Remote Sensing for Natural Hazards Assessment and Control

Remote Sensing for Natural Hazards Assessment and Control

Editors

Paolo Mazzanti
Saverio Romeo

MDPI • Basel • Beijing • Wuhan • Barcelona • Belgrade • Manchester • Tokyo • Cluj • Tianjin

Editors
Paolo Mazzanti
Sapienza University of Rome
Italy

Saverio Romeo
Geological Survey of Italy
Italy

Editorial Office
MDPI
St. Alban-Anlage 66
4052 Basel, Switzerland

This is a reprint of articles from the Special Issue published online in the open access journal *Remote Sensing* (ISSN 2072-4292) (available at: https://www.mdpi.com/journal/remotesensing/special_issues/RS_Natural_Hazards_Assessment_and_Control).

For citation purposes, cite each article independently as indicated on the article page online and as indicated below:

LastName, A.A.; LastName, B.B.; LastName, C.C. Article Title. *Journal Name* **Year**, *Volume Number*, Page Range.

ISBN 978-3-0365-6832-4 (Hbk)
ISBN 978-3-0365-6833-1 (PDF)

Cover image courtesy of Saverio Romeo

Contents

About the Editors

Paolo Mazzanti

Paolo Mazzanti, Prof. Ph.D., is Professor of Remote Sensing and Geological Risks at the Department of Earth Sciences of "Sapienza" University of Rome and member of CERI Research Centre for Geological Risk. Co-Founder of NHAZCA S.r.l., startup of "Sapienza" University of Rome. He is the organizer of several professional training courses, including the annual "International Course on Geotechnical and Structural Monitoring" and invited speaker at several Universities and International Research Centers. He is also member of the TRB (Transportation Research Board) Engineering Geology Committee, of the Technical Committee 220 "Field Monitoring in Geomechanics" of the International Society of Soil Mechanics and Geotechnical Engineering, of the Technical Committee 4.3 Earthworks of PIARC (World Road Association) and auditor of AIGA (Italian Association of Applied and Environmental Geology).

Saverio Romeo

Saverio Romeo, Ph.D., is Engineering Geologist at the Department of the Geological Survey of Italy of the Italian Institute for Environmental Protection and Research (ISPRA), with strong expertise in the fields of geological remote sensing (e.g., Terrestrial SAR Interferometry, Laser Scanning, Imaging, etc.) and design of monitoring systems for geotechnical and structural assets. He is a Licensed Engineering Geologist and former President of LARES—Italian Union of Civil Protection Experts, with international experience, as a technical consultant, on both infrastructure and research projects. He is also a member of EGU (European Geoscience Union) and AIGA (Italian Association of Applied and Environmental Geology).

remote sensing MDPI

Editorial

Introduction to a Thematic Set of Papers on Remote Sensing for Natural Hazards Assessment and Control

Paolo Mazzanti [1,2] **and Saverio Romeo** [3,*]

1 Department of Earth Sciences & CERI Research Centre, Sapienza University of Rome, P.le Aldo Moro, 5, 00185 Rome, Italy
2 NHAZCA s.r.l., Via Vittorio Bachelet, 12, 00185 Rome, Italy
3 Italian Institute for Environmental Protection and Research (ISPRA), Geological Survey of Italy, 00144 Rome, Italy
* Correspondence: saverio.romeo@isprambiente.it

Abstract: Remote sensing is currently showing high potential to provide valuable information at various spatial and temporal scales concerning natural hazards and their associated risks. Recent advances in technology and processing methods have strongly contributed to the development of disaster risk reduction research. In this Special Issue titled "Remote Sensing for Natural Hazards Assessment and Control", we propose state-of-the-art research that specifically addresses multiple aspects of the use of remote sensing for natural hazards. The aim was to collect innovative methodologies, expertise, and capabilities to detect, assess monitor, and model natural hazards. In this regard, 18 open-access papers showcase scientific studies based on the exploitation of a broad range of remote sensing data and techniques, as well as focusing on a well-assorted sample of natural hazard types.

Keywords: remote sensing; natural hazards; hazard; vulnerability; risk assessment

1. Overview of the Special Issue

Each year, natural hazards, such as earthquakes, landslides, avalanches, tsunamis, floods, wildfires, severe storms, and drought, globally affect humans through deaths, suffering, and economic losses. According to the insurance broker Aon, 2010–2019 was the worst decade on record for economic losses due to disasters triggered by natural hazards, amounting to $3 trillion: a $ trillion more than the 2000–2009 decade. In 2019, economic losses from disasters caused by natural hazards were estimated to be over $200 billion (UNDRR Annual Report, 2019).

In this context, remote sensing demonstrates a high potential to provide valuable information, at various spatial and temporal scales, concerning natural processes and their associated risks. Recent advances in remote sensing technologies and analysis, in terms of sensors, platforms, and techniques, have strongly contributed to the development of natural hazards research.

In this Special Issue titled "Remote Sensing for Natural Hazards Assessment and Control", we propose state-of-the-art research that specifically addresses multiple aspects of the use of remote sensing (RS) for natural hazards (NH). The aim was to collect innovative methodologies, expertise, and capabilities to detect, assess monitor, and model natural hazards.

The present Special Issue of the *Remote Sensing* journal encompasses 18 open-access papers that present scientific studies based on the exploitation of a broad range of RS data and techniques, as well as a well-assorted sample of NH types (Figure 1). Table 1 summarizes the RS data, the processing techniques used in each paper, and the general purpose of the presented works.

Citation: Mazzanti, P.; Romeo, S. Introduction to a Thematic Set of Papers on Remote Sensing for Natural Hazards Assessment and Control. *Remote Sens.* **2023**, *15*, 1048. https://doi.org/10.3390/rs15041048

Received: 7 February 2023
Accepted: 13 February 2023
Published: 15 February 2023

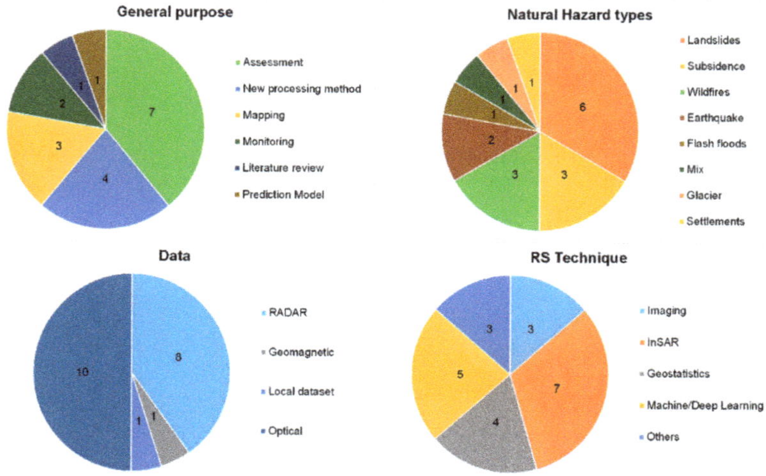

Figure 1. Pie charts of general purpose, natural hazard types, data, and RS technique of published papers.

Table 1. Overview of RS data, techniques, purposes, and NH types that are presented in the papers comprising the SI. Access links to each paper are also provided together with DOI numbers.

Paper Reference and DOI with Access Link	RS Data	Processing Technique	General Purpose	Natural Hazard Types
Chen et al. [1] https://doi.org/10.3390/rs14195059 (accessed on 6 February 2023)	optical, radar	InSAR	assessment	landslide
Wang et al. [2] https://doi.org/10.3390/rs14184562 (accessed on 6 February 2023)	radar	InSAR	new processing method	subsidence
Ma et al. [3] https://doi.org/10.3390/rs14174257 (accessed on 6 February 2023)	optical, radar	InSAR, TRIGRS model	mapping	landslide
Wang et al. [4] https://doi.org/10.3390/rs14153832 (accessed on 6 February 2023)	radar	InSAR	new processing method	subsidence
Xiong et al. [5] https://doi.org/10.3390/rs14133081 (accessed on 6 February 2023)	radar	InSAR, exponential model	new processing method	settlements
Wangcai et al. [6] https://doi.org/10.3390/rs14092131 (accessed on 6 February 2023)	radar	InSAR, random forest	assessment	landslide
Hermle et al. [7] https://doi.org/10.3390/rs14030455 (accessed on 6 February 2023)	optical	Imaging (CD, DIC)	monitoring	landslide

Table 1. *Cont.*

Paper Reference and DOI with Access Link	RS Data	Processing Technique	General Purpose	Natural Hazard Types
Li et al. [8] https://doi.org/10.3390/rs14010030 (accessed on 6 February 2023)	local dataset	Machine learning	prediction model	earthquake
Seydi et al. [9] https://doi.org/10.3390/rs13245138 (accessed on 6 February 2023)	multispectral and hyperspectral	Deep Learning	mapping	wildfires
Nolde et al. [10] https://doi.org/10.3390/rs13244975 (accessed on 6 February 2023)	optical (red and NIR)	Imaging (NDVI)	assessment	wildfires
Kos et al. [11] https://doi.org/10.3390/rs13142694 (accessed on 6 February 2023)	optical, radar	SAR offset tracking	monitoring	glacier
Ding et al. [12] https://doi.org/10.3390/rs13091818 (accessed on 6 February 2023)			review of the literature	flash floods
Cheng et al. [13] https://doi.org/10.3390/rs13091775 (accessed on 6 February 2023)	optical	Imaging (NDWI, SI)	assessment	hazard chain (dam failure, mud and hyperc. flow)
Pacheco et al. [14] https://doi.org/10.3390/rs13071345 (accessed on 6 February 2023)	multispectral	k-Nearest neighbor, random forest	assessment	wildfires
Ranjgar et al. [15] https://doi.org/10.3390/rs13071326 (accessed on 6 February 2023)	radar	InSAR, Machine Learning	mapping	subsidence
Wang et al. [16] https://doi.org/10.3390/rs13050938 (accessed on 6 February 2023)	optical	Geostatistics	assessment	rockfall
Yang et al. [17] https://doi.org/10.3390/rs12223805 (accessed on 6 February 2023)	multispectral	Geostatistics, RUSLE, NBR	new processing method	hillslope erosion
Piersanti et al. [18] https://doi.org/10.3390/rs13142839 (accessed on 6 February 2023)	geomagnetic	Geostatistics	assessment	earthquake

1.1. Overview of the Presented Papers

The 18 papers published in the current Special Issue belong to the section "Environmental Remote Sensing" and cover a wide range of applications in terms of the RS data exploited, processing techniques used, and NH addressed.

Chen et al. [1] applied multi-source remote sensing (InSAR from ALOS PALSAR-1 and -2) and field investigation to study the activity and kinematics of two adjacent landslides along the Datong River in the Qilian Mountains of the Qinghai-Tibet Plateau (China).

Wang et al. [2] proposed a data partition strategy to solve typical limitations due to traditional multi-temporal interferometric synthetic aperture radar (MT-InSAR) methods which require a large computer memory and time when processing full-resolution data. They validated such a strategy in Changzhou City and in Chongqing City (China).

Ma et al. [3] adopted a new open-source tool named MAT.TRIGRS(V1.0) to establish the landslide susceptibility map in landslide abundance areas and to back-analyze the response of the rainfall process to the change in landslide stability. The prediction results were roughly consistent with the actual landslide distributions in Longchuan County (China).

Wang et al. [4] proposed a wide-area InSAR variable-scale deformation detection strategy that combined stacking technology for fast ground-deformation rate calculations and advanced TS–InSAR technology to obtain a fine deformation time series. This new strategy was tested in the Turpan–Hami basin (China).

Xiong et al. [5] presented a new strategy based on the Multitemporal Interferometric Synthetic Aperture Radar (MT-InSAR) method to overcome limitations due to an inaccurate settlement prediction using traditional methods. The Xiamen Xiang'an International Airport (China) was chosen as the test site.

Wangcai et al. [6] assessed landslide susceptibility, hazard, and risk in Yan'an City (China) using a random forest machine learning classifier and eight environmental factors influencing landslides. Additionally, Differential Synthetic Aperture Radar Interferometry (DInSAR) was used for a hazard assessment.

Hermle et al. [7], with the aim of reducing noise from decorrelation in ground motion detection by imaging, applied, for the first time, the optical flow-time series for fast landslides. The debris flows from the Sattelkar area (Austria) was selected as a benchmark site.

Li et al. [8], in order to obtain a precise casualty prediction method that could be applied globally, a spatial division method based on regional differences and a zoning casualty prediction method based on support vector regression (SVR) were proposed in their paper. A selection of 30 historical earthquakes that occurred in China's mainland was chosen.

Seydi et al. [9] presented a novel framework for burned area mapping based on the deep Siamese morphological neural network (DSMNN-Net) and heterogeneous datasets. Two case study areas in Australian forests were selected.

Nolde et al. [10] exploited the possibilities of a recent EO dataset published by the German Aerospace Center (DLR) by exemplarily analyzing fire severity trends on the Australian East coast for the past 20 years.

Kos et al. [11] used SAR offset tracking to reconstruct a unique record of ice surface velocities for a 3.2-year period for the Palcaraju glacier located above Laguna Palcacocha, Cordillera Blanca (Peru).

Ding et al. [12] carried out a review of the literature related to the application of RS and GIS in the study of flash floods. They analyzed more than 200 articles published in the last 20 years, performing keyword co-occurrence, time zone chart, keyword burst, and the literature co-citation analysis.

Cheng et al. [13] presented a detailed analysis to investigate the disaster conditions of the Brumadinho dam failure (Brasil) using satellite images. Their in-depth analysis revealed a hazard chain containing three stages, namely dam failure, mud-, and hyper-concentrated flow.

Pacheco et al. [14] used RS to detect, map, and monitor areas that were affected by forest fires in central Portugal. For this purpose, the study analyzed the performance of the k-nearest neighbor (kNN) and random forest (RF) classifiers.

Ranjgar et al. [15] assessed land subsidence susceptibility for Shahryar County (Iran) using the adaptive neuro-fuzzy inference system (ANFIS) machine learning algorithm. Additionally, they assessed if ensembles of ANFIS with two meta-heuristic algorithms could yield a better prediction performance.

Wang et al. [16] proposed a new approach using the relief–slope angle relationship to identify rockfall source areas controlled by rock mass strength. By using data from helicopter-based RS imagery, a 10m-DEM, and fieldwork, historical rockfalls in the Wolong study area of Tibet (China) were identified.

Yang et al. [17] developed a rapid and innovative approach to estimate post-fire hillslope erosion using weather radar, RS, Google Earth Engine (GEE), GIS, and the revised universal soil loss equation (RUSLE). They assessed the Sydney drinking water catchment area and the Warragamba Dam (Australia).

Lastly, Piersanti et al. [18] presented the first evidence, via observation and modeling, of changes in magnetospheric field line resonance (FLR) eigenfrequency, which was associated with the earthquake occurrence, and demonstrated a causal connection between seismic phenomena and space-based observables.

The Editors expect that these studies will lead to fruitful discussions and scientific progress, which should ultimately help to improve the overall quality and reliability of remote sensing as a now indispensable tool for approaching natural hazards.

1.2. Statistics

The total number of researchers and technologists who contributed to the papers was 104, with an average of 5.8 contributors per article. As shown in Figure 2, most of them worked in China, at least in terms of affiliation, followed by Germany, Italy, Australia, and Iran. Overall, Universities and Institutions from 16 different countries were involved in the present Special Issue. Most of the papers described work with practical applications tested around the world.

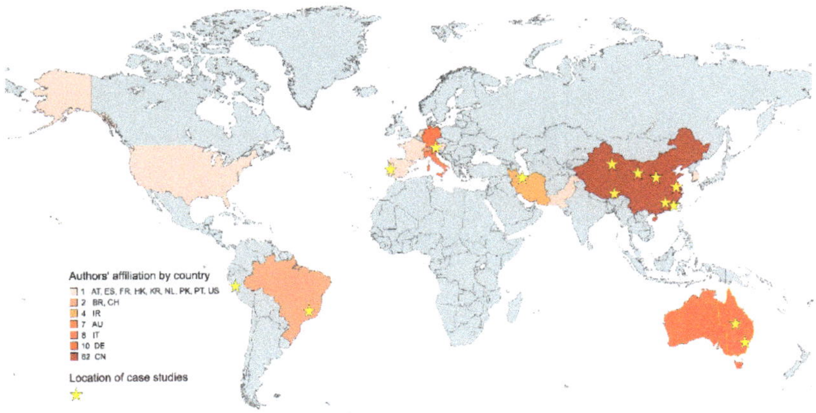

Figure 2. Overview of the authors' affiliation by country together with the location of case studies discussed in the present Special Issue.

The most recurring words among the keywords chosen by the authors are shown in the word cloud in Figure 3. Among them, "InSAR" was selected six times, followed by "landslide" (4 times), "burned area", "sentinel", and "wildfires" with three occurrences.

Figure 3. Word clouds (also known as text clouds or tag clouds) generated from the keywords of all contributions to the present Special Issue. The more a word appears as a keyword, the bigger and bolder it appears in the word cloud.

1.3. Bibliometrics and Impact

The 18 papers were published in the current Special Issue, over 2 years, between November 2020 and October 2022. Each manuscript was assessed via rigorous peer-reviewing from two or more esteemed experts in their respective fields. Based on MDPI's article metrics, this Special Issue has received, up to now (2 February 2023), more than 34,000 total views. A worldwide geographic distribution of readers was also noted. Overall, the published papers already received 95 citations in the indexed literature indicating the high scientific quality of the Special Issue. In detail, the work from Ranjgar et al. [15] reached 22 citations and 3299 online views, followed by the papers from Pacheco et al. [14] and Ding et al. [12] with 16 and 12 citations, respectively.

2. Further Reading

Readers who are interested in the use of remote sensing data and methods for the assessment and control of natural hazards, in addition to this Special Issue, can also refer to manuscripts published in other recent Special Issues of the *Remote Sensing* journal, such as the "Remote Sensing of Natural Hazards" issued in 2019–2020 (https://www.mdpi.com/si/32980 (accessed on 2 February 2023)), "Remote Sensing in Engineering Geology" published in 2020–2021 (https://www.mdpi.com/si/28775 (accessed on 2 February 2023)), and "Natural Hazard Assessment and Disaster Management Using Remote Sensing" available from 2021 (https://www.mdpi.com/si/64420 (accessed on 2 February 2023)). In these Issues, several applications are offered, ranging from GIS-based hazard assessment to the use of multi-sensor data for hazard detection and mapping. Even in these examples, a wide range of natural hazards is covered, including wildfires, earthquakes, landslides, and floods.

Author Contributions: The authors contributed equally to this work. All authors have read and agreed to the published version of the manuscript.

Acknowledgments: The Guest Editors would like to acknowledge all the authors for contributing to this Special Issue and all the involved reviewers for their thoughtful comments and efforts toward improving the submitted manuscripts. Sincere gratitude goes to the Editorial Board and Office of Remote Sensing.

Conflicts of Interest: The authors declare no conflict of interest.

References

1. Chen, J.; Zhang, J.; Wu, T.; Hao, J.; Wu, X.; Ma, X.; Zhu, X.; Lou, P.; Zhang, L. Activity and Kinematics of Two Adjacent Freeze–Thaw-Related Landslides Revealed by Multisource Remote Sensing of Qilian Mountain. *Remote Sens.* **2022**, *14*, 5059. [CrossRef]
2. Wang, Y.; Feng, G.; Feng, Z.; Wang, Y.; Wang, X.; Luo, S.; Zhao, Y.; Lu, H. An MT-InSAR Data Partition Strategy for Sentinel-1A/B TOPS Data. *Remote Sens.* **2022**, *14*, 4562. [CrossRef]
3. Ma, S.; Shao, X.; Xu, C. Characterizing the Distribution Pattern and a Physically Based Susceptibility Assessment of Shallow Landslides Triggered by the 2019 Heavy Rainfall Event in Longchuan County, Guangdong Province, China. *Remote Sens.* **2022**, *14*, 4257. [CrossRef]
4. Wang, Y.; Feng, G.; Li, Z.; Luo, S.; Wang, H.; Xiong, Z.; Zhu, J.; Hu, J. A Strategy for Variable-Scale InSAR Deformation Monitoring in a Wide Area: A Case Study in the Turpan–Hami Basin, China. *Remote Sens.* **2022**, *14*, 3832. [CrossRef]
5. Xiong, Z.; Deng, K.; Feng, G.; Miao, L.; Li, K.; He, C.; He, Y. Settlement Prediction of Reclaimed Coastal Airports with InSAR Observation: A Case Study of the Xiamen Xiang'an International Airport, China. *Remote Sens.* **2022**, *14*, 3081. [CrossRef]
6. Liu, W.; Zhang, Y.; Liang, Y.; Sun, P.; Li, Y.; Su, X.; Wang, A.; Meng, X. Landslide Risk Assessment Using a Combined Approach Based on InSAR and Random Forest. *Remote Sens.* **2022**, *14*, 2131. [CrossRef]
7. Hermle, D.; Gaeta, M.; Krautblatter, M.; Mazzanti, P.; Keuschnig, M. Performance Testing of Optical Flow Time Series Analyses Based on a Fast, High-Alpine Landslide. *Remote Sens.* **2022**, *14*, 455. [CrossRef]
8. Li, B.; Gong, A.; Zeng, T.; Bao, W.; Xu, C.; Huang, Z. A Zoning Earthquake Casualty Prediction Model Based on Machine Learning. *Remote Sens.* **2022**, *14*, 30. [CrossRef]
9. Seydi, S.T.; Hasanlou, M.; Chanussot, J. DSMNN-Net: A Deep Siamese Morphological Neural Network Model for Burned Area Mapping Using Multispectral Sentinel-2 and Hyperspectral PRISMA Images. *Remote Sens.* **2021**, *13*, 5138. [CrossRef]
10. Nolde, M.; Mueller, N.; Strunz, G.; Riedlinger, T. Assessment of Wildfire Activity Development Trends for Eastern Australia Using Multi-Sensor Earth Observation Data. *Remote Sens.* **2021**, *13*, 4975. [CrossRef]
11. Kos, A.; Amann, F.; Strozzi, T.; Osten, J.; Wellmann, F.; Jalali, M.; Dufresne, A. The Surface Velocity Response of a Tropical Glacier to Intra and Inter Annual Forcing, Cordillera Blanca, Peru. *Remote Sens.* **2021**, *13*, 2694. [CrossRef]
12. Ding, L.; Ma, L.; Li, L.; Liu, C.; Li, N.; Yang, Z.; Yao, Y.; Lu, H. A Survey of Remote Sensing and Geographic Information System Applications for Flash Floods. *Remote Sens.* **2021**, *13*, 1818. [CrossRef]
13. Cheng, D.; Cui, Y.; Li, Z.; Iqbal, J. Watch Out for the Tailings Pond, a Sharp Edge Hanging over Our Heads: Lessons Learned and Perceptions from the Brumadinho Tailings Dam Failure Disaster. *Remote Sens.* **2021**, *13*, 1775. [CrossRef]
14. Pacheco, A.d.P.; Junior, J.A.d.S.; Ruiz-Armenteros, A.M.; Henriques, R.F.F. Assessment of k-Nearest Neighbor and Random Forest Classifiers for Mapping Forest Fire Areas in Central Portugal Using Landsat-8, Sentinel-2, and Terra Imagery. *Remote Sens.* **2021**, *13*, 1345. [CrossRef]
15. Ranjgar, B.; Razavi-Termeh, S.V.; Foroughnia, F.; Sadeghi-Niaraki, A.; Perissin, D. Land Subsidence Susceptibility Mapping Using Persistent Scatterer SAR Interferometry Technique and Optimized Hybrid Machine Learning Algorithms. *Remote Sens.* **2021**, *13*, 1326. [CrossRef]
16. Wang, X.; Liu, H.; Sun, J. A New Approach for Identification of Potential Rockfall Source Areas Controlled by Rock Mass Strength at a Regional Scale. *Remote Sens.* **2021**, *13*, 938. [CrossRef]
17. Yang, X.; Zhang, M.; Oliveira, L.; Ollivier, Q.R.; Faulkner, S.; Roff, A. Rapid Assessment of Hillslope Erosion Risk after the 2019–2020 Wildfires and Storm Events in Sydney Drinking Water Catchment. *Remote Sens.* **2020**, *12*, 3805. [CrossRef]
18. Piersanti, M.; Burger, W.J.; Carbone, V.; Battiston, R.; Iuppa, R.; Ubertini, P. On the Geomagnetic Field Line Resonance Eigenfrequency Variations during Seismic Event. *Remote Sens.* **2021**, *13*, 2839. [CrossRef]

remote sensing

MDPI

Article

Activity and Kinematics of Two Adjacent Freeze–Thaw-Related Landslides Revealed by Multisource Remote Sensing of Qilian Mountain

Jie Chen [1,†], Jing Zhang [2,3,†], Tonghua Wu [1,*], Junming Hao [4], Xiaodong Wu [1], Xuyan Ma [5], Xiaofan Zhu [1], Peiqing Lou [1] and Lina Zhang [6]

1 Cryosphere Research Station on the Qinghai-Tibet Plateau, State Key Laboratory of Cryospheric Science, Northwest Institute of Eco-Environment and Resources, Chinese Academy of Sciences, Lanzhou 730000, China
2 Key Laboratory of Seismic and Volcanic Hazards, China Earthquake Administration, Beijing 100029, China
3 Institute of Geology, China Earthquake Administration, Beijing 100029, China
4 School of Civil Engineering, Lanzhou University of Technology, Lanzhou 730050, China
5 Geomatics Technology and Application Key Laboratory of Qinghai Province, Qinghai Remote Sensing Center for Natural Resources, Qinghai 810001, China
6 School of Geography Science and Geomatics Engineering, Suzhou University of Science and Technology, Suzhou 215009, China
* Correspondence: thuawu@lzb.ac.cn
† These authors contributed equally to this work.

Citation: Chen, J.; Zhang, J.; Wu, T.; Hao, J.; Wu, X.; Ma, X.; Zhu, X.; Lou, P.; Zhang, L. Activity and Kinematics of Two Adjacent Freeze–Thaw-Related Landslides Revealed by Multisource Remote Sensing of Qilian Mountain. *Remote Sens.* **2022**, *14*, 5059. https://doi.org/10.3390/rs14195059

Academic Editors: Paolo Mazzanti and Saverio Romeo

Received: 31 August 2022
Accepted: 30 September 2022
Published: 10 October 2022

Publisher's Note: MDPI stays neutral with regard to jurisdictional claims in published maps and institutional affiliations.

Abstract: The increase in temperatures and changing precipitation patterns resulting from climate change are accelerating the occurrence and development of landslides in cold regions, especially in permafrost environments. Although the boundary regions between permafrost and seasonally frozen ground are very sensitive to climate warming, slope failures and their kinematics remain barely characterized or understood in these regions. Here, we apply multisource remote sensing and field investigation to study the activity and kinematics of two adjacent landslides (hereafter referred to as "twin landslides") along the Datong River in the Qilian Mountains of the Qinghai-Tibet Plateau. After failure, there is no obvious change in the area corresponding to the twin landslides. Based on InSAR measurements derived from ALOS PALSAR-1 and -2, we observe significant downslope movements of up to 15 mm/day within the twin landslides and up to 5 mm/day in their surrounding slopes. We show that the downslope movements exhibit distinct seasonality; during the late thaw and early freeze season, a mean velocity of about 4 mm/day is observed, while during the late freeze and early thaw season the downslope velocity is nearly inactive. The pronounced seasonality of downslope movements during both pre- and post-failure stages suggest that the occurrence and development of the twin landslide are strongly influenced by freeze–thaw processes. Based on meteorological data, we infer that the occurrence of twin landslides are related to extensive precipitation and warm winters. Based on risk assessment, InSAR measurements, and field investigation, we infer that new slope failure or collapse may occur in the near future, which will probably block the Datong River and cause catastrophic disasters. Our study provides new insight into the failure mechanisms of slopes at the boundaries of permafrost and seasonally frozen ground.

Keywords: landslides; Gaofen-2; Interferometric synthetic aperture radar (InSAR); freeze–thaw processes; permafrost; Qilian Mountains

1. Introduction

A landslide is the downslope movement of soil, rock, and debris under the action of gravity and the landform that results from such movement [1]. The factors triggering slope failure mainly include rainfall, earthquakes, fluvial erosion, excavation, and construction activities [1]. Slope failure occurs frequently in rainy and mountainous areas, often resulting in casualties and property damage [1,2]. Slope failure at high latitudes and high altitudes in cold regions has attracted significant attention for decades. Climate warming and the consequent varied freeze–thaw states in cold regions may accelerate the occurrence and development of landslides [3,4]. Thaw-related landslide activities (such as retrogressive thaw slumps and active layer detachment) are increasing extensively in the permafrost regions of both the Arctic [5–8] and the Qinghai-Tibet Plateau (QTP) [9–11], a phenomenon that is mainly related to extremely warm summers and extensive summer precipitation. Even in very cold permafrost regions, extreme warming events can cause slope failure due to the thawing of ice-rich permafrost [12]. However, slope failures on the boundaries of permafrost and seasonally frozen ground remain little characterized or understood.

Optical and radar remote sensing techniques are becoming one of the most important tools for investigating the activity and kinematics of landslides, especially in rural and mountainous regions [13,14]. The occurrence and development of landslides can be obtained from optical remotely sensed images [15–19]. Rapid and abrupt slope movements can be quantified according to the differences in DEM from two successive orthorectified images [20–22] or from single-pass InSAR observations [23,24]. However, optical images are prone to the effects of atmospheric conditions and revisit time, which impede the quantification of landslide activities. In addition, optical-based slope movement is insensitive to gradual slope movement or creep, which is often a sign of slope stability. Interferometric synthetic aperture radar (InSAR) has been increasingly explored and successfully applied to identify and monitor gradual slope deformation with centimeter-to-millimeter accuracy [13,25–28]. The advantage of InSAR is that it is less affected by cloudy weather and works regardless of whether day or night conditions. Recently, the development of unmanned aerial vehicle (UAV) cameras allows the generation of very-high-resolution optical images and digital surface models (DSMs). The combination of InSAR and UAV data has advanced the understanding of crack development, landslide evolution, and susceptibility [29–34].

In this study, we combine multisource optical and radar remote sensing images to investigate the activity and kinematics of two adjacent landslides (hereafter referred to as "twin landslides") located in the boundary regions of permafrost and seasonally frozen ground on Qilian Mountain in the QTP. High-resolution optical satellite and UAV-based images are used to investigate the activity of the twin landslides. InSAR measurements are used to map and quantify the gradual slope movement before and after the failure of the twin landslides. The triggering mechanisms of the twin landslides and their potential risks are analyzed.

2. Study Area

In this study, we investigate two adjacent landslides occurring on the slope of the middle and lower reaches of Datong River on the southeast slope of Qilian Mountain. For convenience, we named the two adjacent landslides as twin landslides and refer to them as "QLDT01" and "QLDT02" throughout the paper (Figure 1C). The study area is located between Tuolai Mountain in the north and Datong Mountain in the south, both of which belong to branches of the Qilian Mountains. The Wari Gaqu River rises from the Tuolai Mountains and ultimately joins the Datong River. A number of brooks from the Datong Mountains = discharge into the Datong River (Figure 1A). Therefore, the study area is a place where many rivers converge. The Tuolaishan and Datongshan faults are distributed in the northwest and southwest of the study area, respectively. The altitude in the study area ranges from 3400 to 3600, which is in the lower boundary of permafrost and seasonal frozen ground on the QTP. The permafrost distribution map is shown in Figure 1B [35].

Qilian Mountain is dominated by a continental alpine semi-humid mountain climate, which is characterized by long, cold, and dry winters and short, cool, and moist summers. Annual average air and ground surface temperatures are approximately −2.4 and 2.9 °C, respectively [36]. The coldest January averages below −11 °C, and the warmest July averages below 25 °C; most of the Qilian Mountains are below 0 °C from December to March, and the highest temperatures range from 4 to 15 °C from April to October. Precipitation ranges from 300 to 500 mm annually and is mainly concentrated during May to September in the form of rainfall [37,38].

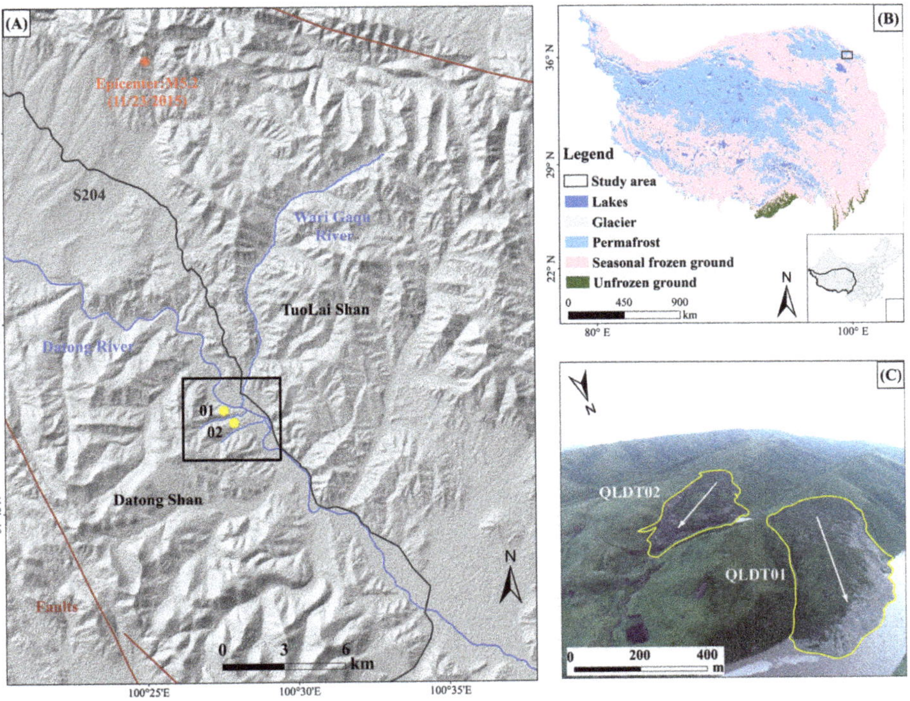

Figure 1. Study area and field photos. (**A**) The locations of the twin landslides (QLDT01 and QLDT02) are marked by yellow dots within the black rectangle. The background map is the hill-shaded DEM. The mountains, rivers, and roads in our study area are marked. (**B**) The distribution map of permafrost and seasonal frozen ground in the QTP [35]. The black rectangle presents the location of our study area. The inner map in the bottom right-hand corner shows the location of the QTP in China. (**C**) An unmanned aerial vehicle (UAV) photograph of the twin landslides taken in April 2021. The white arrow denotes the direction of the landslide movement. The yellow polygons are the boundaries of the twin landslides.

3. Data and Methods

3.1. Data Sources

To investigate the activity and kinematics of the twin landslides, multisource high-resolution optical and radar data were acquired during 2009–2020. Google images with a spatial resolution of 0.65 m were acquired using Map Tile Downloader (version number: release 2.3, developed by Centmap Co., LTD., located in Hefei, Anhui Province, China). The details of the remotely sensed images are listed in Table 1. Gaofen-2 is a Chinese high-resolution optical satellite that was launched in August 2014 and carries two panchromatic and multispectral charge-coupled device camera sensors. We used panchromatic images with a spatial resolution of 0.8–1 m, which were obtained from the China Centre for Resources Satellite Data and Application (http://www.cresda.com/CN/, accessed on 10 October 2021). Based on the selection of images free of cloud and snow cover, four Google and three Gaofen-2 images obtained during 2019–2020 were used.

The L-band Phased Array type L-band Synthetic Aperture Radar (PALSAR) sensor is mounted on the Advanced Land Observation Satellite (ALOS)-1/2. Six ALOS-1 (Path: 477, Frame: 750, incidence angle: 38.7°, heading angle: −10.1°, range resolution: 4.66 m, azimuth resolution: 3.16 m) and thirteen ALOS-2 (Path: 147, Frame: 750, incidence angle: 36.3°, heading angle: −10.4°, range resolution: 4.29 m, azimuth resolution: 3.77 m) SAR images were chosen to map and quantify the ground movements of the twin landslides. We expected better performance from the L-band PALSAR data with a wavelength of about 24 cm. This is because L-band electromagnetic waves can penetrate deeper into snow and vegetated surfaces [39], leading to higher interferometric coherence [40].

To investigate potential unstable zones, we conducted a UAV survey and used structure-from-motion/multiview stereo photogrammetry to map the twin landslides and their surroundings on 29 April 2021. We used the DJI Phantom 4 RTK flying platform with an altitude of 100 m above the ground surface. The heading and lateral overlap of flying were 85% and 60%, respectively. We obtained the very-high-resolution (VHR) images and a DEM with a resolution of around 5 cm/pixel using Agisoft PhotoScan software. The uncertainty of the relative positions was estimated to be around 2–3 cm.

The temperature and precipitation data in the Qilian Mountains from 2000 to 2019 obtained by the National Meteorological Station of China were used to describe the climatic conditions (http://www.cma.gov.cn/, accessed on 10 November 2021).

Table 1. Summary of the remotely sensed dataset used in this study. The acquisition dates of ALOS PALSAR-1/2 can be found in Table 2.

Data	Resolution (m)	Date (YYYYMM)	Number of Scenes
Google	0.65	200910, 201006, 201712, 202007	4
Gaofen-2	0.8	201512, 201811, 202008	3
ALOS PALSAR-1	7	2008–2010	6
ALOS PALSAR-2	7	2015–2020	13
UAV	0.05	202104	1

3.2. *Mapping of Twin Landslides from Optical Remote Sensing*

Google, Gaofen-2, and PALSAR-1/2 satellite images were used to determine the occurrence and development of the twin landslides. The Gaofen-2 images were geometrically corrected using ENVI5.3 software. Then, the one arc-second Shuttle Radar Topography Mission (SRTM) DEM product was used for image orthorectification. Due to destruction of the integrity of the original stratum, landslide features such as changed vegetation and soil collapse can be identified from high-resolution optical images [16,41]. The boundaries of the twin landslides and adjacent shorelines were outlined based on visual inspection by three experienced researchers. To further evaluate the change characteristics, we estimated the rate of landslide areal growth ΔA_{area} [6]:

$$\Delta A_{\text{rate}} = \frac{A_2 - A_1}{t_2 - t_1} \tag{1}$$

where A_1 and A_2 are the total area (m^2) of landslide in different timeframes and t_1 and t_2 (year) are the corresponding time points.

VHR optical images have proven useful in identifying landslide features such as small cracks or ground discontinuities [42,43]. In this study, to map the potential unstable zones surrounding the twin landslides, we identified cracks through visual inspection of the VHR UAV optical images. As the spatial resolution of UAV images is 5 cm, cracks with width larger than 5 cm were very likely to be identified.

3.3. *InSAR for Ground Deformation Monitoring*

The InSAR technique detects ground movements by comparing the phase differences between SAR images acquired from slightly different positions at different times [44]. Differential InSAR (DInSAR) and multi-temporal InSAR (MTInSAR) have frequently been used to measure slope movements in both permafrost and nonpermafrost regions [27,28,45–47]. As there are very limited descending PALSAR-1/2 images, all the archived ascending PALSAR-1/2 images available that covered our study area were examined for ground deformation monitoring. The interferometric coherence decreases rapidly in the thaw season (May–October) and is slower in the freeze season (November–next April). To mitigate the decorrelation impact, we selected image pairs with temporal spans of less than 150 days and perpendicular baselines shorter than 500 m. Considering the accuracy of SRTM DEM and the maximum perpendicular baseline (591 m in our case), we estimated that the residual topographic phase would be about 0.8 radians, corresponding to 1.5 cm in the InSAR measurements. Relying on the interferometric coherence and phase quality, only six ALOS PALSAR-1 images and thirteen ALOS PALSAR-2 images, taken during 2009–2010 and 2015–2020, respectively, were selected. This causes severe disconnection between SAR images and does not allow the use of MTInSAR approaches such as small-baseline subset InSAR [48].

We calculated ground movement using the DInSAR technique, which was conducted using the commercial GAMMA software [49]. We constructed three and nine interferograms for PALSAR-1 and -2, respectively. The range and azimuth look numbers were 2 and 5 for PALSAR-1 and 2 and 4 for PALSAR-2, generating ground pixels of approximately 15 m × 15 m. The one arc-second SRTM DEM product was used to remove the topographic phase of each interferogram. The temporal and perpendicular baselines are presented in Table 2. We applied a power spectrum adaptive filter to mitigate the phase noise and mask out decorrelation areas with a coherence threshold of 0.6 [50]. We unwrapped all the interferograms using the minimum cost flow approach [51]. To compare the deformation between PALSAR-1 and -2, a local reference point with high coherence nearby the twin landslides was selected for calibration of the unwrapped phase. Tropospheric artifacts may contaminate the ground deformation in mountainous regions. As our study area was very small, we mitigated tropospheric artifacts by fitting the topographic-related components [52]. Residual atmospheric and orbital errors were mitigated using a linear deramping approach.

Table 2. The interferogram pairs from ALOS PALSAR-1/2 and their temporal and perpendicular spatial baselines.

ALOS PALSAR-1				ALOS PALSAR-2			
ID	**Master–Slave (YYYYMMDD)**	**Time Span (Days)**	**$B_\perp$ (m)**	**ID**	**Master–Slave (YYYYMMDD)**	**Time Span (Days)**	**$B_\perp$ (m)**
1	20090630–20090815	46	−32	1	20151009–20151218	70	200
2	20091231–20100215	46	491	2	20171201–20180209	70	−98
3	20100703–20100818	46	116	3	20180209–20180601	112	−44
				4	20180601–20180727	56	−6
				5	20181116–20190125	70	15
				6	20190125–20190531	126	−60
				7	20190531–20190726	56	78
				8	20190726–20190906	42	−10
				9	20200306–20200529	84	169

We calculated the light-of-sight (LOS) movement from each interferogram. By dividing the time interval between the interferogram pairs, we calculated the deformation velocities along the LOS direction. Assuming the slopes move purely along the downslope direction, the InSAR-estimated LOS velocities (V_{los}) can be projected into the downslope velocities (V_{ds}) with the following equation [47,53]:

$$V_{ds} = \frac{V_{los}}{\sin(\alpha_{aspect} - \beta)\sin\theta_{inc}\cos\alpha_{slope} + \cos\theta_{inc}\sin\alpha_{slope}} \quad (2)$$

where α_{aspect} and α_{slope} are the aspect and slope angles, respectively, which can be calculated using the SRTM DEM data; β is the flight direction of the SAR satellite; and θ_{inc} is the local incidence angle, which can be calculated using the SAR geometry and SRTM DEM data. To reduce the noise in the calculation of slope, aspect, and local incidence angles, we applied a Gaussian filter with a 7×7 window (around 200 m) to the SRTM DEM.

3.4. Climatic Factors

Air temperature and precipitation data from 2000 to 2019 were used to analyze their impacts on the evolution of the twin landslides. We calculated four temperature indicators: mean annual air temperature (MAAT), thawing index, warming days, and average temperature in the coldest month of the year. To account for warming days, we calculated the number of days with a daily temperature higher than 10 °C. The thawing index TI is the cumulative number of degree days above 0 °C for a given thaw season, which can be calculated by [54]:

$$TI = \sum_{i=1}^{N_T} T_i, \qquad T_i > 0 \quad (3)$$

where T_i is the daily temperature on day i and N_T is the number of days in a year with a temperature greater than 0 °C.

We calculated the annual total precipitation, precipitation intensity, extreme precipitation, and the number of consecutive drought days in a year. The precipitation intensity is the ratio between the total precipitation and the duration of precipitation days, and represents the average amount of precipitation in a certain duration. Daily precipitation of between 10 and 25 mm is defined as moderate rainfall by the World Meteorological Organization. However, because the annual precipitation is about 450 mm, we consider a daily precipitation of higher than 15 mm to be extreme precipitation in our study. The consecutive drought days is the number of days without precipitation.

3.5. Risk Assessment

We evaluated the potential risks related to the twin landslides and their surroundings in the same slopes. A landslide dam forms when a landslide reaches the bottom of a valley and causes partial or complete blockage of a river [55]. The sudden collapse of landslide dams and the rapid release of water storage poses a great risk of flooding downstream [56]. The dimensionless blockage index (DBI) has been developed for the prediction of potential risks of a landslide dam by linking the stability of a landslide dam to three geomorphic parameters [57]. The dam volume V_d controls the dam height H_d, and is considered as the main stabilizing factor. The watershed area A_b indirectly controls the channel flow and flow power, and is the main factor influencing dam instability. The dam height is an important variable for evaluating the stability of landslide dams against overtopping and pipeline failure. Thus, the DBI can be expressed as [57]:

$$\mathrm{DBI} = \log\left(\frac{A_b \times H_d}{V_d}\right) \quad (4)$$

As only QLDT01 has caused the formation of a landslide dam, we calculated the DBI only for QLDT01. The dam height was obtained from UAV-based DEM data. The volume of the landslide dam was calculated from high-resolution UAV-based DEM using the cut-and-fill volume tool in the Global Mapper software. The hydrological analysis tool was used to calculate the catchment area from UAV-based DEM in ArcGIS (version number: release 10.7, developed by Environmental Systems Research Institute, Inc., located in RedLands, CA, USA).

4. Results

4.1. Spatiotemporal Variations of the Twin Landslides

The occurrence and development of the twin landslides are shown below (QLDT01 in Figure 2 and QLDT02 in Figure 3). Based on visual interpretation of optical and PALSAR-1/2 SAR backscatter images, we infer that QLDT01 occurred between October and December of 2009, whereas QLDT02 occurred sometime between October and November of 2015. The failure of QLDT01 caused the mud and rubble to slide into the river channel and almost blocked the Datong River (Figure 2B). Crack features can be found in the headwall regions for both QLDT01 and QLDT02 as far back as 2009 (Figures 2A and 3A). As there are no high-resolution images in our study before 2009, we cannot precisely determine the exact initialization time of these cracks. Compared to 2009, the cracks in QLDT02 had significantly enlarged during 2009–2015 before its failure. The development of QLDT01 is slow, and its slide into the Datong River almost stopped during 2010–2018. Contrary to QLDT01, the headwall of QLDT02 continued to slowly retreat during 2009–2015. The mass of QLDT02 slid along the northwest side of the headwall region and caused the formation of a dammed lake at the foot of the slope.

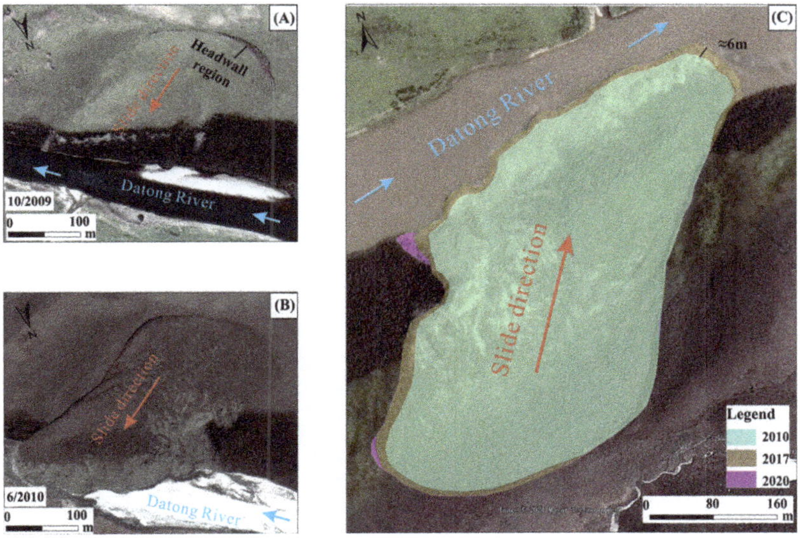

Figure 2. Temporal variations of landslide QLDT01: (**A**,**B**) Google satellite images of landslide boundary changes in 2009 and 2010 and (**C**) landslide evolution in 2010, 2017, and 2020.

The total area of the QLDT01 slide is about 76.5×10^3 m^2 following the slope failure in 2009. The landslide body slid into and dammed the Datong River. QLDT01 slowly expanded at an areal growth rate of 0.5×10^3 m^2 during 2011–2018 (Figure 2C). The total area for slope failure of QLDT02 is about 131×10^3 m^2, which is about double that of QLDT01 (Figure 3C). A small dammed lake has formed at the toe of QLDT02. The areal growth rate of QLDT01 is 10.7×10^3 m^2 during 2016–2018, which has slowed to 5.5×10^3 m^2 during 2018–2020.

The slope failure of QLDT01 completely dammed the Datong River and rerouted its flow (Figure 4). The width of the Datong River beneath QLDT01 was 66 m before the slope failure in 2009. The landslide body slid into the river and reached to about 4 m beyond the northern bank when the slope collapsed. The river quickly expanded towards the northern bank, whereas the river's width changed to 16 m in 2010. Under continuous fluvial erosion, the northern bank expanded northward by about 30 m during 2010–2017, whereas the river's width changed to 48 m in 2017. In other words, the average bank erosion rate was about 4 m/year during 2010–2017.

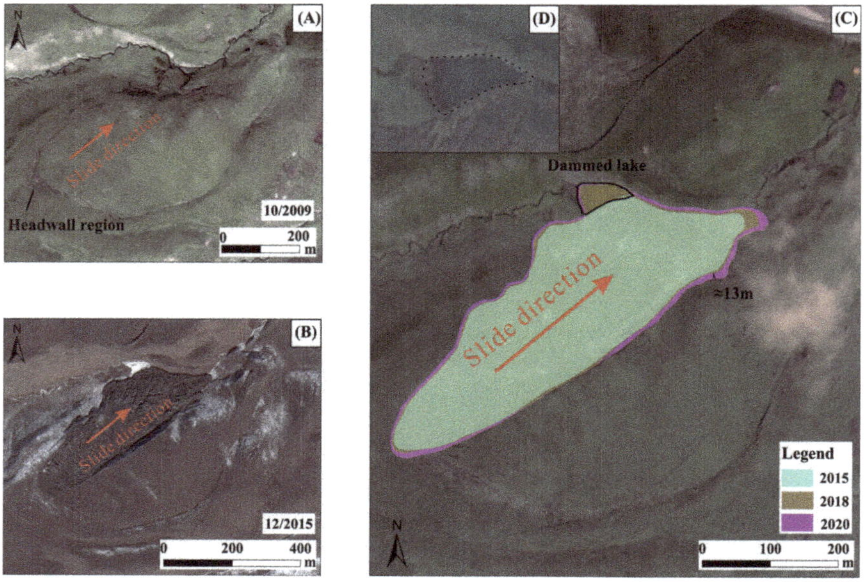

Figure 3. Temporal variations of landslide QLDT02: (**A**,**B**) landslide in Google satellite and Gaofen-2 satellite images from 2009 and 2015; the yellow arrow is the direction of movement of the landslide. (**C**) Landslide characteristics recorded by Gaofen-2 satellite images in 2015, 2018, and 2020. A dammed lake is formed at the toe of the slope, as shown in (**D**).

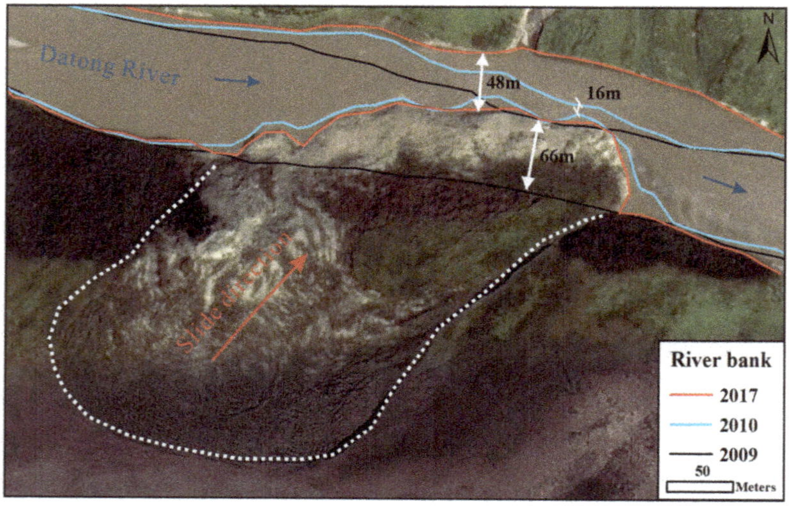

Figure 4. River bank changes of Datong River beneath the twin landslides in 2009, 2010, and 2017. In 2009, the thinnest section of the Datong River was about 66 m. In 2010, the southern river bank expanded northward due to the collapse of Landslide QLDT01, and the width was about 16 m. In 2017, the north bank continued to expand northward, and the thinnest section of the Datong River was 48 m.

4.2. *InSAR-Derived Downslope Movement of the Twin Landslides*

We derive the downslope movement of twin landslides before and after their failure from PALSAR-1/2 InSAR measurements during 2009–2010 and 2015–2020 (Figure 5).

Significant downslope movement is observed within the twin landslides, whereas the maximum displacement rate reaches up to 15 mm/day. We observe strong displacement of up to 5 mm/day outside the twin landslides. To evaluate the potential risks related to the twin landslides and their surroundings, we outline one polygon adjacent to the QLDT02 based on the phenomenological features from UAV images (potential risk zone (PRZ) in Figure 5L).

During the summer before the failure of QLDT01, significant downslope velocities up to 15 mm/day are observed in the boundary and central regions of the landslide body, whereas the mean value is about 4 mm/day (Figure 5A). Five years later, after the failure of the slope, the mean downslope velocities are smaller than 0.5 mm/day in both the summer and winter seasons during 2015–2020 (Figure 5D–F).

We observe that for QLDT02 the mean downslope velocities are about 1.6 mm/day with a maximum value of 5 mm/day from July 2009 to August 2010, i.e., the periods just before and after the failure of QLTD01 (Figure 5A–C). A distinct scarp can be observed in the high-resolution optical image at the head of the landslide body (Figure 3A), which may cause severe InSAR decorrelation and result in no measurements in these regions. The failure of QLDT02 occurred during October and December of 2015; however, there are no valid InSAR measurements due to this severe decorrelation. We observe that significant downslope velocities with mean values of about 2.3 mm/day are pronounced in QLDT02 during July–February of 2016–2020 after slope failure (Figure 5E,H,K). On the contrary, QLDT02 is inactive during March–June (Figure 5F,G,I,J,L).

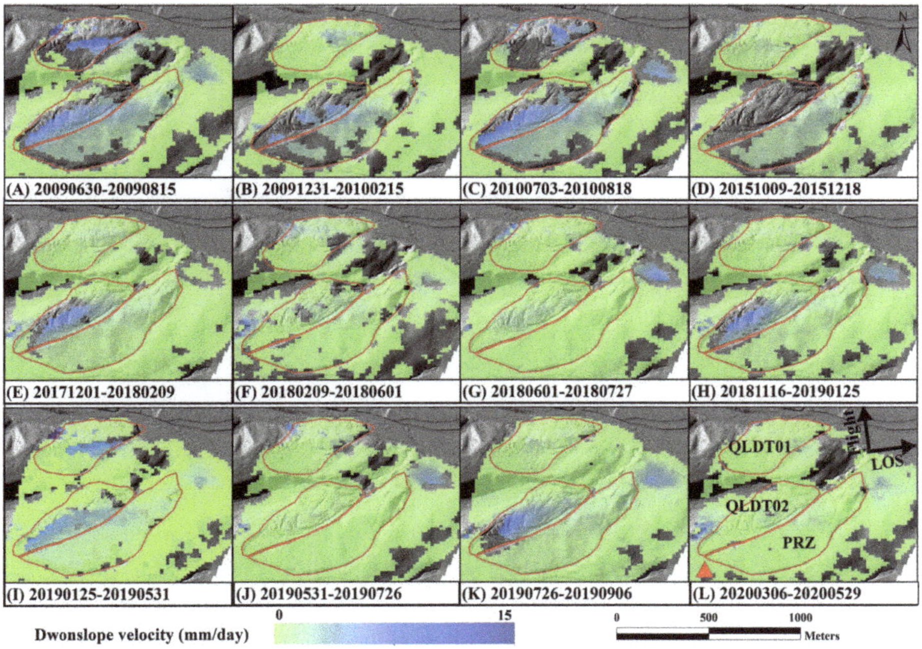

Figure 5. The downslope velocity is derived from line-of-sight (LOS) deformation using Equation (2). The background map is the shaded relief map derived from UAV DEM. The twin landslides (QLDT01 and QLDT02) and the potential risk zone (PRZ) are marked by red polygons in the bottom right-hand corner of the subfigures. The positive values refer to the movement in the downslope direction. The red triangle denotes the location of the reference point.

5. Discussion

5.1. Triggering Mechanisms

5.1.1. Precipitation

In general, changing precipitation patterns increase subsurface saturation and pore pressure, which increase the likelihood of slope failure [58]. Extensive or extreme precipitation and rapid snow/ice melt are therefore likely to increase the frequency and magnitude of landslides [59,60]. The precipitation data near the twin landslides show fluctuating upward trends during 2000–2019 (Figure 6A). The annual precipitation is 451 mm in 2009 and 448 mm in 2015, which is significantly higher than the mean annual precipitation since 2000. The annual precipitation in the preceding year is higher than in the year of slope failure.

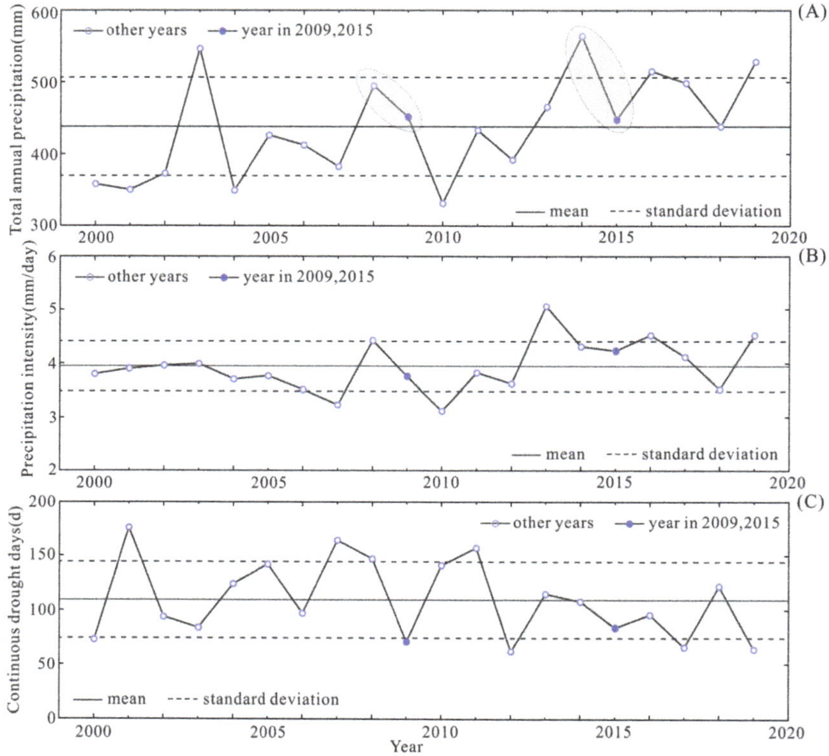

Figure 6. Variations of precipitation during 2000–2019: (**A**) variations in annual precipitation, (**B**) precipitation intensity, (**C**) number of consecutive drought days in a year.

To evaluate the impacts of precipitation events on the occurrence of the twin landslides, we graphed the daily precipitation before slope failure (Figure 7). An extreme precipitation event of 35.8 mm is recorded in August 2009. The accumulated precipitation was about 193 mm during August–September 2019 (Figure 7). The number of consecutive drought days is among the lowest during 2000–2009 (Figure 6C). Extensive rainwater may increase the pore water pressure and reduce the shear strength in weak soil layers. Thus, we infer that the occurrence of QLDT01 may have been primarily triggered by extensive precipitation. In 2015, there was no extreme precipitation event such as that in 2009, however, the annual precipitation was higher than the 20-year average (Figure 6A). The number of consecutive drought days was below the average (Figure 6C). Therefore, we presume that increased precipitation is likely one of the triggering factors of landslide QLDT02.

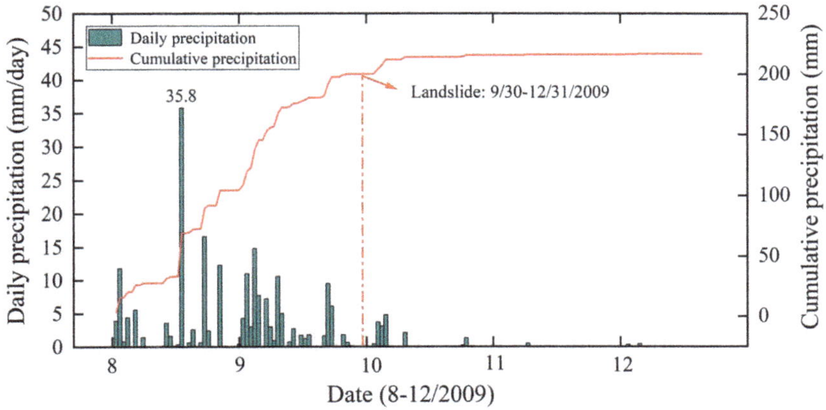

Figure 7. Daily and cumulative precipitation during August–December, 2019. A maximum daily precipitation of 35.8 mm was recorded in August 2009.

5.1.2. Freeze–Thaw Processes

Climate warming and disturbance may have strong impacts on slope stability in cold environments [3,4]. In permafrost areas, rocks are glued together by ice filling their cracks and crevices. Freeze–thaw processes are characterized by variability in subsurface temperature and moisture content, which results in substantial fluctuations of shear strength (cohesion and friction angle) and drives landslide initiation [61]. The transition from perennially frozen to seasonally frozen ground accelerates the effect of freeze–thaw processes on both bedrock and unconsolidated material [62]. As the air temperature increases, the warming and thawing of permafrost may weaken rock faces and the inherent stability of permafrost, leading to slope failure [63]. In mountain permafrost regions, e.g., the European Alps, Canada, and the Tibetan plateau, researchers have recorded an increasing tendency of landslide activities due to the warming climate [9,64,65].

The MAAT shows an obvious warming trend during 2000–2019 (Figure 8A). The MAAT in 2009 and 2015 is 2.32 and 2.36 °C, respectively, which is about 0.3 °C above the 20-year average (Figure 8A). The MAAT values in the preceding years (2008 and 2014) are about 0.7 °C lower than in the failure years. The strong fluctuations in air temperature may amplify freeze–thaw processes and thus affect slope stability. In 2009, the warming days (the number of days with air temperature above 10 °C), the thawing index, and the average temperature in the coldest month were all above their 20-year averages (Figure 8B–D). This suggests that the warming events in 2009 might have been one of the triggers of QLDT01 failure. On the contrary, the warming days and thawing index in 2015 are lower than their averages. However, the mean temperature of the coldest month in 2015 is about 0.7 °C) above the 20-year average, which suggests a warm winter. Warm winters may slow down freezing processes, allow the soil water to remain in an unfrozen state for a longer time, thereby increasing the risk of slope failure.

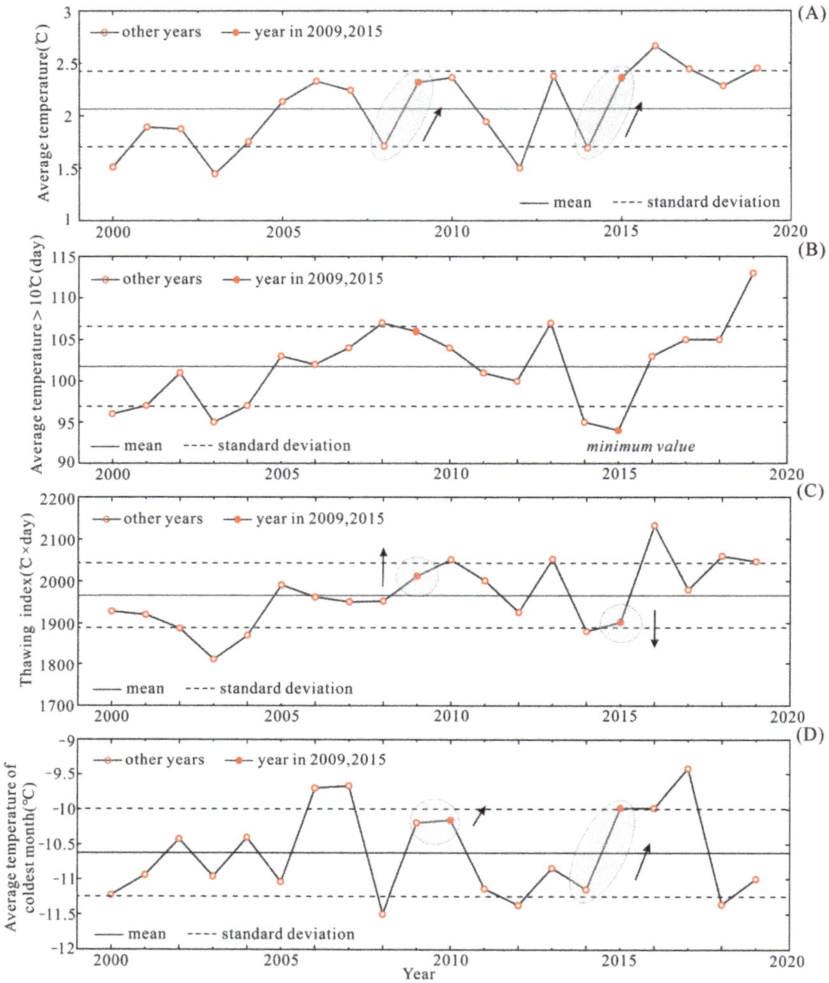

Figure 8. Variations in air temperature during 2000–2019: (**A**) mean annual air temperature (MAAT), (**B**) warming days with an annual average temperature greater than 10 °C, (**C**) thawing index, (**D**) average temperature of the coldest month.

The activities and kinematic patterns in the twin landslides and their surroundings have been derived from InSAR measurements. We observe that the average downslope velocities in QLDT01 and QLDT02 exhibit distinct seasonality (Figure 9). During the early thawing periods from May to early July, the slopes are in an inactive state. In this stage, soil thawing is shallow and does not reach the sliding surface, resulting in limited downslope movement. During the late thawing and early freezing periods from late July to the next January the slopes are in an active state, with average downslope velocities up to 4 mm/day. In the late thawing stage, the sliding surface is thawed, which results in significant downslope movement. Despite the shallow soil being frozen during the early freezing season, downslope movement remains significant, as the sliding surface is in a thawed state. During the early freezing period from February to April the slopes become inactive, as the sliding surface is in a frozen state.

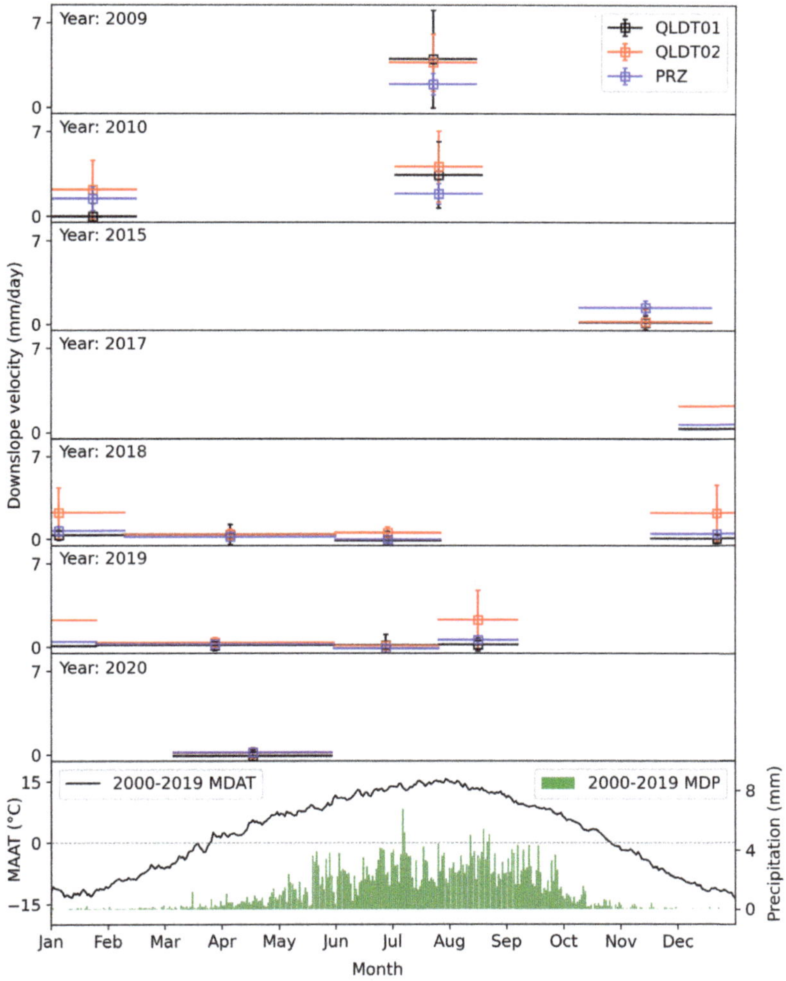

Figure 9. Temporal variations for 2000–2019 in mean daily air temperature (MDAT), mean daily precipitation (MDP), and average downslope velocities of QLDT01, QLDT02, and potential risk zone (PRZ) (Figure 5L). The squares present the average downslope velocities, whereas the corresponding lines show the start and end date of the SAR image pair. The error bars denote the standard deviation of downslope velocities within the red polygons. The bottom panel plots the MDAT and MDP.

The seasonality of significant downslope movement during both the pre- and post-failure stages suggest that the occurrence and development of the twin landslides were strongly influenced by freeze–thaw processes. The seasonal pattern is different from the seasonal deformation corresponding to freeze uplift and thaw subsidence due to ice–water phase change constraints in the active layer [66,67]. To put this work in a spatial context, we compare our study with several freeze–thaw-related slope instability studies on the QTP. Meng et al. and Hao et al. [34,45] observed deformation velocity up to 100 mm/year with a linear trend assumption using the multi-temporal InSAR technique on an earthflow in Yushu, QTP. Dini et al. [68] characterized different magnitudes of LOS deformation over different types of slope instability in the eastern Himalayas. Hu et al. [53] found similar seasonal patterns of downslope velocity up to about 3 mm/day during the active stage in several rock glaciers in the East Kunlun Mountains. The less pronounced downslope velocity may be primarily related to the kinematic behaviors of rock glaciers.

5.1.3. Other Triggering Factors

The slope failure of QLDT01 may be partially attributed to fluvial erosion at the slope toe and its geomorphological characteristics. QLDT01 is situated at the confluence of several rivers. The north bank of the Datong River facing landslide QLDT01 is the Wari Gaqu River, which flows into the Datong River. This results in high runoff flow of the Datong River, which is usually accompanied by transverse expansion when it is scoured downward along the river. Under continuous erosion by river flow or streams, the slope toe becomes too steep to hold itself, consequently resulting in slope failure [69]. Moreover, QLDT01 has a slope of about 17.5 degrees on average, with a slope height of about 66 m, which make it prone to slope failure.

Shaking from earthquakes may be a direct triggering factor of the QLDTL02 failure. According to data from the China Earthquake Networks Center (http://www.ceic.ac.cn, accessed on 10 November 2021), a Mw 5.2 earthquake occurred on 23 November 2015, with a focal depth of 10 km and a direct distance of 24 km from QLDT02 (Figure 1A). The Tuolaishan fault is the seismogenic fault of this earthquake according to the Qinghai Earthquake Administration, China (www.qhdzj.gov.cn, accessed on 10 November 2021). Earthquakes increase the occurrence of landslides due to ground shaking, liquefaction of susceptible sediments, and swelling of soil materials caused by shaking, which allows water to seep in rapidly. In addition, earthquakes can alter friction at the base of landslides, thus accelerating their movement over several days or weeks [70,71].

5.2. Hazard Analysis

We evaluate the stability of the landslide dam of QLDT01 based on the DBI calculation. According to the DBI criterion proposed by Ermini and Casagli [57], the state of landslide dam can be categorized as a stable domain (DBI < 2.75), an uncertain domain (2.75 < DBI < 3.08), and an unstable domain (DBI > 3.08). The height of the landslide dam ranges from 0 m at the toe of the landslide to about 40 m at the south bank of the Datong River. The landslide dam volume and catchment area are 69×10^3 m^3 and 4.4×10^3 m^2, respectively. Relying on different dam heights, the calculated DBI ranges from 2.55 to 2.97, with an average of 2.78 (Figure 10). We find that the toe ($H_d < 3.5$ m) and top ($H_d > 29$ m) of the slope in QLDTL01 can be considered as a stable domain, as their DBI is lower than 2.75. The landslide dam is in the uncertain domain in the middle of the slope (3.5 m < H_d < 29 m), which accounts for 70% of the entire slope. Thus, we infer that the QLDT01 is at risk of further slope collapse.

While only a portion of the slope (QLDT02) has collapsed, we evaluate the stability of the noncollapsed regions of the slope and the potential risks. Two long cracks (about 300–400 m) could be observed as of 2009. While one crack (QLDT01) collapsed in 2009, only a small portion of another crack developed into a landslide (QLDT02) in 2015. Based on the high-resolution UAV DEM (Figure 11B), we find that the slope height varies significantly and the slope gradient is large, providing geomorphological conditions for slope creep. In addition, many new cracks are found in the noncollapsed regions of the slope (Figure 11A), suggesting the occurrence of strong internal movement. Moreover, continuous InSAR-derived downslope movements are observed, further confirming the instability of the noncollapsed slope (PRZ in Figures 5 and 9). The volume of the PRZ region is about 12×10^5 m^3, which is 1.6 times larger than that of landslide QLDT01. In addition, there are temporary houses in the area for locals to graze animals. A potential slope failure may completely block the Datong River and cause a catastrophic disaster.

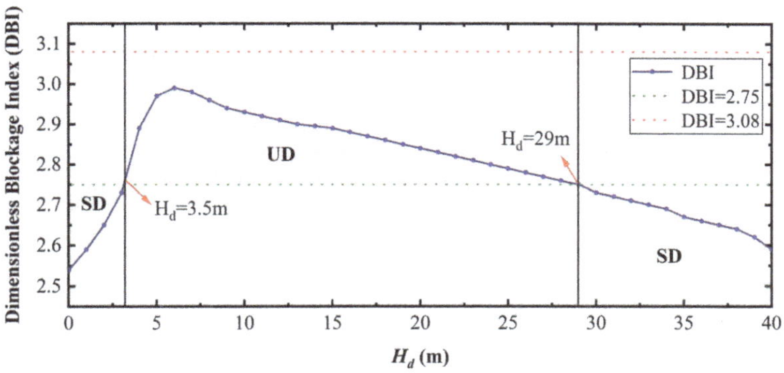

Figure 10. The dimensionless blockage index (DBI) diagram of landslide QLDT01. When the landslide dam is at the height of 3.5 to 29 m, the mean value of DBI is 2.87, which is in the uncertain domain (UD). When the height of landslide dam is lower than 3.5 m and higher than 29 m, the average DBI is 2.66, which is in the stable domain (SD).

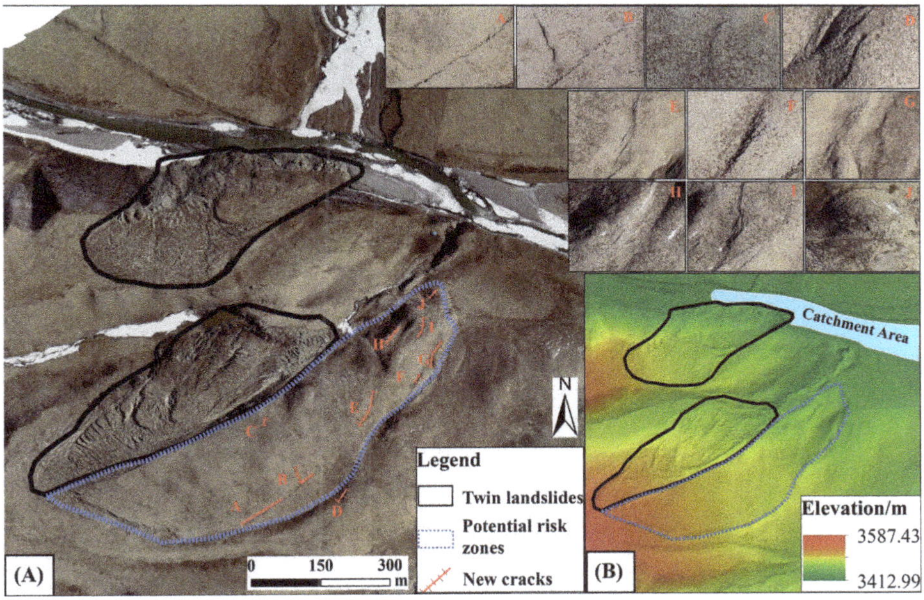

Figure 11. (**A**) The orthophoto map acquired by UAV in April 2021; the red lines denote the cracks. (**B**) The UAV-derived DEM map.

6. Conclusions

We have documented the spatiotemporal evolution of two adjacent landslides on the southeast slope of Qilian Mountain during their pre-failure and post-failure stages from 2008 to 2020 by integrating multisource optical and radar remote sensing techniques. The main conclusions are as follows:

1. The occurrence and development of the twin landslides and the adjacent river bank have been determined using high-resolution optical and radar images.
2. Significant downslope movements are observed within the twin landslides (up to 15 mm/day) and their surrounding slopes (up to 5 mm/day). The downslope movement exhibits distinct seasonality. During February-early July, the downslope velocity is nearly inactive; during late July-next January, mean velocity of about 4 mm/day is

observed. The seasonality of downslope movement during both pre- and post-failure stages suggests that the occurrence and development of the twin landslide are strongly influenced by freeze–thaw processes.

3. Combining data on precipitation, air temperature, and InSAR-based deformation history, we infer that the occurrence of the twin landslides is mainly related to extensive precipitation, warm winters, and freeze–thaw processes.
4. From the UAV-based geomorphological features, InSAR-derived downslope movements, and dimensionless blockage index, we infer that new collapse and slope failure events may occur within the twin landslides and their surroundings, which may completely block the Datong River and cause a catastrophic disaster.

Our study demonstrates the capability of multisource high-resolution remote sensing techniques to monitor landslide activities in cold regions. As the impacts of climate warming becoming more extensive, freeze–thaw-related slope instability in climate-sensitive regions (the boundary regions of permafrost and seasonally frozen ground, in this case) should be afforded greater attention.

Author Contributions: Conceptualization, T.W., J.C. and J.Z.; methodology, J.C. and J.Z.; validation, J.C., J.Z., J.H., X.Z., P.L. and L.Z.; investigation, T.W., J.C., X.W. and X.Z.; field work, J.C., J.Z., J.H., X.Z., X.M. and P.L.; writing—original draft preparation, J.C., J.Z. and T.W.; writing—review and editing, J.C., J.Z., T.W., J.H., X.W., X.M., X.Z.; visualization, J.Z. and J.C.; supervision, T.W.; funding acquisition, J.C. and T.W. All authors have read and agreed to the published version of the manuscript.

Funding: This research was supported by the National Natural Science Foundation of China (Grant No. 42001072 and 41771076), the fellowship of China Postdoctoral Science Foundation (2021T140702 and 2021M693374), the State Key Laboratory of Cryospheric Science (Grant No. sklcs-zz-2022), the National Cryosphere Desert Data Center Program (Grant No. E01Z790208), the Natural Science Foundation of Gansu Province of China (Grant No. 21JR7RA242), and the CAS "Special Research Assistant program" (Jie Chen).

Data Availability Statement: Not applicable.

Conflicts of Interest: The authors declare no conflicts of interest.

References

1. Highland, L.; Bobrowsky, P.T. *The Landslide Handbook: A Guide to Understanding Landslides*; US Geological Survey: Reston, VA, USA, 2008.
2. Anbalagan, R. Landslide hazard evaluation and zonation mapping in mountainous terrain. *Eng. Geol.* **1992**, *32*, 269–277. [CrossRef]
3. Patton, A.I.; Rathburn, S.L.; Capps, D.M. Landslide response to climate change in permafrost regions. *Geomorphology* **2019**, *340*, 116–128. [CrossRef]
4. Gariano, S.L.; Guzzetti, F. Landslides in a changing climate. *Earth-Sci. Rev.* **2016**, *162*, 227–252. 2016.08.011. [CrossRef]
5. Lantz, T.C.; Kokelj, S.V. Increasing rates of retrogressive thaw slump activity in the Mackenzie Delta region, N.W.T., Canada. *Geophys. Res. Lett.* **2008**, *35*, L06502. [CrossRef]
6. Segal, R.A.; Lantz, T.C.; Kokelj, S.V. Acceleration of thaw slump activity in glaciated landscapes of the Western Canadian Arctic. *Environ. Res. Lett.* **2016**, *11*, 034025. [CrossRef]
7. Ward Jones, M.K.; Pollard, W.H.; Jones, B.M. Rapid initialization of retrogressive thaw slumps in the Canadian high Arctic and their response to climate and terrain factors. *Environ. Res. Lett.* **2019**, *14*, 055006. [CrossRef]
8. Lewkowicz, A.G.; Way, R.G. Extremes of summer climate trigger thousands of thermokarst landslides in a High Arctic environment. *Nat. Commun.* **2019**, *10*, 1329. [CrossRef]
9. Huang, L.; Luo, J.; Lin, Z.; Niu, F.; Liu, L. Using deep learning to map retrogressive thaw slumps in the Beiluhe region (Tibetan Plateau) from CubeSat images. *Remote Sens. Environ.* **2020**, *237*, 111534. [CrossRef]
10. Mu, C.; Shang, J.; Zhang, T.; Fan, C.; Wang, S.; Peng, X.; Zhong, W.; Zhang, F.; Mu, M.; Jia, L. Acceleration of thaw slump during 1997–2017 in the Qilian Mountains of the northern Qinghai-Tibetan plateau. *Landslides* **2020**, *17*, 1051–1062. [CrossRef]
11. Luo, J.; Niu, F.; Lin, Z.; Liu, M.; Yin, G. Recent acceleration of thaw slumping in permafrost terrain of Qinghai-Tibet Plateau: An example from the Beiluhe Region. *Geomorphology* **2019**, *341*, 79–85. [CrossRef]
12. Farquharson, L.M.; Romanovsky, V.E.; Cable, W.L.; Walker, D.A.; Kokelj, S.V.; Nicolsky, D.D. Climate Change Drives Widespread and Rapid Thermokarst Development in Very Cold Permafrost in the Canadian High Arctic. *Geophys. Res. Lett.* **2019**, *46*, 6681–6689. [CrossRef]

13. Mondini, A.C.; Guzzetti, F.; Chang, K.T.; Monserrat, O.; Martha, T.R.; Manconi, A. Landslide failures detection and mapping using Synthetic Aperture Radar: Past, present and future. *Earth-Sci. Rev.* **2021**, *216*, 103574. [CrossRef]
14. Zhao, D.; Qu, C.; Shan, X.; Zuo, R.; Liu, Y.; Gong, W.; Zhang, G. Broadscale postseismic deformation and lower crustal relaxation in the central Bayankala Block (central Tibetan Plateau) observed using InSAR data. *J. Asian Earth Sci.* **2018**, *154*, 26–41. [CrossRef]
15. Guzzetti, F.; Mondini, A.C.; Cardinali, M.; Fiorucci, F.; Santangelo, M.; Chang, K.T. Landslide inventory maps: New tools for an old problem. *Earth-Sci. Rev.* **2012**, *112*, 42–66. [CrossRef]
16. Martha, T.R.; Kerle, N.; Van Westen, C.J.; Jetten, V.; Kumar, K.V. Segment optimization and data-driven thresholding for knowledge-based landslide detection by object-based image analysis. *IEEE Trans. Geosci. Remote Sens.* **2011**, *49*, 4928–4943. [CrossRef]
17. Hölbling, D.; Füreder, P.; Antolini, F.; Cigna, F.; Casagli, N.; Lang, S. A Semi-Automated Object-Based Approach for Landslide Detection Validated by Persistent Scatterer Interferometry Measures and Landslide Inventories. *Remote Sens.* **2012**, *4*, 1310–1336. [CrossRef]
18. Lu, P.; Stumpf, A.; Kerle, N.; Casagli, N. Object-oriented change detection for landslide rapid mapping. *IEEE Geosci. Remote Sens. Lett.* **2011**, *8*, 701–705. [CrossRef]
19. Catani, F. Landslide detection by deep learning of non-nadiral and crowdsourced optical images. *Landslides* **2021**, *18*, 1025–1044. [CrossRef]
20. Booth, A.M.; McCarley, J.; Hinkle, J.; Shaw, S.; Ampuero, J.; Lamb, M.P. Transient reactivation of a deep-seated landslide by undrained loading captured with repeat airborne and terrestrial lidar. *Geophys. Res. Lett.* **2018**, *45*, 4841–4850. 77812. [CrossRef]
21. Rossi, G.; Tanteri, L.; Tofani, V.; Vannocci, P.; Moretti, S.; Casagli, N. Multitemporal UAV surveys for landslide mapping and characterization. *Landslides* **2018**, *15*, 1045–1052. [CrossRef]
22. Niethammer, U.; James, M.R.; Rothmund, S.; Travelletti, J.; Joswig, M. UAV-based remote sensing of the Super-Sauze landslide: Evaluation and results. *Eng. Geol.* **2012**, *128*, 2–11. [CrossRef]
23. Bernhard, P.; Zwieback, S.; Leinss, S.; Hajinsek, I. Mapping retrogressive thaw slumps using single-pass TanDEM-X observations. *IEEE J. Sel. Top. Appl. Earth Obs. Remote. Sens.* **2020**, *13*, 3263–3280. [CrossRef]
24. Bernhard, P.; Zwieback, S.; Bergner, N.; Hajnsek, I. Assessing volumetric change distributions and scaling relations of retrogressive thaw slumps across the Arctic. *Cryosphere* **2022**, *16*, 1–15. [CrossRef]
25. Colesanti, C.; Ferretti, A.; Prati, C.; Rocca, F. Monitoring landslides and tectonic motions with the Permanent Scatterers Technique. *Eng. Geol.* **2003**, *68*, 3–14. [CrossRef]
26. Strozzi, T.; Delaloye, R.; Kääb, A.; Ambrosi, C.; Perruchoud, E.; Wegmüller, U. Combined observations of rock mass movements using satellite SAR interferometry, differential GPS, airborne digital photogrammetry, and airborne photography interpretation. *J. Geophys. Res. Earth Surf.* **2010**, *115*. [CrossRef]
27. Hu, X.; Bürgmann, R.; Schulz, W.H.; Fielding, E.J. Four-dimensional surface motions of the Slumgullion landslide and quantification of hydrometeorological forcing. *Nat. Commun.* **2020**, *11*, 2792. [CrossRef]
28. Bekaert, D.P.; Handwerger, A.L.; Agram, P.; Kirschbaum, D.B. InSAR-based detection method for mapping and monitoring slow-moving landslides in remote regions with steep and mountainous terrain: An application to Nepal. *Remote Sens. Environ.* **2020**, *249*, 111983. [CrossRef]
29. Meng, Q.; Li, W.; Raspini, F.; Xu, Q.; Peng, Y.; Ju, Y.; Zheng, Y.; Casagli, N. Time-series analysis of the evolution of large-scale loess landslides using InSAR and UAV photogrammetry techniques: A case study in Hongheyan, Gansu Province, Northwest China. *Landslides* **2020**, *18*, 251–265. [CrossRef]
30. Xie, M.; Zhao, W.; Ju, N.; He, C.; Huang, H.; Cui, Q. Landslide evolution assessment based on InSAR and real-time monitoring of a large reactivated landslide, Wenchuan, China. *Eng. Geol.* **2020**, *277*, 105781. [CrossRef]
31. Zhu, Y.; Qiu, H.; Yang, D.; Liu, Z.; Ma, S.; Pei, Y.; He, J.; Du, C.; Sun, H. Pre- and post-failure spatiotemporal evolution of loess landslides: a case study of the Jiangou landslide in Ledu, China. *Landslides* **2021**, *18*, 3475–3484. [CrossRef]
32. Eker, R.; Aydın, A. Long-term retrospective investigation of a large, deep-seated, and slow-moving landslide using InSAR time series, historical aerial photographs, and UAV data: The case of Devrek landslide (NW Turkey). *Catena* **2021**, *196*, 104895. [CrossRef]
33. Cao, C.; Zhu, K.; Xu, P.; Shan, B.; Yang, G.; Song, S. Refined landslide susceptibility analysis based on InSAR technology and UAV multi-source data. *J. Clean. Prod.* **2022**, *368*, 133146. [CrossRef]
34. Meng, Q.; Intrieri, E.; Raspini, F.; Peng, Y.; Liu, H.; Casagli, N. Satellite-based interferometric monitoring of deformation characteristics and their relationship with internal hydrothermal structures of an earthflow in Zhimei, Yushu, Qinghai-Tibet Plateau. *Remote Sens. Environ.* **2022**, *273*, 112987. [CrossRef]
35. Zou, D.; Zhao, L.; Sheng, Y.; Chen, J.; Hu, G.; Wu, T.; Wu, J.; Xie, C.; Wu, X.; Pang, Q.; et al. A new map of permafrost distribution on the Tibetan Plateau. *Cryosphere* **2017**, *11*, 2527–2542. [CrossRef]
36. Wang, Q.; Jin, H.; Zhang, T.; Cao, B.; Peng, X.; Wang, K.; Xiao, X.; Guo, H.; Mu, C.; Li, L. Hydro-thermal processes and thermal offsets of peat soils in the active layer in an alpine permafrost region, NE Qinghai-Tibet plateau. *Glob. Planet. Chang.* **2017**, *156*, 1–12. [CrossRef]
37. Chen, R.; Han, C.; Liu, J.; Yang, Y.; Liu, Z.; Wang, L.; Kang, E. Maximum precipitation altitude on the northern flank of the Qilian Mountains, northwest China. *Hydrol. Res.* **2018**, *49*, 1696–1710. [CrossRef]

38. Liu, J.; Chen, R. Discriminating types of precipitation in Qilian Mountains, Tibetan Plateau. *J. Hydrol. Reg. Stud.* **2016**, *5*, 20–32. [CrossRef]
39. Rignot, E.; Echelmeyer, K.; Krabill, W. Penetration depth of interferometric synthetic-aperture radar signals in snow and ice. *Geophys. Res. Lett.* **2001**, *28*, 3501–3504. [CrossRef]
40. Zebker, H.; Shankar, P.; Hooper, A. InSAR remote sensing over decorrelating terrains: Persistent scattering methods. In Proceedings of the 2007 IEEE Radar Conference, Waltham, MA, USA, 17–20 April 2007; pp. 717–722.
41. Malamud, B.D.; Turcotte, D.L.; Guzzetti, F.; Reichenbach, P. Landslide inventories and their statistical properties. *Earth Surf. Process. Landforms* **2004**, *29*, 687–711. [CrossRef]
42. Stumpf, A.; Malet, J.P.; Delacourt, C. Correlation of satellite image time-series for the detection and monitoring of slow-moving landslides. *Remote Sens. Environ.* **2017**, *189*, 40–55. [CrossRef]
43. Fan, X.; Xu, Q.; Alonso-Rodriguez, A.; Subramanian, S.S.; Li, W.; Zheng, G.; Dong, X.; Huang, R. Successive landsliding and damming of the Jinsha River in eastern Tibet, China: Prime investigation, early warning, and emergency response. *Landslides* **2019**, *16*, 1003–1020. [CrossRef]
44. Rosen, P.A.; Hensley, S.; Joughin, I.R.; Fuk, K.L.; Madsen, S.N.; Rodriguez, E.; Goldstein, R.M. Synthetic aperture radar interferometry. *Proc. IEEE* **2000**, *88*, 333–382. [CrossRef]
45. Hao, J.; Wu, T.; Wu, X.; Hu, G.; Zou, D.; Zhu, X.; Zhao, L.; Li, R.; Xie, C.; Ni, J.; et al. Investigation of a Small Landslide in the Qinghai-Tibet Plateau by InSAR and Absolute Deformation Model. *Remote Sens.* **2019**, *11*, 2126. [CrossRef]
46. Dini, B.; Daout, S.; Manconi, A.; Loew, S. Classification of slope processes based on multitemporal DInSAR analyses in the Himalaya of NW Bhutan. *Remote Sens. Environ.* **2019**, *233*, 111408. [CrossRef]
47. Liu, L.; Millar, C.I.; Westfall, R.D.; Zebker, H.A. Surface motion of active rock glaciers in the Sierra Nevada, California, USA: Inventory and a case study using InSAR. *Cryosphere* **2013**, *7*, 1109–1119. [CrossRef]
48. Berardino, P.; Fornaro, G.; Lanari, R.; Sansosti, E. A new algorithm for surface deformation monitoring based on small baseline differential SAR interferograms. *IEEE Trans. Geosci. Remote Sens.* **2002**, *40*, 2375–2383. [CrossRef]
49. Werner, C.; Wegmüller, U.; Strozzi, T.; Wiesmann, A. Gamma SAR and interferometric processing software. In Proceedings of the ERS-ENVISAT Symposium, Gothenburg, Sweden, 16–20 October 2000; Volume 1620, p. 1620.
50. Goldstein, R.M.; Werner, C.L. Radar interferogram filtering for geophysical applications. *Geophys. Res. Lett.* **1998**, *25*, 4035–4038. [CrossRef]
51. Chen, C.W. Statistical-Cost Network-Flow Approaches to Two-Dimensional Phase Unwrapping for Radar Interferometry. Ph.D. Thesis, Stanford University, Stanford, CA, USA, 2001.
52. Delacourt, C.; Briole, P.; Achache, J.A. Tropospheric corrections of SAR interferograms with strong topography. Application to Etna. *Geophys. Res. Lett.* **1998**, *25*, 2849–2852. [CrossRef]
53. Hu, Y.; Liu, L.; Wang, X.; Zhao, L.; Wu, T.; Cai, J.; Zhu, X.; Hao, J. Quantification of permafrost creep provides kinematic evidence for classifying a puzzling periglacial landform. *Earth Surf. Process. Landforms* **2021**, *46*, 465–477. [CrossRef]
54. Nelson, F.E.; Shiklomanov, N.I.; Mueller, G.R.; Hinkel, K.M.; Walker, D.A.; Bockheim, J.G. Estimating Active-Layer Thickness over a Large Region: Kuparuk River Basin, Alaska, U.S.A. *Arct. Alp. Res.* **1997**, *29*, 367–378. [CrossRef]
55. Costa, J.E.; Schuster, R.L. Formation and Failure of Natural Dams. *Bull. Geol. Soc. Am.* **1988**, *100*, 1054–1068.<1054:TFAFON>2.3.CO;2. [CrossRef]
56. Casagli, N.; Ermini, L.; Rosati, G. Determining grain size distribution of the material composing landslide dams in the Northern Apennines: Sampling and processing methods. *Eng. Geol.* **2003**, *69*, 83–97. [CrossRef]
57. Ermini, L.; Casagli, N. Prediction of the behaviour of landslide dams using a geomorphological dimensionless index. *Earth Surf. Process. Landforms* **2003**, *28*, 31–47. [CrossRef]
58. Orlowsky, B.; Seneviratne, S.I. Global changes in extreme events: Regional and seasonal dimension. *Clim. Chang.* **2012**, *110*, 669–696. [CrossRef]
59. Huggel, C.; Clague, J.J.; Korup, O. Is climate change responsible for changing landslide activity in high mountains? *Earth Surf. Process. Landforms* **2012**, *37*, 77–91. [CrossRef]
60. Owczarek, P.; Opała-Owczarek, M.; Boudreau, S.; Lajeunesse, P.; Stachnik, Ł. Re-activation of landslide in sub-Arctic areas due to extreme rainfall and discharge events (the mouth of the Great Whale River, Nunavik, Canada). *Sci. Total Environ.* **2020**, *744*, 140991. [CrossRef]
61. Pavlova, I.; Jomelli, V.; Brunstein, D.; Grancher, D.; Martin, E.; Déqué, M. Debris flow activity related to recent climate conditions in the French Alps: A regional investigation. *Geomorphology* **2014**, *219*, 248–259. [CrossRef]
62. McRoberts, E.C.; Morgenstern, N.R. The Stability of Thawing Slopes. *Can. Geotech. J.* **1974**, *11*, 447–469. [CrossRef]
63. Ding, Y.; Zhang, S.; Zhao, L.; Li, Z.; Kang, S. Global warming weakening the inherent stability of glaciers and permafrost. *Sci. Bull.* **2019**, *64*, 245–253. [CrossRef]
64. Lacelle, D.; Brooker, A.; Fraser, R.H.; Kokelj, S.V. Distribution and growth of thaw slumps in the Richardson Mountains-Peel Plateau region, northwestern Canada. *Geomorphology* **2015**, *235*, 40–51. [CrossRef]
65. Savi, S.; Comiti, F.; Strecker, M.R. Pronounced increase in slope instability linked to global warming: A case study from the eastern European Alps. *Earth Surface Process Landforms* **2021**, *46*, 1328–1347. [CrossRef]
66. Liu, L.; Zhang, T.; Wahr, J. InSAR measurements of surface deformation over permafrost on the North Slope of Alaska. *J. Geophys. Res. Earth Surf.* **2010**, *115*. [CrossRef]

67. Chen, J.; Wu, T.; Zou, D.; Liu, L.; Wu, X.; Gong, W.; Zhu, X.; Li, R.; Hao, J.; Hu, G.; et al. Magnitudes and patterns of large-scale permafrost ground deformation inferred from Sentinel-1 InSAR on the central Qinghai-Tibet Plateau. *Remote Sens. Environ.* **2022**, *268*, 112778. [CrossRef]
68. Dini, B.; Manconi, A.; Loew, S. Investigation of slope instabilities in NW Bhutan as derived from systematic DInSAR analyses. *Eng. Geol.* **2019**, *259*, 105111. [CrossRef]
69. Lacroix, P.; Berthier, E.; Maquerhua, E.T. Earthquake-driven acceleration of slow-moving landslides in the Colca valley, Peru, detected from Pléiades images. *Remote Sens. Environ.* **2015**, *165*, 148–158. [CrossRef]
70. Moro, M.; Chini, M.; Saroli, M.; Atzori, S.; Stramondo, S.; Salvi, S. Analysis of large, seismically induced, gravitational deformations imaged by high-resolution COSMO-SkyMed synthetic aperture radar. *Geology* **2011**, *39*, 527–530. [CrossRef]
71. Bontemps, N.; Lacroix, P.; Doin, M.P. Inversion of deformation fields time-series from optical images, and application to the long term kinematics of slow-moving landslides in Peru. *Remote Sens. Environ.* **2018**, *210*, 144–158. [CrossRef]

remote sensing

MDPI

Article

An MT-InSAR Data Partition Strategy for Sentinel-1A/B TOPS Data

Yuexin Wang [1], Guangcai Feng [1,*], Zhixiong Feng [2], Yuedong Wang [1], Xiuhua Wang [1], Shuran Luo [3], Yinggang Zhao [1] and Hao Lu [1]

[1] School of Geosciences and Info-Physics, Central South University, Changsha 410083, China
[2] Guangzhou Urban Planning and Design Survey Research Institute, Guangzhou 510060, China
[3] Guangdong Research Institute of Water Resources and Hydropower, Guangzhou 510635, China
* Correspondence: fredgps@csu.edu.cn; Tel.: +86-182-7486-7449

Abstract: The Sentinel-1A/B satellite launched by European Space Agency (ESA) in 2014 provides a huge amount of free Terrain Observation by Progressive Scans (TOPS) data with global coverage to the public. The TOPS data have a frame width of 250 km and have been widely used in surface deformation monitoring. However, traditional Multi-Temporal Interferometric Synthetic Aperture Radar (MT-InSAR) methods require large computer memory and time when processing full resolution data with large width and long strips. In addition, they hardly correct atmospheric delays and orbital errors accurately over a large area. In order to solve these problems, this study proposes a data partition strategy based on MT-InSAR methods. We first process the partitioned images over a large area by traditional MT-InSAR method, then stitch the deformation results into a complete deformation result by correcting the offsets of adjacent partitioned images. This strategy is validated in a flat urban area (Changzhou City in Jiangsu province, China), and a mountainous region (Qijiang in Chongqing City, China). Compared with traditional MT-InSAR methods, the precision of the results obtained by the new strategy is improved by about 5% for Changzhou city and about 15% for Qijiang because of its advantage in atmospheric delay correction. Furthermore, the proposed strategy needs much less memory and time than traditional methods. The total time needed by the traditional method is about 20 h, and by the proposed method, is about 8.7 h, when the number of parallel processing is 5 in the Changzhou city case. The time will be further reduced when the number of parallel processes increases.

Keywords: MT-InSAR; ground deformation monitoring; Sentinel-1A/B; image partition; block adjustment

Citation: Wang, Y.; Feng, G.; Feng, Z.; Wang, Y.; Wang, X.; Luo, S.; Zhao, Y.; Lu, H. An MT-InSAR Data Partition Strategy for Sentinel-1A/B TOPS Data. *Remote Sens.* **2022**, *14*, 4562. https://doi.org/10.3390/rs14184562

Academic Editors: Paolo Mazzanti and Saverio Romeo

Received: 26 July 2022
Accepted: 2 September 2022
Published: 13 September 2022

1. Introduction

Due to large coverage and high-precision, Interferometric Synthetic Aperture Radar (InSAR) has been widely used for mapping surface deformation, such as urban surface deformation [1,2], seismic deformation [3,4], landslide displacement [5–9], and mining subsidence [10]. With the fast development of SAR satellite technology [11], the observation range and frequency are both improved [12–14], providing cycle monitoring for a large-scale or national wide area. However, the traditional processing strategies for Multi-Temporal Interferometric Synthetic Aperture Radar (MT-InSAR) cannot efficiently process the huge number of images with large spatial and temporal coverage. Furthermore, the possible atmospheric phase screen and orbital errors exist in the SAR images with wide spatial coverage are difficult to be corrected. Therefore, optimizing the InSAR processing strategy and parameters is crucial for the application of wide-area InSAR data.

Using supercomputers or distributed computing systems, such as CASearth Cloud Infrastructure Platform [15], and ESA's G-POD environment [16], is a way to improve the data processing efficiency, but it is too expensive to be popularized. Another way is to segment a large image into small blocks, which can significantly reduce the computation burden and complexity in one block and improve the efficiency of data processing. Currently, data partition strategies are applied in some steps of SAR data processing, such as phase unwrapping [17–20], orbital error correction [21], atmospheric correction [22,23], and PS point decomposition [24,25], but not the whole data process. GAMMA software provides a well-known patch-based point target analysis method, Interferometric Point Target Analysis (IPTA) [26]. However, the method of using local reference points between neighboring patches to extend the results to adjacent blocks is highly affected by unstable connections, resulting in errors propagating in the result easily [27]. StaMPS method and software [28] also provide a block strategy to select permanent scatterers, but it is time-consuming for large-scale areas [29]. Data block processing often introduces systematic errors, such as reference basis errors [29–31]. To remove these errors, external data (GNSS and leveling) and modeling [32,33] are needed. Furthermore, the correction efficacy and precision strongly depend on the precision and spatial distribution of external data.

To address the above problems, we propose a strategy to divide the original data into small blocks by GAMMA software and process these blocks independently by the traditional MT-InSAR method. Then, we use the least square method to estimate the basis between each block and mosaic the corrected block results to obtain the overall results. To validate our strategy, we selected the Sentinel-1 Terrain Observation by Progressive Scans (TOPS) data of a city in the plain area (Changzhou City, Jiangsu Province) and a city in the mountainous area (Qijiang, Chongqing City) in China for the experiment. The results obtained by the traditional and the proposed methods are compared in terms of precision, memory consumption, and time consumption. We also discuss the optimal overlap ratio of blocks and the application of the proposed method.

This study is organized as follows: Section 2 describes the proposed method in detail. The study area and the datasets are introduced in Section 3. In Section 4, we compare the precision and time consumption of the proposed method and traditional method. The block approach and the applicability scenarios of our method is also discussed in Section 5. Finally, some conclusions are drawn in Section 6.

2. The Block MT-InSAR Data Processing Strategy

In order to solve the great calculation burden caused by Sentinel-1A/B TOPS data of large spatial and temporal coverage, this paper proposes a data partition strategy based on the MT-InSAR data method, referred to as the block MT-InSAR algorithm. First, the TOPS data are co-registered in the study area to obtain registered single look complex (RSLC), and then the RSLC data are partitioned and processed separately by the traditional MT-InSAR algorithm. Then, the results are corrected by the adjustment model based on the spatial consistency of homonymy points (the same ground deformation points located in different blocks within the overlap areas.). Finally, the results are spliced to obtain the continuous overall deformation results. The general flow of the method is shown in Figures 1 and A1.

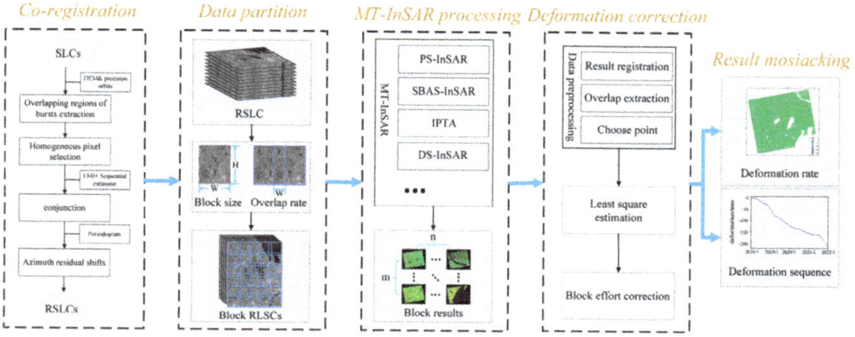

Figure 1. Flowchart of the block MT-InSAR algorithm.

2.1. Data Partition and Block Processing

Even partition [29], quadtree partition [22], and clustering algorithm partition [19,34] have been used in some parts of data processing, such as atmospheric delay removal, PS network construction, and phase unwrapping. In order to facilitate the splicing process, this paper uses even partition to divide the original data into small blocks, in which the block size and the overlap ratio should be considered.

Block size affects the precision and the processing efficiency of the phase unwrapping, atmospheric delay, and orbit error partition. If the block interferograms are highly coherent and easy to unwrap, the block size has little effect on phase unwrapping precision, but a small block size would lead to high unwrapping efficiency [17]. Additionally, the atmospheric delay in the MT-InSAR consists of a short-scale (few kilometers) and a long-scale (tens of kilometers) component [35], so the block size smaller than these scales is conducive to removing atmospheric delay. However, the too-small block size may remove the long-wavelength deformation signal. The block width and height should be larger than 1/3 ALOS-2 data in range and azimuth for ALOS-2 (70 km) datasets [21]. Therefore, we set the initial block size as ~30 km in length and width for S1A/B TOPS data.

The overlap ratio between blocks also affects the reliability of results and data processing efficiency. The larger the overlap ratio the higher the reliability. For example, if a block area is overlapped with the surrounding blocks in four directions by 10%, 36% of the small block is overlapped with the surrounding blocks, and the overlap area will become 96% when the overlap ratio is 40% in width and height directions. (Figure 2a) However, increasing the overlap ratio will lead to a lot of repeated calculations and reduce the data processing efficiency. In order to improve the result reliability (>50% overlap area), we take off the balance between the reliability of results and choose the overlap ratio of 20~40% for further experiments.

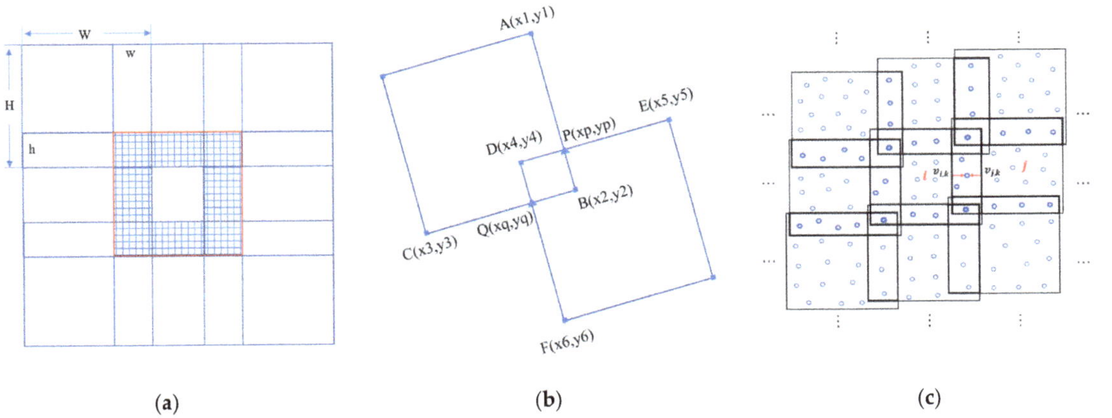

Figure 2. (**a**) Diagram of overlap ratio and overlap area. w and h are the width and height of the overlap region, respectively, and W and H are the width and height of the image, respectively. So, the overlap ratio is w/W or h/H. The shadow area is the overlap area in the block (red line) with the surrounding blocks, and the overlap area is $1-(1-2\times w/W)\times(1-2\times h/H)$. (**b**) Diagram of the coordinates acquirement of corner points in the overlap region. (**c**) Diagram of the adjustment model. The circles represent the deformation points. The thicker the circles, the more times the regions are overlapped. v represents the deformation rate of each point.

2.2. Results Correction Based on Least Square Estimation

After data partition, we process each block of data to obtain the deformation results using the improved IPTA-InSAR method [36,37]. The obtained deformation results of all the blocks are preprocessed through three steps. (1) Co-registration. Due to the location errors caused by orbital errors and low resolution of DEM; the location of deformation results may have a systematic deviation of about 1–2 pixels after geocoding. Such deviation can be solved by an overall offset correction using some feature points on the ground. (2) Automatic extraction of overlapping regions. The deformation rates of the homonymy points in the overlapping region determine the correction precision, so it is necessary to identify the overlapping regions between the deformation results first. We use the topological relationship between image overlays (quadrangles) to find the coordinates of corner points in the overlap region (Figure 2b). (3) High-quality homonymy point selection. We select homonymy points with high coherence and small uncertainty. After these operations, the deformation results can be corrected by adjustment.

Errors can be removed during data processing. However, data partition makes each block have a local reference point, and the benchmarks of these reference points may be different, resulting in discrepancies between the results of adjacent blocks and affecting the precision of the overall results. The difference in the deformation rates of the homonymy points in the overlapping area is described as

$$vel_{i,k} - vel_{j,k} = \delta_{i,k} - \delta_{j,k} \quad (1)$$

where $vel_{i,k}$ and $vel_{j,k}$ denote the deformation rate at deformation point k in image i and j, respectively, and $\delta_{i,k}$, and $\delta_{j,k}$ denote the error of the corresponding points. Since the error contains mainly the difference in benchmarks, this value can be assumed as a constant.

The matrix form of Equation (1) is:

$$V = B\hat{X} - L \quad (2)$$

where $V = \begin{bmatrix}\delta v_1 & \delta v_2 & \cdots & \delta v_i & \cdots & \delta v_M\end{bmatrix}^T$ is the residual of the calculated values and the observations, $\hat{X} = \begin{bmatrix}\hat{x}_1 & \hat{x}_2 & \cdots & \hat{x}_M\end{bmatrix}^T$ is the difference between the reference points of adjacent images estimated by the least square method. B is the coefficient matrix. $L = \begin{bmatrix}vel_{1,k} - vel_{2,k} & vel_{i,k} - vel_{j,k} & \cdots\end{bmatrix}^T$.

To solve Equation (2), we have to determine the weights of the blocks according to the quality of the data involved in the adjustment.

$$D(L) = \sigma_0^2 P^{-1} \quad (3)$$

σ_0^2 denotes the variance of unit weight and P is the weight matrix. Assume that the uncertainty of point i is given by δ. Then, the weight of the point is

$$p_i = \frac{C}{\delta_i} \quad (4)$$

In this study, the data are partitioned into small blocks, which are processed independently. The Helmert variance component estimation for multiple data classes is applied to optimize the solution weights of each data set.

$$S\hat{\theta} = W_\theta \quad (5)$$

where $\hat{\theta} = \begin{bmatrix}\hat{\sigma}_{01}^2 & \hat{\sigma}_{02}^2 & \cdots & \hat{\sigma}_{0M}^2\end{bmatrix}^T$ is the estimated variance of unit weight. W_θ is the square sum of the corrected values, $W_\theta = \begin{bmatrix}V_1^T P V_1 & V_2^T P V_2 & \cdots & V_M^T P V_M\end{bmatrix}^T$. S is the coefficient matrix. After obtaining $\hat{\theta}$, $\hat{X}$ is solved using the least square method. Repeat the above process until $\hat{\theta}$ satisfies the given threshold $T = 3\delta_0$, and the corresponding solution is the optimal $\hat{X}$ for each SAR image block. The corrected deformation rate is obtained by Equation (6).

$$\hat{vel}_{i,k} = vel_{i,k} - \hat{x}_i \quad (6)$$

The posteriori variance of unit weight and the covariance array are used to evaluate the adjustment observation. They can be obtained by

$$\hat{\sigma}_0^2 = \frac{V^T P V}{r} \tag{7}$$

$$Q_{\hat{x}_i} = B_i^T P B_i \tag{8}$$

In Equation (7), r is the number of redundant observations, and it can be referred to as the number of degrees of freedom, $r = N - M$, with N denoting the number of observations, and M denoting the row number of $\hat{X}$. According to the error propagation, the covariance of the estimate of the homonymy points in the overlap region can be obtained by Equation (9).

$$Q_{\hat{L}\hat{L}} = \left(BQ_{\hat{x}}^{-1}B^T P\right) Q (BQ_{\hat{x}}^{-1}B^T P)^T = BQ_{\hat{x}}^{-1}B^T \tag{9}$$

To verify the precision of the adjustment results, the deformation difference of the homonymy points before and after correction are compared. The block processing results are verified by comparing with that of the traditional processing method (the result without partitioning processing).

2.3. Result Mosaicking

The final step is to mosaic the corrected deformation results of all blocks. After geocoding, the block results are horizontally mosaicked. After correction, the deformation of the homonymy points in the overlapping area may still have differences, due to the different errors distribution. We adopt the weighted average method to merge the deformation of the homonymy points.

After correcting the deformation rate, we correct the deformation sequence. Assuming that the deformation is linear, $T(T_1, T_2 \cdots T_a \cdots T_t)$ is a deformation sequence, and the corrected deformation sequence at time T_a can be obtained by

$$\hat{S}_a = \int_{T_1}^{T_a} (v_a + \hat{x}_a) dT = S_a + \hat{x}_a (T_a - T_1) \tag{10}$$

where $\hat{S}_a$ is the accumulated deformation after correction, v_a is the deformation rate. $\hat{x}_a$ is the correction of deformation rate, but cannot be calculated, S_a is the accumulated deformation before correction. If the deformation is linear, $\hat{x}_a$ is equal to the correction of average deformation rate in the deformation sequences T. Variant $\hat{x}$ can be calculated by Equation (2), so the equation can be instead of Equation (11). Figure 3 is the diagram of deformation sequence correction.

$$\hat{S}_a = S_a + \hat{x}(T_a - T_1) \tag{11}$$

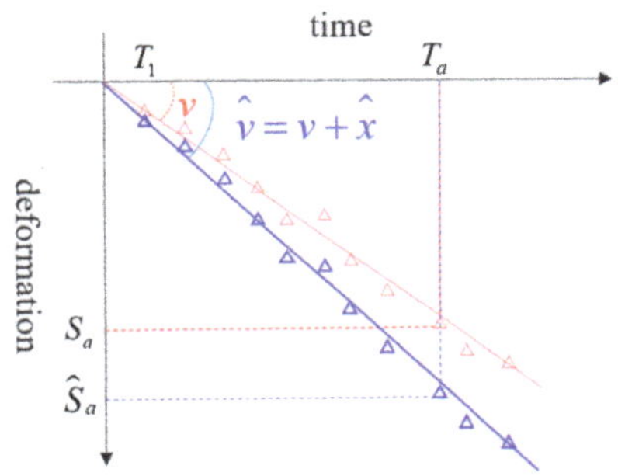

Figure 3. Deformation sequence correction diagram.

3. Experiment and Data Processing

3.1. Study Area and Datasets

Two study areas are selected to validate the proposed method. One is in Changzhou City (31°09′–32°04′N, 119°08′–120°12′E), a coastal city in eastern China. This area is a plain with an elevation of about 10 m [38]. It has a highly developed economy and urban industry. The continuous expansion of urban and engineering projects has changed the geological environment and led to frequent geological hazards. So, surface subsidence monitoring in the area is necessary.

The other study area is Qijiang (28°27′–29°11′N, 106°23′–107°03′E), in western China. It is in the transition zone from the southeastern edge of the Sichuan basin to the Yunnan-Guizhou plateau. The topography is undulating. The mountainous area accounts for 67.6% of the total area and the hills account for 32.4%. The average elevation of this area is 254.8 m [39].

These two regions are used to test the applicability of the proposed method under different error conditions.

We collected 110 Sentinel-1A/B TOPS images covering Changzhou City from path T69 and frame 99, between 5 January 2018 and 31 December 2020, and acquired 115 Sentinel-1 images over Qijiang from path T55 and frame 92, between 9 January 2018 and 31 December 2021. Specific image parameters and image acquisition time are shown in Tables 1 and A1 in Appendix A.

Table 1. The image parameters of the study areas.

Study Area	Direction	Path	Heading	Incidence	Pixel Spacing (Rg × Az)	Num of Images
Changzhou	Ascending	T69	−12.79°	36.65°	2.33 × 13.98 m	110
Qijiang		T55	−12.65°	43.64°	2.33 × 13.96 m	115

3.2. Data Processing

Using the method described in Section 2, we partitioned the acquired single look complex (SLC) images after co-registration and obtained 30 small blocks with overlapping regions (Figure 4). The block size in Changzhou City is about 7000 × 1400 (pixels), and the overlap rate is about 30%; the block size in Qijiang is about 6400 × 1600 (pixels), and the overlap rate is about 25%.

The spatial baselines of Sentinel-1 images are short, so we connected each image with two (temporally) adjacent images to form a network, only considering the temporal baselines. A multi-look operation (range: azimuth = 5:1) was applied to reduce the noise. After the multi-look operation, the image size of Changzhou city was reduced to 1600 × 1400 (pixels) and that of Qijiang was reduced to 1500 × 1600 (pixels). The data were processed by minimum cost flow (MCF) for phase unwrapping, and Goldstein filtering for noise mitigation. The PS points were selected considering the phase coherence threshold and the amplitude dispersion threshold of the amplitude map. Orbital error phases were removed by polynomial fitting. Most atmospheric phases were removed by differencing between neighboring PS points, and the remaining was removed by spatial-temporal filtering. The topographic residual phases were then removed using linear regression. Finally, the deformation sequence was solved from the remaining phases using Singular Value Decomposition (SVD). The obtained time series of deformation was corrected by the method introduced in Section 2.2. The average deformation of the high-quality homonymy points in the overlapping areas was used for correction. Finally, the result of traditional processing and block processing were obtained. Figure 5 shows the deformation results of Changzhou and Qijiang.

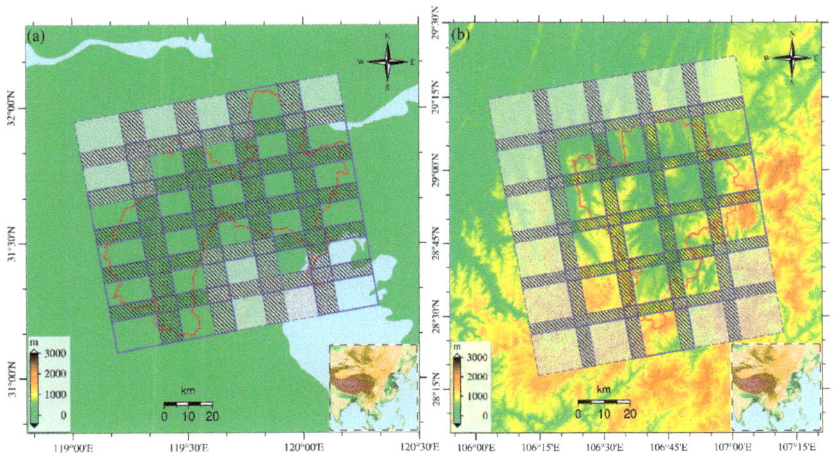

Figure 4. Data coverage for (**a**) Changzhou City and (**b**) Qijiang City. The red line is the administrative division boundary. The blue frame is the image coverage after partition and the shaded part is the overlapping area. The gray blocks are not in the study area.

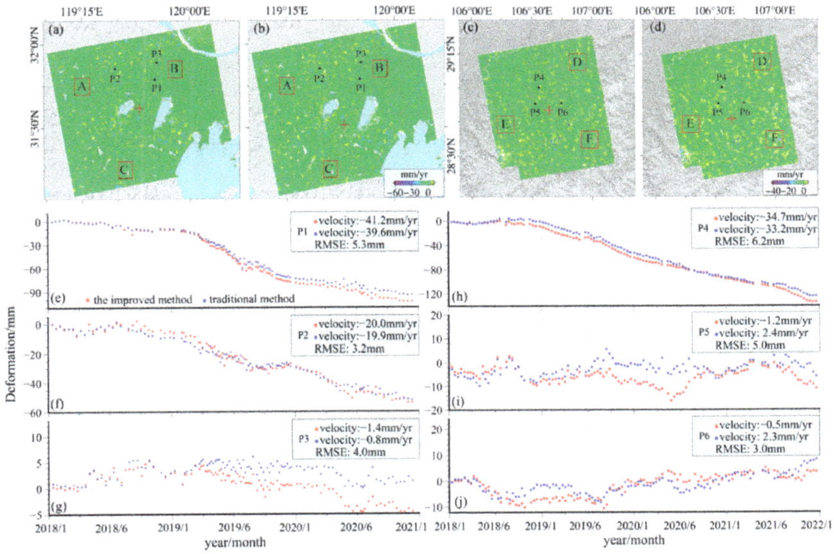

Figure 5. Deformation results of the study areas. The results of Changzhou found by (**a**) partition method and (**b**) traditional method. The results of Qijiang found by (**c**) partition method and (**d**) traditional method. (**e–j**) Are time series results of the selected points. The red "+" is the reference point. The reference points in (**a**,**c**) are virtual reference points after free net adjustment because of there are reference points in each small block before adjustment, and they are the center of gravity of the image coverage. The reference points of (**b**,**d**) are the real reference points in data processing.

4. Result Analysis

When dealing with the deformation time series of a large area, most conventional algorithms use one reference point for phase unwrapping and solve for PS point deformation rates. If the distance between the PS point and the reference point is large, the precision of the results is low. Reducing the size of image coverage by partition can improve the precision of PS points. However, partition leads to different reference points for different blocks, so the deformation results should be corrected to follow one benchmark. In this section, the partition and the traditional methods are compared in terms of precision and time consumption.

4.1. Precision of the Deformation Rate

We compare the deformation results in Changzhou found by the partition method and the traditional method in Figure 5. The two results show a similar distribution of deformation, but a slight difference in details. We manually selected three regions A, B, and C for analysis. The size of these regions is 1000 × 1000 (pixels) These three regions are stable and outside the deformation region, so we assumed the deformation as 0. The statistical analysis shows that the standard deviation (STD) of the partition results and the traditional results in region A is 3.4 mm/yr and 3.7 mm/yr, respectively, in region B is 4.1 mm/yr and 4.2 mm/yr, respectively, and in region C is 4.6 mm/yr and 4.6 mm/yr, respectively. On the whole, the precision of the results obtained by the partition strategy is slightly higher than that of the traditional processing results.

Figure 5c,d show the deformation results of Qijiang found by the two methods, which generally agreed with each other but some local areas have some slight differences, especially in the circled areas D, E, F (the selection criteria is the same as A, B, C). The traditional results contain a large number of uplift signals, which are not deformation signals but residual errors. These errors are significantly less in the partition results. The STD of the partition results in regions D, E, and F are 3.2 mm/yr, 3.1 mm/yr, and 3.6 mm/yr, respectively, and the correspondence of the traditional results are 3.8 mm/yr, 3.9 mm/yr, and 4.0 mm/yr, respectively. Therefore, the partition method outperforms the traditional method in error removal.

The comparison results in Table 2 show that the partition method has higher precision than the traditional method. In the Changzhou experiment, the former obtained a precision of about 5% higher than the traditional method, and in the Qijiang experiment, the precision improvement is about 15%.

Table 2. Precision of the deformation velocity in Changzhou and Qijiang found by the two methods.

Study Area	Strategy	Area	Number of Points	Std /(mm/yr)	Mean /(mm/yr)	Difference /(mm/yr)	Precision Improvement
Changzhou	Partition	A	564,867	3.4	4.0	0.2	5%
		B	589,430	4.1			
		C	593,479	4.6			
	Traditional	A	561,071	3.7	4.2		
		B	582,648	4.2			
		C	591,199	4.6			
Qijiang	Partition	D	992,679	3.2	3.3	0.6	15%
		E	916,098	3.1			
		F	922,811	3.6			
	Traditional	D	986,163	3.8	3.9		
		E	897,871	3.9			
		F	923,017	4.0			

4.2. Precision of the Deformation Sequence

After correcting the deformation rates, we corrected the corresponding deformation sequences. We selected the deformation time series of 3 points in each of Changzhou city (P1, P2, P3) and Qijiang (P4, P5, P6) to test the result precision. These points are in the overlapping regions, and they have different deformation magnitudes. P1 and P4 have large deformation rate, P2 and P5 have medium deformation rate, and P3 and P6 have small deformation rate. The results are shown in Figure 5. In Changzhou City, the difference between the deformation sequences obtained by our method and that obtained by the traditional method is not significant. At P2, the deformation sequences obtained by the two methods almost coincide (Figure 5f), and the deformation rate difference at the three selected points is less than 1 mm/yr. The RMSE between the two results is 5.3 mm for P1, 3.2 mm for P2, and 4.0 mm for P3. We also selected the time series deformation of the three points in Qijiang. The overall deformation trend and deformation magnitude obtained by the two methods are basically the same. The RMSEs between the two results at P4, P5, and P6 are 6.2 mm, 5.0 mm, and 3.0 mm, respectively.

In Figure 5e,h, the annual average deformation rates of these two points are more than 30 mm/yr, the difference between the deformation rates of these two points was about 1.5 mm/yr, and the deformation monitoring precision of InSAR was also basically in this range. A simple proportional function model is used for correcting the time series. If the deformation is nonlinear, the correction of this model might not be appropriate.

4.3. Time and Memory Consumption

The partition and parallel processing strategy can reduce the memory consumption of every single process. Additionally, increasing the number of parallel processing can reduce the time consumed by the whole data processing. The partition strategy can be roughly divided into three stages: image partition, MT-InSAR processing, and correction. The total time consumed by partition depends mainly on the processing times of image blocks, that is, the total number of blocks divided by the number of blocks processed in a single parallel session. The traditional processing spends all its time on MT-InSAR processing.

The program running time and memory consumption of the two strategies are listed in Table 3. The traditional method costs about 20 h. The total time required for block processing is 46.6 h when the number of parallel processing is 1. However, when the number of parallel processing is greater than 2, the block processing needs less time than the traditional processing. Additionally, it only needs 8.7 h when the number of parallel processing is 5. In addition, the computer memory occupied by block processing is much lower than that of traditional processing. In this experiment, the single memory occupied by partitioning is only 1/20 of that occupied by the traditional processing. When the image coverage is large or the computer memory is small, the traditional processing may cause memory overflows and the data cannot be processed successfully, but this problem will not happen to our data partition processing.

Table 3. Time and memory consumption of traditional processing and block processing with the number of parallel processing 5.

	Changzhou		Qijiang	
	Traditional	**Partition**	**Traditional**	**Partition**
Original size (pixels)	29,739 × 6892	29,739 × 6892	28,104 × 7648	28,104 × 7648
Partition strategy	\	6 × 5 ~30% overlap	\	6 × 5 ~25% overlap
Size of block (pixels)	\	7147 × 1373	\	6374 × 1574
Platform	CPU: AMD Ryzen 9 5900X 12-Core/RAM:64 G			
Multi-look	5:1			
Average number of points in a block	\	2,941,200	\	2,493,200
Total number of points	50,941,512	51,178,814	51,620,120	51,883,633
Memory Usage	27.2 G	1.3 G	27.9 G	1.3 G
Time of partition	\	~1 h	\	~1 h
Time of InSAR processing	~20 h	~1.2 h	~20 h	~1.2 h
Time of correction	\	~0.1 h	\	~0.1 h
Total time	~20 h	~8.7 h	~20 h	~8.7 h

5. Discussion

5.1. Space Consistency Correction

The precision of the deformation in each block can be calculated by Formula (9). In the results of Changzhou, the unit weight mean error after adjustment is 0.13 mm/yr, the precision of the block with the highest adjustment precision is 0.27 mm/yr, and the precision of the block with the lowest adjustment precision is 0.59 mm/yr. In the results of Qijiang, the unit weight mean error after adjustment is 0.28 mm/yr, the precision of the block with the highest adjustment precision is 0.40 mm/yr, and the precision of the block with the lowest adjustment precision is 0.81 mm/yr. The adjustment precision is plotted in Figures 6 and 7, which show that the precision of the center blocks is higher than that of the edge blocks.

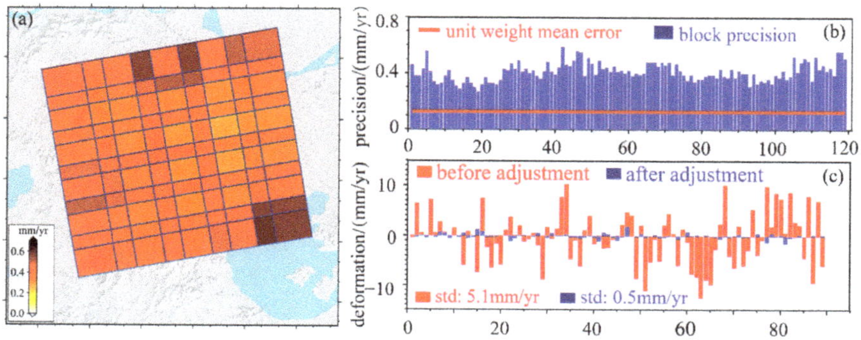

Figure 6. (**a**) Deformation precision of each block after adjustment in Changzhou. (**b**) The histogram of the adjustment precision of each block after adjustment, (**c**) the histogram of the difference of the mean value of the overlapping area before and after adjustment of each block.

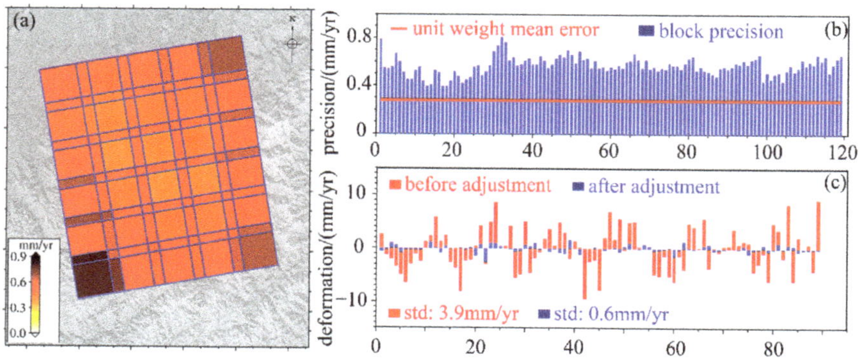

Figure 7. The same as Figure 6, but for Qijiang.

In theory, after adjustment, the deformation rates of the homonymy points in the overlapping region should be the same. As Figure 6c shows, before the adjustment, the difference between homonymy points in Changzhou is more than 5 mm/yr, with an STD of 5.1 mm/yr, and after the adjustment, the difference almost converges to 0, with an STD of 0.5 mm/yr. In Qijiang, the difference between homonymy points is more than 7 mm before adjustment, with an STD of 3.9 mm/yr, and it is reduced to 0.6 mm/yr after the adjustment. The precision of the block processing results in the two study areas was greatly improved by adjustment, indicating that adjustment can improve the consistency of the block deformation results.

We selected 4 deformation areas (A, B, C, D) in the overlapping area in Changzhou (Figure A2), and compared their deformation results before and after adjustment in Figure 8. The results of the two image blocks in region A have little difference, so the improvement of the result is not significant after the adjustment (Figure 8b,f). However, the results of regions B, C and D are improved obviously after adjustment. The mean values of the differences of deformation in these three regions change from 4.1 mm/yr, −5.6 mm/yr, and 5.6 mm/yr to 0.3 mm/yr, 0.1 mm/yr, and 0.2 mm/yr after correction, and the spatial consistency of the results improves more significantly.

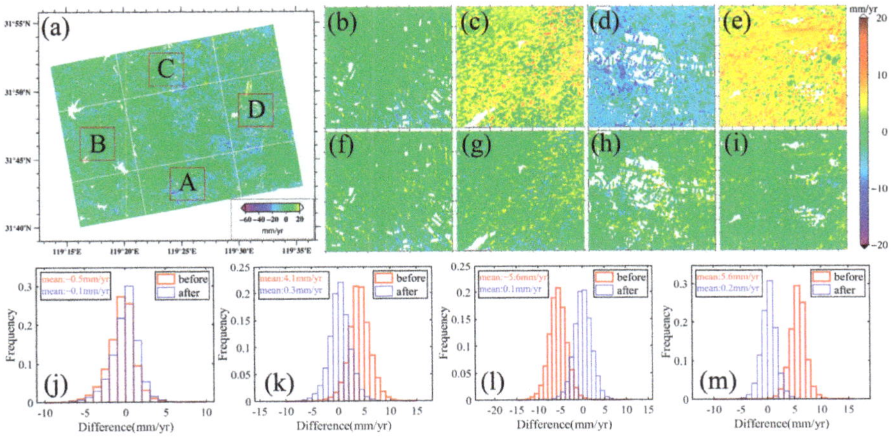

Figure 8. The difference in the homonymy points before and after adjustment in regions A, B, C, and D of Changzhou. (**a**) shows the location of the four regions. (**b–e**) Are the difference of the overlapping areas before adjustment in regions A, B, C, and D, and (**f–i**) are the difference after adjustment. (**j–m**) Shows the statistical histograms of the four regions.

5.2. Effects of Overlap Rate on Result Precision and Time Consumption

In the proposed partition strategy, partitioning the image is the first and most critical step. Different partition strategies provide different precision results, and different data processing efficiency. Because we adopt the even partition strategy, the size and overlap ratio of the blocks have a great impact on the results. To analyze the effects of the overlap ratio on the partition results, we set one block size and obtained the temporal deformation results of Changzhou using the overlap ratios of 10%, 20%, 30%, and 40%, separately. The results are shown in Figure A2. We evaluated the result precision (Table 4).

The precision of the deformation rates obtained by different overlap ratios is similar because the mean values of the high-quality homonymy points in the overlap region are used in adjustment, which is slightly influenced by the overlap region. As long as the block size is the same, the precision of the results obtained by different overlap ratios are similar. However, the time consumed by different overlap ratios is different. The larger the overlap ratio, the larger the number of blocks and the longer the processing. When the number of parallel processing is 5, the total time consumed by the four overlap ratios are 6.6 h, 7.6 h, 9.5 h, and 10.7 h, indicating that the consuming time increases with the increase of overlap ratio. The increase in the overlapping area brings larger double-counted areas and raises the reliability of the results. Considering the precision, reliability, and time consumption, we choose 20% as the best overlap ratio. The experiments show that overlap ratio 20% has similar result precision and time consumption with that of overlap ratio 10%, but it leads to more than 64% overlap area, which contributes to a significantly higher reliability.

Table 4. Results of using different block overlap ratios.

Overlap Ratio	Total Overlap	Block Size (Amount)	Precision /(mm/yr)		Time of Each Block	Total Time
Traditional	0%	29,739 × 6892 (1)	A B C	3.7 4.2 4.6	20 h	20 h
10%	36%	7000 × 1500 (20) 4539 × 1500 (5)	A1 B1 C1	3.5 4.2 4.6	1.2 h 0.8 h	6.7 h
20%	64%	7000 × 1500 (25) 7000 × 892 (5)	A2 B2 C2	3.5 4.1 4.6	1.2 h 0.7 h	7.8 h
30%	84%	7000 × 1500 (30) 5239 × 1500 (6)	A3 B3 C3	3.4 4.1 4.6	1.2 h 0.9 h	10.1 h
40%	96%	7000 × 1500 (42) 4539 × 1500 (7)	A4 B4 C4	3.3 4.0 4.5	1.2 h 0.8 h	12.7 h

5.3. Implications of Data Partition Strategy for MT-InSAR

The administrative boundary of a city is usually an irregular polygon, but the image coverage is a regular quadrangle. Thus, the image coverage contains many data unrelated to the study area. The traditional method will also process these data. If such data accounts for a large proportion of the image, the data processing will waste a lot of time. The proposed method only processes the block data inside the study area, which can improve the data processing efficiency. In Figure 4, the gray blocks do not need processing. Furthermore, the proposed method can refine the data processing for only the blocks with deformation, which further improves the efficiency of data processing.

The difference in the image coverage will definitely lead to the difference in the results. In this paper, we divide data into blocks, and process, correct, and splice the results of all blocks. The atmospheric delay in small range data is easier to remove than that in the data with large range. Studies have shown that the atmospheric phase in InSAR data measurements has a close correlation with spatial scale [35]. The atmospheric phase difference between two PS points with a distance less than 1 km is less than 0.1 rad2 [40], so the smaller the area, the better the atmospheric error removal according to the error propagation law. However, for large deformation areas, long-wavelength deformation may be removed as orbital errors, due to the polynomial fitting [21]. Thus, the proposed method is not fully applicable to the study area with long-wavelength deformation, such as interseismic deformation.

The correction of the partition results is based on the assumption that the deformation rates of the homonymy points are the same. However, the deformation acquired by InSAR is the line-of-sight (LOS) deformation, and the deformation direction at each point is related to the incidence angle. When the deformation of the homonymy point is obtained from the same orbit and has the same incidence angles, the deformation rates should be the same. If the partitioned data are acquired under different imaging geometries, there will be inconsistency in the incidence angles, resulting in different LOS deformation. Therefore, the incidence angle variation of the results should be considered when the partition data are acquired from different orbits.

Finally, the method does not use control points for the adjustment. Although the benchmarks between image blocks are unified, there may be a deviation between the unified benchmark and the real deformation result datum. We only make a simple correction to the result but using external data as control points may improve the correction.

6. Conclusions

In this paper, we propose to partition the data into blocks before obtaining the deformation, to save memory and time for large-scale data processing. To validate this method, we used the Sentinel-1 TOPS data covering Changzhou, a plain area, and Qijiang, a mountainous area in China. The time series deformation results were obtained in these two regions using the traditional processing method (the improved IPTA) and the partition processing method. The latter outperforms the former in precision, time consumption, and memory occupation. Taking Changzhou City as an example, the memory occupation of the traditional processing method is about 27.2 G, and the total time consumed by processing is about 20 h. During partition processing, the memory occupation of each block is only 1.3 G, and the consumed time is 8.7 h when the parallel number is 5. We also compared the precision of the results obtained by the two methods. The results obtained by the partition processing in Changzhou is as about 4.0 mm/yr, while the precision of the traditional processing is about 4.2 mm/yr. The correspondence in Qijiang is about 3.3 mm/yr and 3.9 mm/yr, respectively. The precision of the results obtained by the proposed method is higher than that obtained by traditional processing.

In general, the proposed method can significantly reduce the memory occupation and time consumption of data processing under the condition of sufficient parallelism, and the precision of the results is higher than that obtained by traditional processing. This method is suitable for monitoring the short-wavelength deformation in a large area, such as large-scale urban deformation monitoring and large-scale landslide deformation detection. However, further research is needed for the result splicing and its application to long-wavelength deformation.

Author Contributions: Conceptualization, Y.W. (Yuexin Wang) and G.F.; data curation, S.L.; formal analysis, Z.F. and Y.W. (Yuedong Wang); funding acquisition, Y.W. (Yuexin Wang) and G.F.; methodology, Y.W. (Yuexin Wang) and G.F.; resources, G.F.; Software, Y.W. (Yuexin Wang); supervision, S.L., Y.Z. and H.L.; validation, X.W.; writing—original draft, Y.W. (Yuexin Wang); writing—review and editing, Y.W. (Yuexin Wang), G.F., Z.F., Y.W. (Yuedong Wang), X.W., Y.Z. and H.L. All authors have read and agreed to the published version of the manuscript.

Funding: This research was funded by the National Natural Science Foundation of China (No. 42174039), and the Fundamental Research Funds for the Central Universities of Central South University (No. 506021741).

Acknowledgments: The authors would like to thank the European Space Agency (ESA) for providing free Sentinel-1 data. The data were additionally retrieved from the Alaska Satellite Facility Distributed Active Archive Center. We also thank the contributors for the Generic Mapping Tools (Wessel et al., 2013) open-source software.

Conflicts of Interest: The authors declare no conflict of interest.

Appendix A

Table A1. Parameters of the images used in this study.

Study Area	Parameters		Acquisition Date (YYYY/MM/DD)			
Changzhou	Direction	Ascending	2018/01/10	2018/01/22	2018/02/03	2018/02/15
	Path	T69	2018/02/27	2018/03/11	2018/03/23	2018/04/04
	Heading	−12.79°	2018/04/16	2018/04/28	2018/05/10	2018/05/22
	Incidence	36.65°	2018/06/03	2018/06/15	2018/06/27	2018/07/09
	Pixel Spacing (Rg × Az)	2.33 × 13.98	2018/07/21	2018/08/02	2018/08/14	2018/09/07
			2018/09/19	2018/10/01	2018/10/13	2018/10/25
			2018/11/06	2018/11/18	2018/12/12	2018/12/24
	Number of images	110	2019/01/05	2019/01/17	2019/02/10	2019/02/16
			2019/02/22	2019/03/06	2019/03/18	2019/03/30
			2019/04/05	2019/04/11	2019/04/23	2019/04/29
			2019/05/05	2019/05/11	2019/05/17	2019/05/23
			2019/05/29	2019/06/04	2019/06/10	2019/06/16

Table A1. *Cont.*

Study Area	Parameters		Acquisition Date (YYYY/MM/DD)			
			2019/06/22	2019/06/28	2019/07/04	2019/07/10
			2019/07/16	2019/07/22	2019/07/28	2019/08/03
			2019/08/09	2019/08/15	2019/08/21	2019/08/27
			2019/09/02	2019/09/08	2019/09/20	2019/09/26
			2019/10/02	2019/10/08	2019/10/14	2019/10/20
			2019/10/26	2019/11/01	2019/11/07	2019/11/19
			2019/11/25	2019/12/01	2019/12/07	2019/12/13
			2019/12/19	2019/12/25	2019/12/31	2020/01/12
			2020/01/24	2020/02/05	2020/02/17	2020/02/29
			2020/03/12	2020/03/24	2020/04/05	2020/04/17
			2020/04/29	2020/05/11	2020/05/23	2020/06/04
			2020/06/16	2020/06/28	2020/07/10	2020/07/22
			2020/07/28	2020/08/03	2020/08/15	2020/08/27
			2020/09/08	2020/09/20	2020/10/02	2020/10/14
			2020/10/26	2020/11/07	2020/11/19	2020/12/01
			2020/12/13	2020/12/25		
Qijiang	Direction	Ascending	2018/01/09	2018/01/21	2018/02/02	2018/02/14
	Path	T55	2018/02/26	2018/03/10	2018/03/22	2018/04/03
	Heading	−12.65°	2018/04/15	2018/04/27	2018/05/09	2018/05/21
	Incidence	43.64°	2018/06/02	2018/06/14	2018/06/26	2018/07/08
	Pixel Spacing (Rg × Az)	2.33 × 13.96	2018/07/20	2018/08/01	2018/08/25	2018/09/06
			2018/09/18	2018/09/30	2018/10/12	2018/10/24
			2018/11/05	2018/11/29	2018/12/11	2018/12/23
			2019/01/04	2019/01/16	2019/01/28	2019/02/09
			2019/02/21	2019/03/05	2019/03/17	2019/03/29
			2019/04/10	2019/04/22	2019/05/04	2019/05/16
			2019/05/28	2019/06/09	2019/07/03	2019/07/15
			2019/07/27	2019/08/08	2019/08/20	2019/09/01
			2019/09/13	2019/09/25	2019/10/07	2019/10/19
			2019/10/31	2019/11/12	2019/11/24	2019/12/06
			2019/12/18	2019/12/30	2020/01/11	2020/01/23
			2020/02/04	2020/02/16	2020/02/28	2020/03/11
			2020/03/23	2020/04/04	2020/04/16	2020/04/28
	Number of images	114	2020/05/22	2020/06/03	2020/06/15	2020/06/27
			2020/07/09	2020/07/21	2020/08/02	2020/08/14
			2020/09/07	2020/09/19	2020/10/01	2020/10/13
			2020/10/25	2020/11/06	2020/11/18	2020/11/30
			2020/12/12	2020/12/24	2021/01/05	2021/01/17
			2021/01/29	2021/02/10	2021/02/22	2021/03/06
			2021/03/18	2021/03/30	2021/04/11	2021/04/23
			2021/05/29	2021/06/10	2021/06/22	2021/07/16
			2021/07/28	2021/08/09	2021/08/21	2021/09/02
			2021/09/14	2021/09/26	2021/10/08	2021/10/20
			2021/11/01	2021/11/13	2021/11/25	2021/12/07
			2021/12/19	2021/12/31		

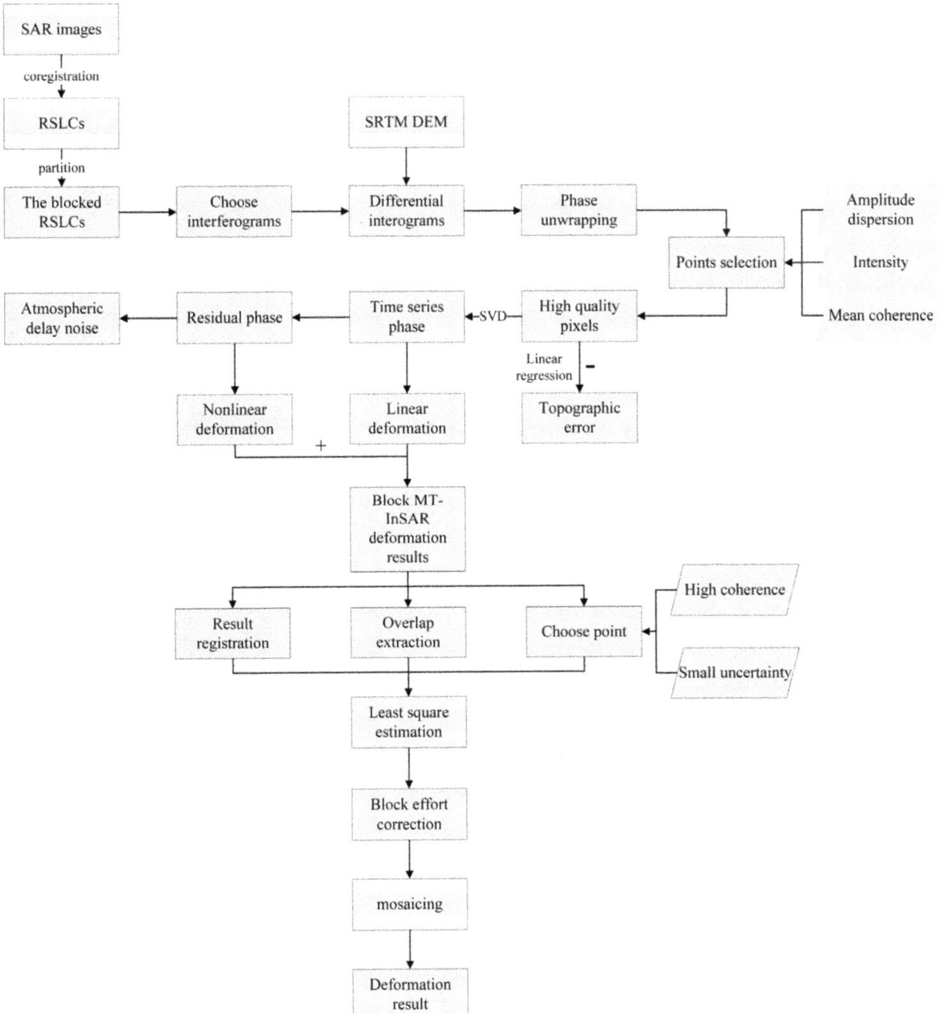

Figure A1. The detailed flowchart of the proposed method.

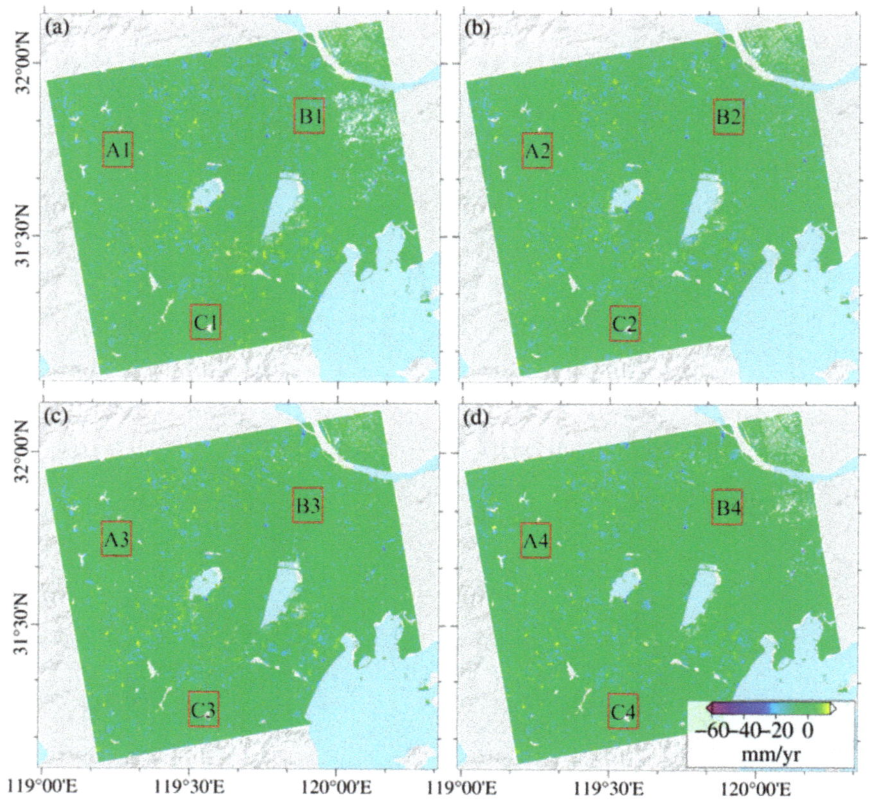

Figure A2. The deformation results in Changzhou obtained using the overlap ratio of (**a**) 10%, (**b**) 20%, (**c**) 30%, and (**d**) 40%. A1-C4 are the same areas as A–C described in Section 4.1.

References

1. Ng, A.H.-M.; Ge, L.; Li, X.; Zhang, K. Monitoring ground deformation in Beijing, China with persistent scatterer SAR interferometry. *J. Geod.* **2011**, *86*, 375–392. [CrossRef]
2. Wang, H.; Wright, T.J.; Yu, Y.; Lin, H.; Jiang, L.; Li, C.; Qiu, G. InSAR reveals coastal subsidence in the Pearl River Delta, China. *Geophys. J. Int.* **2012**, *191*, 1119–1128. [CrossRef]
3. Feng, G.; Li, Z.; Shan, X.; Xu, B.; Du, Y. Source parameters of the 2014 Mw 6.1 South Napa earthquake estimated from the Sentinel 1A, COSMO-SkyMed and GPS data. *Tectonophysics* **2015**, *655*, 139–146. [CrossRef]
4. Chaussard, E.; Johnson, C.W.; Fattahi, H.; Bürgmann, R. Potential and limits of InSAR to characterize interseismic deformation independently of GPS data: Application to the southern San Andreas Fault system. *Geochem. Geophys. Geosystems* **2016**, *17*, 1214–1229. [CrossRef]
5. Dong, J.; Zhang, L.; Tang, M.; Liao, M.; Xu, Q.; Gong, J.; Ao, M. Mapping landslide surface displacements with time series SAR interferometry by combining persistent and distributed scatterers: A case study of Jiaju landslide in Danba, China. *Remote Sens. Environ.* **2018**, *205*, 180–198. [CrossRef]
6. Xiong, Z.; Feng, G.; Feng, Z.; Miao, L.; Wang, Y.; Yang, D.; Luo, S. Pre- and post-failure spatial-temporal deformation pattern of the Baige landslide retrieved from multiple radar and optical satellite images. *Eng. Geol.* **2020**, *279*, 105580. [CrossRef]
7. Novellino, A.; Cesarano, M.; Cappelletti, P.; di Martire, D.; di Napoli, M.; Ramondini, M.; Sowter, A.; Calcaterra, D. Slow-moving landslide risk assessment combining Machine Learning and InSAR techniques. *CATENA* **2021**, *203*, 105317. [CrossRef]
8. Meng, Q.; Confuorto, P.; Peng, Y.; Raspini, F.; Bianchini, S.; Han, S.; Liu, H.; Casagli, N. Regional Recognition and Classification of Active Loess Landslides Using Two-Dimensional Deformation Derived from Sentinel-1 Interferometric Radar Data. *Remote Sens.* **2020**, *12*, 1541. [CrossRef]
9. Miele, P.; Di Napoli, M.; Novellino, A.; Calcaterra, D.; Mallorqui, J.J.; Di Martire, D. SAR data and field surveys combination to update rainfall-induced shallow landslide inventory. *Remote Sens. Appl. Soc. Environ.* **2022**, *26*, 100755. [CrossRef]

10. Yang, Z.; Li, Z.; Zhu, J.; Preusse, A.; Hu, J.; Feng, G.; Papst, M. High-Resolution Three-Dimensional Displacement Retrieval of Mining Areas From a Single SAR Amplitude Pair Using the SPIKE Algorithm. *IEEE J. Sel. Top. Appl. Earth Obs. Remote Sens.* **2018**, *11*, 3782–3793. [CrossRef]
11. Ghasemloo, N.; Matkan, A.A.; Alimohammadi, A.; Aghighi, H.; Mirbagheri, B. Estimating the Agricultural Farm Soil Moisture Using Spectral Indices of Landsat 8, and Sentinel-1, and Artificial Neural Networks. *J. Geov. Spat. Anal.* **2022**, *6*, 19. [CrossRef]
12. Kellogg, K.; Hoffman, P.; Standley, S.; Shaffer, S.; Rosen, P.; Edelstein, W.; Dunn, C.; Baker, C.; Barela, P.; Shen, Y.; et al. NASA-ISRO Synthetic Aperture Radar (NISAR) Mission. In Proceedings of the 2020 IEEE Aerospace Conference, Big Sky, MT, USA, 7–14 March 2020; pp. 1–21. [CrossRef]
13. Fan, W.; Pan, G.; Wang, L. Development and Application of a Networked Automatic Deformation Monitoring System. *J. Geovisualization Spat. Anal.* **2020**, *4*, 11. [CrossRef]
14. Torres, R.; Snoeij, P.; Geudtner, D.; Bibby, D.; Davidson, M.; Attema, E.; Potin, P.; Rommen, B.; Floury, N.; Brown, M.; et al. GMES Sentinel-1 mission. *Remote Sens. Environ.* **2012**, *120*, 9–24. [CrossRef]
15. Wang, C.; Tang, Y.; Zhang, H.; You, H.; Zhang, W.; Duan, W.; Wang, J.; Dong, L.; Zhang, B. First mapping of China surface movement using supercomputing interferometric SAR technique. *Sci. Bull.* **2021**, *66*, 1608–1610. [CrossRef]
16. De Luca, C.; Cuccu, R.; Elefante, S.; Zinno, I.; Manunta, M.; Casola, V.; Rivolta, G.; Lanari, R.; Casu, F. An On-Demand Web Tool for the Unsupervised Retrieval of Earth's Surface Deformation from SAR Data: The P-SBAS Service within the ESA G-POD Environment. *Remote Sens.* **2015**, *7*, 15630–15650. [CrossRef]
17. Chen, C.W.; Zebker, H.A. Phase unwrapping for large SAR interferograms: Statistical segmentation and generalized network models. *IEEE Trans. Geosci. Remote Sens.* **2002**, *40*, 11. [CrossRef]
18. Zhang, K.; Ge, L.; Hu, Z.; Ng, A.H.-M.; Li, X.; Rizos, C. Phase Unwrapping for Very Large Interferometric Data Sets. *IEEE Trans. Geosci. Remote Sens.* **2011**, *49*, 4048–4061. [CrossRef]
19. Yu, H.; Xing, M.; Bao, Z. A Fast Phase Unwrapping Method for Large-Scale Interferograms. *IEEE Trans. Geosci. Remote Sens.* **2013**, *51*, 4240–4248. [CrossRef]
20. Yuan, Z.; Chen, T.; Xing, X.; Peng, W.; Chen, L. BM3D Denoising for a Cluster-Analysis-Based Multibaseline InSAR Phase-Unwrapping Method. *Remote Sens.* **2022**, *14*, 1836. [CrossRef]
21. Du, Y.; Fu, H.; Liu, L.; Feng, G.; Peng, X.; Wen, D. Orbit error removal in InSAR/MTInSAR with a patch-based polynomial model. *Int. J. Appl. Earth Obs. Geoinf.* **2021**, *102*, 102438. [CrossRef]
22. Liang, H.; Zhang, L.; Ding, X.; Lu, Z.; Li, X. Toward Mitigating Stratified Tropospheric Delays in Multitemporal InSAR: A Quadtree Aided Joint Model. *IEEE Trans. Geosci. Remote Sens.* **2019**, *57*, 291–303. [CrossRef]
23. Shi, M.; Peng, J.; Chen, X.; Zheng, Y.; Yang, H.; Su, Y.; Wang, G.; Wang, W. An Improved Method for InSAR Atmospheric Phase Correction in Mountainous Areas. *IEEE J. Sel. Top. Appl. Earth Obs. Remote Sens.* **2021**, *14*, 10509–10519. [CrossRef]
24. Goel, K.; Adam, N.; Shau, R.; Rodriguez-Gonzalez, F. Improving the reference network in wide-area Persistent Scatterer Interferometry for non-urban areas. In Proceedings of the 2016 IEEE International Geoscience and Remote Sensing Symposium (IGARSS), Beijing, China, 10–15 July 2016; pp. 1448–1451. [CrossRef]
25. Xue, F.; Lv, X.; Dou, F.; Yun, Y. A Review of Time-Series Interferometric SAR Techniques: A Tutorial for Surface Deformation Analysis. *IEEE Geosci. Remote Sens. Mag.* **2020**, *8*, 22–42. [CrossRef]
26. Werner, C.; Wegmuller, U.; Strozzi, T.; Wiesmann, A. Interferometric point target analysis for deformation mapping. In Proceedings of the IGARSS 2003 IEEE International Geoscience and Remote Sensing Symposium, Toulouse, France, 21–25 July 2003; Volume 4367, pp. 4362–4364. [CrossRef]
27. Li, T.; Liu, G.; Lin, H.; Jia, H.; Zhang, R.; Yu, B.; Luo, Q. A Hierarchical Multi-Temporal InSAR Method for Increasing the Spatial Density of Deformation Measurements. *Remote Sens.* **2014**, *6*, 3349–3368. [CrossRef]
28. Hooper, A. A multi-temporal InSAR method incorporating both persistent scatterer and small baseline approaches. *Geophys. Res. Lett.* **2008**, *35*, 16. [CrossRef]
29. Hou, J.; Xu, B.; Li, Z.; Zhu, Y.; Feng, G. Block PS-InSAR ground deformation estimation for large-scale areas based on network adjustment. *J. Geod.* **2021**, *95*, 111. [CrossRef]
30. Ge, D. Research on Key Technologies for Regional Ground Subsidence InSAR Monitoring. Ph.D. Thesis, China University of Geosciences, Beijing, China, 2013.
31. Liu, H.; Yu, J.; Zhao, Z.; Jezek, K.C. Calibrating and mosaicking surface velocity measurements from interferometric SAR data with a simultaneous least-squares adjustment approach. *Int. J. Remote Sens.* **2007**, *28*, 1217–1230. [CrossRef]
32. Hu, J. Theory and Methodology of InSAR Three-Dimensional Deformation Estimation Based on Modern Measurement Leveling. Ph.D. Thesis, Central South University, Changsha, China, 2013.
33. Kalia, A.C.; Frei, M.; Lege, T. A Copernicus downstream-service for the nationwide monitoring of surface displacements in Germany. *Remote Sens. Environ.* **2017**, *202*, 234–249. [CrossRef]
34. Murray, K.D.; Lohman, R.B.; Bekaert, D.P.S. Cluster-Based Empirical Tropospheric Corrections Applied to InSAR Time Series Analysis. *IEEE Trans. Geosci. Remote Sens.* **2021**, *59*, 2204–2212. [CrossRef]
35. Li, Z.; Cao, Y.; Wei, J.; Duan, M.; Wu, L.; Hou, J.; Zhu, J. Time-series InSAR ground deformation monitoring: Atmospheric delay modeling and estimating. *Earth-Sci. Rev.* **2019**, *192*, 258–284. [CrossRef]
36. Wang, Y.; Feng, G.; Li, Z.; Luo, S.; Wang, H.; Xiong, Z.; Zhu, J.; Hu, J. A Strategy for Variable-Scale InSAR Deformation Monitoring in a Wide Area: A Case Study in the Turpan–Hami Basin, China. *Remote Sens.* **2022**, *14*, 3832. [CrossRef]

37. Luo, S.; Feng, G.; Xiong, Z.; Wang, H.; Zhao, Y.; Li, K.; Deng, K.; Wang, Y. An Improved Method for Automatic Identification and Assessment of Potential Geohazards Based on MT-InSAR Measurements. *Remote Sens.* **2021**, *13*, 3490. [CrossRef]
38. Zhong, Y.; Dong, S.; Hu, H. Monitoring and Analysis of Ground Settlement in Changzhou City Based on Time-Series InSAR Technology. *Geol. J. China Univ.* **2019**, *25*, 131–143.
39. Wang, J.; Wei, Z.; Liu, W.; Zhong, S.; Zhang, G. Functional zoning of land consolidation in mountainous and hilly areas based on "Production-ecological" perspective: A case study of Qijiang District, Chongqing. *Areal Res. Dev.* **2018**, *37*, 155–159.
40. Williams, S.; Bock, Y.; Fang, P. Integrated satellite interferometry: Tropospheric noise, GPS estimates and implications for interferometric synthetic aperture radar products. *J. Geophys. Res. Solid Earth* **1998**, *103*, 27051–27067. [CrossRef]

remote sensing

MDPI

Article

Characterizing the Distribution Pattern and a Physically Based Susceptibility Assessment of Shallow Landslides Triggered by the 2019 Heavy Rainfall Event in Longchuan County, Guangdong Province, China

Siyuan Ma [1,2], Xiaoyi Shao [3,4] and Chong Xu [3,4,*]

1 Institute of Geology, China Earthquake Administration, Beijing 100029, China
2 Key Laboratory of Seismic and Volcanic Hazards, Institute of Geology, China Earthquake Administration, Beijing 100029, China
3 National Institute of Natural Hazards, Ministry of Emergency Management of China, Beijing 100085, China
4 Key Laboratory of Compound and Chained Natural Hazards Dynamics, Ministry of Emergency Management of China, Beijing 100085, China
* Correspondence: chongxu@ninhm.ac.cn

Abstract: Rainfall-induced landslides pose a significant threat to the lives and property of residents in the southeast mountainous and hilly area; hence, characterizing the distribution pattern and effective susceptibility mapping for rainfall-induced landslides are regarded as important and necessary measures to remediate the damage and loss resulting from landslides. From 10 June 2019 to 13 June 2019, continuous heavy rainfall occurred in Longchuan County, Guangdong Province; this event triggered extensive landslide disasters in the villages of Longchuan County. Based on high-resolution satellite images, a landslide inventory of the affected area was compiled, comprising a total of 667 rainfall-induced landslides over an area of 108 km^2. These landslides consisted of a large number of shallow landslides with a few flowslides, rockfalls, and debris flows, and the majority of them occurred in Mibei and Yanhua villages. The inventory was used to analyze the distribution pattern of the landslides and their relationship with topographical, geological, and hydrological factors. The results showed that landslide abundance was closely related to slope angle, TWI, and road density. The landslide area density (LAD) increased with the increase in the above three influencing factors and was described by an exponential or linear relationship. In addition, southeast and south aspect hillslopes were more prone to collapse than the northwest–north aspect ones because of the influence of the summer southeast monsoon. A new open-source tool named MAT.TRIGRS(V1.0) was adopted to establish the landslide susceptibility map in landslide abundance areas and to back-analyze the response of the rainfall process to the change in landslide stability. The prediction results were roughly consistent with the actual landslide distribution, and most areas with high susceptibility were located on both sides of the river valley; that is, the areas with relatively steep slopes. The slope stability changes in different periods revealed that the onset of heavy rain on 10 June 2019 was the main triggering factor of these group-occurring landslides, and the subsequent rainfall with low intensity had little impact on slope stability.

Keywords: heavy rainfall; shallow landslides; TRIGRS model; spatial distribution; susceptibility assessment; Longchuan County; Guangdong Province

Citation: Ma, S.; Shao, X.; Xu, C. Characterizing the Distribution Pattern and a Physically Based Susceptibility Assessment of Shallow Landslides Triggered by the 2019 Heavy Rainfall Event in Longchuan County, Guangdong Province, China. *Remote Sens.* **2022**, *14*, 4257. https://doi.org/10.3390/rs14174257

Academic Editors: Paolo Mazzanti and Saverio Romeo

Received: 29 July 2022
Accepted: 24 August 2022
Published: 29 August 2022

1. Introduction

In recent years, an increasing number of intense rainfall events have occurred in mountainous areas due to the impact of global climate change, which has dramatically increased the frequency of global rainfall-induced landslides [1,2]. Rainfall-induced landslides are not only widely distributed in the world but also occur frequently and cause significant damage to humanity [3–6]. Therefore, a good understanding of the fundamentals of rainfall-induced landslide occurrence, distribution patterns, and susceptibility assessments can provide useful guidance for regional disaster prevention and mitigation, and landscape evolution [7–9].

A new landslide inventory that is generated after a major triggering event (e.g., an earthquake, volcanic eruptions, or heavy rainfall) is referred to as an event-based landslide inventory. Owing to the advancements in earth observation technology, such as multi-temporal high-resolution optical satellite remote sensing, more high-quality earthquake-induced landslide inventories have been developed. In particular, since the 2008 Wenchuan earthquake, the establishment of coseisimic landslide inventories has made great progress. At present, there are roughly 46 detailed coseismic landslide databases mapped as polygons [10–13]. However, unlike earthquake events, the construction speed of landslide inventories triggered by heavy rainfall events is still relatively slow, and currently there are only a few heavy rainfall-induced landslide inventories [14–16]. The main reason is that clouds are often a major obstacle in the affected areas, which may limit the visibility of satellite images and thus affect the visual interpretation of rainfall-induced landslides [15]. At present, there are 16 public landslide inventories triggered by heavy rainfall events around the world, with the majority of these landslide databases being on a small scale. The southeast coastal region in China is economically developed and densely populated. Influenced by monsoon rainfall, this area is also considered a landslide-prone zone [17,18]. Once landslides occur, the social and economic losses in this area will be huge. A comprehensive rainfall-induced landslide database not only contributes to a deeper understanding of the event occurrence but also provides data support for the subsequent in-depth analysis of the formation and evolution of the geological disaster chain[15,19]. However, there are few rainfall-induced landslide inventories for a single event in the southeast coastal region, and thus more analyses are needed for rainfall-induced landslide inventories in this area.

Rainfall-induced landslide susceptibility can provide valuable information for landslide risk assessment. Currently, there are two quantitative methods for assessing the susceptibility for rainfall-induced landslides, which include the data-driven methods based on mathematical methods and physical-based methods that couple the hydrological models and infinite slope stability models. For the data-driven method, the relationship between the influencing factors and the landslide occurrence are analyzed by mathematical models [20–22]. Currently, many models have been widely used in landslide susceptibility mapping, particularly with the development of machine learning technology, such as logistic regression [23,24], random forest [25], artificial neural network [26], convolutional neural network (CNN) [27], support vector machine (SVM) [28], and decision tree [29]. However, the outcomes of landslide susceptibility mapping based on the data-driven method could be subject to considerable uncertainties due to errors and variability in model choice, data selection, system understanding of weighting factors, and human judgment [30,31]. Meanwhile, the data-driven model does not possess the timeliness of emergency assessment for a single triggering event, because it requires sufficient landslide data to establish the susceptibility assessment model. As a consequence, assessment results frequently lag behind practical application and cannot serve the emergency assessment in a short time [32,33]. Otherwise, due to the fact that the majority of these models are trained by regional landslide data and are thus limited by regional geological and geomorphic characteristics [14,34,35], the data-driven model's applicability in different areas is greatly diminished. However, the physically based landslide susceptibility assessment can better solve the above problems.

Unlike the data-driven method, the physically based method does not take into account actual landslide data, but rather simulates the physical process of rainfall-induced landslide occurrence by coupling the hydrological and infinite slope models [36]. The physically based method has been pervasively used because of its high predictive capability and the most acknowledged feasibility for a quantitative assessment of the effects of the individual parameters that contribute to landslide initiation [37] and it is a useful tool for determining the susceptibility zonation of rainfall-induced shallow landslides [38].

In addition, the wide application of GIS technology facilitates the wide application of physical models in large areas [39,40]. Due to its preferable practicability and wide regional applicability, physically based models are popular in the spatial prediction of regional rainfall-induced landslides [41–45]. In recent years, some physically based models for rainfall-induced landslide susceptibility mapping have been developed, such as the TRIGRS model [40], the Slip model [46–48], the GIS-TiVaSS model [45,49], the GIS-TISSA model [50], the CRESTSLIDE model [51,52], and the HIRESSS model [53–55]. Among them, the TRIGRS model, which accounts for transient pore water pressure, can predict the impact of heavy rainfall on groundwater changes in a short period. At present, it is the most widely used physically-based model of slope instability [41,56–58], and has been used in many countries around the world, including Italy, the United States, China, South Korea, and Southeast Asia [59–64]. However, the application of the TRIGRS model in China's southeast area is limited so it is necessary to investigate the applicability of the model in the spatio-temporal prediction of rainfall-induced landslides in the southeast mountainous area.

Longchuan County experienced continuous heavy rainfall from 10 June 2019 to 13 June 2019. Extensive landslides, collapses, and debris flows occurred in the villages of Longchuan county. A total of 352 villages of Longchuan County were devastated to varied degrees, of which Mibei village in Beiling town was the most severely hit with 1571 individuals affected, 120 buildings fully collapsed, and more than 100 houses damaged. The direct economic loss of this event reached CNY 110 million, exerting a significant impact on the normal productivity and lives of local residents. Thus, the objectives of this study are: (1) establishing a landslide inventory including landslides induced by the 2019 Longchuan heavy rainfall event and analyzing the spatial distribution of landslides with topographical, geological, and hydrological factors; (2) conducting the physically based susceptibility assessment based on a new open-source tool named MAT.TRIGRS(V1.0) for predicting the spatiotemporal distribution of rainfall-induced landslides and back-analyzing the response of the rainfall process on the change of landslide stability.

2. Study Area

Longchuan County is situated in the northeast of Guangdong Province, spanning from 23.8°N to 24.7°N of latitude and from 115.0° to 115.6°E of longitude, and covers an area of approximately 3089 km^2 on the surface. The study area is Beiling Town, which is located in the north of Longchuan County and the upper reaches of Dongjiang River. The climate in the region is subtropical monsoon with abundant rainfall and sunshine. The annual rainfall is 1500 mm, and the average temperature is about 22 °C. The study area experiences the most rainfall in May, June, and July. The geomorphic unit of the study area is a hilly landform with an elevation range from 100 m to 1100 m (Figure 1). The mountains are steep, and the peaks are conical due to the relatively developed hydrographic nets and strong topographic cutting in this area. As a result, numerous "V" shaped valleys developed in this area, with slope angles ranging from 20 to 50 degrees. The main lithology of the study area is acid intrusive rock of Ordovician and Silurian, mainly monzogranite (O_3-S_1), which accounts for more than 70% of the rock in the whole study area (Figure 2a). In addition, tuff of Yousheng formation of Middle Cretaceous (K_{2ys}) and quartz mica schist of Daganshan formation of Sinian(Z_{2djs}) also developed in this area (Figure 2a). The main land use type is forest, which accounts for 80% of the whole study area, followed by cropland, accounting for more than 10% (Figure 2b).

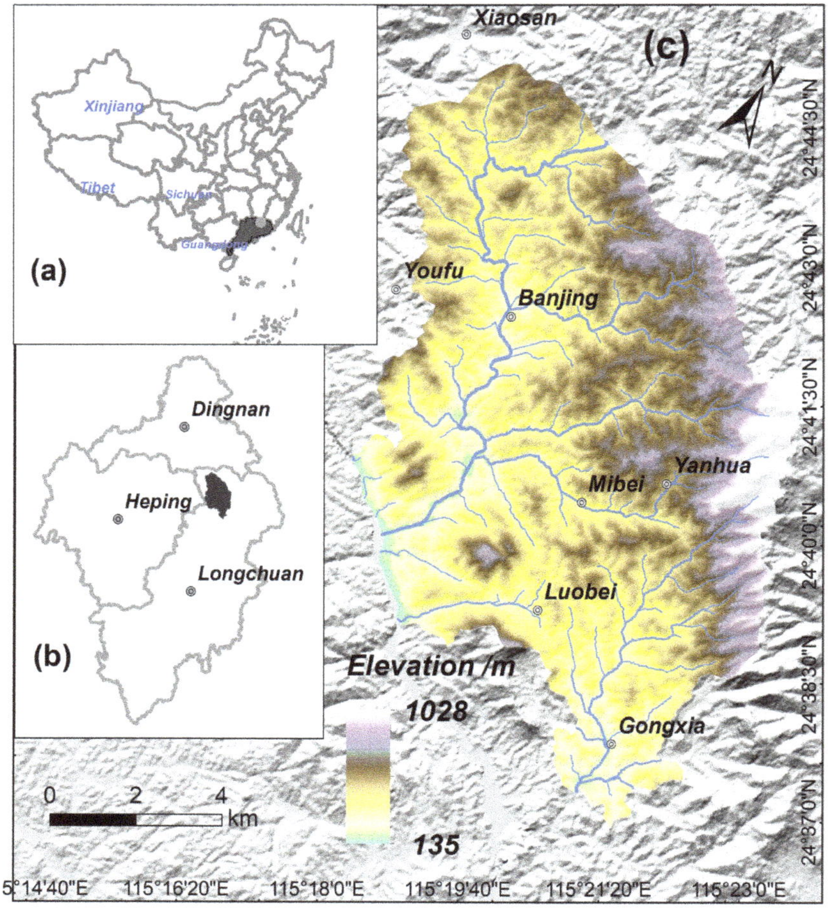

Figure 1. Mapping shows the location and elevation of the study area; (**a**) Guangdong Province; (**b**) location of Longchuan county; (**c**) the elevation and water net distribution of the study area.

Due to the unique geographical and climatic conditions, Longchuan area experiences several large or small rainstorms every year, making it one of most vulnerable zones to geological disasters. From 10 June 2019 to 13 June 2019, Longchuan County suffered continuous heavy rainfall; this rainfall event triggered a large number of landslides. As far as local people can recollect, since the evening of the 10 June 2019, transportation has been disrupted, communication has been lost, and electricity has been cut off. Meanwhile, the settlement below the mountain was engulfed in mist, and the sound of collapses and landslides was constant.

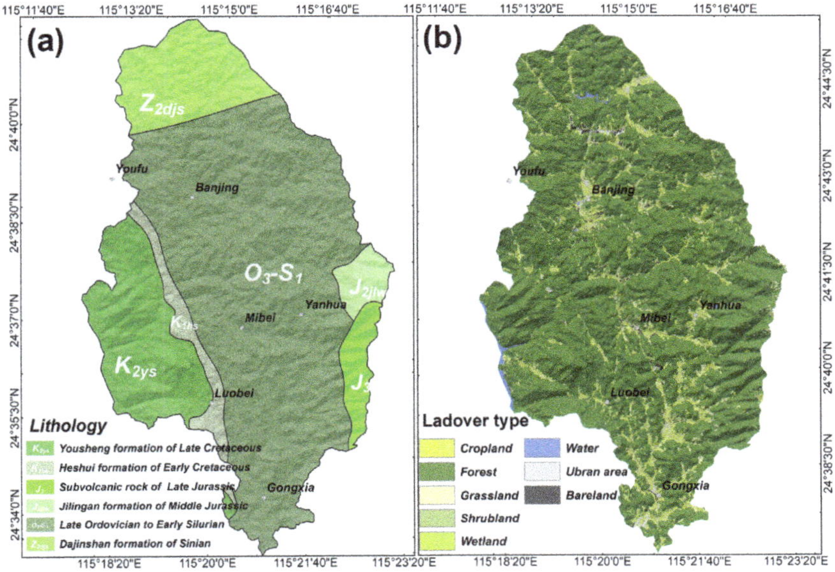

Figure 2. (**a**) Geological map of the study area obtained from 1:200,000 geological maps published by China Geological Survey (http://dcc.cgs.gov.cn/, accessed on 1 July 2022); (**b**) the land use type map of the study area derived from the 10-m resolution global land cover results [65].

3. Data and Method

3.1. Landslide Mapping

The availability of high-resolution satellite images on the Google Earth (GE) platform allowed us to conduct a detailed visual interpretation of landslides [66,67]. The remote sensing images used for landslide interpretation in this study are based on the GE platform. It was important that the high-resolution satellite image covered the entire study area, and the dates of images before and after the rainfall event were mainly in January 2019 and January 2021. Meanwhile, given the relatively long interval between the images before and after the rainfall event, we obtained the Sentinel-2 images with 10 m resolution as a supplementary (the pre- and post-events images were 17 April 2019 and 24 September 2019, respectively) (Figure 3). The landslide inventory was checked by Sentinel-2 images to ensure that the interpreted landslides were caused by the 2019 rainfall event. The reason for selecting these two images was that they had the closest interval between rainfall events without cloud cover in the study area. Landslides were identified by visual interpretation and mapped as polygons. Since the study area has high vegetation coverage, landslides can be better delineated by satellite images before and after this event. Figure 3 depicts the Sentinel-2 satellite images before and after the rainfall. According to remote sensing images, most landslides triggered by this event were small and medium-scale shallow landslides, and a majority of them were located near Mibei village, showing obvious group-occurring characteristics (Figure 4).

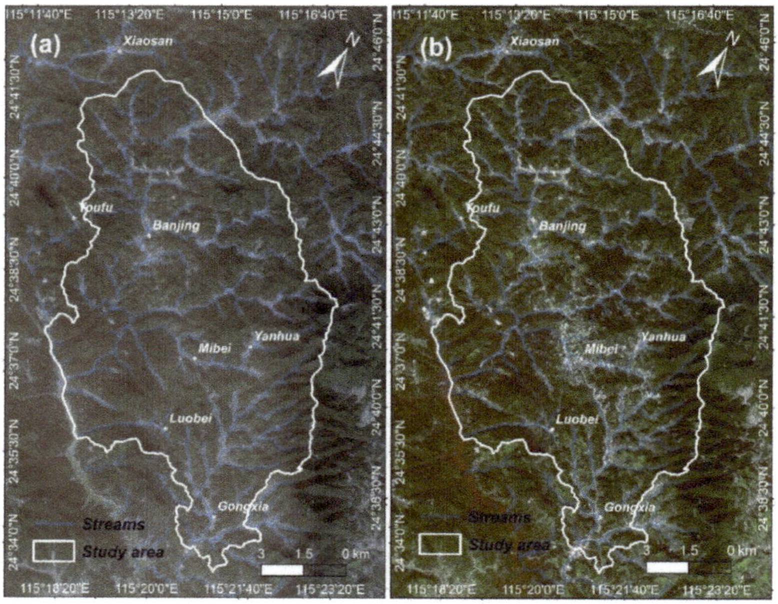

Figure 3. Mapping shows the Sentinel-2 images before and after the rainfall event; (**a**) satellite image before rainfall event taken on 17 April 2019; (**b**) satellite image before rainfall event taken on 24 September 2019.

Figure 4. (**a**) Aerial photograph of Mibei village after the rainfall event, houses are damaged by rainfall-induced landslides; (**b**) group-occurring shallow landslides; (**c**) the landslide damaged the hillside residences, and the floors on the second floor crashed on the first floor; (**d**) road damage caused by landslides (Picture source: http://www.gdlctv.com/Pc/index/new_detalis.html?id=3320, accessed on 25 June 2022).

3.2. Rainfall Data

We collected the precipitation data over the past two decades from 2000 to 2020 in Longchuan County (Figure 5). The results show that the average rainfall remained between 1200 and 2400 mm, with periodic fluctuations. The annual rainfall in 2006 and 2017 was unusually high, reaching almost 2300 mm or more. In comparison, the annual rainfall in 2019 was low with 1500 mm, which was roughly the same as the recent 20-year average (Figure 5a). Comparing the monthly rainfall in 2019 with the average value over the past two decades (Figure 5b), we also found that the rainfall from March to June in 2019 was higher than the monthly average rainfall in the last 20 years. The precipitation in June of 2019 was 300 mm, slightly higher than the monthly average rainfall of 250 mm in previous years.

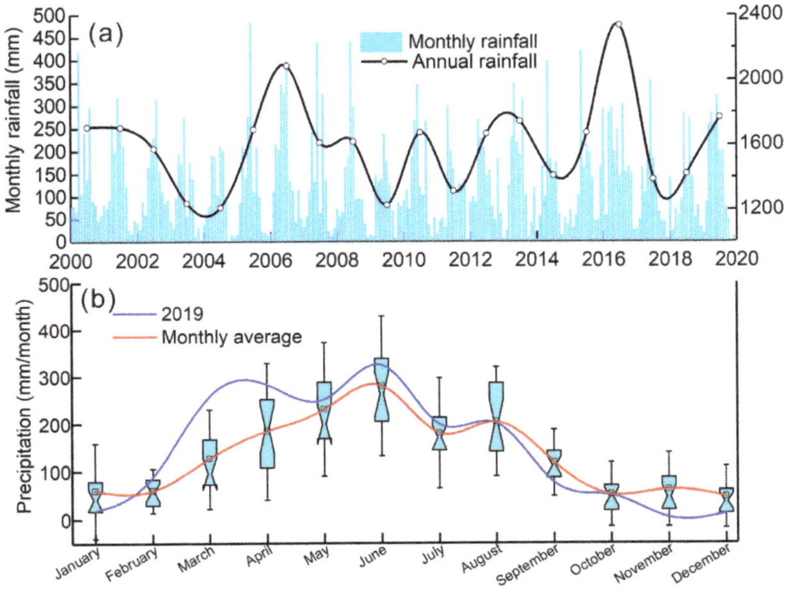

Figure 5. Monthly rainfall data of Longchuan County in the past 20 years from 2000 to 2020; (**a**) monthly and annual average rainfall data over the last 20 years; (**b**) comparing the monthly rainfall in 2019 with the average value over the last two decades.

We obtained the data for the rainfall every 12 h based on the rainfall stations of China Meteorological Administration. Eight national rainfall stations within 50 km of the study area were utilized for interpolation, and the most popular Kriging interpolation algorithm was used to obtain the spatial distribution of rainfall (Figure 6). The results show that this rainfall event occurred primarily from 10 June 2019 to 13 June 2019 (Figure 7). The cumulative rainfall was basically the same, remaining at 210 to 220 mm, with rainfall in the west slightly higher than that in the east (Figure 6). Figure 8 shows the distribution of daily rainfall from 10th to 13th of June during this rainfall event. The rainfall on 10 June 2019 was the heaviest, peaking at around 120 mm, accounting for more than half of this rainfall event. The rainfall for the next three days was expected to be around 20–40 mm. Otherwise, the spatial change of daily rainfall in the study area from June 10 to 13 was relatively small, and the difference of daily rainfall of the study area was essentially maintained within 10 mm.

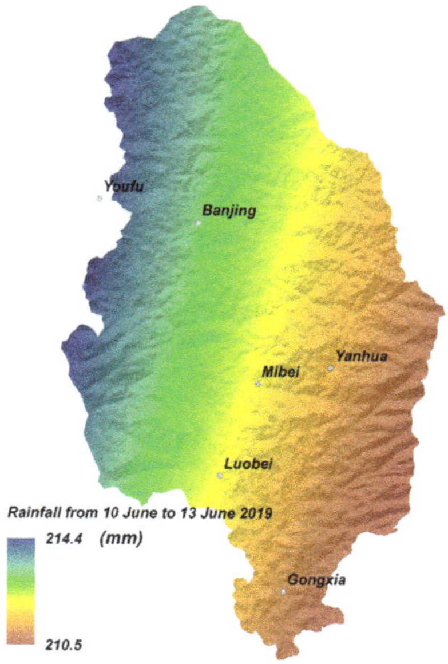

Figure 6. Mapping shows the spatial distribution of total rainfall from 10 June 2019 to 13 June 2019.

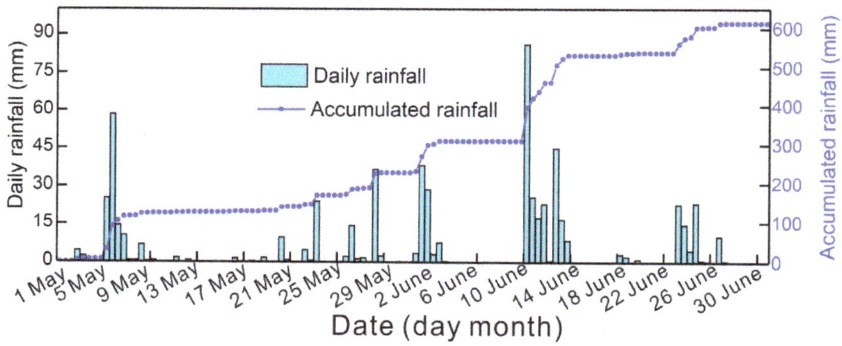

Figure 7. Data of the rainfall every 12 h from the national rainfall stations in the study area from 1 May to 30 June.

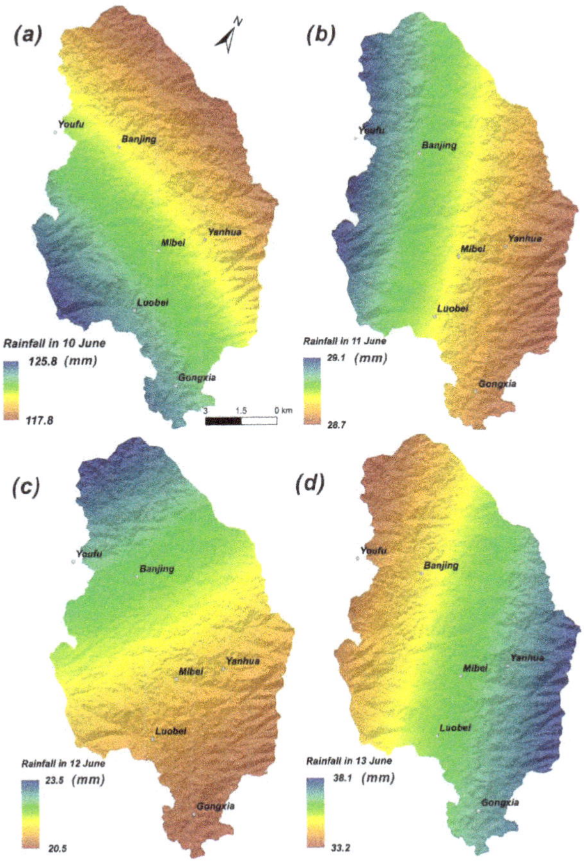

Figure 8. Mapping shows the distribution of daily rainfall from the 10th to 13th of June during this rainfall event; (**a**) 10 June 2019; (**b**) 11 June 2019; (**c**) 12 June 2019; (**d**) 13 June 2019.

3.3. Data Related to Other Influencing Factors

To assess the role of topographic, geologic, and hydrologic factors on the distribution of rainfall-induced landslides, we obtained several terrain metrics (i.e., elevation, hillslope gradient, and topographic relief) and lithologic and hydrological data. The elevation data were derived from ALOS PALSAR DEM with 12.5 m resolution, which were then resampled into a 5 m resolution based on the bilinear algorithm. The hillslope gradient and slope aspect were derived from the DEM data. In addition, we estimated the topographic relief from the elevation range within a 1.0 km radius. TWI was computed using GRASS GIS and the DEM data. Drainages were also derived from DEM by AcrGIS. The road data were downloaded from the OpenStreetMap Data (https://master.apis.dev.openstreetmap.org/export#map=11/35.2510/103.4308, accessed on 5 June 2022). The lithology data are obtained from 1:200,000 geological maps published by China Geological Survey (http://dcc.cgs.gov.cn/, accessed on 1 July 2022). The land use type data were derived from the 10 m resolution global land cover results [65]. The spatial distribution of the above influencing factors was converted into a raster format with a grid cell size of 5 m. Finally, seven influencing factors were considered for the statistical analysis, including the elevation, hillslope gradient, relief, slope aspect, land use type, road density, and distance to river (Figure 9). Meanwhile, the relationship between different influencing factors and the occurrence of landslides were analyzed by the polygon feature.

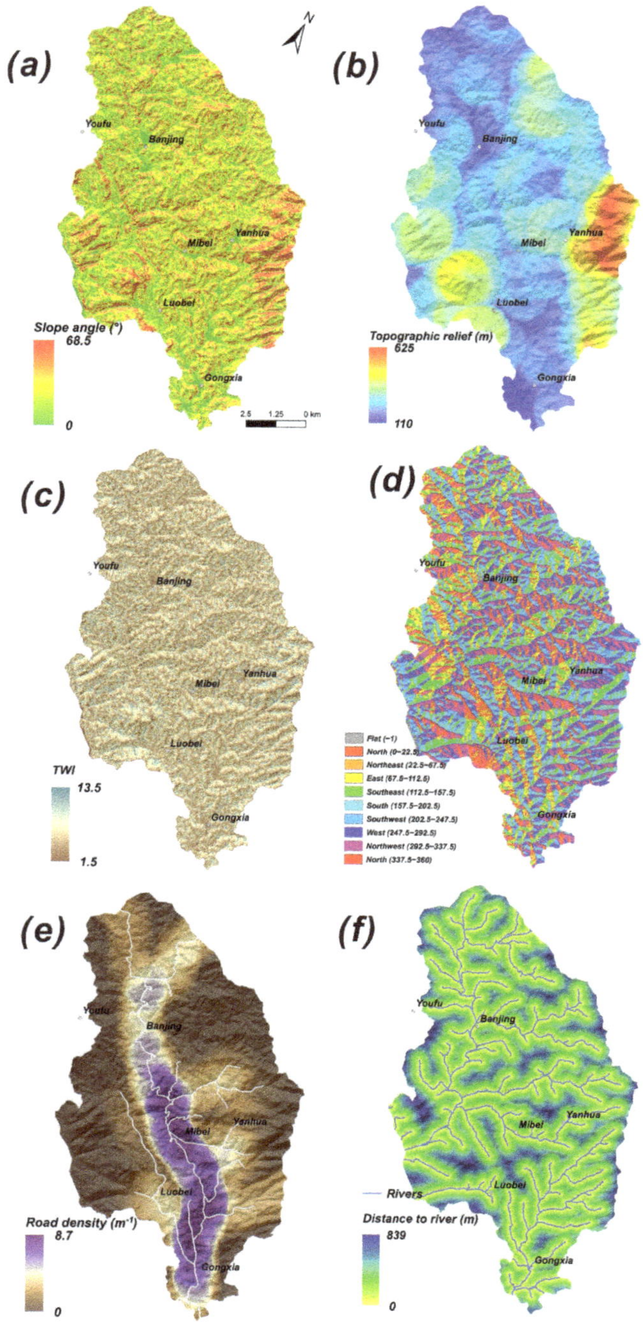

Figure 9. Mapping showing the distribution of the influencing factors in the study area; (**a**) slope angle; (**b**) topographic relief; (**c**) topographic wetness index; (**d**) aspect; (**e**) road density; (**f**) distance to river.

3.4. TRIGRS Model

The TRIGRS model (Transient rainfall infiltration and grid-based regional slope-stability model) is a widely used and effective evaluation model of rainfall-induced shallow landslide susceptibility [68,69]; the model was developed by the United States Geological Survey (USGS) [40,70] and written by Baum et al in FORTRAN [40], and it needs specific input parameters, mainly including rainfall parameters, soil mechanics parameters, and hydrological parameters of the study area. Following the determination of the parameters, the grid stability caused by the change of transient pore water pressure of each grid during the rainfall period is calculated based on the GIS platform for the purpose of evaluating the slope stability of all grids in a certain rainfall period.

Iverson [36] linearized the solution of the Richards equation and this serves as the foundation for the infiltration models for wet initial conditions. It consists of a steady component and a transient component of seepage. The steady seepage is determined by the initial depth of the water table and steady infiltration rate. Under steady infiltration, the slope is stable. Transient infiltration is the short-term change in pore water pressure caused by rainfall. The infinite slope model is then applied using the computed transient pore water pressure. The generalized solution in TRIGRS is:

$$\begin{gathered}\psi(Z,t) = (Z-d)\beta + 2\sum_{n=1}^{N}\frac{I_{nz}}{K_s}H(t-t_n)[D_1(t-t_n)]^{\frac{1}{2}}\\ \sum_{m=1}^{\infty}\left\{ierfc\left[\frac{(2m-1)d_{LZ}-(d_{LZ}-Z)}{2[D_1(t-t_n)]^{\frac{1}{2}}}\right]+ierfc\left[\frac{(2m-1)d_{LZ}+(d_{LZ}-Z)}{2[D_1(t-t_n)]^{\frac{1}{2}}}\right]\right\}-\\ 2\sum_{n=1}^{N}\frac{I_{nz}}{K_s}H(t-t_{n+1})[D_1(t-t_{n+1})]^{\frac{1}{2}}\sum_{m=1}^{\infty}\left\{ierfc\left[\frac{(2m-1)d_{LZ}-(d_{LZ}-Z)}{2[D_1(t-t_{n+1})]^{\frac{1}{2}}}\right]+ierfc\left[\frac{(2m-1)d_{LZ}+(d_{LZ}-Z)}{2[D_1(t-t_{n+1})]^{\frac{1}{2}}}\right]\right\}\end{gathered} \tag{1}$$

where ψ is the groundwater pressure head; t is time; N is the total number of time intervals; Z is depth below the ground surface in the vertical coordinate direction; d is the depth of steady-state water table; d_{LZ} is the depth of the impermeable basal boundary; $\beta = \cos^2\delta - (I_{ZLT}/K_s)$, δ is the slope angle; I_{ZLT} is the steady surface flux; Ks is the saturated hydraulic conductivity; I_{nZ} is the the surface flux or rainfall intensity for the nth time interval; $D_1 = D_0/\cos^2\delta$, D_0 is the saturated hydraulic diffusivity; and $H(t-t_n)$ is the Heaviside step function in which t_n is the time at the nth time interval in the rainfall infiltration sequence.

$$ierfc(\eta) = \frac{1}{\sqrt{\pi}}\exp\left(-\eta^2\right) - \eta erfc(\eta) \tag{2}$$

where $erfc(\eta)$ is the complementary error function.

The model calculates infiltration (*I*) at each cell as the sum of precipitation (*P*) and any runoff from upslope cells (*Ru*), with the caveat that infiltration cannot exceed the saturated hydraulic conductivity (*Ks*):

$$I = P + R_u,\ if\ P + R_u \le K_s \tag{3}$$

$$I = K_s,\ if\ P + R_u > K_s \tag{4}$$

When *P* + *Ru* exceeds *Ks* in a cell, the excess is considered runoff (*Rd*) and is diverted to nearby downslope cells.

$$R_d = P + R_u - K_s,\ if\ P + R_u - K_s \ge 0 \tag{5}$$

$$R_d = 0,\ if\ P + R_u - K_s < 0 \tag{6}$$

The TRIGRS model calculates the slope stability using an infinite-slope stability analysis (Equation (7)), as described in Iverson [36]. The ratio of resistant basal coulomb friction to gravitationally induced downslope basal driving stress characterizes the instability of an infinite slope in the failure analysis [71]. This ratio F_s, is computed at depth Z by

$$F_s(Z,t) = \frac{\tan\varphi'}{\tan\delta} + \frac{c' - \psi(Z,t)\gamma_w \tan\varphi'}{\gamma_s Z \sin\delta\cos\delta} \quad (7)$$

where c' is the cohesion of the soil, φ' is the friction angle, γ_s is the unit weight, and γ_w is unit weight of groundwater.

The flow chart of this study is shown in Figure 10.

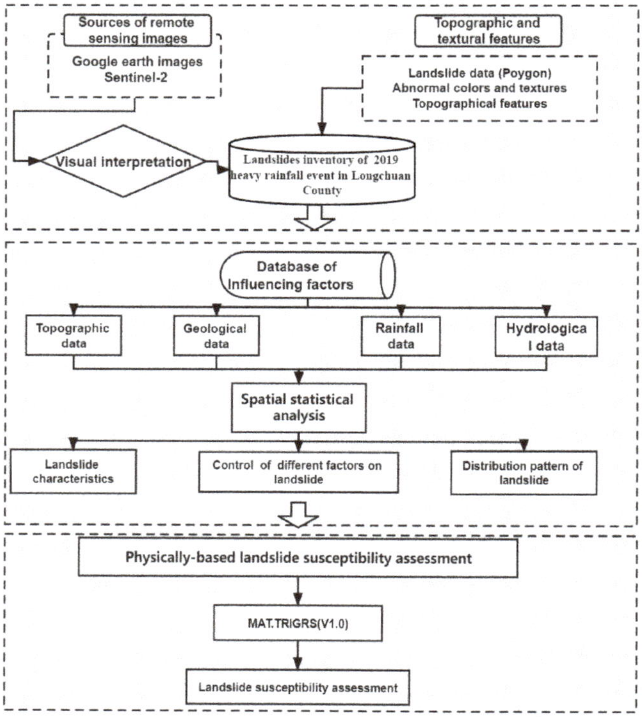

Figure 10. Flow chart of this study.

4. Rainfall-Induced Landslide Inventory

4.1. Basic Characteristics

This rainfall event triggered 667 landslides over an area of 108 km^2, and the majority of them (552 landslides) occurred in the Mibei and Yanhua villages (Figure 11a,b). The types of landslides were various, including shallow landslides combined with a small number of flowslides, rockfalls, and debris flows with a total landslide area of 0.75 km^2. The largest landslide area was approximately 20,000 m^2, the smallest area was 50 m^2, and the average landslide area was about 1100 m^2. According to the statistics, there were 288 landslides with an area of less than 500 m^2, accounting for approximately 43% of all landslides. There were 291 landslides with an area of 500~2000 m^2, accounting for approximately 44%. The number of landslides with an area of 2000~5000 m^2 and > 5000 m^2 was 71 and 17, respectively (Figure 10).

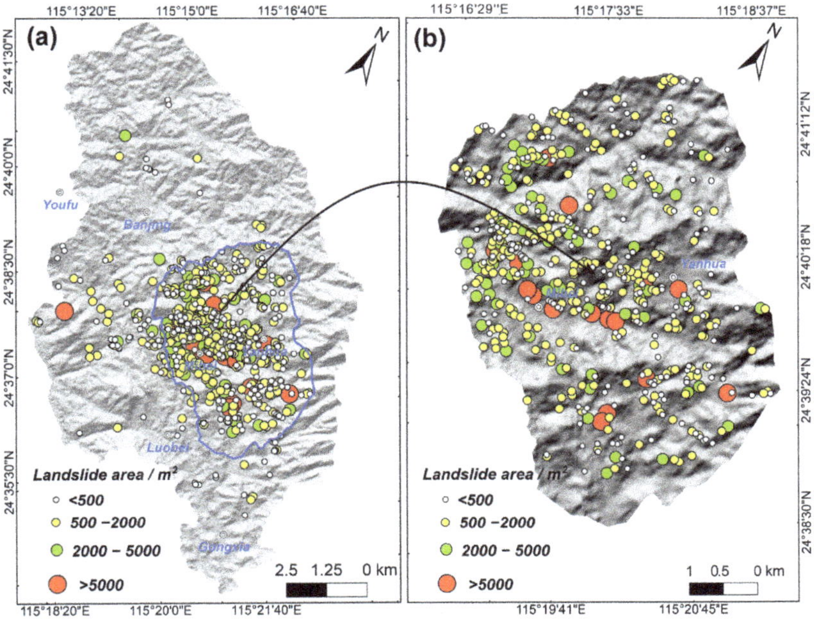

Figure 11. (**a**) Rainfall-induced landslide inventory of the 2019 heavy rainfall event in Longchuan County, the landslide abundance area of this event; (**b**) map showing the zooming of the landslide abundance area.

We calculated the landslide number density (LND) and landslide area density (LAD) within a 1.5 km-radius moving window using a Gaussian density kernel function. The LND and LAD maps indicated that the maximum LAD and LND of the study area were 9.5% and 78/km^2, respectively (Figure 11). Landslides had obvious cluster distribution characteristics, and a large number of landslides were concentrated within 2 km of the Mibei Village (Figure 12).

4.2. Factor Analysis

In order to analyze the relationship between different influencing factors and the occurrence of landslides, we calculated the frequency distribution of landslides and landscape (i.e., non-landslide area) and the LAD of different influencing factors. Figure 13 shows the frequency density distribution of landslide and non-landslide areas, and Figure 14 shows the LAD in different intervals of six influencing factors (the higher the LAD, the more likely the landsliding will occur). For elevation, the frequency density distribution of landslide area and non-landslide area was basically the same (Figure 13a), the peak LAD was situated at elevations from 300 to 450 m, indicating that landslides were more likely to occur within this elevation range (Figure 14a). For the slope angle, the landscape area was clustered between 5 and 20°, while most of the landslides occurred on slopes with the inclination of 15–35° (Figure 13b). Overall, the LAD increased with the increase in the slope angle and was described by an exponential relationship of $y = e^{(0.21+0.08x)}$, (where x is the slope angle and y is the LAD, Figure 14b), suggesting that with the increase in the slope angle, the landslide occurrence possibility also increased. In terms of topographic relief, the relief of non-landslide area was primarily concentrated in the 200~250 m range, whereas the landslide area was primarily clustered in the range of 250~350 m (Figure 13c). Overall, there was a negative logarithmic relationship between the LAD and relief, indicating that the LAD decreased with the increase in relief (Figure 14c). On the part of TWI, landslides were most commonly seen in the range with TWI values between

4 and 6, and there was a positive exponential relationship between the LAD and TWI of $y = e^{(-12.16+1.18x)}$ (where x is the TWI and y is the LAD), and the LAD increased as the TWI increased (Figure 14d). For road density, landslides were primarily distributed in the road density interval between 2 and 4 (Figure 13d). In general, LAD and road density had a linear relationship of $y = 0.14x + 0.45$ (where x is the road density and y is the LAD), which shows that landslides were more likely to occur in areas with a high road density (Figure 14e). For the distance to river, landslides were more likely to occur in the range of 100~400 m, and there was no obvious correlation between the LAD and the river distance (Figure 14f).

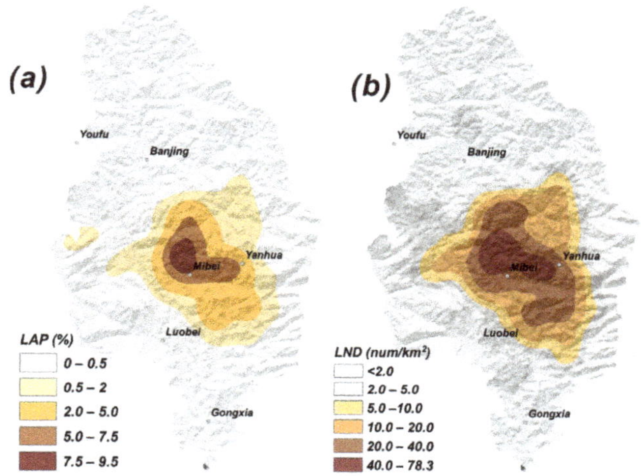

Figure 12. Map showing spatial density of landslides triggered by this rainfall event. (**a**) landslide number density (LND); (**b**); landslide areal density (LAD).

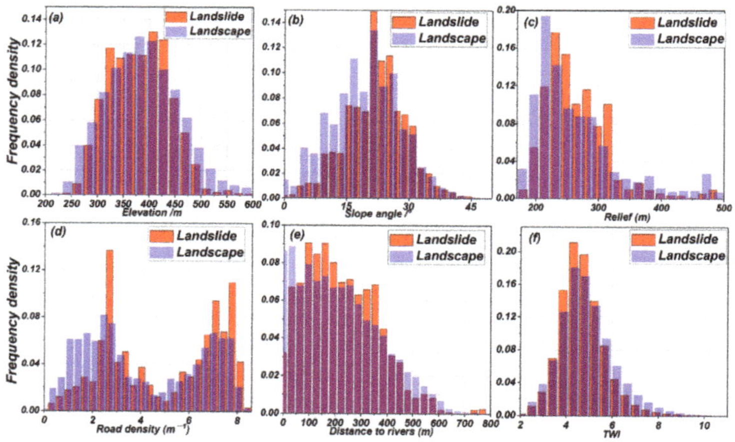

Figure 13. Frequency density estimates of landslides and landscape area for different influencing factors; (**a**) slope angle; (**b**) elevation; (**c**) relief; (**d**) road density; (**e**) distance to rivers; (**f**) TWI.

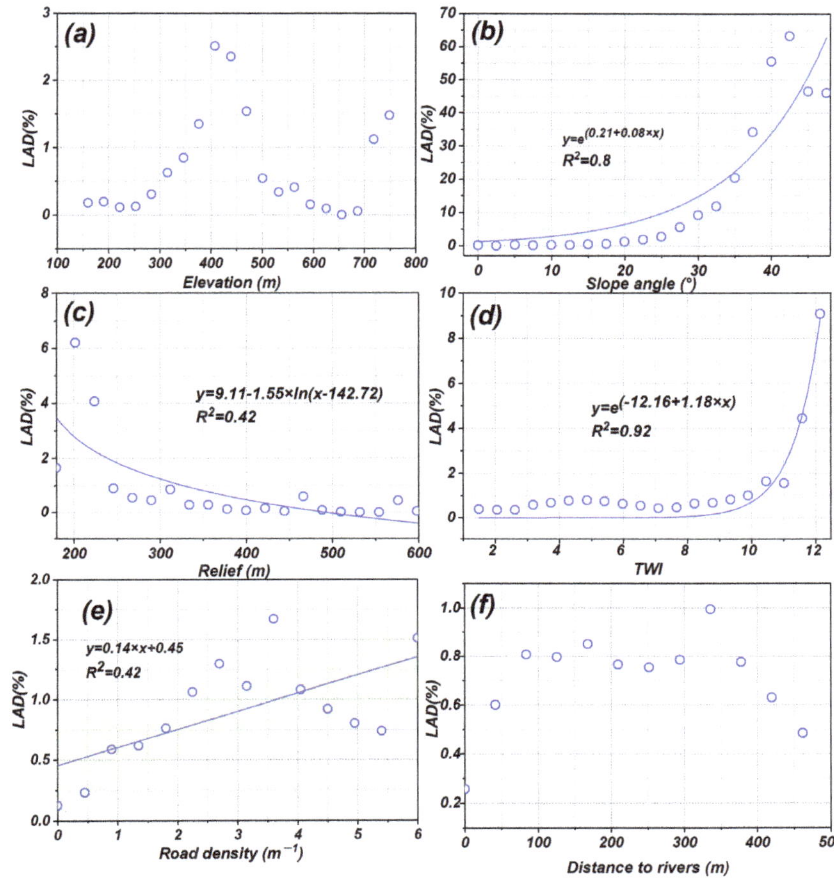

Figure 14. Map showing the relationship between the six influencing factors and the landslide areal density (LAD); (**a**) elevation; (**b**) slope angle; (**c**) topographic relief; (**d**) TWI; (**e**) road density; (**f**) distance to rivers.

Figure 15 shows the statistical results of the landslides and the slope aspect. Figure 15a shows the frequency density of the landslides and landscape (i.e., non-landslide area) on different slope aspects. The result demonstrates that the non-landslide area was evenly distributed in all aspects, but most of the landslide area was concentrated in the aspect of 110°~180° (SE to S). The statistical results of LAD show that the peak LAD of 1.4% was present at the aspects from SE to S for the landslides.

Figure 16 shows the distribution of the landslide and non-landslide area, and the average landslide area in each land use unit. The result shows that the predominant land type was forest, which accounts for 80% of the study area, followed by cropland land, which accounts for more than 10%. The area of urban area and bare land was less than 1%. Among all land types, shrubland was the most prone to landslides, with roughly 10% of landslides occurring in the 5% area. Landslides were the least developed in cropland, maybe due to the relatively gentle slope of this unit. Furthermore, statistics on the average landslide area of different units suggest that bare land had the largest average landslide area, with more than 1600 m^2, followed by forest land, which had an average landslide area of 1200 m^2, and cropland had a relatively small average landslide area, only 600 m^2.

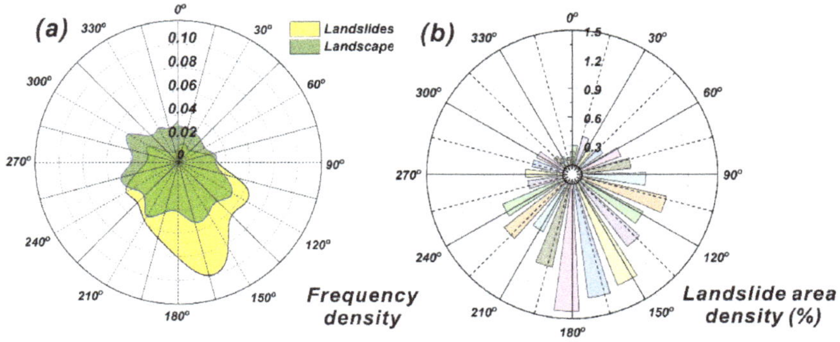

Figure 15. (**a**) The distribution of aspect within landslides and landscape; (**b**) correlations between aspect and landslide area density (LAD).

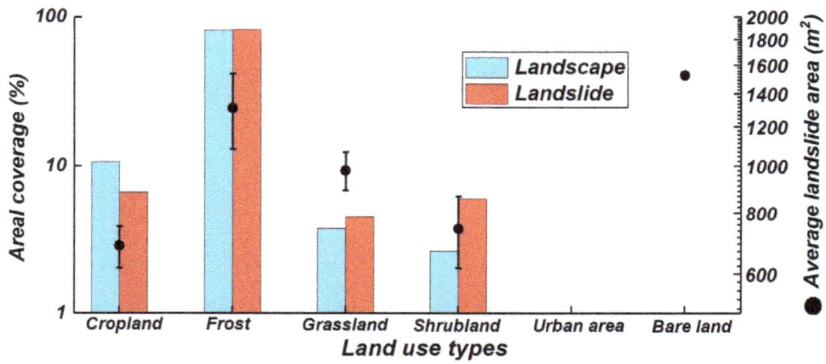

Figure 16. Areal coverage (%) of different land use types for both landslide and landscape overlaid by average landslide area calculated per each unit.

5. Physically Based Landslide Susceptibility Assessment

5.1. Brief Description of MAT.TRIGRS(V1.0)

To address the issues of the manual modification of plentiful model parameters and complex data processing in the traditional TRIGRS model, Ma, et al. [72] proposed a new TRIGRS model using Matlab®programming. It can directly read the grid data of TIF format as the input, and then directly exports the prediction results of grid files, which greatly simplifies data preparation and parameter setting. It includes the script files INPUT DATA.m and TRIGRS.m. The INPUT DATA.m file is used to read the TIF input files, and TRIGRS.m is the executable program that can be used to calculate the pressure head and Fs. The minimum Fs and the corresponding pressure head are generated in the TIF format by calculating the pressure head and Fs at various soil depths. More description can be obtained in [72].

In the physically based model, in order to obtain accurate landslide prediction results, sufficient and accurate input data are required [68,73–75]. For the soil thickness distribution, the Z-model developed by Saulnier, et al. [76] was used to evaluate the soil thickness. We assumed that the maximum thickness of the soil in the study area was 5 m and the minimum thickness was 0.5 m based on previous studies [17,77]. Soil thickness can be estimated and calculated by Equation (8). The bedrock in the study area is monzogranite (O_3-S_1), and the landslide occurred primarily in the weathered soil layer on the bedrock's surface. The soil type of the weathered soil layer is sandy clay loam. Therefore, combined with previous studies [17,52,78], we assigned the corresponding values to mechanical and hydrological parameters including cohesion, internal friction angle, and soil weight of this soil type (Figure 17 and Table 1). Based on previous experience [57,79], saturated hydraulic diffusivity D0 was set to $D_0 = 200K_s$ and the initial surface flux (I_{ZLT}) was generally less than the Ks to one power or more and was set to $I_{ZLT} = 0.01K_s$.

$$h_i = h_{max} - \left(\frac{Z_i - Z_{min}}{Z_{max} - Z_{min}}\right)(h_{max} - h_{min}) \tag{8}$$

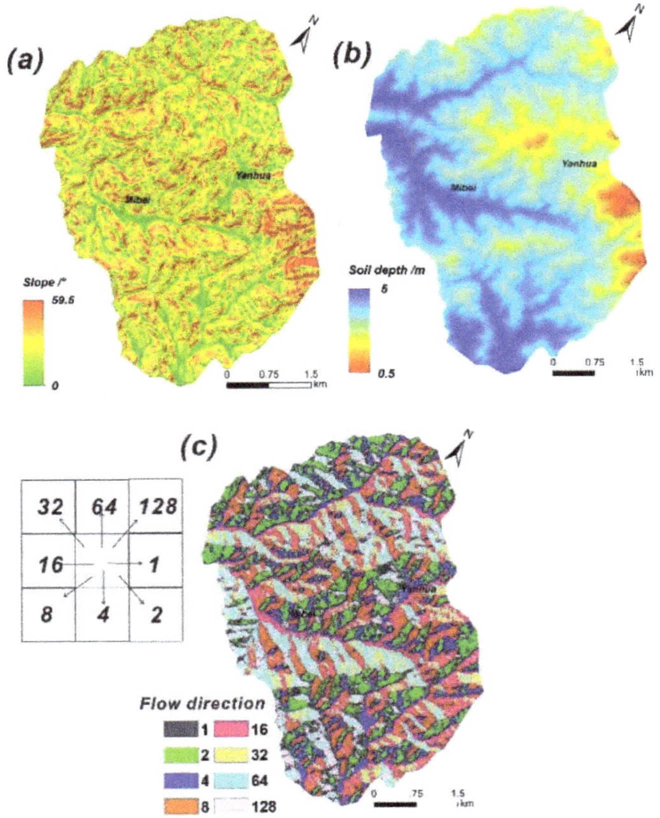

Figure 17. Maps showing the distribution of slope angle (**a**), soil thickness (**b**) and flow direction (**c**).

Table 1. Mechanical properties of the soil.

Input Parameters	Cohesion (Kpa)	Friction Angle (°)	Unit Weight (kN/m^3)	Saturated Hydraulic Conductivity (m/s)
	29	20	15	6.59×10^{-6}

Simultaneously, in order to account for the uncertainties in the physical process that lead to slope failure, the Monte Carlo simulation, which is a robust and well-known approach in applications concerning probability analyses and reliability studies, was used in this study [56,80]. We considered the uncertainties of two main parameters (cohesion and internal friction angle) that primarily influence the slope failure. To characterize the probability density function (PDF) of the two random variables, the normal PDF was adopted. We assumed that the average and standard deviation of the cohesion were 29 kPa and 9 kPa and those of the internal friction angle were 20° and 6°. Based on the Monte Carlo simulation, the input data were calculated by the TRIGRS model, yielding 1000 predicted pictures of potential landslides in the study area. Finally, the slope failure probability (Pf) of the study area was obtained.

5.2. Landslide Susceptibility Assessment

Figure 18 shows the distribution of the average value of 1000 predicted pictures calculated by rainfall data over different time periods. From the calculation results, we can observe that the Fs of all raster cells were greater than that before the rainfall event, indicating that all slopes were stable (Figure 17a). In addition, after 12 h of the rainfall (at 8:00 on 10 June 2019) 12-h rainfall reached 86 mm), the Fs of some grid cells in the study area decreased. Particularly, some grid cells with a large slope angle began to fail (Figure 17b). Then, although continuous rainfall occurred in the subsequent stage (after 11 June 2019), the change of Fs in the study area was relatively small, and few new grid units became unstable.

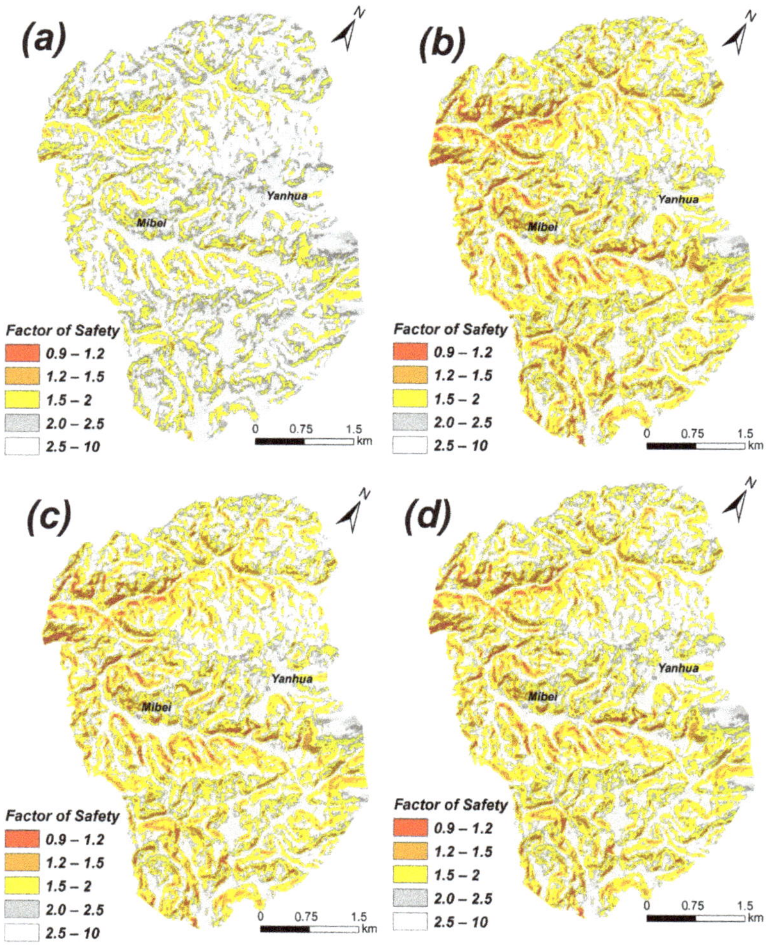

Figure 18. Slope stability conditions, expressed in terms of Safety of Factor (FS) in different time periods of this rainfall event; (**a**) 20:00 on 9 June 2019 (UTC + 8, before rainfall event); (**b**) 8:00 on 10 June 2019 (UTC + 8); (**c**) 20:00 on 11 June 2019(UTC + 8); (**d**) 8:00 on 13 June 2019(UTC + 8).

We calculated the Fs results in the various slope interval over different time periods (Figure 19). The result shows that the Fs of the grids with slope angles between 30 and 40° was mostly distributed between 1.3 and 2.5, with an average value of around 1.6. After the onset of heavy rain on 10 June 2019, the Fs of raster cells rapidly decreased, and the Fs of most grids ranged between 0.9 and 1.7, with an average value of about 1.2. From 8:00 on 10 June 2019, although there was rainfall every day at a subsequent stage, the average rainfall was less than 2 mm/h. The low rainfall intensity had a little impact on the slope stability. Rainfall increased to some extent on 12 June 2019, reaching 45 mm in 12 h, and the Fs decreased slightly. For grids with a slope larger than 40°, we also found the same trend that the Fs of most grid units decreased rapidly after heavy rainfall, and then basically remained unchanged. Overall, the Fs of grids with a slope greater than 40 degrees was much smaller than grids with a slope between 30 and 40 degrees.

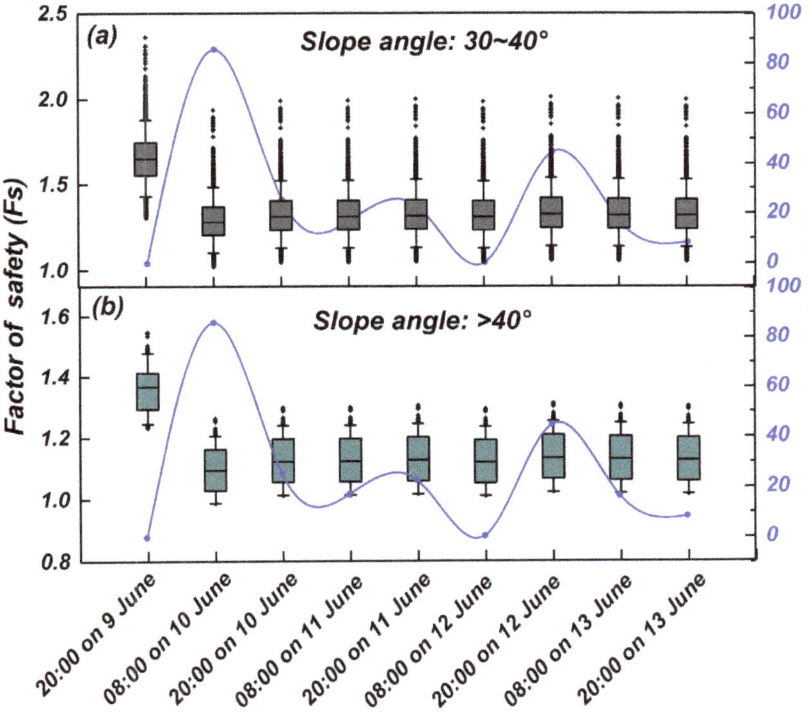

Figure 19. The statistical results of Factor of safety (Fs) in the various slope interval at different rainfall times; (**a**) slope angle: 30~40°; (**b**) slope angle: >40°.

Figure 20 shows the probability distribution of slope failure in different time periods. Obviously, the prediction results of Pf were roughly consistent with the actual landslide distribution. Most areas with a high probability (blue areas) were located on both sides of the river valley, that is, the areas with relatively steep slopes. Before rainfall, almost all the grids in the study area were less than 0.1, indicating that the slope before rainfall was stable. After 12 h of rainfall (at 8:00 on 10 June 2019), the area with steep slopes began to show the instability phenomenon, and the Pf of some grids reached more than 0.6. In the following continuous rainfall, with the decrease in rainfall intensity, there was a slight increase in the area with a high probability of failure.

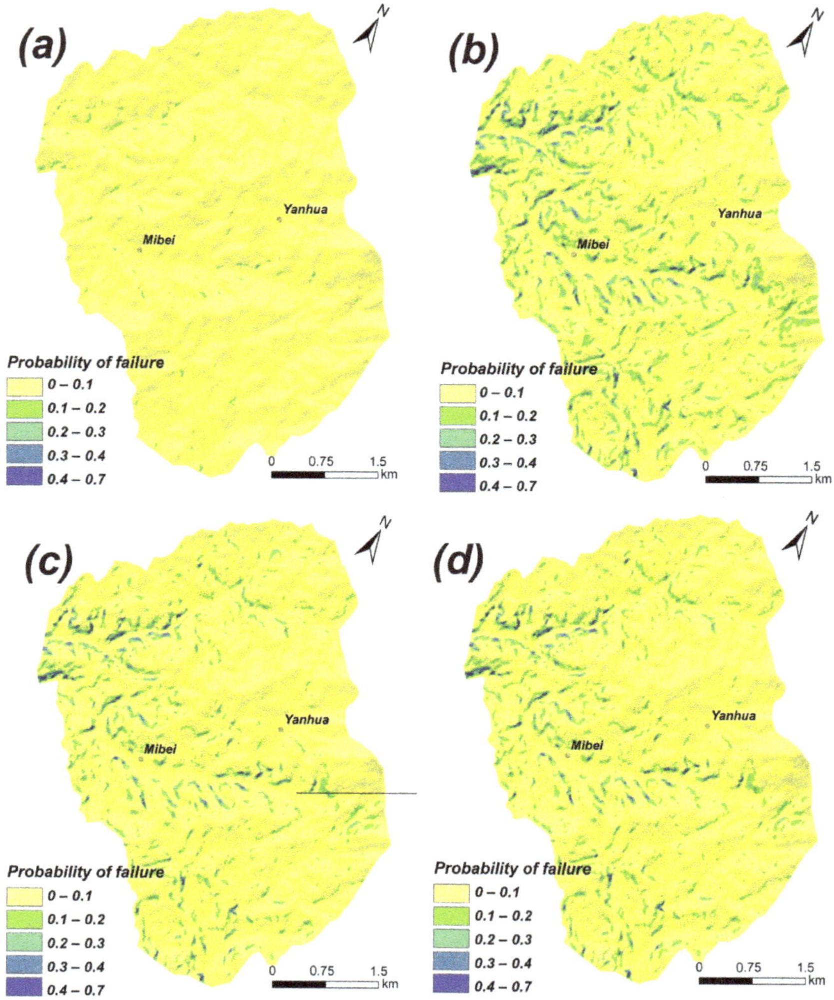

Figure 20. Probability of slope failure (Pf) in different time periods of this rainfall event; (**a**) before rainfall event; (**b**) 8:00 on 10 June 2019; (**c**) 20:00 on 11 June 2019; (**d**) 8:00 on 13 June 2019.

To quantitatively analyze the susceptibility results, we counted the class area, landslide area, and the corresponding LAD of different susceptibility classes before and after rainfall (Figure 21). Based on the natural breaks, the susceptibility level was divided into four classes (i.e., very low, low, moderate, and high). The result shows that before the occurrence of rainfall, most areas belonged to the low susceptibility area, and the majority of landslides were concentrated in very low and low susceptibility areas. With the occurrence of rainfall, the area of low susceptibility areas decreased, while the area of high susceptibility areas increased. The statistical result reveals that 12.1% of the total landslides occurred in the 25.0% of the area which were classified as moderate and high. Meanwhile, the LAD increased with the increase in the susceptibility level, which also shows that the model can effectively predict the potential landslide-prone zone.

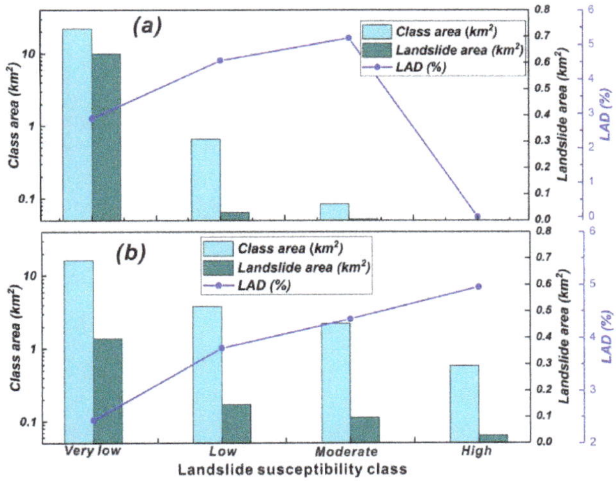

Figure 21. Susceptibility class distribution and the occurrence of landslides within the study area; (**a**) before rainfall; (**b**) after rainfall.

6. Discussion

China's southeast area is situated in a subtropical monsoon climate zone with frequent typhoons and rainstorms. The most common types of geological hazards in this area are landslides and debris flows caused by rainfall, which have the characteristics of a small scale of individual hazard point, a large number of groups, and a wide distribution range [18,19]. In mountainous areas, the effect of the orographic amplification of rainfall and the projection of rainfall-vector on hillslopes [81,82] might result in the windward hill-slope receiving more rainfall, leading to more landslides on the hillslope scale [83]. Due to the influence of the monsoon depression and tropical cyclone, the southeast monsoon prevails in the Longchuan area during the summer (June and July). The landslide distribution of this rainfall event indicates that the southeast and south aspect hillslopes are more prone to collapse than the northwest-north aspect ones (Figure 14). The main reason for this phenomenon is that the south slope is mostly windward, which causes more rainfall and splash erosion in the area. Otherwise, the bedrock weathering degree of the south slope will also be high due to the influence of environmental factors such as soil moisture content, surface temperature, light time, and so forth, leading to relatively weak mechanical parameters of rock and soil mass. Therefore, under the condition of heavy rainfall, the south slope is more prone to landsliding.

Slope angle is an important topographic factor affecting the occurrence of landslides. From the spatial distribution of the landslides, we can observe that the landslides were mainly distributed in low mountainous areas, with the sections at elevations within 300~450 m and slopes ranging from 15 to 35° (Figure 12). The LAD increased with the increase in slope angle and was described by an exponential relationship, indicating that the landslides of this event more easily occurred in areas with steep slopes (Figure 13b). TWI reflects how surface morphology affects soil groundwater level and moisture content, which is represented by a theoretical measure of the accumulation of flow [84,85]. According to the statistical results, there was an exponential relationship between the LAD and TWI, and the LAD increased as the TWI value increased. Especially when the TWI was greater than 10, the LAD increased rapidly (Figure 14f). Higher soil moisture causes higher pore water pressure and reduces the strength of rock and soil mass. As a result, when it rains, the pore water pressure in these areas rises rapidly, resulting in slope failure.

Anthropogenic factors (such as land-use change, deforestation, hill cutting, etc.) play a significant role in the initiation of landslides in active mountain ranges [86–88]. The construction of roads has significantly altered the slope stability of mountainous areas, making them prone to landslides. When a road is built, the toe of the slope is excavated or the weight of the slope is increased, and the overall stability above the slope is reduced, resulting in the occurrence of new landslides or the reactivation of old landslides[89]. From Figure 13d, we can observe that the landslides of this event were more likely to occur in areas with high road density, illustrating that anthropogenic factors have accelerated the instability of the slope in this area. Furthermore, in the Longchuan area, the majority of local residents have excavated mountains to build houses, leading to a number of nearly vertical artificial slopes. Meanwhile, human activities will fragment surrounding natural slopes and increase the degree of rock weathering, which will also exacerbate slope instability in mountain areas.

The formation lithology of the slope is the material basis of landslides. Granite layers are one of the most common strata in China's southeast coastal regions. Long-term weathering of granite results in widely distributed residual soil layers. For the Longchuan area, the bedrock is monzogranite (O_3-S_1), and the landslides occurred primarily in the weathered soil layer on the bedrock surface [17,77]. The major influence depth of heavy rainfall was limited to the superficial zone of slopes due to the difference in rainfall intensity and permeability of granite residual soil. This is why the shallow surface zone was severely affected by landslides [17]. A saturated seepage field was formed in the shallow surface zone of slopes as a result of prolonged heavy rainfall. The mechanical strength of saturated soil diminished, and slide failure occurred at the shallow surface saturation zone.

7. Conclusions

In this work, we established a landslide inventory including all the landslides induced by the 2019 Longchuan heavy rainfall event in Guangdong Province, China. We described the topographical, geological, and hydrological control of landslide hazards. Furthermore, we conducted the physically based susceptibility assessment of shallow landslides based on the MAT.TRIGRS (V1.0) tool. The following conclusions can be drawn: (1) This rainfall event triggered about 670 landslides with a total area of 0.75 km^2; the landslides had obvious cluster distribution characteristics, and a large number of landslides were concentrated within 2 km of the Mibei village. (2) The landslide abundance was closely related to slope angle, TWI, and road density but had a low correlation with elevation and distance to rivers. Among them, the LAD increased with the increase in the slope angle and TWI and was described by an exponential relationship. Otherwise, the statistical results of the landslides and the slope aspect showed that most of the landslide area was concentrated in the aspect of 110°~180° (SE to S). (3) The physically based susceptibility assessment results indicated that the prediction results were roughly consistent with the actual landslide distribution, and most areas with a high susceptibility were located on both sides of the river valley. The onset of heavy rain on 10 June 2019 was the main triggering factor of this group-occurring landslides. Our study will be beneficial for understanding the distribution pattern and cause of rainfall-induced shallow landslides in the Longchuan area, and it can provide data and technical support for the prevention of rainfall-induced geological disasters in the southeast mountainous area of China.

Author Contributions: Conceptualization, C.X. and S.M.; methodology, S.M.; software, S.M. and X.S.; validation, S.M. and X.S.; formal analysis, S.M. and X.S.; data curation, S.M. and X.S.; writing—original draft preparation, S.M.; writing—review and editing, C.X. and S.M.; supervision, C.X.; project administration, C.X.; funding acquisition, C.X. All authors have read and agreed to the published version of the manuscript.

Funding: This study was supported by the National Institute of Natural Hazards, Ministry of Emergency Management of China (ZDJ2020-10 and ZDJ2021-12).

Acknowledgments: We thank Google Earth and Sentinel-2 satellite images for the free access satellite images used in this study. We also express our appreciation of the constructive comments provided by the anonymous reviewers, which is beneficial for the quality of the manuscript a lot.

Conflicts of Interest: The authors declare that they have no known competing financial interests or personal relationships that could have appeared to influence he work reported in this paper.

References

1. Kirschbaum, D.; Kapnick, S.B.; Stanley, T.; Pascale, S. Changes in Extreme Precipitation and Landslides Over High Mountain Asia. *Geophys. Res. Lett.* **2020**, *47*, e2019GL085347. [CrossRef]
2. Emberson, R.; Kirschbaum, D.; Stanley, T. Global connections between El Nino and landslide impacts. *Nat. Commun.* **2021**, *12*, 2262. [CrossRef]
3. Petley, D. Global patterns of loss of life from landslides. *Geology* **2012**, *40*, 927–930. [CrossRef]
4. Lin, Q.; Wang, Y. Spatial and temporal analysis of a fatal landslide inventory in China from 1950 to 2016. *Landslides* **2018**, *15*, 2357–2372. [CrossRef]
5. Gariano, S.L.; Guzzetti, F. Landslides in a changing climate. *Earth Sci. Rev.* **2016**, *162*, 227–252. [CrossRef]
6. Huang, Y.; Xu, C.; Zhang, X.; Li, L. Bibliometric analysis of landslide research based on the WOS database. *Nat. Hazards Res.* **2022**, *2*, 49–61. [CrossRef]
7. Dai, L.; Scaringi, G.; Fan, X.; Yunus, A.P.; Liu-Zeng, J.; Xu, Q.; Huang, R. Coseismic Debris Remains in the Orogen Despite a Decade of Enhanced Landsliding. *Geophys. Res. Lett.* **2021**, *48*, e2021GL095850. [CrossRef]
8. Fan, X.; Yunus, A.P.; Scaringi, G.; Catani, F.; Siva Subramanian, S.; Huang, R. Rapidly Evolving Controls of Landslides After a Strong Earthquake and Implications for Hazard Assessments. *Geophys. Res. Lett.* **2020**, *48*, e2020GL090509. [CrossRef]
9. Fan, X.; Scaringi, G.; Korup, O.; West, A.J.; van Westen, C.J.; Tanyas, H.; Hovius, N.; Hales, T.C.; Jibson, R.W.; Allstadt, K.E.; et al. Earthquake-induced chains of geologic hazards: Patterns, mechanisms, and impacts. *Rev. Geophys.* **2019**, *57*, 421–503. [CrossRef]
10. Schmitt, R.; Tanyas, H.; Jessee, A.; Zhu, J.; Biegel, K.; Allstadt, K.; Jibson, R.; Thompson, E.; Westen, C.J.; Sato, H.; et al. *An Open Repository of Earthquake-Triggered Ground-Failure Inventories*; U.S. Geological Survey: Reston, VA, USA, 2017. [CrossRef]
11. Tian, Y.; Xu, C.; Yuan, R.-M. Earthquake-Triggered Landslides. *Treatise Geomorphol.* **2021**, *2*, 583–614. [CrossRef]
12. Xu, C.; Xu, X.; Zhou, B.; Shen, L. Probability of coseimic landslides: A new generation of earthquake-triggered landslide hazard model. *J. Eng. Geol.* **2019**, *27*, 1122. [CrossRef]
13. Shao, X.; Xu, C. Earthquake-induced landslides susceptibility assessment: A review of the state-of-the-art. *Nat. Hazards Res.* **2022**, in press. [CrossRef]
14. Kirschbaum, D.; Stanley, T. Satellite-Based Assessment of Rainfall-Triggered Landslide Hazard for Situational Awareness. *Earths Future* **2018**, *6*, 505–523. [CrossRef] [PubMed]
15. Emberson, R.; Kirschbaum, D.; Amatya, P.M.; Tanyas, H.; Marc, O. Insights from the topographic characteristics of a large global catalog of rainfall-induced landslide event inventories. *Nat. Hazards Earth Syst. Sci.* **2022**, *22*, 1129–1149. [CrossRef]
16. Marc, O.; Stumpf, A.; Malet, J.P.; Gosset, M.; Uchida, T.; Chiang, S.-H. Initial insights from a global database of rainfall-induced landslide inventories: The weak influence of slope and strong influence of total storm rainfall. *Earth Surf. Dyn.* **2018**, *6*, 903–922. [CrossRef]
17. Feng, W.; Bai, H.; Lan, B.; Wu, Y.; Wu, Z.; Yan, L.; Ma, X. Spatial–temporal distribution and failure mechanism of group-occurring landslides in Mibei village, Longchuan County, Guangdong, China. *Landslides* **2022**, *19*, 1957–1970. [CrossRef]
18. Ma, T.; Li, C.; Lu, Z.; Bao, Q. Rainfall intensity–duration thresholds for the initiation of landslides in Zhejiang Province, China. *Geomorphology* **2015**, *245*, 193–206. [CrossRef]
19. Chen, H.; Dadson, S.; Chi, Y.-G. Recent rainfall-induced landslides and debris flow in northern Taiwan. *Geomorphology* **2006**, *77*, 112–125. [CrossRef]
20. Merghadi, A.; Yunus, A.P.; Dou, J.; Whiteley, J.; ThaiPham, B.; Bui, D.T.; Avtar, R.; Abderrahmane, B. Machine learning methods for landslide susceptibility studies: A comparative overview of algorithm performance. *Earth Sci. Rev.* **2020**, *207*, 103225. [CrossRef]
21. Reichenbach, P.; Rossi, M.; Malamud, B.D.; Mihir, M.; Guzzetti, F. A review of statistically-based landslide susceptibility models. *Earth Sci. Rev.* **2018**, *180*, 60–91. [CrossRef]
22. Hong, H.; Pradhan, B.; Xu, C.; Tien Bui, D. Spatial prediction of landslide hazard at the Yihuang area (China) using two-class kernel logistic regression, alternating decision tree and support vector machines. *Catena* **2015**, *133*, 266–281. [CrossRef]
23. Shao, X.; Ma, S.; Xu, C.; Zhang, P.; Wen, B.; Tian, Y.; Zhou, Q.; Cui, Y. Planet Image-Based Inventorying and Machine Learning-Based Susceptibility Mapping for the Landslides Triggered by the 2018 Mw6.6 Tomakomai, Japan Earthquake. *Remote Sens.* **2019**, *11*, 978. [CrossRef]
24. Shao, X.; Ma, S.; Xu, C.; Zhou, Q. Effects of sampling intensity and non-slide/slide sample ratio on the occurrence probability of coseismic landslides. *Geomorphology* **2020**, *363*, 107222. [CrossRef]
25. Tanyu, B.F.; Abbaspour, A.; Alimohammadlou, Y.; Tecuci, G. Landslide susceptibility analyses using Random Forest, C4.5, and C5.0 with balanced and unbalanced datasets. *CATENA* **2021**, *203*, 105355. [CrossRef]
26. Kavzoglu, T.; Sahin, E.K.; Colkesen, I. An assessment of multivariate and bivariate approaches in landslide susceptibility mapping: A case study of Duzkoy district. *Nat. Hazards* **2015**, *76*, 471–496. [CrossRef]
27. Wang, Y.; Fang, Z.; Hong, H. Comparison of convolutional neural networks for landslide susceptibility mapping in Yanshan County, China. *Sci. Total Environ.* **2019**, *666*, 975–993. [CrossRef]
28. Xu, C.; Xu, X.; Dai, F.; Saraf, A.K. Comparison of different models for susceptibility mapping of earthquake triggered landslides related with the 2008 Wenchuan earthquake in China. *Comput. Geosci.* **2012**, *46*, 317–329. [CrossRef]

29. Arabameri, A.; Chandra Pal, S.; Rezaie, F.; Chakrabortty, R.; Saha, A.; Blaschke, T.; Di Napoli, M.; Ghorbanzadeh, O.; Thi Ngo, P.T. Decision tree based ensemble machine learning approaches for landslide susceptibility mapping. *Geocarto Int.* **2021**, *37*, 1–35. [CrossRef]
30. Adnan, M.S.G.; Rahman, M.S.; Ahmed, N.; Ahmed, B.; Rabbi, M.F.; Rahman, R.M. Improving Spatial Agreement in Machine Learning-Based Landslide Susceptibility Mapping. *Remote Sens.* **2020**, *12*, 3347. [CrossRef]
31. Rossi, M.; Guzzetti, F.; Reichenbach, P.; Mondini, A.C.; Peruccacci, S. Optimal landslide susceptibility zonation based on multiple forecasts. *Geomorphology* **2010**, *114*, 129–142. [CrossRef]
32. Ma, S.; Xu, C.; Shao, X. Spatial prediction strategy for landslides triggered by large earthquakes oriented to emergency response, mid-term resettlement and later reconstruction. *Int. J. Disaster Risk Reduct.* **2020**, *43*, 101362. [CrossRef]
33. Gutiérrez-Martín, A. A GIS-physically-based emergency methodology for predicting rainfall-induced shallow landslide zonation. *Geomorphology* **2020**, *359*, 107121. [CrossRef]
34. Hong, Y.; Adler, R.; Huffman, G. Evaluation of the potential of NASA multi-satellite precipitation analysis in global landslide hazard assessment. *Geophys. Res. Lett.* **2006**, *33*, L22402. [CrossRef]
35. Lin, Q.; Lima, P.; Steger, S.; Glade, T.; Jiang, T.; Zhang, J.; Liu, T.; Wang, Y. National-scale data-driven rainfall induced landslide susceptibility mapping for China by accounting for incomplete landslide data. *Geosci. Front.* **2021**, *12*, 101248. [CrossRef]
36. Iverson, R.M. Landslide triggering by rain infiltration. *Water Resour. Res.* **2000**, *36*, 1897–1910. [CrossRef]
37. Corominas, J.; van Westen, C.; Frattini, P.; Cascini, L.; Malet, J.P.; Fotopoulou, S.; Catani, F.; van Den Eeckhaut, M.; Mavrouli, O.; Agliardi, F.; et al. Recommendations for the quantitative analysis of landslide risk. *Bull. Eng. Geol. Environ.* **2014**, *73*, 209–263. [CrossRef]
38. Fell, R.; Corominas, J.; Bonnard, C.; Cascini, L.; Leroi, E.; Savage, W. Guidelines for landslide susceptibility, hazard and risk zoning for land use planning. *Eng. Geol.* **2008**, *102*. [CrossRef]
39. Sorbino, G.; Sica, C.; Cascini, L. Susceptibility analysis of shallow landslides source areas using physically based models. *Nat. Hazards* **2010**, *53*, 313–332. [CrossRef]
40. Baum, R.L.; Savage, W.Z.; Godt, J.W. *TRIGRS-A Fortran Program for Transient Rainfall Infiltration and Grid-Based Regional Slope-Stability Analysis, Version 2.0*; US Geological Survey: Reston, VA, USA, 2008; pp. 1159–2008.
41. Baum, R.L.; Godt, J.W.; Savage, W.Z. Estimating the timing and location of shallow rainfall-induced landslides using a model for transient, unsaturated infiltration. *J. Geophys. Res. F Earth Surf.* **2010**, *115*, F03013. [CrossRef]
42. Domènech, G.; Alvioli, M.; Corominas, J. Preparing first-time slope failures hazard maps: From pixel-based to slope unit-based. *Landslides* **2019**, *17*, 249–265. [CrossRef]
43. Schiliro, L.; Montrasio, L.; Scarascia Mugnozza, G. Prediction of shallow landslide occurrence: Validation of a physically-based approach through a real case study. *Sci. Total Environ.* **2016**, *569-570*, 134–144. [CrossRef] [PubMed]
44. Sarkar, S.; Roy, A.K.; Raha, P. Deterministic approach for susceptibility assessment of shallow debris slide in the Darjeeling Himalayas, India. *Catena* **2016**, *142*, 36–46. [CrossRef]
45. An, H.; Viet, T.T.; Lee, G.; Kim, Y.; Kim, M.; Noh, S.; Noh, J. Development of time-variant landslide-prediction software considering three-dimensional subsurface unsaturated flow. *Environ. Model. Softw.* **2016**, *85*, 172–183. [CrossRef]
46. Montrasio, L.; Valentino, R.; Losi, G.L. Rainfall-induced shallow landslides: A model for the triggering mechanism of some case studies in Northern Italy. *Landslides* **2009**, *6*, 241–251. [CrossRef]
47. Montrasio, L.; Valentino, R.; Corina, A.; Rossi, L.; Rudari, R. A prototype system for space–time assessment of rainfall-induced shallow landslides in Italy. *Nat. Hazards* **2014**, *74*, 1263–1290. [CrossRef]
48. Montrasio, L.; Valentino, R.; Meisina, C. Soil Saturation and Stability Analysis of a Test Site Slope Using the Shallow Landslide Instability Prediction (SLIP) Model. *Geotech. Geol. Eng.* **2018**, *36*, 2331–2342. [CrossRef]
49. Tran, T.V.; Lee, G.; An, H.; Kim, M. Comparing the performance of TRIGRS and TiVaSS in spatial and temporal prediction of rainfall-induced shallow landslides. *Environ. Earth Sci.* **2017**, *76*, 315. [CrossRef]
50. Escobar-Wolf, R.; Sanders, J.D.; Vishnu, C.L.; Oommen, T.; Sajinkumar, K.S. A GIS Tool for Infinite Slope Stability Analysis (GIS-TISSA). *Geosci. Front.* **2021**, *12*, 756–768. [CrossRef]
51. He, X.; Hong, Y.; Vergara, H.; Zhang, K.; Kirstetter, P.-E.; Gourley, J.J.; Zhang, Y.; Qiao, G.; Liu, C. Development of a coupled hydrological-geotechnical framework for rainfall-induced landslides prediction. *J. Hydrol.* **2016**, *543*, 395–405. [CrossRef]
52. Zhang, K.; Xue, X.; Hong, Y.; Gourley, J.J.; Lu, N.; Wan, Z.; Hong, Z.; Wooten, R. iCRESTRIGRS: A coupled modeling system for cascading flood–landslide disaster forecasting. *Hydrol. Earth Syst. Sci.* **2016**, *20*, 5035–5048. [CrossRef]
53. Rossi, G.; Catani, F.; Leoni, L.; Segoni, S.; Tofani, V. HIRESSS: A physically based slope stability simulator for HPC applications. *Nat. Hazards Earth Syst. Sci.* **2013**, *13*, 151–166. [CrossRef]
54. Tofani, V.; Bicocchi, G.; Rossi, G.; Segoni, S.; D'Ambrosio, M.; Casagli, N.; Catani, F. Soil characterization for shallow landslides modeling: A case study in the Northern Apennines (Central Italy). *Landslides* **2017**, *14*, 755–770. [CrossRef]
55. Salvatici, T.; Tofani, V.; Rossi, G.; D'Ambrosio, M.; Tacconi Stefanelli, C.; Masi, E.B.; Rosi, A.; Pazzi, V.; Vannocci, P.; Petrolo, M.; et al. Application of a physically based model to forecast shallow landslides at a regional scale. *Nat. Hazards Earth Syst. Sci.* **2018**, *18*, 1919–1935. [CrossRef]
56. Salciarini, D.; Fanelli, G.; Tamagnini, C. A probabilistic model for rainfall—induced shallow landslide prediction at the regional scale. *Landslides* **2017**, *14*, 1731–1746. [CrossRef]

57. Park, D.W.; Nikhil, N.V.; Lee, S.R. Landslide and debris flow susceptibility zonation using TRIGRS for the 2011 Seoul landslide event. *Nat. Hazards Earth Syst. Sci.* **2013**, *13*, 2833–2849. [CrossRef]
58. Liao, Z.; Hong, Y.; Kirschbaum, D.; Adler, R.F.; Gourley, J.J.; Wooten, R. Evaluation of TRIGRS (transient rainfall infiltration and grid-based regional slope-stability analysis)'s predictive skill for hurricane-triggered landslides: A case study in Macon County, North Carolina. *Natural Hazards* **2011**, *58*, 325–339. [CrossRef]
59. Lee, J.H.; Park, H.J. Assessment of shallow landslide susceptibility using the transient infiltration flow model and GIS-based probabilistic approach. *Landslides* **2015**, *13*, 885–903. [CrossRef]
60. Li, H.-C.; Wu, T.; Wei, H.-P.; Shih, H.-J.; Chao, Y.-C. Basinwide disaster loss assessments under extreme climate scenarios: A case study of the Kaoping River basin. *Nat. Hazards* **2016**, *86*, 1039–1058. [CrossRef]
61. Zhuang, J.; Peng, J.; Wang, G.; Iqbal, J.; Wang, Y.; Li, W.; Zhu, X. Prediction of rainfall-induced shallow landslides in the Loess Plateau, Yan'an, China, using the TRIGRS model. *Earth Surf. Processes Landf.* **2016**, *42*, 915–927. [CrossRef]
62. An, K.; Kim, S.; Chae, T.; Park, D. Developing an Accessible Landslide Susceptibility Model Using Open-Source Resources. *Sustainability* **2018**, *10*, 293. [CrossRef]
63. Hsu, Y.-C.; Liu, K.-F. Combining TRIGRS and DEBRIS-2D Models for the Simulation of a Rainfall Infiltration Induced Shallow Landslide and Subsequent Debris Flow. *Water* **2019**, *11*, 890. [CrossRef]
64. Saadatkhah, N.; Mansor, S.; Kassim, A.; Lee, L.M.; Saadatkhah, R.; Sobhanmanesh, A. Regional modeling of rainfall-induced landslides using TRIGRS model by incorporating plant cover effects: Case study in Hulu Kelang, Malaysia. *Environ. Earth Sci.* **2016**, *75*, 445. [CrossRef]
65. Gong, P.; Liu, H.; Zhang, M.; Li, C.; Wang, J.; Huang, H.; Clinton, N.; Ji, L.; Li, W.; Bai, Y.; et al. Stable classification with limited sample: Transferring a 30-m resolution sample set collected in 2015 to mapping 10-m resolution global land cover in 2017. *Sci. Bull.* **2019**, *64*, 370–373. [CrossRef]
66. Sato, H.; Harp, E. Interpretation of earthquake-induced landslides triggered by the 12 May 2008, M7.9 Wenchuan earthquake in the Beichuan area, Sichuan Province, China using satellite imagery and Google Earth. *Landslides* **2009**, *6*, 153–159. [CrossRef]
67. Tian, Y.; Owen, L.A.; Xu, C.; Ma, S.; Li, K.; Xu, X.; Figueiredo, P.M.; Kang, W.; Guo, P.; Wang, S.; et al. Landslide development within 3 years after the 2015 Mw 7.8 Gorkha earthquake, Nepal. *Landslides* **2020**, *17*, 1251–1267. [CrossRef]
68. Weidner, L.; Oommen, T.; Escobar-Wolf, R.; Sajinkumar, K.S.; Samuel, R.A. Regional-scale back-analysis using TRIGRS: An approach to advance landslide hazard modeling and prediction in sparse data regions. *Landslides* **2018**, *15*, 2343–2356. [CrossRef]
69. Tran, T.V.; Alvioli, M.; Lee, G.; An, H.U. Three-dimensional, time-dependent modeling of rainfall-induced landslides over a digital landscape: A case study. *Landslides* **2018**, *15*, 1071–1084. [CrossRef]
70. Alvioli, M.; Baum, R.L. Parallelization of the TRIGRS model for rainfall-induced landslides using the message passing interface. *Environ. Model. Softw.* **2016**, *81*, 122–135. [CrossRef]
71. Godt, J.; Baum, R.L.; Savage, W.Z.; Salciarini, D.; Schulz, W.; Harp, E.L. Transient deterministic shallow landslide modeling: Requirements for susceptibility and hazard assessments in a GIS framework. *Eng. Geol.* **2008**, *102*, 214–226. [CrossRef]
72. Ma, S.; Shao, X.; Xu, C.; He, X.; Zhang, P. MAT.TRIGRS (V1.0): A new open-source tool for predicting spatiotemporal distribution of rainfall-induced landslides. *Nat. Hazards Res.* **2021**, *1*, 161–170. [CrossRef]
73. Harp, E.; Michael, J.; Laprade, W. Shallow landslide hazard map of Seattle, Washington. *Rev. Eng. Geol.* **2008**, *20*, 67–82. [CrossRef]
74. Baumann, V.; Bonadonna, C.; Cuomo, S.; Moscariello, M.; Biass, S.; Pistolesi, M.; Gattuso, A. Mapping the susceptibility of rain-triggered lahars at Vulcano island (Italy) combining field characterization, geotechnical analysis, and numerical modelling. *Nat. Hazards Earth Syst. Sci.* **2019**, *19*, 2421–2449. [CrossRef]
75. Vieira, B.C.; Fernandes, N.F.; Augusto Filho, O.; Martins, T.D.; Montgomery, D.R. Assessing shallow landslide hazards using the TRIGRS and SHALSTAB models, Serra do Mar, Brazil. *Environ. Earth Sci.* **2018**, *77*, 260. [CrossRef]
76. Saulnier, G.-M.; Beven, K.; Obled, C. Including spatially variable soil depths in TOPMODEL. *J. Hydrol.* **1997**, *202*, 158–172. [CrossRef]
77. Bai, H.; Feng, W.; Yi, X.; Fang, H.; Wu, Y.; Deng, P.; Dai, H.; Hu, R. Group-occurring landslides and debris flows caused by the continuous heavy rainfall in June 2019 in Mibei Village, Longchuan County, Guangdong Province, China. *Nat. Hazards* **2021**, *108*, 3181–3201. [CrossRef]
78. Das, B. *Advanced Soil Mechanics*; Taylor & Francis: London, UK; New York, NY, USA, 2008.
79. He, J.; Qiu, H.; Qu, F.; Hu, S.; Yang, D.; Shen, Y.; Zhang, Y.; Sun, H.; Cao, M. Prediction of spatiotemporal stability and rainfall threshold of shallow landslides using the TRIGRS and Scoops3D models. *Catena* **2021**, *197*, 104999. [CrossRef]
80. Medina, V.; Hürlimann, M.; Guo, Z.; Lloret, A.; Vaunat, J. Fast physically-based model for rainfall-induced landslide susceptibility assessment at regional scale. *Catena* **2021**, *201*, 105213. [CrossRef]
81. Yu, C.-K.; Cheng, L.-W. Distribution and Mechanisms of Orographic Precipitation Associated with Typhoon Morakot (2009). *J. Atmos. Sci.* **2013**, *70*, 2894–2915. [CrossRef]
82. Liu, J.-K.; Shih, P.T.-Y. Topographic Correction of Wind-Driven Rainfall for Landslide Analysis in Central Taiwan with Validation from Aerial and Satellite Optical Images. *Remote Sens.* **2013**, *5*, 2571–2589. [CrossRef]
83. Chen, Y.-C.; Chang, K.-T.; Wang, S.F.; Huang, J.-C.; Yu, C.-K.; Tu, J.-Y.; Chu, H.-J.; Liu, C.-C. Controls of preferential orientation of earthquake- and rainfall-triggered landslides in Taiwan's orogenic mountain belt. *Earth Surf. Processes Landf.* **2019**, *44*, 1661–1674. [CrossRef]

84. Vorpahl, P.; Elsenbeer, H.; Märker, M.; Schröder, B. How can statistical models help to determine driving factors of landslides? *Ecol. Model.* **2012**, *239*, 27–39. [CrossRef]
85. Kavzoglu, T.; Sahin, E.K.; Colkesen, I. Landslide susceptibility mapping using GIS-based multi-criteria decision analysis, support vector machines, and logistic regression. *Landslides* **2014**, *11*, 425–439. [CrossRef]
86. Laimer, H.J. Anthropogenically induced landslides–A challenge for railway infrastructure in mountainous regions. *Eng. Geol.* **2017**, *222*, 92–101. [CrossRef]
87. Vuillez, C.; Tonini, M.; Sudmeier-Rieux, K.; Devkota, S.; Derron, M.-H.; Jaboyedoff, M. Land use changes, landslides and roads in the Phewa Watershed, Western Nepal from 1979 to 2016. *Appl. Geogr.* **2018**, *94*, 30–40. [CrossRef]
88. Li, Y.; Wang, X.; Mao, H. Influence of human activity on landslide susceptibility development in the Three Gorges area. *Nat. Hazards* **2020**, *104*, 2115–2151. [CrossRef]
89. Tanyaş, H.; Görüm, T.; Kirschbaum, D.; Lombardo, L. Could road constructions be more hazardous than an earthquake in terms of mass movement? *Nat. Hazards* **2022**, *112*, 639–663. [CrossRef]

remote sensing

MDPI

Article

A Strategy for Variable-Scale InSAR Deformation Monitoring in a Wide Area: A Case Study in the Turpan–Hami Basin, China

Yuedong Wang [1,2], Guangcai Feng [1,2,*], Zhiwei Li [1,2], Shuran Luo [1], Haiyan Wang [3], Zhiqiang Xiong [1], Jianjun Zhu [1,2] and Jun Hu [1,2]

1 School of Geosciences and Info-Physics, Central South University, Changsha 410083, China
2 Key Laboratory of Metallogenic Prediction of Nonferrous Metals and Geological Environment Monitoring Ministry of Education, Changsha 410083, China
3 Chongqing 208 Geological Environment Research Institute Co., Ltd., Chongqing 400700, China
* Correspondence: fredgps@csu.edu.cn; Tel.: +86-182-7486-7449

Abstract: In recent years, increasing available synthetic aperture radar (SAR) satellite data and gradually developing interferometric SAR (InSAR) technology have provided the possibility for wide-scale ground-deformation monitoring using InSAR. Traditionally, the InSAR data are processed by the existing time-series InSAR (TS–InSAR) technology, which has inefficient calculation and redundant results. In this study, we propose a wide-area InSAR variable-scale deformation detection strategy (hereafter referred to as the *WAVS–InSAR strategy*). The strategy combines stacking technology for fast ground-deformation rate calculation and advanced TS–InSAR technology for obtaining fine deformation time series. It adopts an adaptive recognition algorithm to identify the spatial distribution and area of deformation regions (regions of interest, ROI) in the wide study area and uses a novel wide-area deformation product organization structure to generate variable-scale deformation products. The Turpan–Hami basin in western China is selected as the wide study area (277,000 km^2) to verify the proposed WAVS–InSAR strategy. The results are as follows: (1) There are 32 deformation regions with an area of $\geq$1 km^2 and a deformation magnitude of greater than $\pm$2 cm/year in the Turpan–Hami basin. The deformation area accounts for 2.4‰ of the total monitoring area. (2) A large area of ground subsidence has occurred in the farmland areas of the ROI, which is caused by groundwater overexploitation. The popularization and application of facility agriculture in the ROI have increased the demand for irrigation water. Due to the influence of the tectonic fault, the water supply of the ROI is mainly dependent on groundwater. Huge water demand has led to a continuous net deficit in aquifers, leading to land subsidence. The WAVS–InSAR strategy will be helpful for InSAR deformation monitoring at a national/regional scale and promoting the engineering application of InSAR technology.

Keywords: wide-area deformation; deformation detection; time-series InSAR; stacking; Turpan–Hami basin

Citation: Wang, Y.; Feng, G.; Li, Z.; Luo, S.; Wang, H.; Xiong, Z.; Zhu, J.; Hu, J. A Strategy for Variable-Scale InSAR Deformation Monitoring in a Wide Area: A Case Study in the Turpan–Hami Basin, China. *Remote Sens.* **2022**, *14*, 3832. https://doi.org/10.3390/rs14153832

Academic Editors: Paolo Mazzanti and Saverio Romeo

Received: 11 July 2022
Accepted: 6 August 2022
Published: 8 August 2022

Publisher's Note: MDPI stays neutral with regard to jurisdictional claims in published maps and institutional affiliations.

1. Introduction

Advanced microwave remote sensing technology can precisely monitor deformation over wide areas, which helps geohazard surveys of phenomena such as underground fluid development, mineral mining, and landslide. In recent years, fast-developing interferometric synthetic aperture radar (InSAR) technology and abundant available synthetic aperture radar (SAR) data [1–4] has laid the foundation for high-precision and wide-scale InSAR ground-deformation monitoring. InSAR technology has been successfully used to monitor ground deformation at a regional [5–9] and national scale [10–13]. Large-scale geodetic technology, such as InSAR, usually describes the spatial characteristics of ground deformation by deformation rate, and shows deformation development over time using a time series of deformation. The deformation region usually accounts for a small part of the monitoring area [11], so the ground deformation we are interested in only accounts for a small part of the wide-area monitoring results. At present, the engineering projects to obtain ground deformation in a wide study area (WSA) usually calculate the deformation time series using InSAR datasets covering the whole WSA, using time-series InSAR (TS–InSAR) technology. Even with multiple spatial resolutions, such schemes require a lot of computing resources and storage space, and even then require repeated calculations and provide redundant results, especially in the non-deformation region [14]. Therefore, it is necessary to develop a set of efficient monitoring methods and procedures for wide-area InSAR deformation and a more feasible multi-scale deformation product organization structure in the WSA.

One way to improve the computational efficiency of the TS–InSAR method is to introduce a parallel processing method, which can be realized using high-performance computers (HPC) [12,15–19]. However, the high cost of HPC equipment hinders the popularization of this strategy. Another way is to improve the TS–InSAR method itself, by introducing sequential adjustment theory for real-time InSAR data processing [20–22], introducing a geological model or time-series filtering algorithm for high-dimensional deformation calculation [23–26], or realizing a high-precision TS–InSAR deformation calculation using block solutions [27,28]. These strategies can improve the efficiency of the TS–InSAR solution to a certain extent. However, for wide-area InSAR deformation monitoring, high-precision independent calculation of all InSAR datasets in the WSA will provide many useless time-series results, especially in the non-deformation area. Therefore, it is necessary to develop a demand-oriented multiple spatio-temporal-scale deformation monitoring method, considering the universality of monitoring strategies, computing resources, measurement accuracy, and the efficiency of deformation calculation and interpretation.

The averaging of multiple interferograms (stacking) method was proposed by the authors in [29], which can obtain the ground-deformation rate by averaging the phase of the multitemporal differential InSAR (DInSAR) dataset. Compared with conventional TS–InSAR technologies, such as persistent scatterer (PS) [30], small-baseline subset (SBAS) [31], and interferometric point target analysis (IPTA) [32], stacking only obtains the deformation rate with lower technical requirements and higher computational efficiency. Stacking has been widely used for deformation identification [33–37]. A wide-area deformation monitoring project usually identifies deformation regions based on the ground-deformation rate [38]. For the deformation region, the corresponding deformation time series is extracted to analyze the spatio-temporal evolution of deformation. The deformation time series in stable zones has less information. Therefore, combining stacking and TS–InSAR may contribute to efficient variable-scale deformation monitoring.

In this study, we propose a wide-area InSAR variable-scale deformation detection strategy (WAVS–InSAR). WAVS–InSAR uses stacking technology to quickly calculate the low-spatial-resolution ground-deformation rate over the WSA. Then, an adaptive intelligent recognition algorithm is used to identify the location and area of the deformation regions and determine the regions of interest (ROI). Advanced TS–InSAR technologies are then used to obtain the high-spatio-temporal-resolution deformation time series in the ROI. Finally, the variable-scale InSAR deformation product in the WSA is obtained by a novel variable-scale deformation product organization structure. To verify the proposed WAVS–InSAR strategy, we applied it to the Turpan–Hami basin (about 277,000 km^2) in Xinjiang, China. The Turpan–Hami basin is the driest place in China, and has the least rainfall in China. Many tectonic faults, as well as agricultural and mining areas, are scattered across the basin. It is of great significance to obtain the spatio-temporal distribution characteristics of ground subsidence and to investigate the surface deformation related to the active agricultural economy and mineral exploitation in the basin.

The remainder of the paper is organized as follows. We introduce the WAVS–InSAR strategy in Section 2. In Section 3, the general situation of the Turpan–Hami basin, InSAR data, and the data-processing details are briefly described. The variable-scale deformation product in the Turpan–Hami basin is shown in Section 4, followed by the discussion in Section 5. Section 6 presents the conclusions.

2. Methodology

We first collect all available InSAR datasets covering the WSA, and preprocess all datasets through registration and DInSAR, to generate the multitemporal DInSAR datasets with the same spatial reference data. Then, we apply the WAVS–InSAR strategy to process the multitemporal DInSAR data to obtain variable-scale deformation products in a wide area. The WAVS–InSAR includes the following four modules (Figure 1).

(1) We obtain the wide-area deformation rate using the stacking method [29]. First, we calculate the deformation rate of each frame using stacking. Then, we mosaic the results of all frames to obtain the wide-area deformation rates.

(2) We detect ROI from the deformation rates. Setting the threshold for the deformation rate, the extension radius, and the minimum clustering area, we calculate the spatial distribution and area of the ROI in the WSA using an adaptive deformation detection method [39].

(3) We obtain the high-spatio-temporal-resolution deformation result of ROI. The high-spatio-temporal-resolution time-series and/or multidimensional deformation of the ROI are calculated using advanced TS–InSAR technologies, such as PS, SBAS, IPTA, and the multidimensional small-baseline subset (MSBAS) [40–42].

(4) We generate the variable-scale deformation product, combining the high-spatio-temporal-resolution results of ROI and wide-area deformation rate to generate the variable-scale deformation product, which can describe deformation in stable areas only with low-spatial-resolution deformation rate, and in the ROI with the high-spatio-temporal-resolution deformation rate and time series.

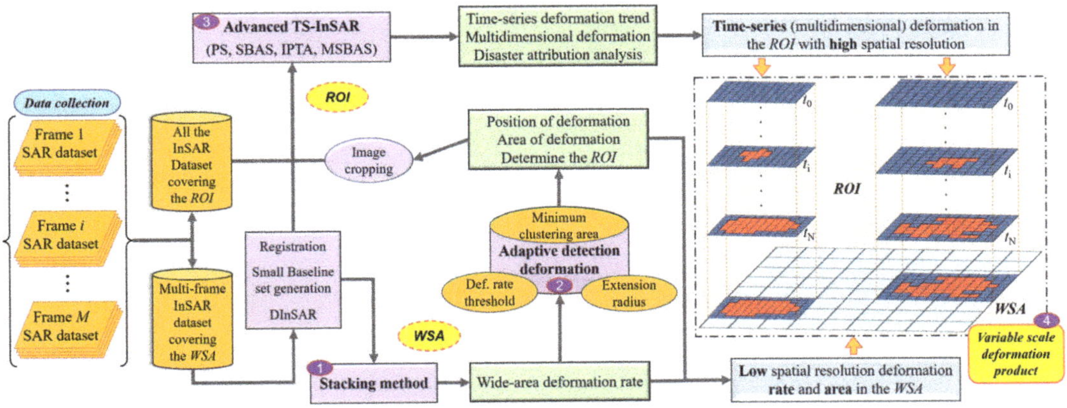

Figure 1. Flowchart of the method for the variable-scale monitoring of deformation in a wide area.

2.1. Wide-Area Deformation Monitoring Using Stacking

The stacking technology can calculate the deformation rate based on weight and average the unwrapped phases of the multitemporal DInSAR dataset. The stacking technology assumes linear ground-deformation changes, and temporal randomly distributed phase noise, such as atmospheric delay phase. Assuming $N+1$ SAR images of one frame covering the WSA constitute M InSAR pairs, the displacement phase can be separated as

$$\overline{\varphi} = \sum_{i=1}^{M} \phi_i \cdot \Delta t_i / \sum_{i=1}^{M} \Delta t_i^2 \quad (1)$$

in which $\overline{\varphi}$ is the rate of deformation phase change. ϕ_i and Δt_i are the interference phase and the time interval of the i-th InSAR pair, respectively.

The rate of the deformation phase change would be converted to the deformation rate (V_{def}) as,

$$V_{def} = \lambda \cdot \overline{\phi}/4\pi \tag{2}$$

where λ is the wavelength of the SAR sensor. The multitemporal DInSAR data in each frame is processed using stacking technology to obtain the ground deformation of the WSA.

2.2. ROI Detection Based on Wide-Area Deformation Rate

Luo et al. [39] proposed an improved method to automatically identify and evaluate geological hazards using TS–InSAR results. By judging and analyzing the deformation rate and time series in the monitoring area, the method can automatically identify the deformation region and evaluate its hazard grade. In this study, we improve this method to accurately delineate the ROI.

To improve the accuracy of ROI detection, we first apply spatial domain filtering to the wide-area monitoring results to obtain deformation results with good spatial consistency. Then, we set the thresholds for deformation rate, extension radius, and minimum clustering area. When the absolute value of the deformation rate is greater than the deformation rate threshold, it is considered to be an active point. Otherwise, it is a stable point. Buffer zones are established around the active points according to the extension radius. The active points are clustered following the principle of spatial proximity relationship [43]. The clustering regions are smoothed to refine the boundary. The robust deformation regions and their area are obtained by removing regions smaller than the minimum clustering area. The ROI can be finally located based on spatial clustering and the area of deformation.

A detailed description of the intelligent recognition part of the method can be found in [39]. It should be noted that InSAR can only obtain one-dimensional (1D) deformation along the line-of-sight (LOS) direction of the SAR sensor, so the InSAR data of one geometry is insensitive to the deformation of some regions, especially landslides [44]. To obtain more reliable deformation detection results, we need to use the above method and InSAR data from different observational geometry. The detection results of multitrack InSAR data are taken together as the final deformation regions. Then, we can adaptively determine the ROI and perform fine monitoring.

2.3. ROI Deformation Refinement Using Advanced TS–InSAR

When calculating the wide-area deformation rate, we select the InSAR data with the same acquisition time from different frames to facilitate the splicing of the results from different frames and to maintain the consistency of the wide-area deformation rate. To accurately monitor the deformation in the ROI, we first crop the registered InSAR datasets. The cropped datasets are used to obtain the time-series and multidimensional ground deformation of the ROI. Detailed steps are as follows.

(1) Deformation time-series calculation. We process the collected InSAR datasets using TS–InSAR technology, with a smaller multi-looking number (a higher spatial resolution). In this study, we use an improved IPTA method to calculate the deformation time series of the ROI [45,46].
(2) Multidimensional deformation rate/time-series calculation. If the ROI has InSAR data with different observation geometry during the same acquisition time, we can obtain the vertical and horizontal displacements using the MSBAS method.

If multi-sensor and multitemporal InSAR data covering the ROI are available, we can collect all data to analyze the long-term deformation and understand the deformation spatio-temporal evolution features based on the data-overlapping and deformation model [47,48].

2.4. Variable-Scale Deformation Product Generation

The low-spatial-resolution deformation rate can be used to detect a stable surface in the WSA, which greatly reduces the task and data volume of wide-area InSAR deformation monitoring. In addition, we obtain the fine results of the deformation time series with a high spatial resolution of ROI using advanced TS–InSAR technology. A variable-scale deformation product organization structure includes low-spatial-resolution deformation rates in stable areas of the WSA and the high-spatio-temporal-resolution deformation in the ROI. Hence, we superimpose the high-spatio-temporal-resolution deformation at the corresponding regions of the ROI on the wide-area deformation rate results to improve the spatial and temporal dimensions of the deformation in the ROI. At this stage, we can obtain variable-scale deformation products in the WSA, which only contain low-spatial-resolution deformation rates in stable regions, and fine monitoring results in the ROI.

3. Study Area and Data Processing

3.1. The Turpan–Hami Basin

The Turpan–Hami basin, consisting of the Turpan and the Hami depressions, is an intermountain basin located in northwest China (Figure 2). Since the end of the Early Permian period, the Turpan–Hami basin has developed following the model of "fault-depression foreland". It is a typical faulted basin, with limited sedimentary range, great lateral variation of sedimentary thickness, and multiple depositions and subsidence centers. The geological conditions and active tectonic motion contribute to oil and gas accumulation and make the Turpan–Hami basin the largest coal-derived petroleum-producing basin in China [49]. Moreover, there are many mineral resources in this basin, e.g., coal, iron, and potassium (sodium) saltpeter. It is the world's largest potassium (sodium) saltpeter resource. Aydingkol Lake, located in the middle of the Turpan depression, is the lowest depression in China, 154.31 m below sea level [50]. Centering on Aydingkol Lake, the Turpan depression presents a roughly three-ring shape. The outermost ring has high snow-capped mountains. The middle ring is the Gobi gravel belt. The inner ring is an oasis plain belt, most of which belongs to a piedmont sloping plain, and accumulates a large area of fine soil alluvium. The water in the basin mainly comes from rainfall and meltwater from the surrounding mountains. The Tianshan mountains, e.g., Bogurda Mountain and Harlick Mountain, are in the north of the Turpan depression. The Flaming Mountains fault zone lies nearly east–west in the Turpan depression, between Turpan city and Shanshan county (Figure 2). Weathered material is transported from the Tianshan mountains to the center of the basin by water flow, but is blocked by the Flaming Mountains fault line and accumulates in the northern part of the mountains. The surface water and groundwater from the Tianshan mountains are also blocked by the Flaming Mountains fault line. The head height of the shallow aquifers is raised on both sides of the Flaming Mountains, creating overflow zones and an oasis in these areas.

The Turpan–Hami basin has a typical continental warm temperate desert climate, with abundant heat and extremely little precipitation. It has 3200 h of sunshine in a year. The hydrogeology, climate, and lighting conditions make it an ideal place for growing cantaloupe, grapes, cotton, and off-season vegetables. Groundwater is the main source of agricultural water in the arid area. Previously, karezes were the predominant underground water conservancy project in this region. A karez uses the principle of water potential artware to divert water from shallow aquifers to the surface for irrigation. There are more than 2000 karezes in the Turpan–Hami basin, accounting for more than 70% of the total number of karezes in Xinjiang [51,52]. However, many electromechanical wells have been built in the Turpan–Hami basin since the 1960s. Groundwater exploitation has increased yearly, with the annual overexploitation reaching 2.48×10^{10} m^3, leading to the continuous decline of groundwater level. Advanced water conservancy facilities have reduced people's dependence on karezes. Meanwhile, the water supply source of karezes is shallow aquifers. The continuous reduction of groundwater level directly leads to the decrease or even drying-up of karezes [53]. The number of water-filled karezes in the Turpan depression decreased from 1237 in 1957 to 214 in 2014 [51]. In addition, the increased demand and excess consumption of water resources in upstream areas have seriously threatened the water supply of Aydingkol Lake, resulting in water area shrinkage. The exploitation of groundwater and mineral resources will make the surface of the Turpan–Hami basin unstable and threatened by potential geohazards.

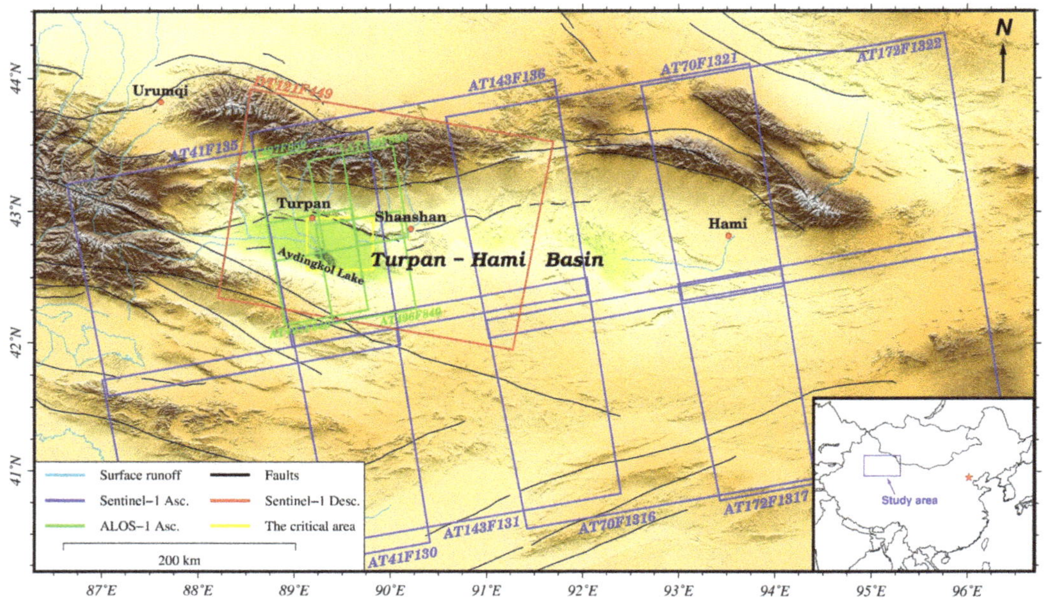

Figure 2. The Turpan–Hami basin and SAR data coverage.

3.2. InSAR Datasets

To monitor wide-area deformation in the Turpan–Hami basin, we collected eight frames of InSAR data covering the whole Turpan–Hami basin from the Sentinel-1 satellite. The Sentinel-1 satellite began operation in April 2014, and has different observation periods in different regions, resulting in inconsistent periods of SAR data in different regions. To ensure the consistency of deformation rates from multiple frames, we selected the images (628 images in total) from the eight frames acquired from October 2017 to May 2020 (Table 1). The spatial coverage of each dataset is shown in Figure 2.

Table 1. Acquisition periods of the datasets.

Sensor	Frame	Time	Number	Frame	Time	Number
Sentinel-1	AT172F1317 AT172F1322	13/10/2017–30/05/2020	77 77	AT143F131 AT143F136	11/10/2007–28/05/2020	81 81
	AT70F1316 AT70F1321	18/10/2017–30/05/2020	78 78	AT41F130 AT41F135	16/10/2017–21/05/2020	78 78
	DT121F449	19/03/2015–27/04/2020	107	AT41F135	25/03/2015–21/05/2020	123
ALOS-1	AT496F840 AT496F850	22/01/2007–14/09/2009	11 11	AT497F840 AT497F850	08/02/2007–04/10/2010	11 11

Wide-area InSAR deformation shows that many subsidence funnels are concentrated in the south part of the Flaming Mountains fault zone in the Turpan depression (hereafter referred to as the SFM–def region). The SFM–def region (the yellow box in Figure 2) was selected as an application demonstration area of ROI to carry out the fine monitoring of the deformation time series. Four frames from the ALOS-1/PALSAR dataset spanning from 2007 to 2010 (green rectangles in Figure 2) and a descending track from the Sentinel-1 dataset (red rectangle in Figure 2) covering the SFM–def region were collected. The common monitoring time of the Sentinel-1 ascending (AT41F135) and descending (DT121F449) tracks data is from 2015 to 2020 (Table 1). These data were used to precisely monitor the long-term and fine deformation in the SFM–def region.

3.3. Data Processing

We preprocessed all InSAR datasets covering the WSA. In each frame, one image was selected as the master image to register and resample the rest images. Multitemporal InSAR pairs were generated from SAR data in the same frame, based on the appropriate spatio-temporal baseline thresholds. All multitemporal DInSAR pairs were processed using GAMMA software [54] and two-pass DInSAR technology [55] to obtain multitemporal deformation signals. The shuttle radar topography mission (SRTM) digital elevation model (DEM) with a resolution of 30 m [56] was employed to remove the topographic phases. The point targets with a coherence lower than 0.3 were eliminated [57]. Least-squares-based filtering and the minimum cost flow method [58] were then applied to further suppress phase noise [59] and unwrap the differential interferogram, respectively.

The eight frames of the Sentinel-1 data were preprocessed with a spatial baseline (perpendicular) and temporal baseline of 100 m and 48 days, respectively, and a multi-looking operation of 20:4. Then, stacking was used to process all the multitemporal deformation signals in each frame, to obtain a wide-area deformation rate of the Turpan–Hami basin. The adaptive deformation detection method proposed in Section 2.2 was used to delineate the deformation regions. The thresholds of the deformation rate, extension radius, and minimum clustering area were set as ± 2 cm/year, 250 m, and 1 km^2, respectively.

For the SFM–def region, we set the multi-looking parameters of ALOS-1/PALSAR and Sentinel-1 data as 3:8 and 8:2, respectively. The improved IPTA method was used to compute the four frames of the ALOS-1/PALSAR data and the ascending/descending tracks from the Sentinel-1 datasets to obtain long-term and high-resolution displacements. Moreover, MSBAS technology was used to obtain multidimensional deformation from the ascending/descending tracks of the Sentinel-1 datasets. Then, we obtained the variable-scale deformation product of the Turpan–Hami basin, which consists of low-spatial-resolution deformation rates in the stable areas and high-spatio-temporal-resolution deformation in the SFM–def region.

4. Results

4.1. Monitoring and Detecting the Wide-Area Deformation in the Turpan–Hami Basin

The wide-area ground subsidence in the Turpan–Hami basin (Figure 3) shows that the surface of the Turpan–Hami basin is generally stable. The regions with deformation account for a small proportion of the whole. The main deformation type is subsidence. Based on the deformation detection threshold set in Section 3.3, we identified 32 deformation areas (the funnel) in the Turpan–Hami basin (the blue lines in Figure 3). The area of each funnel is shown in Table 2. The detected deformation area accounts for about 2.4‰ of the total monitoring area.

Analyzing the hydrogeology and land cover of the deformation areas, we divided the ground deformation in the Turpan–Hami basin into three types:

(1) Ground subsidence in agricultural areas caused by groundwater overexploitation. This kind of subsidence has the largest area and is concentrated in the oasis plain south of the Flaming Mountains fault zone (Figure 3a).
(2) Ground subsidence associated with mineral mining. This kind of deformation is sporadically distributed over the Turpan–Hami basin. Such deformation regions have a small area but large deformation magnitude, e.g., Figure 3b.
(3) Ground uplift associated with the lake water withdrawal, resulting in saline–alkali lands. This kind of deformation is mainly distributed around Aydingkol Lake, characterized by small magnitude and mainly horizontal movement (Figure 4e,f).

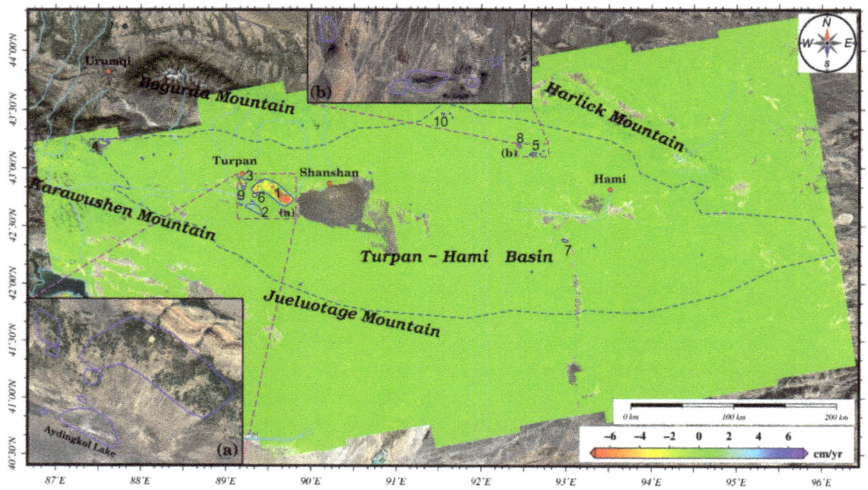

Figure 3. Wide-area subsidence rate map and the detected deformation regions. The numbers identify the location of the top 10 deformation regions. (**a**) The SFM–def region in Figure 4. (**b**) One of the major mining areas. Background image: Google Maps satellite image.

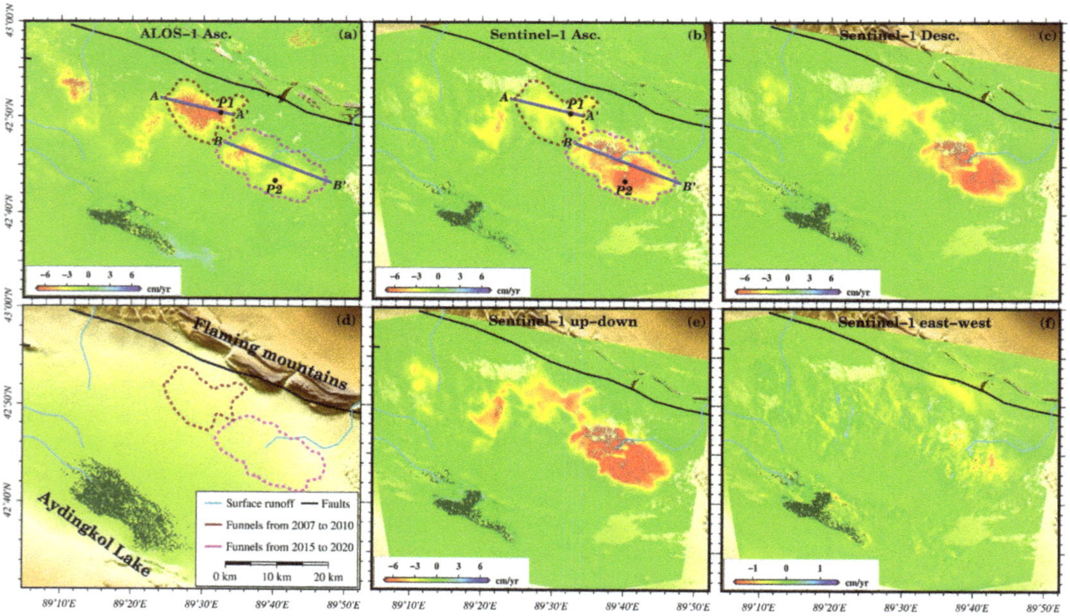

Figure 4. Deformation rate along LOS directions from (**a**) ALOS-1/PALSAR, (**b**) ascending track Sentinel-1, and (**c**) descending track Sentinel-1 data. Negative values indicate the direction away from the SAR satellite, while positive values indicate the opposite. (**d**) The hydrogeology of this area. (**e**,**f**) The deformation rate along the up–down and east-west directions calculated from ascending/descending tracks Sentinel-1 data. The red dotted line delineates the central area of the subsidence funnels from 2007 to 2010. The magenta dotted line delineates the central area of the subsidence funnels from 2015 to 2020.

Table 2. The area of the deformation funnels.

Num.	Area (km^2)	Num.	Area (km^2)	Num.	Area (km^2)	Num.	Area (km^2)
1	437.6	9	5.5	17	2.3	25	1.6
2	61.2	10	5.1	18	2.3	26	1.5
3	42.6	11	4.8	19	2.2	27	1.5
4	26.1	12	4.1	20	2.2	28	1.5
5	16.1	13	3.4	21	2.1	29	1.4
6	11.1	14	3.1	22	2.1	30	1.2
7	9.1	15	2.6	23	2.0	31	1.1
8	6.8	16	2.5	24	1.9	32	1.0
Total area (km^2)				669.6			

The largest deformation funnel is distributed in the SFM–def region, with an area of 437.6 km^2, surrounded by small funnels (Figure 3a). The optical images show that the subsidence funnels in the SFM–def region are highly correlated with the location of agricultural areas. Aydingkol Lake is in the south of the SFM–def region (Figure 3a). In recent years, the area of the lake has continuously shrunk, and a large area of saline–alkali land has appeared. There is obvious ground uplift in these saline–alkali regions. In addition, multiple subsidence funnels are observed close to some mines, e.g., the funnel cluster in Figure 3b. The wide-area deformation results are discussed in detail in Section 5.

4.2. Deformation Time Series of the SFM–Def Region from 2007 to 2020

4.2.1. Long-Term Deformation in the Spatial Dimension

The long-term (2007–2010 and 2015–2020), multidimensional (along with up–down and east–west directions), and high-spatial-resolution displacements are obtained from the four frames of the ALOS-1/PALSAR data and the ascending/descending tracks from the Sentinel-1 data, using advanced IPTA and MSBAS technologies (Figure 4). The ground deformation is mainly distributed in a plain area south of the Flaming Mountains fault line (Figure 4d). There are large areas of farmland in this region (Figure 3a), and the irrigation relies heavily on groundwater. The ground deformation is mainly vertical, with small horizontal movement (Figure 4e,f), which is typical for displacements caused by groundwater extraction [60–62]. The red and magenta dotted lines in Figure 4 delineate the settlement funnel centers in 2007–2010 and 2015–2020, respectively. The area and magnitude of the subsidence in the northwest of the SFM–def region gradually decrease, but the subsidence area in the southeast gradually expands and becomes connected. The center of the funnel shifts from the northwest to the southeast, and form a giant funnel with a larger subsidence rate and area in the southeast region. See Section 5 for detailed analysis and discussion.

4.2.2. Long-Term Deformation in the Time Dimension

The long-term deformation rate can reflect the spatial distribution and evolution characteristics of ground deformation. We select two profiles (AA′ and BB′) and two points (P1 and P2) in the SFM–def region (Figure 4) to investigate the variation characteristics of deformation in the time domain. The long-term time-series displacements at the corresponding position in the two monitoring periods, i.e., 2007–2010 and 2015–2020, are shown in Figures 5 and 6. The time-series cumulative deformation at AA′ and P1, BB′ and P2 can represent the deformation characteristics of the central region of the subsidence funnels during the two monitoring periods.

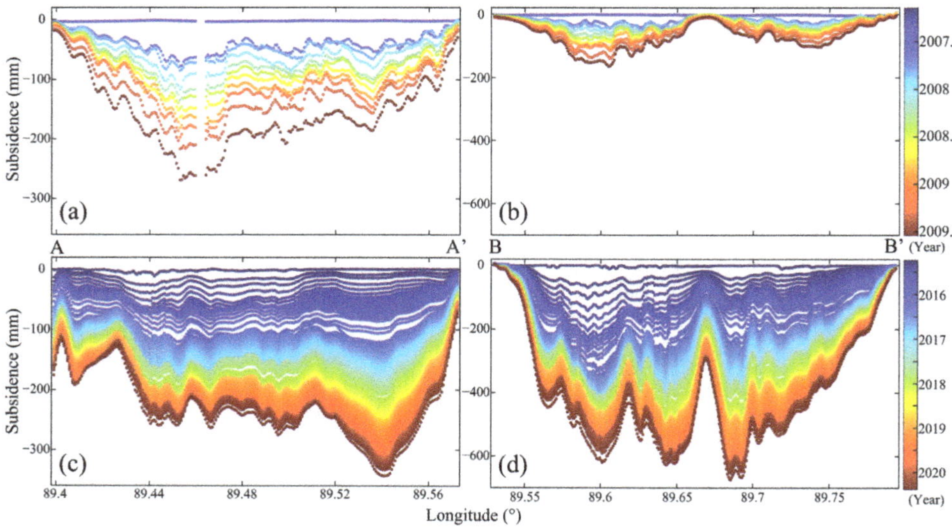

Figure 5. The long-term cumulative deformation at the profiles (**a**,**c**) A-A′ and (**b**,**d**) B-B′ in Figure 4.

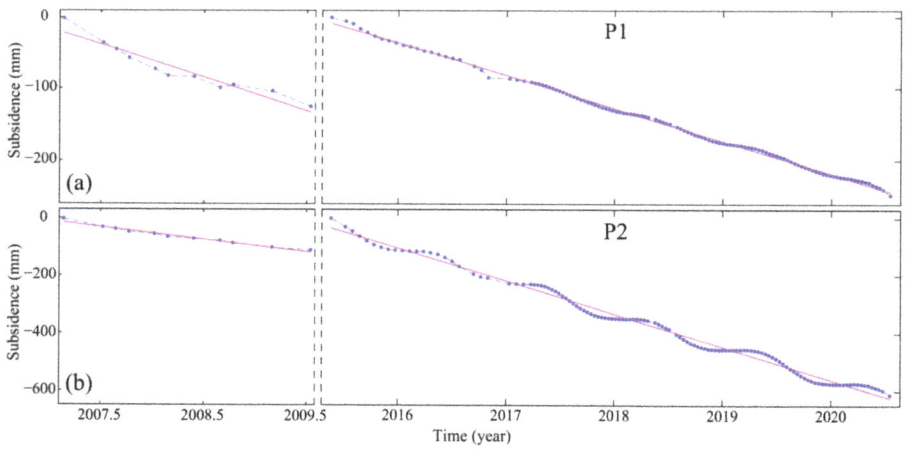

Figure 6. The time-series cumulative deformation at P1 and P2 in Figure 4. The blue dots represent the InSAR observations. The magenta lines are the linear fitting results of the corresponding InSAR observations.

The long-term deformation at AA′ and P1 shows that the subsidence of the area with the most significant subsidence in the first monitoring period tends to be stable, and slows down in the second period. In the first period, the subsidence rate of the section northwest of AA′ is higher than that of the southeast section. However, in the second period, this phenomenon is reversed. The subsidence center moves from northwest to southeast, which is consistent with the spatial evolution of the global subsidence funnel. In the first period, the subsidence rate of BB′ is small, and presents two separate funnels. In the latter period, the two funnels merge into a giant funnel. The subsidence area and rate increases significantly.

Both ALOS-1/PALSAR and Sentinel-1 data can reflect the overall change characteristics of the subsidence in time and space well (Figures 4 and 5). However, compared with the ALOS-1/PALSAR data, which have a revisit period of ≥46 days, the Sentinel-1 data can capture more detailed changes to the deformation signals with obvious periodicities in the time dimension due to its higher temporal resolution (≥12 days) (Figure 6). The subsidence mainly occurs in summer. The ground tends to be stable or slightly uplifted in winter. See Section 5 for detailed analysis and discussion.

4.2.3. Reliability Assessment

As can be seen from Figure 4, the deformation results of the SFM–def region from the ALOS-1 data of different frames have good consistency. The deformation obtained by the Sentinel-1 data of ascending and descending tracks also has good consistency in spatial distribution and magnitude. This indicates that the TS–InSAR results have a good consistency. To quantitatively assess the reliability of the TS–InSAR results, we compare the average subsidence rates extracted from the overlapped areas of two adjacent InSAR frames acquired at the same period, e.g., ALOS-1/PALSAR datasets from AT496F840 and AT496F850, and Sentinel-1 datasets from AT143F136 and DT121F449 (Figure 2). Due to the different observation geometry of each monitoring point in different frames, we convert the LOS deformation to the vertical direction for comparison. The correlation between the results at AT496F840 and AT496F850, and Sentinel-1 results at AT143F136 and DT121F449, are 0.98 and 0.99, respectively. The root-mean-square errors (RMSEs) between them are 0.02 and 0.01 mm/year, respectively. These results show good consistency, and the differences at most points are smaller than three times the RMSE (between the red dotted lines in Figure 7).

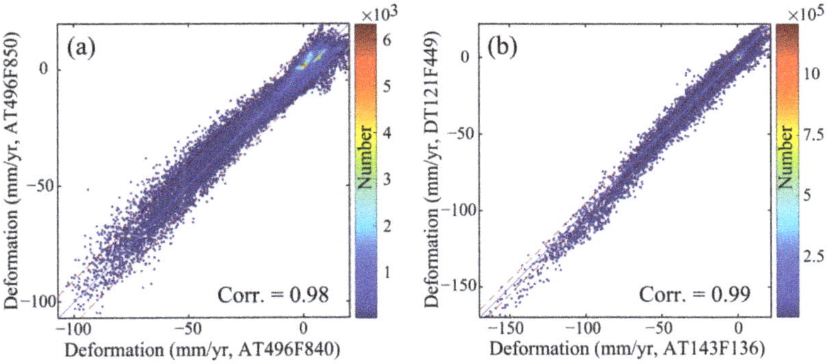

Figure 7. Comparison between the results obtained by (**a**) ALOS-1/PALSAR AT496F840 and AT496F850 data, and (**b**) the Sentinel-1 AT143F136 and DT121F449 data. The red dotted lines denote the value three times the root-mean-square error.

5. Discussion

5.1. Anthropogenic Factors of Ground Deformation in the Turpan–Hami Basin

We obtained variable-scale deformation products in the Turpan–Hami basin using the proposed WAVS–InSAR method. The distribution of most detected deformation funnels (Section 4.1) is highly consistent with human activity, such as agriculture cultivation and mineral mining. The agricultural area in the SFM–def region has a funnel cluster with the largest deformation area and magnitude in the Turpan–Hami basin. We obtained the long-term and multidimensional ground deformation in the SFM–def region in Section 4.2. The subsidence center of the first period (2007–2010) shifted from the northwest to the southeast in the second period (2015–2020).

We collect optical images of the SFM–def region in 2007 and 2018 (Figure 8a,b), corresponding to the two monitoring periods. The green lines mark the locations of the greenhouses that appeared in the latter period. As the optical images show, the majority of farmland in the SFM–def region in 2007 was open-air farmland. However, in 2018, there was a large area of greenhouses, especially in the farmland far from the Flaming Mountains fault line. Many open-air farmlands in 2007 had been changed to greenhouses (Figure 8). In traditional open-air farmland, the crops are mainly grain and cotton, which are planted in spring, managed in summer, and harvested in autumn. However, in greenhouse farmland, the expected proportion of fruit and vegetable cultivation is more than 70% [63]. After 2009, many greenhouses were built in Turpan, especially in the agricultural areas far from the southern margin of the Flaming Mountains fault line (Figure 8b). Advanced agricultural planting technologies have brought huge economic benefits to Turpan, but also increased the environmental burden, especially the demand for water [53]. Irrigation water in the SFM–def region is mainly groundwater. Hence, ground subsidence caused by groundwater overexploitation is more significant in the greenhouse areas of the SFM–def region, resulting in aquifers carrying net deficit and the subsidence center shifting to the southeast (Figure 8a,b).

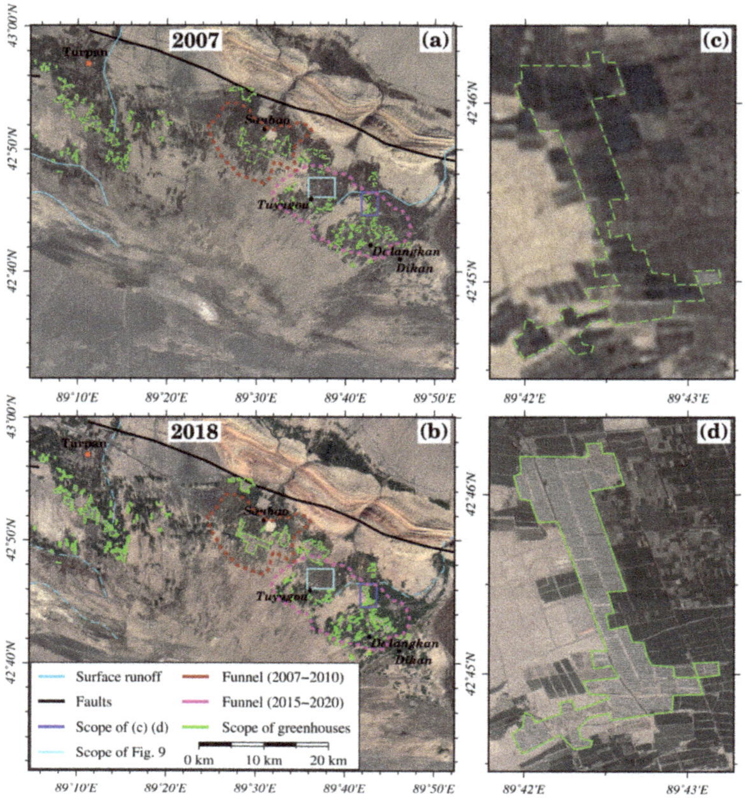

Figure 8. (**a**,**b**) Optical images of the SFM–def region in 2007 and 2018. Green lines delineate greenhouse planting areas. (**c**,**d**) Zoom-ins of the blue rectangular area in (**a**,**b**) in 2007 and 2018. Background image: Google Maps satellite image.

Karezes are an important water supply in arid agricultural areas, known as "the fountains of life". In China, karezes are mainly distributed in the Turpan–Hami basin (Section 3.1). A karez is composed of vertical shafts, culverts, water outlets, open channels, and waterlogging dams, with length ranging from several to dozens of kilometers [51]. The number and distribution of karezes can reflect the changes to the ecological environment in the Turpan–Hami basin. It is important to evaluate the health of the aquifer. We compared two high-resolution (0.44 m) optical images covering the blue rectangular region of Figure 8a,b in July 2003 and May 2013 (Figure 9), where ground subsidence funnels in the second period (Figure 9a,b) were developed. In July 2003, lots of small mounds—the shaft part of a karez—are linearly distributed in this area (Figure 9c). However, a lot of small mounds have disappeared in the optical image taken in May 2013, indicating the karezes in the area were severely damaged (Figure 9d). Some areas that karezes passed through were turned into farmland (Figure 9e,f). The water supply of karezes was destroyed. The water supply in this area will depend mainly on the extraction of groundwater using electromechanical wells. The karezes may have ceased to function, and dried up.

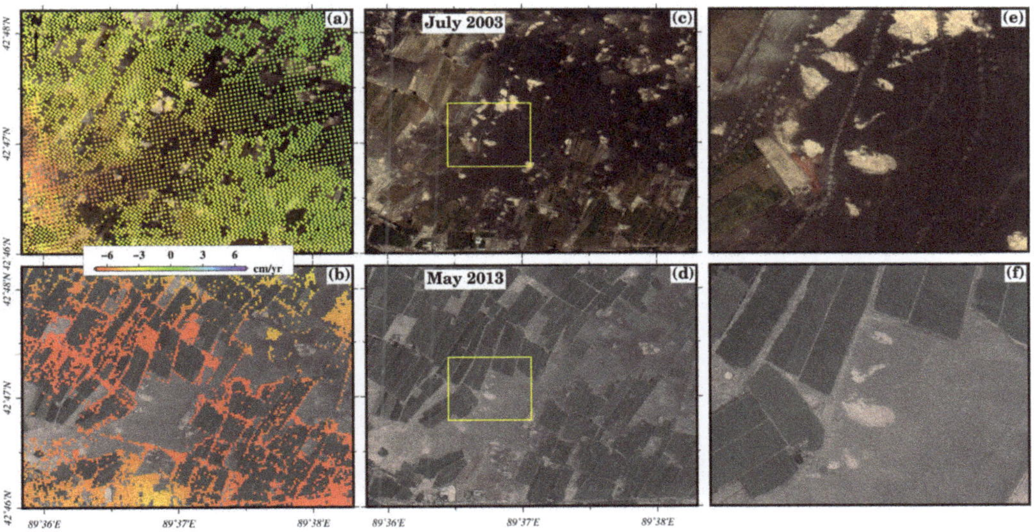

Figure 9. Deformation rate in the blue rectangular area of Figure 8 from (**a**) ALOS-1/PALSAR, (**b**) ascending track Sentinel-1 data. (**c**,**d**) Optical images of this area in July 2003 and May 2013, respectively. (**e**,**f**) Zoom-ins of the yellow rectangular area in (**c**,**d**). Background image: Google Maps satellite image.

In addition, we collected land cover data of the SFM–def region in 2000, 2010, and 2020 (Figure 10) (data from global Land Cover Data Product and Service website of National Basic Geographic Information Center of China (http://www.globallandcover.com/, accessed on 10 July 2022)), and the corresponding area of land cover type in each period (Table 3). The agricultural area has continuously expanded in the past two decades. Artificial areas have expanded rapidly in the past decade, more than 10 times the rate of the previous decade. Water and wetland areas have decreased in the last decade. In 2000 and 2010, the lake area and the surrounding wetland area of Aydingkol Lake was stable, indicating that surface runoff and groundwater are still effective for supply of the lake. These water sources can also partially alleviate the overexploitation of groundwater for agricultural use. However, in 2020, land cover data showed that the waters and wetland of Aydingkol Lake had almost disappeared. Farmland area is in the inner ring of Aydingkol Lake (Section 3.1). The excessive use of surface and underground water in farmland areas has seriously reduced the water supply of the lake, resulting in the shrinkage of water and wetland, which will seriously endanger the ecological environment. The transformation of local agriculture and the economy has upset the ecological balance in the SFM–def region and the balance of aquifers. Conflicts between the development of the local agricultural economy and ecological environment should arouse the attention of local governments.

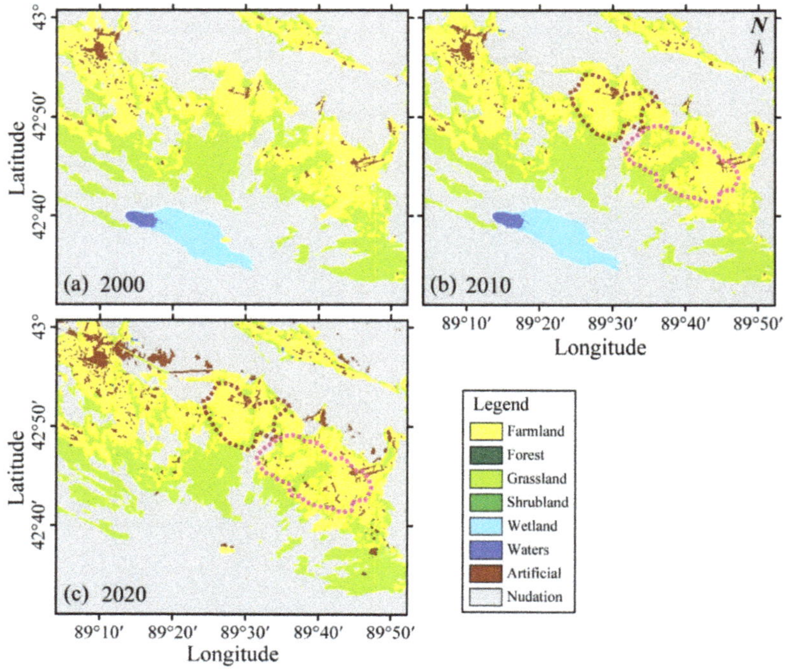

Figure 10. Spatio-temporal evolution of the land covers in the SFM–def region in 2000, 2010, and 2020. The red and magenta dotted lines delineate the central area of the subsidence funnels during the periods 2007–2010 and 2015–2020, respectively.

Table 3. The area of different land covers in the SFM–def region in 2000, 2010, and 2020, obtained from Globeland 30.

Time \ Sort	Farmland	Grassland	Wetland	Waters	Artificial	Nudation
2000	574.33	652.74	99.90	13.68	43.81	2288.24
2010	657.46	669.62	99.90	14.09	48.50	2181.47
2020	705.65	658.64	0.16	1.53	110.14	2194.59
Percentage 1 [a]	14.5%	2.6%	0	3.0%	10.7%	−4.7%
Percentage 2 [b]	7.3%	−1.6%	−99.8%	−89.1%	127.1%	0.6%
Percentage 3 [c]	22.9%	0.9%	−99.8%	−88.8%	151.4%	−4.1%

Unit: km^2. [a]: Percentage of numerical growth in 2010 compared with 2000. [b]: Percentage of numerical growth in 2020 over 2010. [c]: Percentage of numerical growth in 2020 compared with 2000.

5.2. Geological Explanation of Ground Deformation in the Turpan–Hami Basin

There are many farmlands in both the Turpan and Hami depressions. Facility agriculture planting areas are also developed in other agricultural areas, e.g., the oasis areas in Hami and the western part of Turpan. However, why is there a large area of ground subsidence funnels in only the agricultural areas of the SFM–def region?

We plotted the deformation results and the corresponding optical images and faults of the oasis areas in the Turpan depression and the Hami depression (Figure 11). Rainfall is scarce in the Turpan–Hami basin. Irrigation water in the oasis agricultural areas depends on rainfall and meltwater from the surrounding mountains (Section 3.1). The Flaming Mountains fault line lies east–west in the Turpan depression, blocking water flowing from the Tianshan mountain to the south. The other areas, e.g., the northern part of the Flaming Mountains fault line and Hami, can directly obtain abundant mountain water. The surplus water in the Hami oasis can even form a river to supply the downstream area in the southwest (Figure 11d). However, the SFM–def region is short of surface water and groundwater, and the only river channel has almost dried up. As the distance from the southern margin of the fault increases, the water supply gradually decreases. The limited surface water cannot meet the continuously increasing demand for irrigation water, resulting in the continuous overexploitation of aquifers, causing the development of many subsidence funnels in this area.

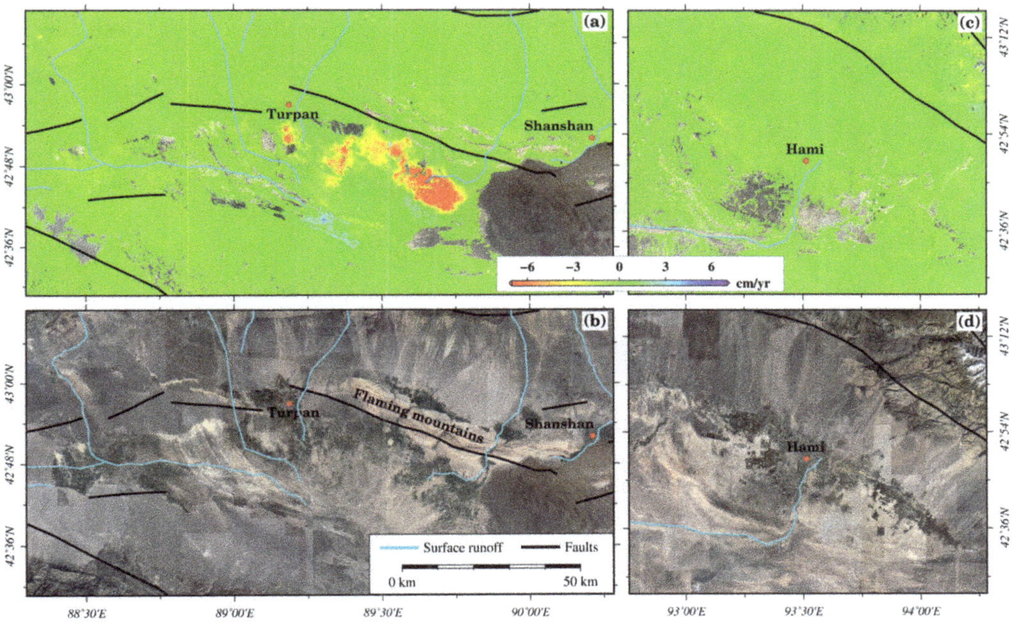

Figure 11. Comparison of ground deformation and optical images in (**a**,**b**) Turpan and (**c**,**d**) Hami oases. Background image: Google Maps satellite image.

The climate in the Turpan–Hami basin is dry and sunny, so evaporation is serious, especially in late spring and summer, when 75% of the year's evaporation occurs. Summer is also the main period of crop growth, which demands more water for irrigation. The agriculture in the SFM–def region relies heavily on groundwater exploitation, which directly leads to the short-term sharp loss of aquifers, and accelerates surface subsidence. From late autumn to early spring, groundwater exploitation intensity in farmland decreases. The aquifers are replenished by surface runoff and groundwater reflux. This explains why the subsidence of the funnels in farmland accelerates in summer and autumn, and slows down or turns to slight uplift in winter and spring (Figure 6).

5.3. Development of InSAR Deformation Monitoring in a Wide Area

At present, most wide-area InSAR deformation monitoring projects use the TS–InSAR algorithm to resolve the deformation time series of all highly coherent monitoring points in each frame [10,11,13,17]. Even though different multi-looking ratios for the WSA and the ROI were used to improve the efficiency of data processing by controlling the spatial resolution of the results [14], these methods are still not out of the scope of time-series deformation calculation. Moreover, if the strategy of reducing spatial resolution is not optimized, it will cause repeated calculations and reduce monitoring efficiency. The deformation rate is usually used to detect potential geohazards in a wide area [38]. Therefore, the deformation time series of some points is unnecessary, especially for stable areas. Hence, calculating the deformation time series at all monitoring points wastes computing resources and labor costs, and produces lots of redundant results. For example, deformation areas account for about 2.4‰ of the total monitoring area in the Turpan–Hami basin. WAVS–InSAR only calculates the deformation rate at each monitoring point in the WSA. Reducing the time dimension of the wide-area deformation results can greatly improve the efficiency of the multitemporal InSAR solution, especially for a lot of InSAR frames in the WSA. Spatial distribution and area of deformation are detected by an adaptive deformation detection method combined with the obtained wide-area deformation rate. After that, high-precision time-series monitoring is only done in the ROI to obtain effective fine deformation results.

For variable-scale deformation results, the WAVS–InSAR strategy proposes a novel variable-scale deformation product organization structure, i.e., it shows the deformation information at the stable surface with low-spatial-resolution deformation rate, while the ROI has a high-spatio-temporal-resolution deformation time series. This structure reduces the amount of deformation results in the stable regions of the WSA, locates the ROI efficiently, and improves the spatial and temporal dimensions of the deformation in the ROI, which is convenient for the calculation, storage, display, and interpretation of the deformation results.

As the SAR satellites and InSAR data increase, InSAR deformation monitoring projects will produce many monitoring results. In the future, wide-area InSAR deformation monitoring projects should be object-oriented, integrating different deformation monitoring to obtain deformation results of multidimensional and high spatio-temporal resolution, and ultimately form a set of universal deformation products. The data-processing strategy and deformation product organization structure proposed in WAVS–InSAR will greatly improve deformation monitoring efficiency and reduce the storage space of massive InSAR monitoring data, which may become a standardized data-processing procedure and data-storage format for future wide-area InSAR deformation products.

6. Conclusions

In this study, we proposed a variable-scale InSAR ground-deformation detection strategy and a deformation product organization structure for wide-area monitoring, namely WAVS–InSAR. This strategy efficiently obtains the deformation rate in the WSA, and uses an adaptive deformation detection method to process the wide-area deformation rate and obtain the spatial distribution and area of the deformation areas (ROI). High-precision time-series monitoring is then only done in the ROI, to obtain effective fine deformation results. Therefore, we can produce variable-scale deformation products in the WSA that consist of low-spatial-resolution deformation rates in stable regions, and fine monitoring results in the ROI.

The proposed WAVS–InSAR was used to monitor wide-area deformation in the Turpan–Hami basin, which has an area of 277,000 km^2. The results show that there are 32 deformation regions with an area of more than 1 km^2 and a deformation magnitude of more than 2 cm/year. The detected deformation areas account for about 2.4‰ of the total monitoring area. The SFM–def region is selected as an application demonstration area of the ROI to carry out fine monitoring of the deformation time series. We obtain the long-term and multidimensional deformation of this area from 2007 to 2010 and from 2015 to 2020 using improved IPTA and MSBAS technologies.

The subsidence funnel center in the SFM–def region moved from northwest to southeast during 2007 to 2020. Based on the variable-scale deformation products and the information regarding hydrogeology, land cover and human activities, we analyze the causes of ground subsidence. Tectonic faults have blocked the water supply in the SFM–def region. The rapid development of facility agriculture has increased the water demand for irrigation. To solve this problem, groundwater has been overexploited. The aquifers in the oasis plain in the SFM–def region are in a state of net deficit. Increased demand for water in the upper reaches of Aydingkol Lake has reduced the lake's water supply. Aydingkol Lake has shrunk dramatically. In addition, there are several deformation areas related to mining in the Turpan–Hami basin.

Author Contributions: Conceptualization, Y.W., G.F. and Z.L.; Formal analysis, H.W. and Z.X.; Funding acquisition, Y.W., G.F. and Z.L.; Investigation, J.Z. and J.H.; Methodology, Y.W. and G.F.; Resources, G.F. and Z.L.; Software, Y.W.; Supervision, Z.L., J.Z. and J.H.; Validation, Y.W., S.L. and H.W.; Writing—original draft, Y.W. and G.F.; Writing—review and editing, all authors. All authors have read and agreed to the published version of the manuscript.

Funding: This research was funded by the Natural Science Foundation of Hunan Province (2021JJ30807), the National Natural Science Foundation of China (Nos. 42174039), the National Science Fund for Distinguished Young Scholars (41925016), the Scientific Research Innovation Project for Graduate Students in Hunan Province (CX20200111), and the Fundamental Research Funds for the Central Universities of Central South University (2020zzts168).

Data Availability Statement: The data used to support the findings of this study are available from the corresponding author upon request.

Acknowledgments: The authors would like to thank the European Space Agency (ESA) for providing free and open Sentinel-1 data, the Japan Aerospace Exploration Agency (JAXA) for providing the ALOS-1/PALSAR images (No: PER2A2N038), and the National Basic Geographic Information Center of China for providing the free and open global Land Cover Data Product (http://www.globallandcover.com/, accessed on 10 July 2022).

Conflicts of Interest: The authors declare no conflict of interest.

References

1. Xue, F.; Lv, X.; Dou, F.; Yun, Y. A Review of Time-Series Interferometric SAR Techniques: A Tutorial for Surface Deformation Analysis. *IEEE Geosci. Remote Sens. Mag.* **2020**, *8*, 22–42. [CrossRef]
2. Even, M.; Schulz, K. InSAR Deformation Analysis with Distributed Scatterers: A Review Complemented by New Advances. *Remote Sens.* **2018**, *10*, 744. [CrossRef]
3. Hu, J.; Li, Z.W.; Ding, X.L.; Zhu, J.J.; Zhang, L.; Sun, Q. Resolving three-dimensional surface displacements from InSAR measurements: A review. *Earth-Sci. Rev.* **2014**, *133*, 1–17. [CrossRef]
4. Liao, M.; Balz, T.; Rocca, F.; Li, D. Paradigm Changes in Surface-Motion Estimation From SAR: Lessons From 16 Years of Sino-European Cooperation in the Dragon Program. *IEEE Geosci. Remote Sens. Mag.* **2020**, *8*, 8–21. [CrossRef]
5. Chang, L.; Dollevoet, R.P.B.J.; Hanssen, R.F. Nationwide Railway Monitoring Using Satellite SAR Interferometry. *IEEE J. Sel. Top. Appl. Earth Obs. Remote Sens.* **2017**, *10*, 596–604. [CrossRef]
6. Ge, D.; Ling, Z.; Yan, W.; Guo, X.; Ye, X. Merging multi-track PSI result for land subsidence mapping over very extended area. In Proceedings of the 2010 IEEE International Geoscience and Remote Sensing Symposium, Honolulu, HI, USA, 25–30 July 2010.
7. Ketelaar, G.; van Leijen, F.; Marinkovic, P.; Hanssen, R. Multi-track PS-InSAR datum connection. In Proceedings of the 2007 IEEE International Geoscience and Remote Sensing Symposium, Barcelona, Spain, 23–28 July 2007; pp. 2481–2484. [CrossRef]
8. Raspini, F.; Bianchini, S.; Ciampalini, A.; Del Soldato, M.; Solari, L.; Novali, F.; Del Conte, S.; Rucci, A.; Ferretti, A.; Casagli, N. Continuous, semi-automatic monitoring of ground deformation using Sentinel-1 satellites. *Sci. Rep.* **2018**, *8*, 7253. [CrossRef]
9. Wang, H.; Wright, T.J.; Yu, Y.; Lin, H.; Jiang, L.; Li, C.; Qiu, G. InSAR reveals coastal subsidence in the Pearl River Delta, China. *Geophys. J. Int.* **2012**, *191*, 1119–1128. [CrossRef]
10. Kalia, A.C.; Frei, M.; Lege, T. A Copernicus downstream-service for the nationwide monitoring of surface displacements in Germany. *Remote Sens. Environ.* **2017**, *202*, 234–249. [CrossRef]
11. Costantini, M.; Ferretti, A.; Minati, F.; Falco, S.; Trillo, F.; Colombo, D.; Novali, F.; Malvarosa, F.; Mammone, C.; Vecchioli, F.; et al. Analysis of surface deformations over the whole Italian territory by interferometric processing of ERS, Envisat and COSMO-SkyMed radar data. *Remote Sens. Environ.* **2017**, *202*, 250–275. [CrossRef]
12. Luca, C.D.; Bonano, M.; Casu, F.; Manunta, M.; Manzo, M.; Onorato, G.; Zinno, I.; Lanari, R. The Parallel SBAS-DInSAR Processing Chain for the Generation of National Scale Sentinel-1 Deformation Time-Series. *Procedia Comput. Sci.* **2018**, *138*, 326–331. [CrossRef]
13. Ferretti, A.; Novali, F.; Giannico, C.; Uttini, A.; Iannicella, I.; Mizuno, T. A Squeesar Database Over the Entire Japanese Territory. In Proceedings of the IGARSS 2019–2019 IEEE International Geoscience and Remote Sensing Symposium, Yokohama, Japan, 28 July–2 August 2019.
14. Dong, J.; Lai, S.; Wang, N.; Wang, Y.; Zhang, L.; Liao, M. Multi-scale deformation monitoring with Sentinel-1 InSAR analyses along the Middle Route of the South-North Water Diversion Project in China. *Int. J. Appl. Earth Obs. Geoinf.* **2021**, *100*, 102324. [CrossRef]
15. Casu, F.; Elefante, S.; Imperatore, P.; Zinno, I.; Manunta, M.; De Luca, C.; Lanari, R. SBAS-DInSAR Parallel Processing for Deformation Time-Series Computation. *IEEE J. Sel. Top. Appl. Earth Obs. Remote Sens.* **2014**, *7*, 3285–3296. [CrossRef]

16. Casu, F.; Manunta, M.; Agram, P.S.; Crippen, R.E. Big Remotely Sensed Data: Tools, applications and experiences. *Remote Sens. Environ.* **2017**, *202*, 1–2. [CrossRef]
17. Lanari, R.; Bonano, M.; Casu, F.; Luca, C.D.; Manunta, M.; Manzo, M.; Onorato, G.; Zinno, I. Automatic Generation of Sentinel-1 Continental Scale DInSAR Deformation Time Series through an Extended P-SBAS Processing Pipeline in a Cloud Computing Environment. *Remote Sens.* **2020**, *12*, 2961. [CrossRef]
18. Wang, C.; Tang, Y.; Zhang, H.; You, H.; Zhang, W.; Duan, W.; Wang, J.; Dong, L.; Zhang, B. First mapping of China surface movement using supercomputing interferometric SAR technique. *Sci. Bull.* **2021**, *66*, 1608–1610. [CrossRef]
19. Duan, W.; Zhang, H.; Wang, C.; Tang, Y. Multi-Temporal InSAR Parallel Processing for Sentinel-1 Large-Scale Surface Deformation Mapping. *Remote Sens.* **2020**, *12*, 3749. [CrossRef]
20. Ansari, H.; De Zan, F.; Bamler, R. Sequential Estimator: Toward Efficient InSAR Time Series Analysis. *IEEE Trans. Geosci. Remote Sens.* **2017**, *55*, 5637–5652. [CrossRef]
21. Wang, B.; Zhao, C.; Zhang, Q.; Peng, M. Sequential InSAR Time Series Deformation Monitoring of Land Subsidence and Rebound in Xi'an, China. *Remote Sens.* **2019**, *11*, 2854. [CrossRef]
22. Palanisamy Vadivel, S.K.; Kim, D.-j.; Jung, J.; Cho, Y.-K.; Han, K.-J. Monitoring the Vertical Land Motion of Tide Gauges and Its Impact on Relative Sea Level Changes in Korean Peninsula Using Sequential SBAS-InSAR Time-Series Analysis. *Remote Sens.* **2020**, *13*, 18. [CrossRef]
23. Wang, Y.; Yang, Z.; Li, Z.; Zhu, J.; Wu, L. Fusing adjacent-track InSAR datasets to densify the temporal resolution of time-series 3-D displacement estimation over mining areas with a prior deformation model and a generalized weighting least-squares method. *J. Geod.* **2020**, *94*, 5. [CrossRef]
24. Liu, J.; Hu, J.; Li, Z.; Sun, Q.; Ma, Z.; Zhu, J.; Wen, Y. Dynamic Estimation of Multi-Dimensional Deformation Time Series from InSAR Based on Kalman Filter and Strain Model. *IEEE Trans. Geosci. Remote Sens.* **2021**, *60*, 1–16. [CrossRef]
25. Liu, J.; Hu, J.; Bürgmann, R.; Li, Z.; Sun, Q.; Ma, Z. A Strain-Model Based InSAR Time Series Method and Its Application to The Geysers Geothermal Field, California. *J. Geophys. Res. Solid Earth* **2021**, *126*, 8. [CrossRef]
26. Samsonov, S.; Dille, A.; Dewitte, O.; Kervyn, F.; d'Oreye, N. Satellite interferometry for mapping surface deformation time series in one, two and three dimensions: A new method illustrated on a slow-moving landslide. *Eng. Geol.* **2020**, *266*, 105471. [CrossRef]
27. Zhang, R.; Liu, G.; Li, Z.; Zhang, G.; Lin, H.; Yu, B.; Wang, X. A Hierarchical Approach to Persistent Scatterer Network Construction and Deformation Time Series Estimation. *Remote Sens.* **2014**, *7*, 211–228. [CrossRef]
28. Hou, J.; Xu, B.; Li, Z.; Zhu, Y.; Feng, G. Block PS-InSAR ground deformation estimation for large-scale areas based on network adjustment. *J. Geod.* **2021**, *95*, 10. [CrossRef]
29. Sandwell, D.T.; Price, E.J. Phase gradient approach to stacking interferograms. *J. Geophys. Res. Solid Earth* **1998**, *103*, 30183–30204. [CrossRef]
30. Ferretti, A.; Prati, C.; Rocca, F. Permanent Scatterers in SAR Interferometry. *IEEE Trans. Geosci. Remote Sens.* **2001**, *39*, 8–20. [CrossRef]
31. Berardino, P.; Fornaro, G.; Lanari, R.; Sansosti, E. A New Algorithm for Surface Deformation Monitoring Based on Small Baseline Differential SAR Interferograms. *IEEE Trans. Geosci. Remote Sens.* **2003**, *40*, 2375–2383. [CrossRef]
32. Werner, C.; Wegmuller, U.; Strozzi, T.; Wiesmann, A. Interferometric point target analysis for deformation mapping. In Proceedings of the IGARSS 2003.,2003 IEEE International Geoscience and Remote Sensing Symposium, (IEEE Cat. No.03CH37477), Toulouse, France, 21–25 July 2003. [CrossRef]
33. Dai, K.; Liu, G.; Li, Z.; Ma, D.; Wang, X.; Zhang, B.; Tang, J.; Li, G. Monitoring Highway Stability in Permafrost Regions with X-band Temporary Scatterers Stacking InSAR. *Sensors* **2018**, *18*, 1876. [CrossRef]
34. Kotsis, I.; Kontoes, C.; Paradissis, D.; Karamitsos, S.; Elias, P.; Papoutsis, I. A Methodology to Validate the InSAR Derived Displacement Field of the September 7(th), 1999 Athens Earthquake Using Terrestrial Surveying. Improvement of the Assessed Deformation Field by Interferometric Stacking. *Sensors* **2008**, *8*, 4119–4134. [CrossRef]
35. McMillan, M.; Shepherd, A.; Gourmelen, N.; Park, J.-W.; Nienow, P.; Rinne, E.; Leeson, A. Mapping ice-shelf flow with interferometric synthetic aperture radar stacking. *J. Glaciol.* **2017**, *58*, 265–277. [CrossRef]
36. Qin, Z.; Agarwal, V.; Gee, D.; Marsh, S.; Grebby, S.; Chen, Y.; Meng, N. Study of Ground Movement in a Mining Area with Geological Faults Using FDM Analysis and a Stacking InSAR Method. *Front. Environ. Sci.* **2021**, *9*, 787053. [CrossRef]
37. Zhang, L.; Dai, K.; Deng, J.; Ge, D.; Liang, R.; Li, W.; Xu, Q. Identifying Potential Landslides by Stacking-InSAR in Southwestern China and Its Performance Comparison with SBAS-InSAR. *Remote Sens.* **2021**, *13*, 3662. [CrossRef]
38. Festa, D.; Bonano, M.; Casagli, N.; Confuorto, P.; De Luca, C.; Del Soldato, M.; Lanari, R.; Lu, P.; Manunta, M.; Manzo, M.; et al. Nation-wide mapping and classification of ground deformation phenomena through the spatial clustering of P-SBAS InSAR measurements: Italy case study. *ISPRS J. Photogramm. Remote Sens.* **2022**, *189*, 1–22. [CrossRef]
39. Luo, S.; Feng, G.; Xiong, Z.; Wang, H.; Zhao, Y.; Li, K.; Deng, K.; Wang, Y. An Improved Method for Automatic Identifi-cation and Assessment of Potential Geohazards Based on MT-InSAR Measurements. *Remote Sens.* **2021**, *13*, 3490. [CrossRef]
40. Samsonov, S.; D'Oreye, N. Multidimensional time-series analysis of ground deformation from multiple InSAR data sets applied to Virunga Volcanic Province. *Geophys. J. Int.* **2012**, *191*, 1095–1108.
41. Samsonov, S.; d'Oreye, N.; Smets, B. Ground deformation associated with post-mining activity at the French–German border revealed by novel InSAR time series method. *Int. J. Appl. Earth Obs. Geoinf.* **2013**, *23*, 142–154. [CrossRef]

42. Samsonov, S.V.; Feng, W.; Peltier, A.; Geirsson, H.; d'Oreye, N.; Tiampo, K.F. Multidimensional Small Baseline Subset (MSBAS) for volcano monitoring in two dimensions: Opportunities and challenges. Case study Piton de la Fournaise volcano. *J. Volcanol. Geotherm. Res.* **2017**, *344*, 121–138. [CrossRef]
43. Solari, L.; Bianchini, S.; Franceschini, R.; Barra, A.; Monserrat, O.; Thuegaz, P.; Bertolo, D.; Crosetto, M.; Catani, F. Satel-lite interferometric data for landslide intensity evaluation in mountainous regions. *Int. J. Appl. Earth Obs. Geoinf.* **2020**, *87*, 102028. [CrossRef]
44. Hu, X.; Burgmann, R.; Schulz, W.H.; Fielding, E.J. Four-dimensional surface motions of the Slumgullion landslide and quantification of hydrometeorological forcing. *Nat. Commun.* **2020**, *11*, 2792. [CrossRef]
45. Liu, Q.; Yue, G.; Ding, X.; Yang, K.; Feng, G.; Xiong, Z. Temporal and characteristics analysis of deformation along Foshan subway using time series InSAR. *Geomat. Inf. Sci. Wuhan Univ.* **2019**, *44*, 1099–1106. [CrossRef]
46. Xiong, Z.; Feng, G.; Feng, Z.; Miao, L.; Wang, Y.; Yang, D.; Luo, S. Pre- and post-failure spatial-temporal deformation pattern of the Baige landslide retrieved from multiple radar and optical satellite images. *Eng. Geol.* **2020**, *279*, 105880. [CrossRef]
47. Wang, Y.; Feng, G.; Li, Z.; Xu, W.; Zhu, J.; He, L.; Xiong, Z.; Qiao, X. Retrieving the displacements of the Hutubi (China) underground gas storage during 2003–2020 from multi-track InSAR. *Remote Sens. Environ.* **2021**, *268*, 112768. [CrossRef]
48. Wu, S.; Yang, Z.; Ding, X.; Zhang, B.; Zhang, L.; Lu, Z. Two decades of settlement of Hong Kong International Airport measured with multi-temporal InSAR. *Remote Sens. Environ.* **2020**, *248*, 111976. [CrossRef]
49. Zhang, M.; Philp, P. Geochemical characterization of aromatic hydrocarbons in crude oils from the Tarim, Qaidam and Turpan Basins, NW China. *Pet. Sci.* **2010**, *7*, 448–457. [CrossRef]
50. Yan, N.; Wu, B.; Zhu, W. Assessment of Agricultural Water Productivity in Arid China. *Water* **2020**, *12*, 1161. [CrossRef]
51. Xiao, L. Study of Protection of Karez in Xinjiang from the Perspective of Ecological Civilization Construction. Master's Thesis, Beijing Forestry University, Beijing, China, 2019.
52. Ye, K. The Water Quantity Monitoring of Karez and Rational Utilization of Water Resources in Turpan. Master's Thesis, Jilin University, Changchun, China, 2016.
53. Rebiyamu, M. Study on the Relationship between the Karez and Turpan Oasis Ecological Environment. Master's Thesis, Xinjiang University, Urumqi, China, 2014.
54. Werner, C.; Wegmüller, U.; Strozzi, T.; Wiesmann, A. Gamma SAR and interferometric processing software. In Proceedings of the Ers-Envisat Symposium, Gothenburg, Sweden, 16–20 October 2000; pp. 211–219.
55. Zebker, H.A.; Rosen, P.A.; Goldstein, R.M.; Gabriel, A.; Werner, C.L. On the derivation of coseismic displacement fields using differential radar interferometry: The Landers earthquake. *J. Geophys. Res. Solid Earth* **1994**, *99*, 19617–19634. [CrossRef]
56. Farr, T.G.; Rosen, P.A.; Caro, E.; Crippen, R.; Duren, R.; Hensley, S.; Kobrick, M.; Paller, M.; Rodriguez, E.; Roth, L.; et al. The Shuttle Radar Topography Mission. *Rev. Geophys.* **2007**, *45*, RG2004. [CrossRef]
57. Chen, C.W.; Zebker, H.A. Two-dimensional phase unwrapping with use of statistical models for cost functions in nonlinear optimization. *JOSA A* **2001**, *18*, 338–351. [CrossRef]
58. Chen, C.W.; Zebker, H.A. Phase unwrapping for large SAR interferograms: Statistical segmentation and generalized network models. *IEEE Trans. Geosci. Remote Sens.* **2002**, *40*, 1709–1719. [CrossRef]
59. Li, Z.W.; Ding, X.L.; Zheng, D.W.; Huang, C. Least Squares-Based Filter for Remote SensingImage Noise Reduction. *IEEE Trans. Geosci. Remote Sens.* **2008**, *46*, 2044–2049. [CrossRef]
60. Smith, R.G.; Knight, R.; Chen, J.; Reeves, J.A.; Zebker, H.A.; Farr, T.; Liu, Z. Estimating the permanent loss of groundwater storage in the southern San Joaquin Valley, California. *Water Resour. Res.* **2017**, *53*, 2133–2148. [CrossRef]
61. Chen, J.; Knight, R.; Zebker, H.A.; Schreüder, W.A. Confined aquifer head measurements and storage properties in the San Luis Valley, Colorado, from spaceborne InSAR observations. *Water Resour. Res.* **2016**, *52*, 3623–3636. [CrossRef]
62. Miller, M.M.; Shirzaei, M.; Argus, D. Aquifer Mechanical Properties and Decelerated Compaction in Tucson, Arizona. *J. Geophys. Res. Solid Earth* **2017**, *122*, 8402–8416. [CrossRef]
63. Xinjiang-Investigation-Team. *Turpan: Building Green Organic Products Brand to Promote the Construction of Facility Agriculture*; National Bureau of Statistics China: Beijing, China, 2010. Available online: http://www.stats.gov.cn/ztjc/ztfx/dfxx/201011/t20101126_35260.html (accessed on 10 July 2022).

remote sensing

Article

Settlement Prediction of Reclaimed Coastal Airports with InSAR Observation: A Case Study of the Xiamen Xiang'an International Airport, China

Zhiqiang Xiong [1], Kailiang Deng [2,*], Guangcai Feng [1], Lu Miao [1], Kaifeng Li [2], Chulu He [3] and Yuanrong He [4]

1 School of Geosciences and Info-Physics, Central South University, Changsha 410083, China; zqxiong_flhs@csu.edu.cn (Z.X.); fredgps@csu.edu.cn (G.F.); insarml@csu.edu.cn (L.M.)
2 Naval Institute of Hydrographic Surveying and Charting, Tianjin 300061, China; kaifeng_li2022@163.com
3 Fujian Provincial Institute of Land Surveying and Mapping, Xiamen 361006, China; chulu_he@163.com
4 Big Data Institute of Digital Natural Disaster Monitoring in Fujian, Xiamen University of Technology, Xiamen 361024, China; 2012112001@xmut.edu.cn
* Correspondence: dkl_hych@163.com

Abstract: Many coastal cities reclaim land from the sea to meet the rapidly growing demand for land caused by population growth and economic development. Settlement in reclaimed land may delay construction and even damage infrastructures, so accurately predicting the settlement over reclaimed areas is important. However, the limited settlement observation and ambiguous final settlement estimation affect accurate settlement prediction in traditional methods. This study proposes a new strategy to solve these problems by using the Multitemporal Interferometric Synthetic Aperture Radar (MT-InSAR) method and takes the Xiamen Xiang'an International Airport, built on reclaimed land, as an example. The MT-InSAR is adopted to process the Sentinel-1 images to obtain the settlement history of the study area. The results show that settlement mainly occurs in the reclaimed areas, with the maximum average settlement rate exceeding 40 mm/y. We use the statistical properties of curve fitting to choose the best curve model from several candidate curve models to predict the settlement time series. The Asaoka method is used to identify the critical state between settlement and stability. We predict the consolidation time of the whole study area and reveal that the deformation rate is positively correlated with the consolidation time. The maximum remaining settlement time is over ten years since 24 December 2019. Therefore, manual compaction operations can be carried out to speed up settlement in the areas that need a long time to consolidate. The proposed method can be used to predict the settlement of similar reclaimed areas, and the predicted results can provide a reference for engineering construction.

Keywords: settlement prediction; reclaimed land; InSAR; exponential model; Asaoka method

Citation: Xiong, Z.; Deng, K.; Feng, G.; Miao, L.; Li, K.; He, C.; He, Y. Settlement Prediction of Reclaimed Coastal Airports with InSAR Observation: A Case Study of the Xiamen Xiang'an International Airport, China. *Remote Sens.* **2022**, *14*, 3081. https://doi.org/10.3390/rs14133081

Academic Editors: Paolo Mazzanti and Saverio Romeo

Received: 12 May 2022
Accepted: 22 June 2022
Published: 27 June 2022

Publisher's Note: MDPI stays neutral with regard to jurisdictional claims in published maps and institutional affiliations.

1. Introduction

Land reclamation from the sea has become an important strategy to promote economic growth and alleviate the population density in coastal areas [1–3]. However, the compaction of the underlying soft soil layers in reclaimed areas often leads to land subsidence, which may cause damage to undergoing construction and infrastructures [4,5]. Settlement monitoring and prediction are important in engineering geology [6], especially for reclaimed areas, where the underlying layers are highly compressible and require a long time for consolidation. Moreover, post-construction settlement prediction based on the foundation displacements can provide a reference for engineering construction, deformation early warning, and future land reclamation.

The methods used to predict the settlement include theoretical estimations based on the soil consolidation theory [7] and curve-fitting or prediction models based on deformation measurements [8]. The application of the former is limited by the few samples used and theoretical deviations due to different soil compositions. The latter relies on deformation observations, which are easy to obtain. Therefore, the second group of methods, such as hyperbolic curve [4,9,10], exponential function [11–13], and grey model [14–16], are well-applied in engineering settlement monitoring and prediction.

Land settlement is often monitored via in situ measurements and remote sensing technology. However, in situ measurements, such as Global Navigation Satellite System (GNSS), leveling, and extensometers, usually have low spatial resolutions due to labor intensity and high cost [17]. Therefore, with in situ measurements, it is difficult to obtain the detailed spatial distribution of deformation, which may lead to misunderstanding settlement behaviors. Interferometric Synthetic Aperture Radar (InSAR) has the ability to obtain the spatial–temporal distribution of deformation with promising accuracy. More importantly, it works in a non-contact style, so the acquisition of those data does not affect engineering construction. Consequently, InSAR has been widely used in measuring ground deformation, such as landslides [18–20], tectonic movements [21–23], volcano dynamics [24,25], and land subsidence [26–29], as well as oil and gas fields [30,31]. Moreover, the ability of long-term deformation time series retrieval of multitemporal InSAR (MT-InSAR) enables InSAR to predict the settlement of the reclaimed areas.

For settlement prediction, most studies have chosen prior functional models (e.g., hyperbolic function, Poisson function, exponential function) to predict the total amount and termination time of settlement. Kim et al. [9] introduced the hyperbolic model and persistent scatterer InSAR (PS-InSAR) to monitor the land subsidence in Mokpo City, South Korea. They showed that the prior hyperbolic model has better performance than the linear model. Hu et al. [11] used multisource remote sensing imagery to characterize landscape changes in Yan'an, China. They chose the exponential curve model to predict the consolidation time of the subsiding area. Deng et al. [15] combined PS-InSAR with Grey system theory to monitor and predict land subsidence in the Beijing Plain. The results indicated that this method can be an alternative to conventional numerical and empirical models for short-term prediction in cases in which there is a lack of detailed geological or hydraulic information. The method of combining InSAR with curve models to monitor and predict land subsidence has also been used in different scenarios [4,12,13,32]. Most studies chose curve models based on experience for settlement monitoring and prediction. The selected prior models, however, are likely not suitable to describe the deformation of the studied areas. Shi et al. [17] reported that different models lead to large variations in settlement predictions. They used exponential curve, hyperbolic curve, and quadratic curve to model the settlement time series. However, the quadratic curve failed to predict the settlement. The maximum difference in the final settlement amount predicted by the two curves can reach 0.2 m. In addition, it is not easy to identify the critical state between deformation and stability by the curve-fitting method, resulting in biased predictions of the total amount and the termination time of settlement.

To solve the abovementioned problems, this study proposes a two-step settlement prediction method. In step one, sufficient deformation points provided via MT-InSAR are used to choose the best curve to model the settlement pattern. In step two, the predicted settlement time series and the Asaoka method [33] are used to predict the final settlement amount and consolidation time. The method is validated using the Xiamen Xiang'an International Airport (referred to as XXIA hereafter), China, a reclaimed airport under construction, as a case study.

This study is organized as follows: The study area and datasets are introduced in Section 2. Section 3 describes the processing flow of MT-InSAR and the deformation results of the XXIA. Section 4 describes the procedure of settlement prediction. In Section 5, we discuss the total settlement and the consolidation time of the study area. Some conclusions are drawn in Section 6.

2. Study Area and Datasets

Sentinel-1 images over the XXIA acquired between July 2015 and December 2019 were processed via MT-InSAR to obtain the deformation of the whole study area. Then, we chose the best curve function to model and predicted the deformation time series of the study area. The study area and datasets are described in subsequent sections.

2.1. Study Area

The XXIA, a planned 4F-class international airport, is located in the southeast of Dadeng Island, Xiamen City, China (Figures 1 and 2). The construction of the XXIA started on 4 January 2022, and about 26 km^2 of the XXIA will be built on reclaimed land. The land reclamation project of the XXIA has three phases (Figure 2); the first two phases have been completed, after which the reclaimed land experienced settlement [2,34]. Settlement monitoring and prediction of the reclaimed land are crucial to the safe construction of the XXIA. However, predictions of the total settlement amount and consolidation time of the XXIA have not yet been reported.

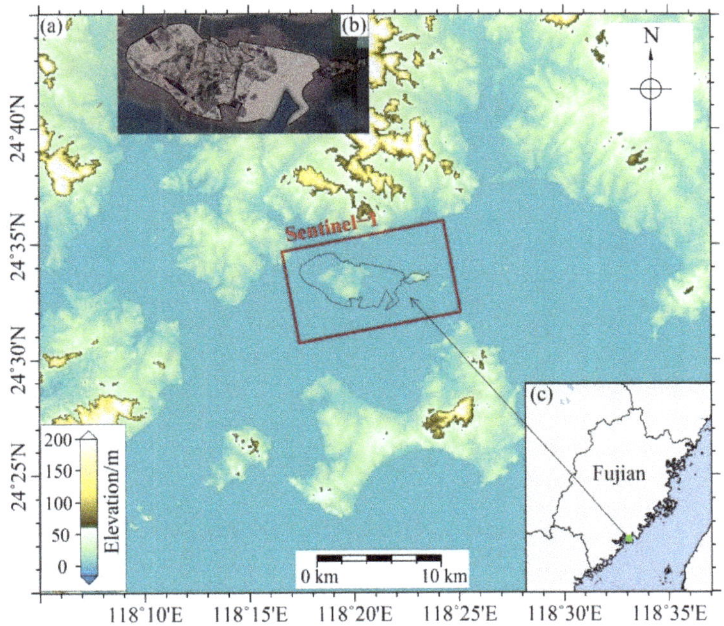

Figure 1. The Xiamen Xiang'an International Airport (XXIA): (**a**) coverage of Sentinel-1 images and the location of the XXIA; (**b**) optical image of Dadeng Island; (**c**) location of the XXIA in Fujian Province, China.

2.2. Datasets

A total of 128 ascending Sentinel-1 images acquired between 6 July 2015 and 24 December 2019 were collected to obtain the deformation time series of the reclaimed land (Table A1). Coverage of the Sentinel-1 data is shown in Figure 1. The main parameters of the Sentinel-1 images are summarized in Table 1. The images were acquired after the first and second reclamation phases were completed. In addition, the SRTM DEM with a resolution of 30 m was collected to simulate and remove the topographic phase in the differential interferograms.

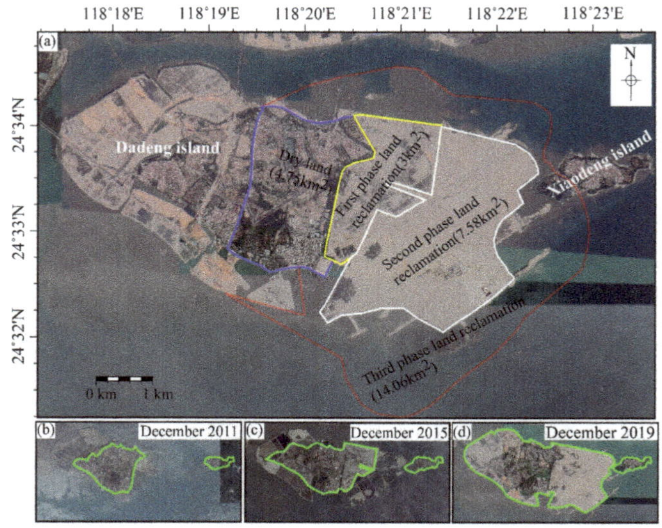

Figure 2. Optical images of the XXIA, Dadeng Island, and Xiaodeng Island: (**a**) illustration of the three phases of land reclamation. The yellow, white, and red lines are the boundaries of the reclaimed land in the first, second, and third phases, respectively. The blue line represents the dry land of Dadeng Island; (**b–d**) optical images of the study area acquired in December 2011, December 2015, and December 2019, respectively. The green lines delineate the boundaries of Dadeng Island and Xiaodeng Island.

Table 1. Acquisition dates and parameters of the collected Sentinel-1 images.

Sensors	Direction	Incidence Angle	Path-Frame	Number of Images	Temporal Coverage
Sentinel-1	Ascending	33.91°	142-75	128	6 July 2015–24 December 2019

3. MT-InSAR Processing and Deformation Results

3.1. MT-InSAR Processing

Sentinel-1 SAR images have small perpendicular baselines, so only the temporal baselines were considered selecting interferometric pairs. The maximum temporal baseline was set as 36 days. The interferometric pairs were processed via a multi-look operation (range × azimuth = 5 × 1) to form the interferograms. The SRTM DEM was used to remove the phase contribution of topography in the interferograms. After interferogram filtering and phase unwrapping, the first-order polynomial function model was used to remove the phase ramps. Some areas in the study area were reclaimed from the sea, so they have no external DEM data. We assumed the elevation of the reclaimed land to be 1 and selected high-quality interferograms to perform linear regression between topographic errors and perpendicular baselines to estimate and remove topographic errors. We used the amplitude dispersion of pixels [35], intensity, and coherence to remove poor-quality pixels. The singular value decomposition (SVD) [36] was used to calculate the phase time series of each pixel. We estimated the linear deformation using the least square method and then subtracted the phase contributions of linear deformation from the phase time series. The residual components in the phase time series mainly include nonlinear deformation, atmospheric artifacts, and noise. The atmospheric phases are highly correlated in space, while lowly correlated in time. Noise is lowly correlated in both space and time domains. Therefore, we used temporal and spatial filtering to extract nonlinear deformations from the residual phases. The final deformation time series of each pixel is the sum of linear deformation and nonlinear deformation. The least-square method was performed in the final deformation time series to estimate the average deformation rate of each pixel.

3.2. Deformation Results

Figure 3 shows the average deformation rate in the line-of-sight (LOS) direction of the study area. The selected pixels were sufficient to show the spatial distribution of the settlement, and the deformation areas obtained in this study are generally consistent with the previous studies [2,34], although the monitoring periods were different. Four areas, areas—A, B, C, and D—suffered severe settlement, and the maximum settlement rate exceeded 40 mm/y. The settlement mainly occurred in the reclaimed areas, including the area reclaimed in the first phase (Figure 3d), the second phase (Figure 3e), and the reclaimed areas in the northwest (Figure 3a) and southwest (Figure 3b) of Dadeng Island. The settlement was particularly severe in the northeast of the area reclaimed in the second phase.

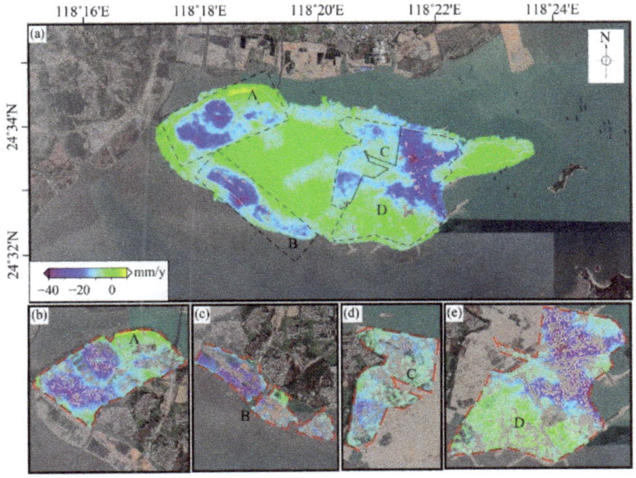

Figure 3. Average deformation rate map of the study area. Positive values denote the deformation toward the satellite. Negative values denote the deformation away from the satellite: (**a**) average deformation rate map of the whole study area; (**b–e**) enlarged view of areas outlined by dotted lines in (**a**).

The first reclamation phase was completed before July 2015, earlier than the monitoring time of InSAR. According to the soil consolidation theory [7], some points (see Section 4) in area C have finished the primary consolidation and entered the second compression stage. The settlement rate in the area reclaimed in the first stage was around 15 mm/y, with a maximum settlement rate of 20 mm/y. For area C, the maximum settlement rate was observed in the southern part, but some points in this area had become stable after the reclamation. Among the four areas, area C was first reclaimed, so it had the minimum deformation rate, due to the longer time of consolidation. It is possible that areas A, B, and D will have a settlement pattern similar to area C in the time domain, as they had similar strategies of reclamation. Area D had the maximum average settlement rate, as it was reclaimed in the second phase, which was accomplished in March 2018 and has just entered the primary consolidation stage. Inside each area, the settlement rates were different. Specifically, in area D, the southern part was stable, but the northern part experienced serious settlement. One explanation is that this area underwent several reclamations at different times (see Section 5).

4. Methods of Settlement Prediction

The considerable number of measured points obtained from MT-InSAR provided a good opportunity for the settlement prediction of the XXIA. In this section, we discuss our methodology, according to which we chose the best model to describe the settlement pattern of the XXIA and then used the Asaoka method to estimate the termination time of settlement for predicting the total amount of settlement. The flowchart is shown in Figure 4.

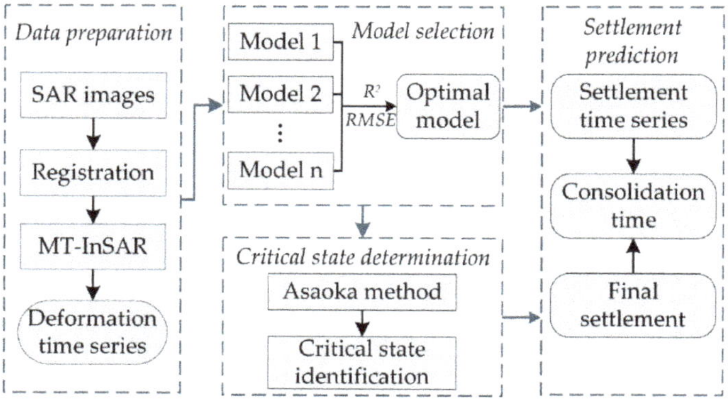

Figure 4. Flowchart of the settlement prediction.

4.1. Function Model Selection

To describe the settlement patterns of the reclaimed areas, we selected the best one from three traditional curve functions—namely, the hyperbolic model, Poisson curve model, and the exponential function model—according to the deformation time series of the XXIA obtained. The three utilized models are introduced below.

4.1.1. Hyperbolic Model

The hyperbolic model [4,9,10] is an empirical curve fitting method, suitable for fitting a large amount of measured data. The amount of settlement S_t at time t can be expressed as

$$S_t = S_0 + \frac{t - t_0}{\alpha + \beta(t - t_0)} \tag{1}$$

where S_0 is the initial amount of settlement, t_0 is the initial time of settlement, and α and β are two unknown parameters. Formula (1) can be rewritten as

$$\frac{t - t_0}{S_t - S_0} = \alpha + \beta(t - t_0) \tag{2}$$

As Formula (2) shows, $(t - t_0)/(S_t - S_0)$ and $(t - t_0)$ are linearly related. α and β can be estimated by linear regression in t/S_t versus t, and then the hyperbolic model of the settlement can be determined.

4.1.2. Poisson Curve Model

The Poisson curve model [32] can be written as

$$S_t = c/(1 + ae^{(-bt)}) \tag{3}$$

where S_t represents the amount of settlement at time t. a, b, and c are three unknown parameters, which can be determined by the three-stage calculation method.

Assume the total number of SAR images is n, and the time interval of each two images is equal. The settlement time series are $y_1, y_2, y_3, \dots, y_n$. The settlement time series is divided into three groups. Each group has $r = n/3$ items. Let d_1, d_2, and d_3 be the reciprocal sum of the settlement in the following three groups:

$$d_1 = \sum_{i=1}^{r} \frac{1}{y_i};\ d_2 = \sum_{i=r+1}^{2r} \frac{1}{y_i};\ d_3 = \sum_{i=2r+1}^{n} \frac{1}{y_i} \quad (4)$$

a, b, and c can be determined by the following formulas:

$$b = \ln \frac{(d_1 - d_2)}{d_2 - d_3} / r \quad (5)$$

$$c = r / [d_1 - \frac{(d_1 - d_2)^2}{(d_1 - d_2) - (d_2 - d_3)}] \quad (6)$$

$$a = \frac{(d_1 - d_2)^2 \left(1 - e^{-b}\right) c}{[(d_1 - d_2) - (d_2 - d_3)] e^{-b} (1 - e^{-rb})} \quad (7)$$

In this study, the SAR images used to retrieve the settlement time series were not acquired at equal intervals, so the settlement time series should be interpolated before applying the Poisson curve model.

4.1.3. Exponential Function Model

The exponential function model [11–13] assumes that the settlement rate decreases along the exponential curve with time. The model is expressed as follows:

$$S_t = S_\infty - (S_\infty - S_0) e^{(t_0 - t)/\eta} \quad (8)$$

where S_t represents the settlement amount at time t, t_0 is the initial time of settlement, η is the unknown parameter, and S_0 and S_∞ represent the initial settlement amount and the final settlement amount, respectively. Nonlinear regression can be used to apply Formula (8).

4.1.4. Optimal Model Selection

Most studies used a few points to evaluate the prediction ability of models. However, these previously selected points may not be able to describe the settlement characteristics of the whole study area. Moreover, settlement prediction based on a small number of points is not very useful in practical engineering. The optimal prediction model based on sufficient measurements can help monitor the whole study area, guide the construction process, and provide a reference for the prediction of settlement in similar cases. The application of MT-InSAR greatly contributes to the optimal model selection.

As mentioned in Section 3.2, the area reclaimed in the first phase has been subsiding for a long time. The settlement pattern can be well-represented by the long-term settlement time series of the monitored points, which helps the prediction model selection. Therefore, the settlement time series of the monitored points (4217 points) in the area reclaimed in the first phase was used to choose the optimal model from the above three models to predict the settlement. Points P1–P6 (Figure A1) were chosen to display the settlement time series and the three fitted curves (Figures 5–7). As shown in Figures 5–7, the three fitted curves were consistent with the settlement time series. In terms of the determination coefficient R^2 and root-mean-square error (RMSE) of points P1–P6 in the curve-fitted models, the Poisson curve had the worst performance among the three models. Thus, the hyperbolic curve and exponential curve were used to predict the settlement.

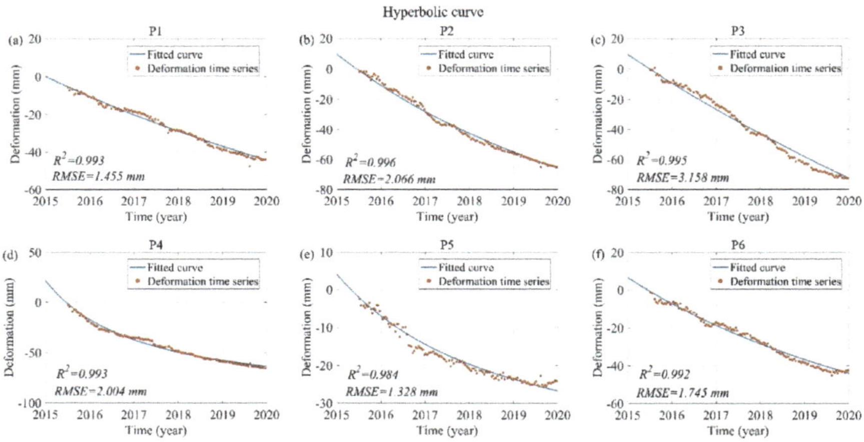

Figure 5. Hyperbolic curve fitting of (**a**) P1, (**b**) P2, (**c**) P3, (**d**) P4, (**e**) P5 and (**f**) P6. Locations of points P1–P6 are shown in Figure A1.

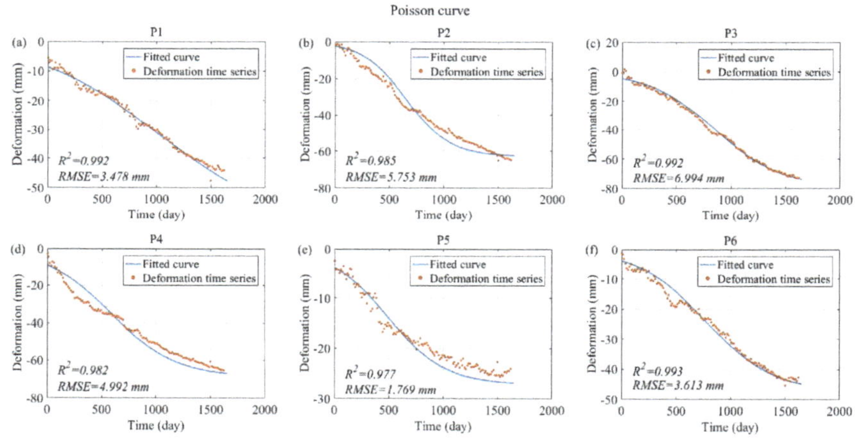

Figure 6. Poisson curve fitting of (**a**) P1, (**b**) P2, (**c**) P3, (**d**) P4, (**e**) P5 and (**f**) P6. Locations of points P1–P6 are shown in Figure A1.

We calculated the R^2 and RMSE values of the 4217 monitored points to qualitatively choose the optimal curve model. Figure 8 shows the distribution of R^2 of the three fitted curves. R^2 of the Poisson curve was mainly concentrated around 0.984, whereas that of the hyperbolic curve was concentrated around 0.992, and the exponential curve had the highest frequency when the correlation coefficient reached 0.996. The statistical histograms of RMSE (Figure 9) showed that the maximum RMSE of the fitted exponential curves was about 7 mm, far less than that of the fitted Poisson and hyperbolic curves. The fitted exponential curves also had the minimum mean RMSE. Therefore, the exponential curve model had the best performance among the three curve models and, therefore, was determined as the optimal settlement prediction model in the study area.

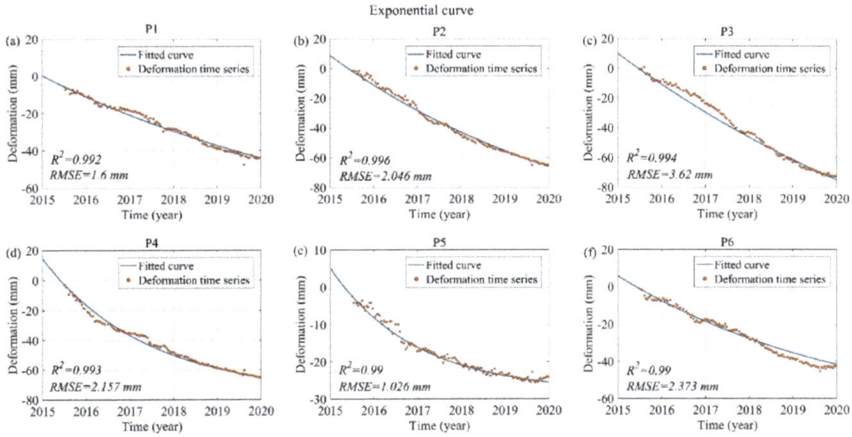

Figure 7. Exponential curve fitting of (**a**) P1, (**b**) P2, (**c**) P3, (**d**) P4, (**e**) P5 and (**f**) P6. Locations of points P1–P6 are shown in Figure A1.

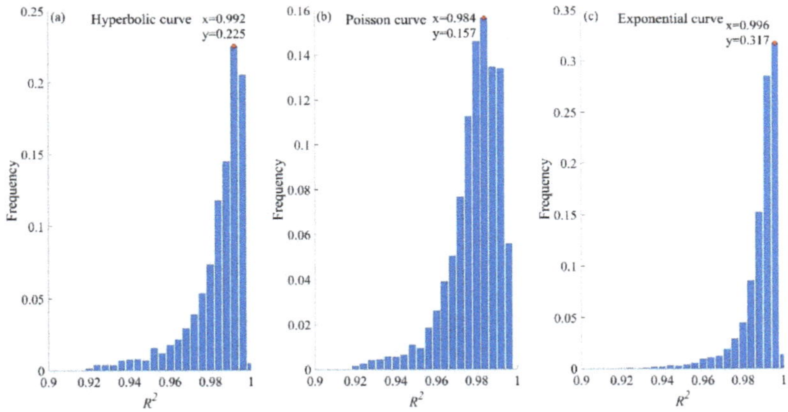

Figure 8. R^2 of (**a**) Hyperbolic curve fitting, (**b**) Poisson curve fitting and (**c**) Exponential curve fitting.

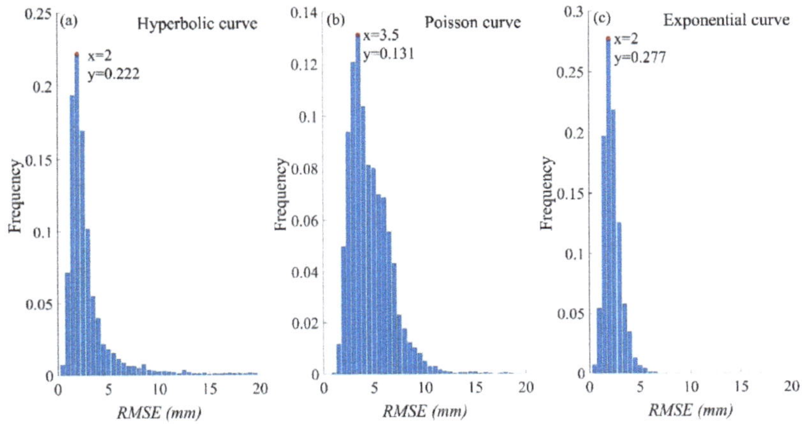

Figure 9. RMSE of (**a**) Hyperbolic curve fitting, (**b**) Poisson curve fitting and (**c**) Exponential curve fitting.

4.2. Method of Total Settlement Prediction

Although curve functions are applicable in settlement prediction, they can hardly determine when the foundation becomes stable. We used the Asaoka method [33] to identify when the settlement ends, which can obtain a reliable final settlement estimation using a large amount of measured data. Then, the total settlement was determined by combing the exponential curve model and the Asaoka method.

The Asaoka method is derived based on the Mikasa one-dimensional consolidation equation [33]. This method approximates the consolidation differential equation by the following equation:

$$S + \lambda_1 \frac{dS}{dt} + \lambda_2 \frac{d^2 s}{dt^2} + \ldots + \lambda_n \frac{d^n s}{dt^n} = \mu \tag{9}$$

where S is the total final settlement, and $\lambda_1, \lambda_2, \ldots \lambda_n, \mu$ are constants.

The first-order differential of Formula (9) provides high accuracy for practical engineering applications, so Formula (9) can be rewritten as

$$S + \lambda_1 \frac{dS}{dt} = \mu \tag{10}$$

The time is divided into j (j = 1, 2, 3, ...) equal parts. For time t_j, the settlement is S_{tj}, and we thus obtain

$$S_{t_{j+\Delta t}} = \gamma_0 + \gamma_1 S_{t_j} \tag{11}$$

where γ_0 is the settlement value, and γ_1 is a constant. Linear fitting is performed on a series of scattered points $(S_{t_j}, S_{t_{j+\Delta t}})$ to obtain the parameters γ_0 and γ_1, and the final settlement of the foundation is expressed as $S_{t\to\infty} = \gamma_0/(1-\gamma_1)$.

5. Discussion

5.1. Settlement Time Series Prediction

The consolidation rate of the foundation is affected by many factors, such as foundation treatment ways, the reclamation materials, and the thickness of the underlying alluvial layers, so it varies over regions [7]. It may take years or even more than ten years for some regions to reach stability under natural conditions. The prediction of the settlement time series helps locate the areas that need a long time to stabilize, so as to apply manual intervention, such as tamping and strengthening, to unstable areas and ensure that the construction is carried out on schedule. In addition, understanding the settlement behaviors can provide prior information for the planning of engineering projects.

Using the exponential curve model and the Asaoka method, we predicted the settlement time series in the XXIA. Figure 10 shows the predicted settlement curves and the settlement termination time for points P7–P12 (locations of points P7–P12 are shown in Figure A2). Their final settlement amount and the settlement termination time were different. As Figure 10 shows, under the current loads, the final settlement amount of P8 will reach about 11 cm in late 2027, and P9 will become stable within one year after December 2019. However, P7, P8, P10, and P11 need a long time for consolidation, which may be longer than ten years, exceeding the planned time of construction. Therefore, an artificial compaction process should be performed to speed up consolidation.

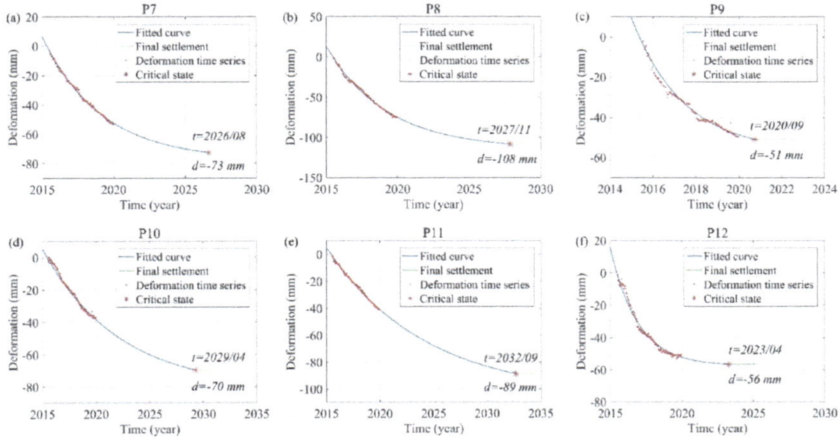

Figure 10. Predicted settlement curves of (**a**) P7, (**b**) P8, (**c**) P9, (**d**) P10, (**e**) P11 and (**f**) P12. Locations of points P7–P12 are shown in Figure A2. *t* is the consolidation time, and *d* is the final settlement.

5.2. Consolidation Time Prediction of the Whole Study Area

The areas reclaimed in the first and second phases have the same geological structure and foundation treatment methods, so the settlement patterns should be similar. Therefore, we extended the optimal prediction model and the final settlement calculation method determined in Section 4 to the whole study area to analyze consolidation. We show the time required for each point to reach stabilization intuitively in Figure 11.

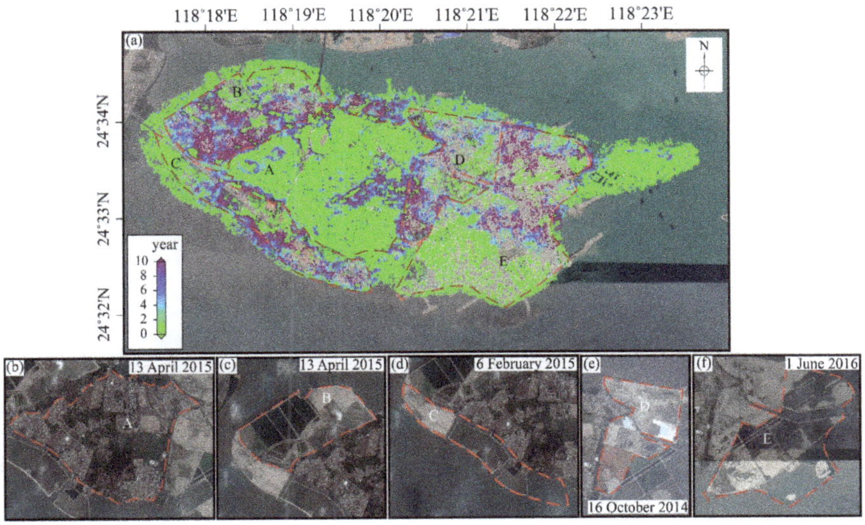

Figure 11. (**a**) The time needed for stabilization since the acquisition date (24 December 2019) of Dadeng Island and its reclamation areas: (**b**) optical image of area A acquired on 13 April 2015; (**c**) optical image of area B acquired on 13 April 2015; (**d**) optical image of area C acquired on 6 February 2015; (**e**) optical image of area D acquired on 16 October 2014; (**f**) optical image of area E acquired on 1 June 2016.

As Figure 11 shows, a large number of points that require more than ten years to stabilize were observed in the southern part of the area reclaimed in the first phase, the northeast part of the area reclaimed in the second phase, the northwest of Dadeng Island,

and the southwest of Dadeng Island. In the interior of Dadeng Island, almost all points reached stability. Combining with the deformation rate map of the study area (Figure 3), we found that the locations of the areas with large deformation rates were coincident with that of the areas needing long settlement duration. The points with long consolidation times were mostly distributed in the reclaimed area. The type and thickness of the reclamation materials, the completion time of reclamation, and the effect of foundation treatments all affect the settlement duration. In this study, Dadeng Island and its reclamation areas were divided into five parts—namely A, B, C, D, and E. In what follows, we discuss their remaining consolidation time separately and analyze the main reasons that affect consolidation in each area.

Area A is the original land of Dadeng Island. As can be seen from Figures 11b and 12, most of this area is covered by buildings, farmland, planting, and fishery farming areas, which had no changes during the InSAR monitoring period, except for a few scattered areas (Figure 3). The deformation results acquired by Liu et al. [2] and Zhuo et al. [34] also showed that there was no large-scale heavy subsidence in area A. By analyzing the optical images (Figure 12), we found that the subsidence in a few scattered areas was caused by city road construction (white dotted rectangle and Figure A3) and farmland. City road construction started between 29 April 2017 and 10 July 2017. It is reasonable that the construction activity may cause land subsidence. Therefore, the prediction results are in line with the actual condition, which further proves the reliability of the proposed method.

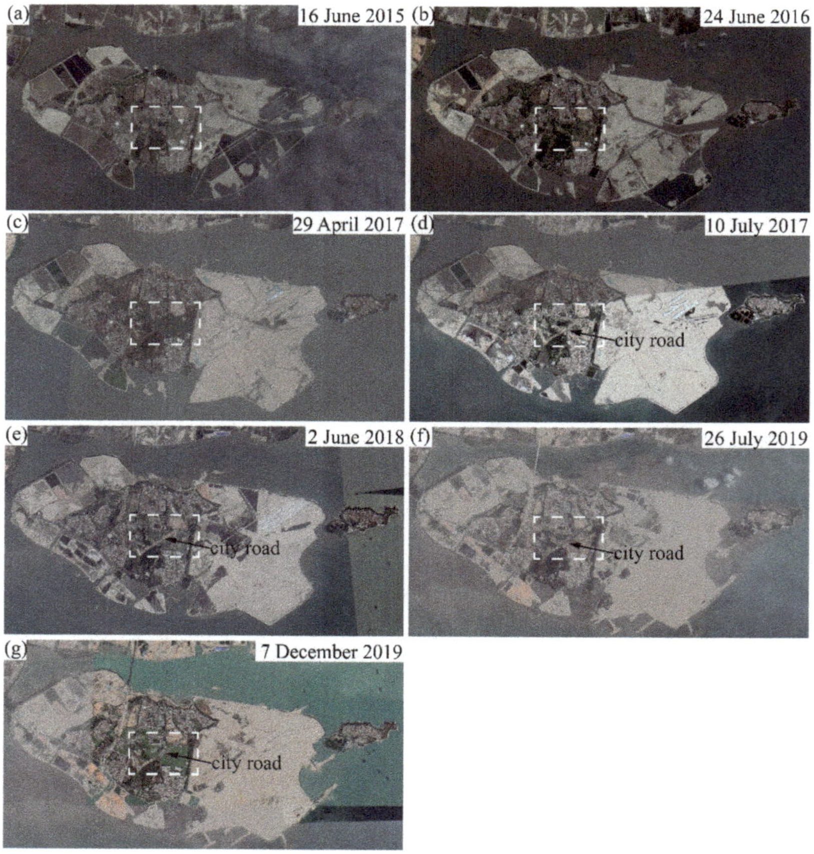

Figure 12. Optical images of the study area acquired on (**a**) 16 June 2015, (**b**) 24 June 2016, (**c**) 29 April 2017, (**d**) 10 July 2017, (**e**) 2 June 2018, (**f**) 26 July 2019, and (**g**) 7 December 2019.

Area B was reclaimed in the second phase. Most of this region, except for the northeast part, needs two to ten years to consolidate. The northeast part was reclaimed before 13 April 2015, when the reclamation of the rest part of area B had not yet started (Figure 12). Therefore, when settlement in the northeast part ends, the remaining part of area B may still be in the primary consolidation stage or secondary compression stage. Similarly, areas C and E have varying consolidation times over different parts.

Area D was reclaimed in the first phase, but its northern and southern parts had different deformation rates (Figure 3). The southern part needs 4 to 10 years for consolidation. The northern part needs a shorter time to consolidate, and some parts have stopped subsiding. The first phase of the reclamation project was from March 2014 to 16 June 2015. As shown in Figure 11e, during the foundation treatment, different types of landfill materials were used in the north and south. The different landfill materials led to the different remaining settlement times.

In foundation treatment and engineering construction, special attention should be paid to the areas that need long consolidation times. However, it is worth noting that the prediction method used in this study may not be appropriate to predict the settlement in some parts of area A. Similar to the abovementioned description, farmland and some construction activity exist in area A (dry land of Dadeng Island). The subsidence patterns of farmland and some construction activity may be different from that of reclaimed areas. Therefore, the interpretation of the predicted settlement should rely on actual deformation patterns.

6. Conclusions

In this study, we focused on settlement time series, final settlement amount, and consolidation time prediction over reclaimed areas using the dense deformation measurements obtained from InSAR. The proposed method consists of two steps: (1) optimal curve model selection; (2) final settlement prediction using the Asaoka method. The Xiamen Xiang'an International Airport, a planned reclaimed area, was chosen as the study site to validate our method. A total of 128 Sentinel-1 images were used to obtain the deformation history of the study area. We analyzed the InSAR-derived deformation results and discussed the predicted settlement. The following main conclusions were drawn from this research:

(1) Settlement mainly occurred in the reclaimed areas, with the maximum average settlement rate exceeding 40 mm/y between 6 July 2015 and 24 December 2019. Different parts in one reclaimed area had different settlement rates, due to the uneven construction progress;
(2) The exponential curve model showed the best performance in fitting the settlement time series obtained from MT-InSAR over the area reclaimed in the first phase. The Asaoka method was effective in the determination of deformation and stability;
(3) The settlement time series and the final settlement of the reclaimed land could be predicted by combining the exponential curve model and the Asaoka method. Predicted consolidation time indicated that some areas need more than ten years to stabilize (since 24 December 2019). Manual consolidation should be applied to those regions to ensure construction speed.

We provided an alternative method to predict the settlement over reclaimed areas that have no in situ measurements and subsurface information. In this study, no in situ measurements or geotechnical models were used to predict the settlement, which may lead to deviation from the actual final settlement. In the future, InSAR, in situ measurements, geotechnical models, and subsurface information can be integrated to conduct precise and wide-coverage settlement predictions.

Author Contributions: Conceptualization, Z.X., K.D., G.F. and L.M.; methodology, G.F., L.M., K.L. and Y.H; writing—original draft preparation, Z.X., K.D. and L.M.; writing—review and editing, Z.X., K.D., G.F., C.H. and Y.H.; funding acquisition, G.F. All authors have read and agreed to the published version of the manuscript.

Funding: This research was supported by the National Natural Science Foundation of China (No. 42174039) and the Natural Science Foundation of Hunan Province (No. 2021JJ30807).

Data Availability Statement: The Sentinel-1 data are archived at the Copernicus Open Access Hub (https://scihub.copernicus.eu (accessed on 3 March 2020.)) and Alaska Satellite Facility (https://search.asf.alaska.edu (accessed on 6 March 2020)).

Acknowledgments: The authors would like to thank three anonymous reviewers for their constructive comments and suggestions that significantly improved this study. The authors acknowledge the GMT open-source software. The authors would like to also thank the European Space Agency (ESA) for providing the Sentinel-1 SAR images.

Conflicts of Interest: The authors declare no conflict of interest.

Appendix A

Table A1. Acquisition dates of Sentinel-1 images.

Satellite	Pixel Spacing (Ran × Azi)	Acquisition Dates (Year/Month/Day)
Sentinel-1	2.3 m × 13.9 m	2015/07/06, 2015/07/18, 2015/08/11, 2015/08/23, 2015/09/04, 2015/09/16, 2015/09/28, 2015/10/10, 2015/10/22, 2015/11/03, 2015/11/15, 2015/11/27,2015/12/09, 2015/12/21, 2016/01/14, 2016/01/26, 2016/02/07, 2016/02/19, 2016/03/02, 2016/03/14, 2016/03/26, 2016/04/07, 2016/04/19, 2016/05/01, 2016/05/13, 2016/05/25, 2016/06/06, 2016/06/30, 2016/07/24, 2016/08/05, 2016/08/17, 2016/08/29,2016/09/10, 2016/09/22, 2016/10/04, 2016/10/16, 2016/10/28, 2016/11/09, 2016/11/21, 2016/12/03, 2016/12/15, 2016/12/27, 2017/01/08, 2017/01/20, 2017/02/01, 2017/02/13, 2017/02/25, 2017/03/09, 2017/03/21, 2017/04/02, 2017/04/14, 2017/04/26, 2017/05/08, 2017/05/20, 2017/06/01, 2017/06/13, 2017/06/25, 2017/07/19, 2017/07/31, 2017/08/12, 2017/08/24, 2017/09/05, 2017/09/17, 2017/10/11, 2017/10/23, 2017/11/04, 2017/11/16, 2017/11/28, 2017/12/10, 2017/12/22, 2018/01/03, 2018/01/15, 2018/01/27, 2018/02/08, 2018/02/20, 2018/03/04, 2018/03/28, 2018/04/09, 2018/04/21, 2018/05/03, 2018/05/15, 2018/05/27, 2018/06/08, 2018/06/20, 2018/07/02, 2018/07/14, 2018/07/26, 2018/08/07, 2018/08/19, 2018/08/31, 2018/09/12, 2018/09/24, 2018/10/06, 2018/10/18, 2018/11/11, 2018/11/23, 2018/12/05, 2018/12/17, 2018/12/29, 2019/01/10, 2019/01/22, 2019/02/03, 2019/02/27, 2019/03/11, 2019/03/23, 2019/04/04, 2019/04/16, 2019/04/28, 2019/05/10, 2019/05/22, 2019/06/03, 2019/06/15, 2019/06/27, 2019/07/09, 2019/07/21, 2019/08/02, 2019/08/14, 2019/08/26, 2019/09/07, 2019/09/19, 2019/10/01, 2019/10/13, 2019/10/25, 2019/11/06, 2019/11/18, 2019/11/30, 2019/12/12, 2019/12/24

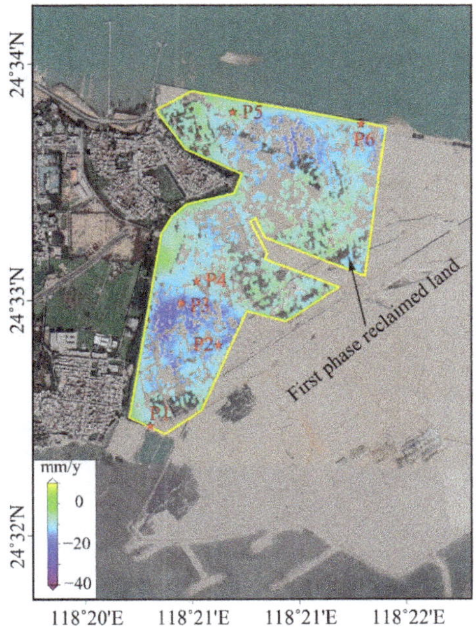

Figure A1. Locations of points P1–P6.

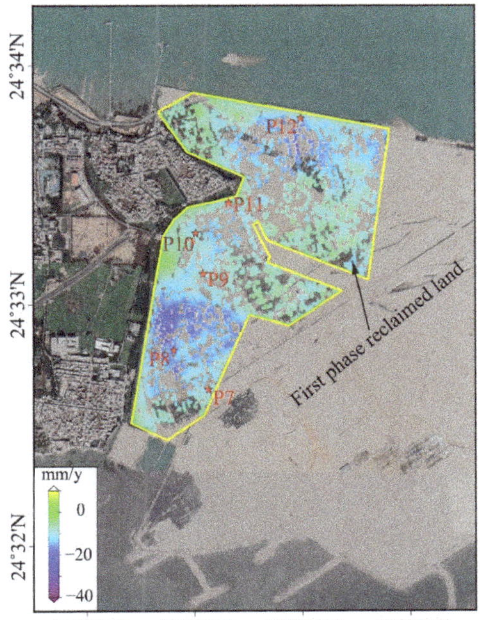

Figure A2. Locations of points P7–P12.

Figure A3. Optical images of the white dotted rectangle area in Figure 12. Optical image acquired on (**a**) 16 June 2015, (**b**) 24 June 2016, (**c**) 10 July 2017, (**d**) 2 June 2018, (**e**) 26 July 2019, and (**f**) 7 December 2019.

References

1. Yu, Q.; Yan, X.; Wang, Q.; Yang, T.; Lu, W.; Yao, M.; Dong, J.; Zhan, J.; Huang, X.; Niu, C.; et al. A Spatial-Scale Evaluation of Soil Consolidation Concerning Land Subsidence and Integrated Mechanism Analysis at Macro-, and Micro-Scale: A Case Study in Chongming East Shoal Reclamation Area, Shanghai, China. *Remote Sens.* **2021**, *13*, 2418. [CrossRef]
2. Liu, X.; Zhao, C.; Zhang, Q.; Yang, C.; Zhang, J. Characterizing and Monitoring Ground Settlement of Marine Reclamation Land of Xiamen New Airport, China with Sentinel-1 SAR Datasets. *Remote Sens.* **2019**, *11*, 585. [CrossRef]
3. Wang, W.; Liu, H.; Li, Y.; Su, J. Development and management of land reclamation in China. *Ocean Coastal Manag.* **2014**, *102*, 415–425. [CrossRef]
4. Park, S.; Hong, S. Nonlinear Modeling of Subsidence from a Decade of InSAR Time Series. *Geophys. Res. Lett.* **2021**, *48*, 2020GL090970. [CrossRef]
5. Pepe, A.; Bonano, M.; Zhao, Q.; Yang, T.; Wang, H. The Use of C-/X-Band Time-Gapped SAR Data and Geotechnical Models for the Study of Shanghai's Ocean-Reclaimed Lands through the SBAS-DInSAR Technique. *Remote Sens.* **2016**, *8*, 911. [CrossRef]
6. Nadeem, M.; Akbar, M.; Pan, H.; Li, X.; Ou, G.; Amin, A. Investigation of the settlement prediction in soft soil by Richards Model: Based on a linear least squares-iteration method. *Arch. Civ. Eng.* **2021**, *67*, 491–506. [CrossRef]
7. Jiang, L.; Lin, H. Integrated analysis of SAR interferometric and geological data for investigating long-term reclamation settlement of Chek Lap Kok Airport, Hong Kong. *Eng. Geol.* **2010**, *110*, 77–92. [CrossRef]
8. Mei, G.; Zhao, X.; Wang, Z. Settlement Prediction Under Linearly Loading Condition. *Mar. Geores. Geotechnol.* **2015**, *33*, 92–97. [CrossRef]
9. Kim, S.; Wdowinski, S.; Dixon, T.; Amelung, F.; Kim, J.; Won, J. Measurements and predictions of subsidence induced by soil consolidation using persistent scatterer InSAR and a hyperbolic model. *Geophys. Res. Lett.* **2010**, *37*. [CrossRef]
10. Al-Shamrani, M. Applying the hyperbolic method and C-alpha/C-c concept for settlement prediction of complex organic-rich soil formations. *Eng. Geol.* **2005**, *77*, 17–34. [CrossRef]
11. Hu, X.; Liang, X.; Yu, Y.; Guo, S.; Cui, Y.; Li, Y.; Qi, S. Remote Sensing Characterization of Mountain Excavation and City Construction in Loess Plateau. *Geophys. Res. Lett.* **2021**, *48*. [CrossRef]

12. Chaussard, E.; Bürgmann, R.; Shizaei, M.; Fielding, E.; Baker, B. Predictability of hydraulic head changes and characterization of aquifer-system and fault properties from InSAR-derived ground deformation. *J. Geophys. Res.-Solid Earth* **2014**, *119*, 6572–6590. [CrossRef]
13. Shi, G.; Ma, P.; Hu, X.; Huang, B.; Lin, H. Surface response and subsurface features during the restriction of groundwater exploitation in Suzhou (China) inferred from decadal SAR interferometry. *Remote Sens. Environ.* **2021**, *256*, 112327. [CrossRef]
14. Ren, X.; Tang, Y.; Li, J.; Yang, Q. A prediction method using grey model for cumulative plastic deformation under cyclic loads. *Nat. Hazards* **2012**, *64*, 441–457. [CrossRef]
15. Deng, Z.; Ke, Y.; Gong, H.; Li, X.; Li, Z. Land subsidence prediction in Beijing based on PS-InSAR technique and improved Grey-Markov model. *GISci. Remote Sens.* **2017**, *54*, 797–818. [CrossRef]
16. Chen, P.; Yu, H. Foundation Settlement Prediction Based on a Novel NGM Model. *Math. Probl. Eng.* **2014**, *2014*, 242809. [CrossRef]
17. Shi, G.; Lin, H.; Bürgmann, R.; Ma, P.; Wang, J.; Liu, Y. Early soil consolidation from magnetic extensometers and full resolution SAR interferometry over highly decorrelated reclaimed lands. *Remote Sens. Environ.* **2019**, *231*, 111231. [CrossRef]
18. Xiong, Z.; Feng, G.; Feng, Z.; Miao, L.; Wang, Y.; Yang, D. Pre- and post-failure spatial-temporal deformation pattern of the Baige landslide retrieved from multiple radar and optical satellite images. *Eng. Geol.* **2020**, *279*, 105880. [CrossRef]
19. Hu, X.; Bürgmann, R.; Schulz, W.; Fielding, E. Four-dimensional surface motions of the Slumgullion landslide and quantification of hydrometeorological forcing. *Nat. Commun.* **2020**, *11*, 2792. [CrossRef]
20. Xu, Q.; Guo, C.; Dong, X.; Li, W.; Lu, H.; Fu, H.; Liu, X. Mapping and Characterizing Displacements of Landslides with InSAR and Airborne LiDAR Technologies: A Case Study of Danba County, South west China. *Remote Sens.* **2021**, *13*, 4234. [CrossRef]
21. He, L.; Feng, G.; Wu, X.; Lu, H.; Xu, W.; Wang, Y.; Liu, J.; Hu, J.; Li, Z. Coseismic and Early Postseismic Slip Models of the 2021 M_W7.4 Maduo Earthquake (Western China) Estimated by Space-Based Geodetic Data. *Geophys. Res. Lett.* **2021**, *48*. [CrossRef]
22. Zhou, Y.; Thomas, M.; Parsons, B.; Walker, R. Time-dependent postseismic slip following the 1978 M_W 7.3 Tabas-e-Golshan, Iran earthquake revealed by over 20 years of ESA InSAR observations. *Earth Planet. Sci. Lett.* **2018**, *483*, 64–75. [CrossRef]
23. Huang, M.; Fielding, E.; Liang, G.; Milillo, P.; Bekaert, D.; Dreger, D.; Salzer, J. Coseismic deformation and triggered landslides of the 2016 M_W 6.2 Amatrice earthquake in Italy. *Geophys. Res. Lett.* **2017**, *44*, 1266–1274. [CrossRef]
24. Xu, W.; Xie, L.; Aoki, Y.; Rivalta, E.; Jónsson, S. Volcano-Wide Deformation After the 2017 Erta Ale Dike Intrusion, Ethiopia, Observed with Radar Interferometry. *J. Geophys. Res.-Solid Earth* **2020**, *125*. [CrossRef]
25. Hooper, A.; Zebker, H.; Segall, P.; Kampes, B. A new method for measuring deformation on volcanoes and other natural terrains using InSAR persistent scatterers. *Geophys. Res. Lett.* **2004**, *31*. [CrossRef]
26. Peng, M.; Lu, Z.; Zhao, C.; Motagh, M.; Bai, L.; Conway, B.; Chen, H. Mapping land subsidence and aquifer system properties of the Willcox Basin, Arizona, from InSAR observations and independent component analysis. *Remote Sens. Environ.* **2022**, *271*, 112894. [CrossRef]
27. Xu, B.; Feng, G.; Li, Z.; Wang, Q.; Wang, C.; Xie, R. Coastal Subsidence Monitoring Associated with Land Reclamation Using the Point Target Based SBAS-InSAR Method: A Case Study of Shenzhen, China. *Remote Sens.* **2016**, *8*, 652. [CrossRef]
28. Costantini, M.; Ferretti, A.; Minati, F.; Falco, S.; Trillo, F.; Colombo, D.; Novali, F.; Malvarosa, F.; Mammone, C.; Vecchioli, F.; et al. Analysis of surface deformation over the whole Italian territory by interferometric processing of ERS, Envisat and COSMO-SkyMed radar data. *Remote Sens. Environ.* **2017**, *202*, 250–275. [CrossRef]
29. Ma, P.; Wang, W.; Zhang, B.; Wang, J.; Shi, G.; Huang, G.; Chen, F.; Jiang, L.; Lin, H. Remotely sensing large- and small-scale ground subsidence: A case study of the Guangdong-Hong Kong-Macao Greater Bay Area of China. *Remote Sens. Environ.* **2019**, *232*, 111282. [CrossRef]
30. Wang, Y.; Feng, G.; Li, Z.; Xu, W.; Zhu, J.; He, L.; Xiong, Z.; Qiao, X. Retrieving the displacements of the Hutubi (China) underground gas storage during 2003–2020 from multi-track InSAR. *Remote Sens. Environ.* **2022**, *268*, 112768. [CrossRef]
31. Yang, C.; Zhang, D.; Zhao, C.; Han, B.; Sun, R.; Du, J.; Chen, L. Ground Deformation Revealed by Sentinel-1 MSBAS-InSAR Time-Series over Karamay Oilfield, China. *Remote Sens.* **2019**, *11*, 2027. [CrossRef]
32. Zhu, L.; Xing, X.; Zhu, Y.; Peng, W.; Yuan, Z.; Xia, Q. An Advanced Time-Series InSAR Approach Based on Poisson Curve for Soft Clay Highway Deformation Monioring. *IEEE J.-STARS* **2021**, *14*, 7682–7698. [CrossRef]
33. Asaoka, A. Observational Procedure of Settlement Prediction. *Soils Found.* **1978**, *18*, 87–101. [CrossRef]
34. Zhuo, G.; Dai, K.; Huang, H.; Li, S.; Shi, X.; Feng, Y.; Li, T.; Dong, X.; Deng, J. Evaluating Potential Ground Subsidence Geo-Hazard of Xiamen Xiang'an New Airport on Reclaimed Land by SAR Interferometry. *Sustainability* **2020**, *12*, 6991. [CrossRef]
35. Ferretti, A.; Prati, C.; Rocca, F. Permanent scatterers in SAR interferometry. *IEEE Trans. Geosci. Remote Sens.* **2001**, *39*, 8–20. [CrossRef]
36. Berardino, P.; Fornaro, G.; Lanari, R.; Sansosti, E. A new algorithm for surface deformation monitoring based on small baseline differential SAR interferograms. *IEEE Trans. Geosci. Remote Sens.* **2002**, *40*, 2375–2383. [CrossRef]

remote sensing

MDPI

Article

Landslide Risk Assessment Using a Combined Approach Based on InSAR and Random Forest

Wangcai Liu [1], Yi Zhang [1,*], Yiwen Liang [1], Pingping Sun [2], Yuanxi Li [1], Xiaojun Su [3], Aijie Wang [1] and Xingmin Meng [1]

1 Technology & Innovation Centre for Environmental Geology and Geohazards Prevention, School of Earth Sciences, Lanzhou University, Lanzhou 730000, China; liuwc20@lzu.edu.cn (W.L.); liangyw19@lzu.edu.cn (Y.L.); liyq2021@lzu.edu.cn (Y.L.); wangaj21@lzu.edu.cn (A.W.); xmmeng@lzu.edu.cn (X.M.)

2 Key Laboratory for Geo-Hazards in Loess Area, Ministry of Natural Resources, Xi'an Center of Geological Survey, China Geological Survey, Xi'an 710054, China; sunpingping203@gmail.com

3 College of Earth and Environmental Sciences, Lanzhou University, Lanzhou 730000, China; suxj19@lzu.edu.cn

* Correspondence: zhangyigeo@lzu.edu.cn

Abstract: Landslide risk assessment is important for risk management and loss–damage reduction. Herein, we assessed landslide susceptibility, hazard, and risk in the urban area of Yan'an City, which is located on the Loess Plateau of China and affected by many loess landslides. Based on 1841 slope units mapped in the study area, a random forest machine learning classifier and eight environmental factors influencing landslides were used for a landslide susceptibility assessment. In addition, differential synthetic aperture radar interferometry (DInSAR) technology was used for a hazard assessment. The accuracy of the random forest is 0.903 and the area under the receiver operating characteristics (ROC) curve is 0.96. The results show that 16% and 22% of the slope units were classified as being at very high and high-susceptibility levels for landslides, respectively, whereas 16% and 24% of the slope units were at very high and high-hazard levels for landslides, respectively. The landslide risk was obtained based on the susceptibility map and hazard map of landslides. The results show that only 26% of the slope units were located at very high and high-risk levels for landslides and these are mainly concentrated in urban centers. Such risk zones should be taken seriously and their dynamics must be monitored. Our landslide risk map is expected to provide information for planners to help them choose appropriate locations for development schemes and improve integrated geohazard mitigation in Yan'an City.

Keywords: landslides; risk assessment; random forest; DInSAR; Yan'an city

Citation: Liu, W.; Zhang, Y.; Liang, Y.; Sun, P.; Li, Y.; Su, X.; Wang, A.; Meng, X. Landslide Risk Assessment Using a Combined Approach Based on InSAR and Random Forest. *Remote Sens.* **2022**, *14*, 2131. https://doi.org/10.3390/rs14092131

Academic Editors: Paolo Mazzanti and Saverio Romeo

Received: 16 February 2022
Accepted: 26 April 2022
Published: 29 April 2022

1. Introduction

Landslides are common natural phenomena on mountains and slopes that can change the geomorphology of the landscape. Thus, the massive destruction caused by landslides is of great concern [1,2]. With global climate change and increasingly intense human engineering, landslides tend to occur more frequently, resulting in huge economic losses and many casualties [3,4]. Therefore, risk assessment is often the focus of research [5–8], especially in populated areas that are prone to landslides. This should help provide the necessary information to governments and decision makers [6,9]. Risk assessment is the basis for risk management. It refers to the possibility and severity of landslides impacting life, health, property, and the environment. In practice, the risk of landslides is computed as the product of landslide hazard and the vulnerability to potential value loss [5]. Quantitative and accurate risk assessment can be effective information for government departments in land and resources planning, engineering construction, the prevention and early warning of landslides, and sustainable development.

It is crucial to select an optimal model and methodology for landslide risk assessment because different assessments may have different results and accuracies for the same areas. In recent decades, numerous landslide susceptibility, hazard, and risk assessment methods have been applied. However, there has been no study showing that a certain model has the optimal solution for all risk assessments [10,11]. Models for landslide susceptibility assessment can be classified as physically deterministic, heuristic based on experts' knowledge, and data-driven quantitative [12]. Physically deterministic models are commonly based on hydrological characterizations combined with infinite-slope stability analyses to estimate the relative stability of slopes [13]. Some research has pointed out that these models are used only for particular hydrological conditions and high model preconditions [14], such as detailed and homogeneous soil mechanical parameters, hydro environmental factors, and simple landslide types. For this reason, they could be effective only for mapping small areas in detail [15,16]. Heuristic models based on experts' knowledge, including the analytical hierarchy process (AHP) [17], expert knowledge systems [18], and gray relational modes [18], mainly rely on constructing a relatively simple ranking method determined by experts' knowledge [16]. Although heuristic models have the advantages of easy application, the assessment results have low accuracy with a certain level of subjectivity [19]. Previous studies show that data-driven quantitative models are preferred and applied more frequently than qualitative evaluation models, such as heuristic or geomorphological mapping [20,21]. Logistic regression (LR) [22,23], frequency ratio (FR) [24,25] and weights of evidence [26] are the most frequently used statistical models. They are based on considered classical statistics; index-based, machine learning; neural networks; and multi-criteria decision analysis. In particular, the use of machine learning for landslide assessment is rapidly increasing [20]. It is a modeling methodology that builds complex relationships between data and target variables through iterative training and learning without assuming additional structural constraints [27,28]. Machine learning is often used to solve nonlinear geological environment problems, such as landslide susceptibility assessment and prediction. For example, Chen et al. [29] introduced a new bivariate statistical-based kernel logistic regression to obtain landslide susceptibility maps by optimizing different kernel functions and two-component statistical correlation analyses. Behnia et al. [30] produced susceptibility maps for debris flows and other geohazards along the Yukon Alaska Highway Corridor, in Canada. Hong et al. [31] built a higher-precision susceptibility map of the Guangchang area in China based on a decision tree model. Furthermore, many studies have compared the accuracy of machine learning with classical statistical models in landslide susceptibility assessment [32–34]. They showed that machine learning models provide more accurate assessments and predictions [35].

Apart from models and methods, selecting appropriate mapping units associated with the research purpose is a key issue for reasonable and accurate assessment maps. Generally, the mapping units fall into several groups: grid cells, terrain units, unique conditional units, topographic units, slope units, complementary geohydrological units and political or administrative units [20,36]. Each type of unit has certain analytical advantages and disadvantages. For this reason, the type of unit needs to be determined at the beginning of a study according to the purpose and scale of the research [36,37]. Landslides tend to show a clear shape and boundary soon after their occurrence so the slope unit is often preferred for representing the form of landslides or unstable slopes. In some studies, the slope unit also performed better than the pixel unit in landslide assessment [38–40].

Yan'an, which is located in the north of Shaanxi Province, on the Loess Plateau, is a typical valley city. Its particular geography and geological environment background, as well as increasing human engineering activities, appear to be the causes of more frequent landslides, collapses, and other geohazards [41]. Several studies have evaluated the susceptibility and stability of landslides in Yan'an City and Baota District based on qualitative methods and physical models [42–44]. However, the evaluation factors in those studies are limited to geological or topographic conditions, and few studies have focused on the deformations which can reflect the activity of slopes through SAR data in the risk assessment of the study area. Interferometric synthetic aperture radar (InSAR) technology can be used to optimize the landslide susceptibility assessment and reduce landslide classification errors [45]. Additionally, a smaller range and larger scale of quantitative assessments are necessary for future urban development in Yan'an City if we want to mitigate the geohazards occurring in current urban constructions. Therefore, this study aims at constructing a detailed landslide risk assessment in Yan'an City using high-resolution aerial images and a digital elevation model (DEM). A detailed investigation and understanding of the characteristics of the geological hazards in the urban area of Yan'an City is considered a critical part of risk assessment. In the process, it becomes necessary to combine ground deformation using InSAR technology with conventional topographic and geomorphic factors for risk analysis. Advanced random forest machine learning classifiers and InSAR technology are used in our study to assess landslide susceptibility, landslide hazards, and the identification of areas exposed to a higher landslide risk in the urban parts of Yan'an City. It is expected that the assessments of urban hazards and risks in urban areas based on the slope units can provide more accurate information for government departments and decision makers in urban planning, construction, and disaster prevention as well as control.

2. Study Area

The present study area is the central urban area of Yan'an City, which is located in the northern part of Shaanxi Province, China, on the Loess Plateau between the latitudes of 36°27′N and 36°41′N and the longitudes of 109°22′E and 109°33′E, covering an area of 185 km^2 (Figure 1). Its landform features typical and complex loess beams, mounds, and gullies. The highest elevation in the study area is 1300 m and the lowest elevation is 927 m, which is in the river valley, so the elevation difference is about 370 m. The climate in the area is semi-humid and semi-arid, with a continental monsoon climate. In the past, the average annual precipitation in Baota District was 537 mm, which occurred mainly from June to September [46].

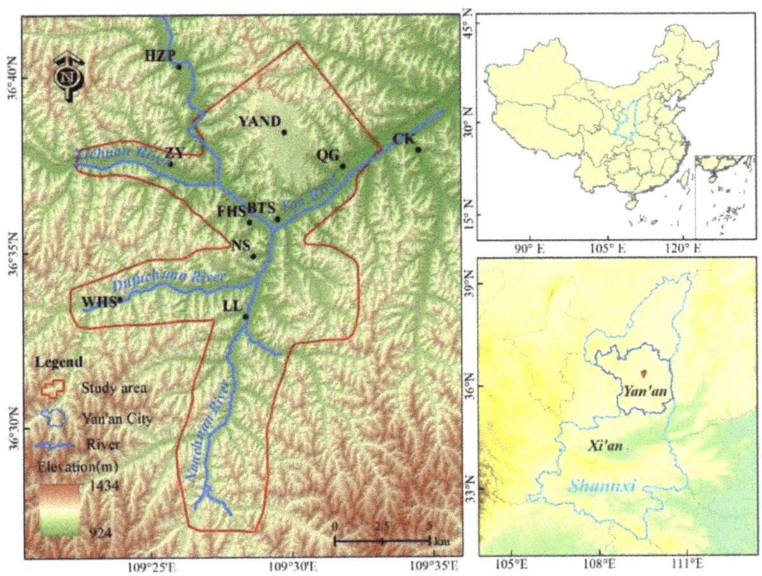

Figure 1. The location, boundary, and geomorphology of the present study area in Yan'an City and Shaanxi Province. YAND—Yan'an New District; HZP—Hezhuangping Town; QG—Qiaogou Street; CK—Chuankou Town; BTS—Baotashan Street; FHS—Fenghuangshan Street; ZY—Zaoyuan Town; NS—Nanshi Street; WHS—Wanhuashan Town; LL—Liulin Town.

From the perspective of regional geology and geotectonics, the study area is located in the middle-eastern part of the Ordos Block in the North China Block. The tectonic movement is slight without strong structural deformation and maintains the characteristics of a stable sedimentary basin. The strata are mainly Mesozoic and Cenozoic, including Triassic, Jurassic, and Quaternary; however, the Quaternary loess is the most widely distributed [47]. Triassic and Jurassic strata are mostly seen along both sides of the valley. Although there is no strong tectonic movement and fault, many landslides have occurred and developed in the area due to the unique physical and mechanical properties of loess. Loess is characterized by high porosity, low bulk density, weak cementation, water sensitivity, collapsibility, structural joints, vertical joints, unloading cracks, and a soft layer structural plane. Under the area's special landform conditions, landslide hazards could be induced by summer rainstorms and human engineering activities, which seriously affect the sustainable development of the local economy and society.

3. Data and Methods

The methodology applied for landslide susceptibility, hazard, and risk assessment is shown as a flowchart in Figure 2. A detailed explanation is provided in the following subsections.

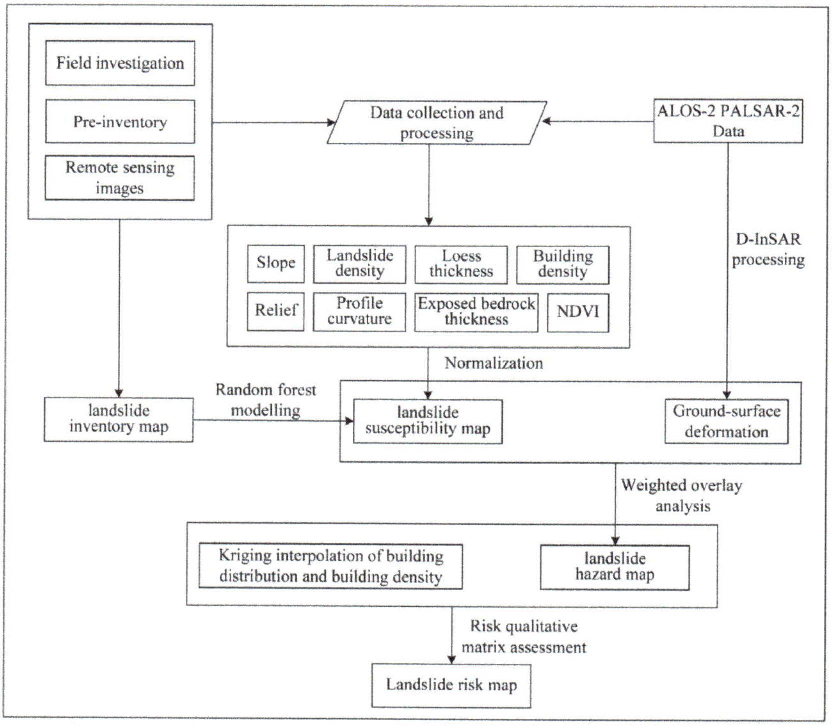

Figure 2. The flowchart of landslide risk assessment in the present study.

3.1. *Landslide Inventory*

A landslide inventory map is the first step to assess susceptibility. It shows information on all historical and active landslides. Combined with field surveys, relevant literature records and news reports of landslide records were used in this study to verify the spatial distribution of landslides using Google Earth high-resolution images and an unmanned aerial vehicle's digital orthophoto map (DOM) (0.1 m) provided by the Xi'an Center of China Geological Survey for visual interpretation.

Generally, slope units are defined by topographic characteristic lines (such as ridgelines and gully lines) and waterway paths, which are closely related to the DEM of the mountain area [36]. Therefore, these topographic characteristic lines and waterway paths are also the basic means to determine slope units in this study. Additionally, to better reflect the terrain of landslides or unstable slopes, we divided slope units according to topographic and geomorphic characteristics in the detailed field surveys. The basic requirement in a field survey is that every gully and slope must be investigated and the results are presented in the form of slope units. In this study, a total of 1841 slope units covering the whole study area were surveyed and the location and boundary of discernible slope units were obtained by the geographic information system (GIS) (Figure 3a). According to the current morphological characteristics and active state of the slopes in the field investigation, the slope units were divided into loess landslides, unstable slopes, and slopes to be evaluated. Loess landslides are the main historical landslides in the study area; unstable slopes show some deformation signs, such as creep slip, collapse, toppling, etc., and are developing toward becoming potential landslides. Finally, the landslide inventory map of the study area was aggregated and shown in Figure 3b, including 344 loess landslides and 411 unstable slopes, detailed in Section 4.1.

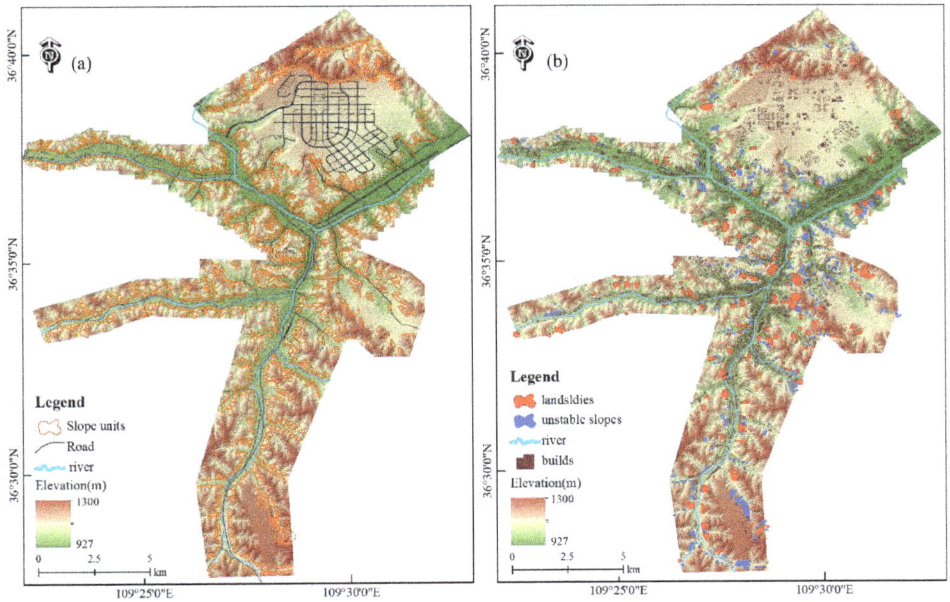

Figure 3. The inventory map of the present study. (**a**) Slope unit mapping and road elements based on topography and field surveys. (**b**) The distribution of landslides and unstable slopes in the study area.

3.2. Factors Influencing Landslides

Selecting appropriate environmental factors and inducing factors is the basis of risk assessment, which depends on data availability, scale, and study area, and affects future predictions [21,48]. Based on the field survey and previous work on landslides in the study area, we considered eight factors: slope, profile curvature, relief, the normalized difference vegetation index (NDVI), landslide density, building density, the thickness of loess, and the thickness of exposed bedrock. They are described below.

- Slope and profile curvature: A slope gradient is the measurement of the steepness of a surface. If the slope is too low, the gravitational potential energy is insufficient, and if the slope is too high, the material accumulation cannot provide the material basis for landslides. A profile curvature is used to describe the complexity of the terrain, which is divided into convex, straight, and concave profiles, and reflects convergent and divergent drainages in addition to variations in erosion rate [49]. In the study, the slope and profile curvature were calculated using ArcGIS and a DEM with a spatial resolution of 2 m (Figure 4a,b).

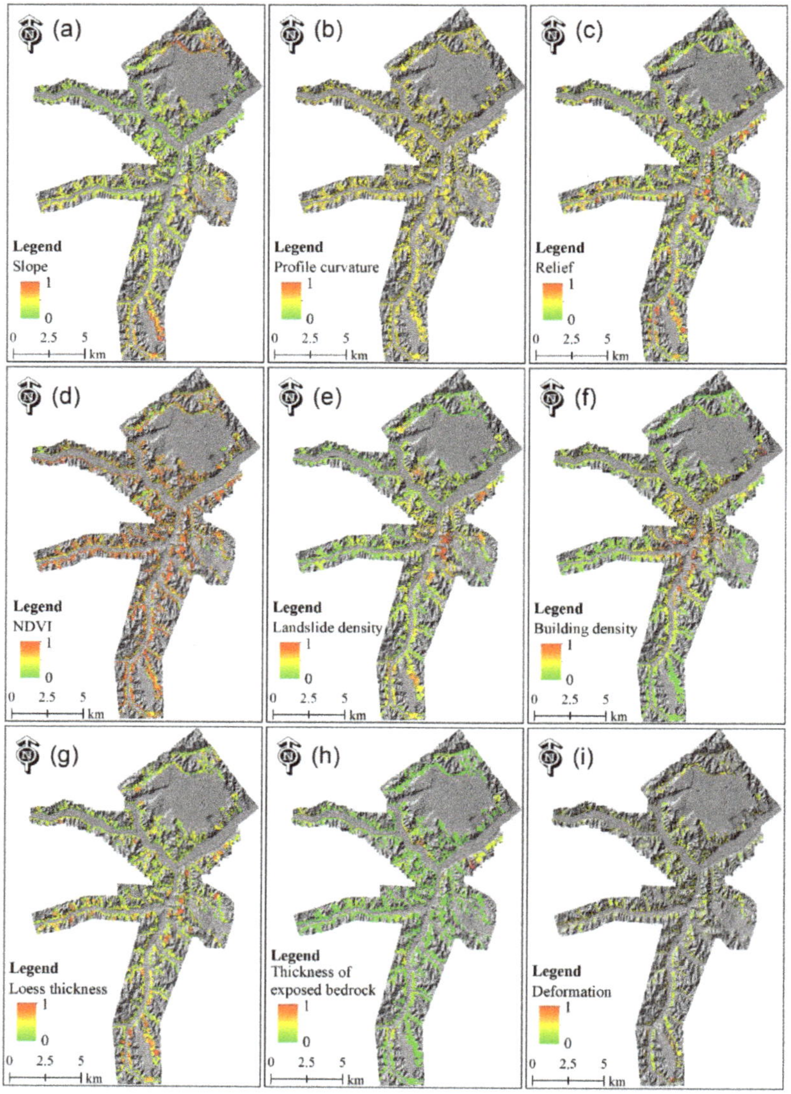

Figure 4. The conditioning factors used in this study: (**a**) slope; (**b**) profile curvature; (**c**) relief; (**d**) NDVI; (**e**) landslide density; (**f**) building density; (**g**) loess thickness; (**h**) thickness of exposed bedrock; and (**i**) deformation. The above parameters have been normalized.

- Relief: A relief represents the elevation difference within a certain range of the slope and determines the gravitational potential energy. Only enough gravitational potential energy can cause landslides (Figure 4c).
- NDVI: The NDVI reflects the vegetation cover in the study area. High vegetation coverage is needed to stabilize the slope by the root system and reduce the devel-

opment of landslides [50]. The NDVI value was calculated using the expression NDVI = (NIR − R)/(NIR + R) from Landsat-8 images, where NIR is the reflectivity of the near-infrared portion of the electromagnetic spectrum and R is the reflectivity of the red portion of the electromagnetic spectrum (Figure 4d).

- Landslide density and building density: The landslide density directly reflects the development quantity of disasters in an area. In urban areas, construction activity is one of the most dominant human activities that cause slope instability. Settlement along slopes in urban areas is an important factor in slope failure. Therefore, we used building density to reflect the effect of human activities on slope stability. Landslide density and building density were calculated for the slope units by vectorizing landslides and building contours and then interpolating them into grid data (Figures 3 and 4e,f).

The thicknesses of the loess and the exposed bedrock were measured during field investigations (Figure 4g,h). The loess thickness on a hillslope, which coincides with the failure depth, is a critical parameter in performing the slope stability analysis. The overlying loess thickness plays an important role in hydrological effects, such as the ratio of the saturated depth to the losses [51]. The thickness of the exposed bedrock of the slope has a great impact on the landslide scale, landslide type, and slope deformation [52]. Due to the undeveloped tectonic activity in the study area, the effect of earthquakes and faults has not been considered in this study. In addition, because precipitation within the relatively small study area is mostly unvarying, precipitation data were excluded from the analysis processes.

A statistical description of the influencing factors is shown in Table 1. To eliminate the dimensional influence of factors, the minimax normalization method was applied [53]. The continuous factor values of each factor were normalized, so all the values fall in the (0, 1) interval, where the normalized data were calculated following the equation below:

$$X_i{}' = \frac{x_i - x_{min}}{x_{max} - x_{min}} \tag{1}$$

where $X_i{}'$ is the normalized input and x_i, x_{min}, and x_{max} are the actual, minimum, and maximum input data, respectively. The results of normalized factors are shown in Figure 4.

Table 1. Statistical description of the influencing factors.

Factors	Min	Max	Standard Deviation
Slope (°)	0	86.8	15.5
Profile curvature	−497.4	499.0	33.6
Relief (m)	8	166.6	29.4
NDVI	−0.23	0.98	0.18
Landslide density	0	0.27	0.06
Build density	0	0.22	0.04
Loess thickness (m)	8.0	160.2	29.3
Bedrock thickness (m)	0	24.5	3.4
Deformation (m)	0.11	−0.09	0.014

3.3. DInSAR

In general, the displacement of a pixel is calculated using the interference phase difference between two SAR images by using the pixel product of a reference image and slave image—this is the basic principle of InSAR [54]. DInSAR is applied to the removal of the topographic phase contribution from the interferogram deformation phase using a two-pass, three-pass, or four-pass technique; however, it is worth noting that the two-pass technique, which imports an external DEM, yields a more reliable and operational outcome [54,55]. Furthermore, several limitations of InSAR technology must be considered at the beginning of use. One limitation is geometric distortion caused by topography, especially in mountainous areas with high elevations, which is affected by the look side of radar observation modes [56]. Another limitation is the poor coherence, even incoherence of interferograms caused by diffuse vegetation, which is very obvious in the C-band Sentinel-1 images of the study area [57].

In this research, two ascending SAR images acquired from ALOS-2 on 5 November 2018 and 20 May 2019 were selected for interference calculations (Figure 5a and Table 2). Due to the relatively flat terrain of the study area, the SAR images from a single orbit can be used to detect and monitor the deformation of most of the slopes. The sensor of ALOS-2 can transmit and receive the L-band with strong penetrating ability and can capture the ground deformation under the dense vegetation. The external DEM for removing the topographic phase and geocoding is the 1-arc-second (~30 m) Shuttle Radar Topography Mission (SRTM) data from NASA. Ground deformation along the LOS (light-of-sight) of the Yan'an City area was obtained after registration and resampling, differential interference, coherence calculation, filtering and phase unwrapping, orbit refining, and reflattening in addition to geocoding. All of these were processed with the DInSAR tool of the SARScape software (Figure 5b). The normalized deformation factor image is shown in Figure 4i.

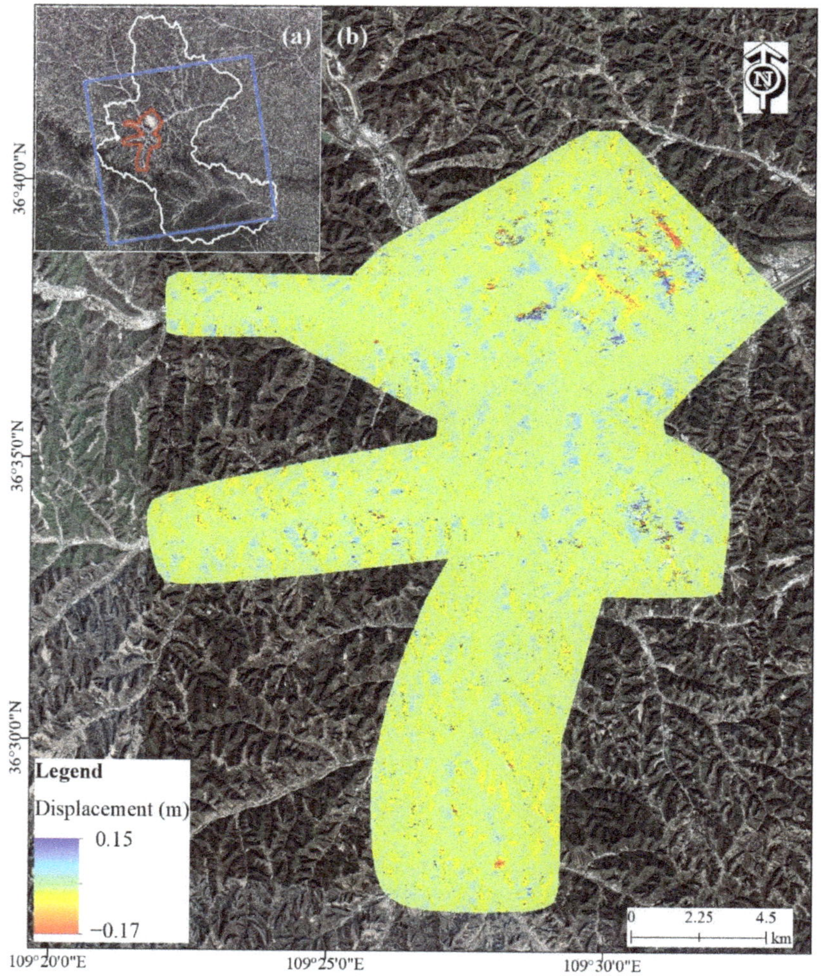

Figure 5. Area covered by two ascending SAR images from ALOS-2 (blue rectangle) in (**a**), and image of the LOS displacement of the study area (red polygon) using DInSAR technology in (**b**).

Table 2. Details of two SAR images and displacement from ALOS-2.

Sensor	PALSAR-2
Wavelength	23 cm
Band	L
Acquired time	5 November 2018 20 May 2019
Orbit direction	Ascending
Angle of incidence	32.5°
Polarization	HH
Observation mode	Fine
Resolution	10 m
Normal baseline	140.798 m
Absolute time baseline	196 days
Max displacement	0.150 m
Min displacement	−0.172 m
Standard deviation	0.018 m

3.4. Random Forest

Random forest has been widely used in data classification and management and has excellent performance in landslide susceptibility mapping. Random forest is an ensemble machine learning algorithm based on a decision tree. The classifier is a recursive process from root nodes to child nodes, which is similar to the combination of a decision tree and the flowchart of a tree structure [58]. The bootstrap method is used to extract multiple samples from the original samples. Starting from the nodes of a tree, the optimal features among different internal nodes are selected, and the corresponding branches are determined based on the test output. Finally, the results are obtained from the leaf nodes of the decision tree.

Random forest has strong generalization ability and can deal with multi-dimensional and large learning sets. Compared with other statistical learning models, random forest does not easily generate overfitting. It improves prediction accuracy without significantly increasing the amount of calculation. It has a higher tolerance for outliers and noise, resulting in data loss and imbalance. In this study, a random forest module was built based on the R language. Before running the random forest module to perform the landslide susceptibility assessment, the training and validation datasets must be selected. In the study, landslide inventories including stable slopes, unstable slopes, stable landslides, and unstable landslides were selected as training sites. Using ArcGIS, these slope units were converted into 109,981 vector points, where 3000 points were selected randomly at each of the landslide sites and non-landslide sites to train and test the classifier. Finally, its performance was evaluated with the ROC curve and confusion matrix.

4. Results

4.1. Characteristics of Landslides

Based on the field survey, the geohazards of Yan'an City were counted as 334 landslides, 411 unstable slopes, and land subsidence locations. Their depths of the sliding surface are mainly shallow (less than 30 m) [47]. The landslides can be classified as loess landslides and loess-bedrock interface landslides because most landslides occur in the loess layer or on the top of the bedrock (Figure 6a,b). The geometric morphology and characteristics of the loess landslides and the unstable slopes including the types, lengths, widths, height, slope angles, and others, such as the longitudinal shape and depth of slide surfaces, were mapped using the GIS and field investigations with high-resolution DEM (~2 m) (Table 3). The length and width of landslides are mainly in the range of 50 to 200 m. The height and slope angle are also condition factors of loess landslides. A higher or steeper slope has a higher degree of stress concentration and tensile stress range so it is more prone to failure and sliding. The study area is located in the loess hilly gully region with dense gullies, and the relative height differences of 60 to 150 m leads to the height of landslides usually being less than 120 m. Since the late Cenozoic era, the Loess Plateau has been in a state of intermittent uplift, with rivers cutting sharply and ravines crisscrossing, creating topographic conditions for loess landslides. The longitudinal shape can control the values and positions of the stress inside the slope body and plays a key role in the stability. For example, flat and convex slopes tend to be more easily destroyed under stress, suggesting an unstable evolutionary trend, whereas sunken and stepped slopes tend to be more stable with less stress concentration. Therefore, the longitudinal shapes of landslides in this study are mainly flat and convex slopes.

Figure 6. Examples of geohazard types mapped in the study area: (**a**) loess–bedrock interface landslide; (**b**) loess landslide; (**c**) soil–bedrock unstable slope; (**d**) soil unstable slope; and (**e**,**f**) cracks and damages in the ground and buildings due to land subsidence. Arrows indicate the direction of the slide and the location of the cracks.

Table 3. Types and characteristics of landslides and unstable slopes. L and U represent landslides and unstable slopes, respectively.

Type			Length (m)		Width (m)		Height (m)		Slope (°)		Area (10^3m^2)	
			<50	26	<50	25	≤30	7	≤20	10	≤5	73
	Loess	243	50–100	108	50–100	107	30–60	113	20–30	95	5–10	73
L			100–150	104	100–150	99	60–90	118	30–40	174	10–15	59
			150–200	50	150–200	41	90–120	69	40–50	55	15–20	46
	Loess-bedrock	91	200–250	22	200–250	31	120–150	24	>50	0	20–25	17
			>250	24	>250	31	>150	3			25	66
	Loess	285	≤50	76	≤50	35	≤30	19	≤20	12	≤5	145
			50–100	218	50–100	160	30–60	184	20–30	59	5–10	143
U	Bedrock	5	100–150	82	100–150	138	60–90	161	30–40	231	10–15	61
			150–200	19	150–200	39	90–120	40	40–50	108	15–20	28
	Loess-bedrock	121	200–250	10	200–250	18	120–150	6	>50	1	20–25	12
			>250	6	>250	21	>150	1			>25	22

Unstable slopes refer to a slope with creep slip, collapse, toppling, lateral tensile fracture, and other deformation characteristics or trends, and that is regarded as a potential geohazard. The 411 unstable slopes from the field survey were divided into three types according to their material composition: (i) soil unstable slope, (ii) rock unstable slope, and (iii) soil–bedrock unstable slope. There were about 285 soil unstable slopes in the study area, accounting for 69% of the total number of unstable slopes. Soil–bedrock unstable slopes and rock unstable slopes are fewer, numbering 125 and five, respectively, accounting for 31% of the total number of unstable slopes and mainly occurring in the Quaternary loess and the Jurassic sandstone strata. The unstable slopes have similar characteristics to landslides in their ranges of length, width, height, and area but the slope angles of unstable slopes are relatively larger. The characteristics of landslides and unstable slopes in the present study are summarized in Table 3.

In addition, under the pressure of population growth and development as well as the preservation of historical and cultural sites, in 2012 the government built a new district called Yan'an New District by cutting mountains and filling ditches. However, because of the special microstructure and complex engineering-geological conditions of the loess, land subsidence in Yan'an New District has become one of the geohazards that requires much attention. The surface deformation along the radar LOS calculated by the DInSAR technique was very similar to that of the small baseline subsets InSAR (SBAS-InSAR) approach from Sentinel-1 images (see [59,60] for more details). DInSAR and field surveys show that the land subsidence area is the ribbon (Figure 7), mainly concentrated in the filling area manifested as wall cracking or collapse, ground subsidence, and cracks (Figure 6e,f).

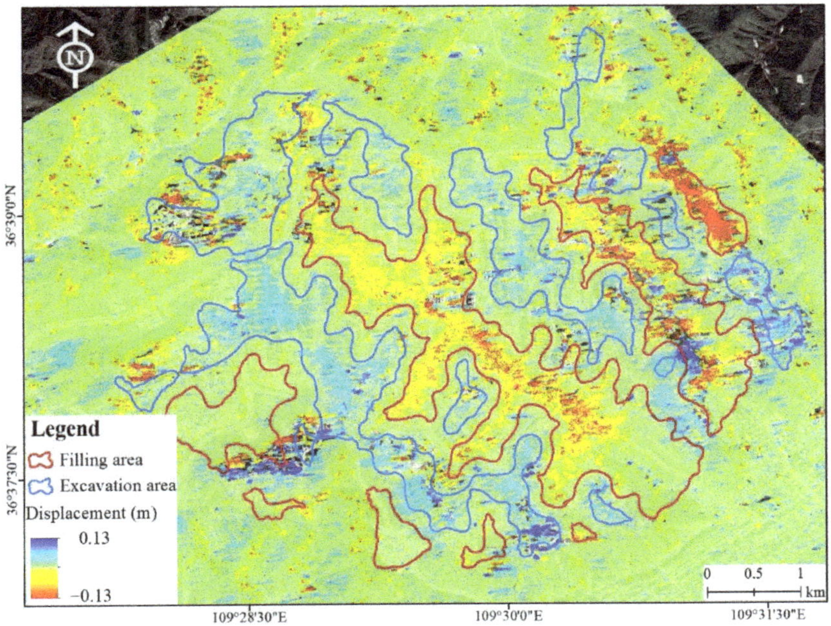

Figure 7. Surface deformation image of Yan'an New District from 2018 to 2019 calculated using DInSAR. Positive values indicate that the ground object deformation is close to the radar along the radar LOS, and negative values indicate that the ground object deformation is far away from the radar along the radar LOS.

4.2. *Landslide Susceptibility Mapping*

The normalized factors were used as the input, and the landslide susceptibility index was the output data. The mean decrease accuracy and mean decrease Gini coefficient of the random forest are used to order the eight variables (Figure 8). The vertical axis represents the eight variables, with the mean decrease accuracy and mean decrease Gini coefficient decreasing from top to bottom. It shows that the importance of the hazard density is the highest and that the build density, thickness of exposed bedrock, loess thickness, relief, and the NDVI are the next most-important. The ROC curve is widely used to evaluate the classification results of the random forest classifier through the area under the ROC curve (AUC) [35]. The vertical axis and horizontal axis represent the true positive rate (TPR) and false-positive rate (FPR) using the random forest classifier, respectively. TPR and FPR, also called the sensitivity and specificity, are the ratio of the landslide sample points correctly detected by the classifier and the ratio of the non-landslide sample points incorrectly classified as landslide sample points, respectively [61]. The larger the AUC, that is, the closer the vertex of the curve is to the upper left corner, the better the classifier's test capability. In this research, the AUC is 0.96, which indicates excellent classification results of the random forest classifier (Figure 9). In addition, the confusion matrix shows that the overall accuracy of the random forest classifier is 0.903 and that the predicted precision of non-landslides and landslides is 0.927 and 0.881, respectively, which is a good method to analyze the prediction accuracy (Table 4). Four levels of susceptibility, i.e., very high (>0.711), high (0.711–0.458), moderate (0.458–0.231), and low (<0.231), were categorized based on the natural breaks classification conducted using the ArcGIS software (Figure 10). The natural breaks classification was determined based on natural groupings inherent in the data. Then, the classification interval was identified to provide an optimum grouping of similar values and maximize the differences between classes [62,63]. Additionally, the distribution of the landslide susceptibility index using the natural breaks is shown in Figure 10b.

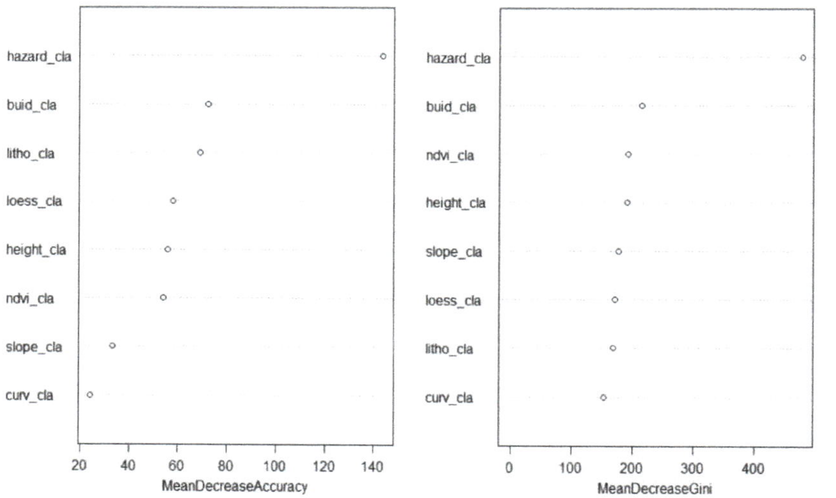

Figure 8. Mean decrease accuracy and mean decrease Gini of variables assigned by the random forest classifier. The vertical axis is the inducing factor variable. hazard—hazard density; build—build density; lithe—thickness of exposed bedrock; loess—loess thickness; height—relief; ndvi—NDVI; curv—profile curvature.

The results show that the distribution of landslides and unstable slopes in the study area is closely related to the susceptibility partitioning (Table 5). Over one-third of the slope units in the study area are in the high- and very high-susceptibility areas, accounting for 21% and 16% of the total, respectively, with a total area of 10.1 km^2; the remaining slope units are in the moderate- and low-susceptibility areas. Moreover, 35% of the landslides and the unstable slopes are located in the very high-susceptibility area, accounting for 16%

of the total number of the slope units; 33% of the landslides and the unstable slopes are located in the high-susceptibility areas, accounting for 21% of the total number of the slopes; 21% of the landslides and the unstable slopes are located in the moderate-susceptibility area, accounting for 27% of the total number of the slopes; and 11% of the landslides and the unstable slopes are located in the low-susceptibility areas, which account for 36% of the total number of slopes.

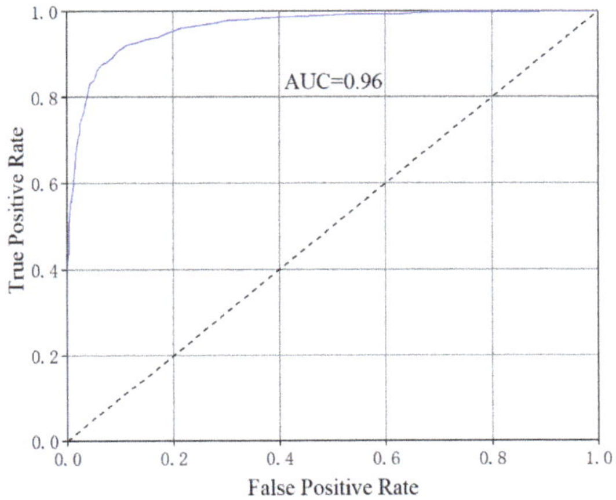

Figure 9. ROC curves and AUC value for evaluating the classification results of landslide susceptibility using the random forest classifier.

Table 4. Confusion matrix of the random forest classifier.

RF		Predicted: Non-Landslide	Predicted: Landslide	Recall
Actual	Non-landslide	543	73	0.881
	Landslide	43	541	0.926
	Precision	0.927	0.881	0.903

The results of the landslide susceptibility assessments show that two regions are highly prone to landslides, one being the urban center of Baota District, and the other Nanniwan Airport (Nnwa) and its surrounding areas. In the urban center of Baota District, including Yangjialing Village (Yjl), Nanshi Street (Ns), Baiping Village (Bp), Hutoumao Village (Htm), Zezigou Village (Zzg), Nanzhaibian Village (Nzb), Majiawan Village (Mjw), Huanghaowa Village (Hhw), Mata Village (Mt), Erzhuangke Village (Ezk), and Shanlangcha Village (Slc), where landslides occur frequently, the landslide susceptibility is high and very high because of the very high-density population and frequent human activities (Figure 10). Due to the effects of road construction, domestic water discharge, crop planting, slope toe excavation, and other activities, landslides, including rock falls, slope failures, unstable slopes, and creep, occur frequently, which poses a great threat to the lives and properties of the local residents. The other highly landslide-prone areas are Nanniwan Airport (Nnwa) and its surrounding areas, including Yangjiawan Village (Yjw), Maozegou Village (Mzg), Sanshipu Village (Ssp), and Yejiagou Village (Yjg). Nnwa is the area of mountain excavation and valley infilling on the Loess Plateau and its construction destroys the stability of the surrounding slopes, resulting in the development of landslides and unstable slopes in the surrounding areas.

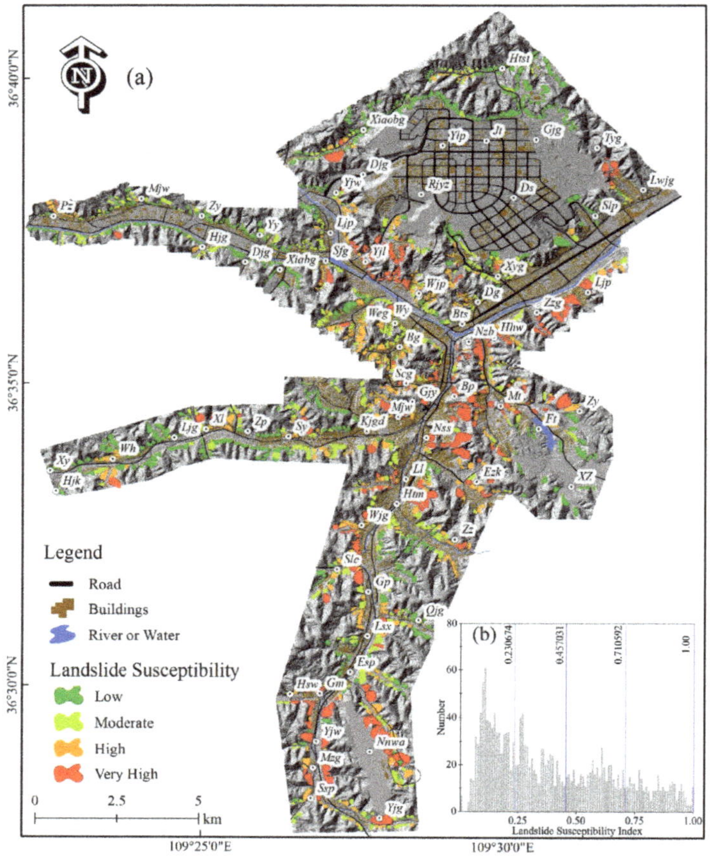

Figure 10. Landslide susceptibility map of the study area in (**a**), and the distribution of the landslide susceptibility index using the natural breaks in (**b**).

Table 5. Slope unit statistics based on landslide susceptibility, hazard, and risk zone.

	Slope Units	VH	H	M	L
Susceptibility zone	Number	291	401	495	654
	Total areas	5.8	4.3	4.6	6.9
	Proportion (N)	16%	22%	27%	36%
	Landslides	122	96	62	54
	Unstable slopes	138	147	96	30
Hazard zone	Number	293	439	583	526
	Total areas	6.4	4.7	6	4.5
	Proportion (N)	16%	24%	32%	29%
	Landslides	123	89	77	45
	Unstable slopes	131	149	93	38
Risk zone	Number	116	377	560	788
	Total areas	2.0	4.4	6.0	9.2
	Proportion (N)	6%	20%	30%	43%
	Landslides	55	78	99	102
	Unstable slopes	41	132	133	105

4.3. Landslide Hazard Mapping

Reliable landslide hazard mapping is crucial for hazard mitigation and risk management. In this study, InSAR technology was used to obtain the landslide hazard assessment, aiming for an ongoing and quantitative practice [63]. The fieldwork showed that the landslide type in the study area was relatively single, mainly loess landslides, and that the geological environment and inducing conditions, such as rainfall, are similar in the small study area, leading to the relatively simple mechanism of loess landslide activity. Therefore, DInSAR was used as a comprehensive indicator to reflect slope displacement, whether caused by rainfall or human activities, in landslide hazard assessment. The spatial probability of landslides (landslide susceptibility) and the intensity of ground surface deformation were used in the weighted overlay model parameters to calculate landslide hazard [64]. The weighted overlay technique is defined to develop a map using the overlays of several raster layers by giving weight to each raster layer according to expert opinions [65]. The weighted overlay analysis was applied to obtain the landslide hazard assessment using the following equation:

$$WX_i = \sum_{j=1}^{m} R(j) \times X(i,j) \quad (2)$$

where m is the total number of factors to assess, WX_i is the hazard index of the assessment units, $R(j)$ is the weight value of each factor, and $X(i,j)$ is the value of the assessment factors. In this study, $X(i,j)$ is the landslide susceptibility index obtained from the random forest and the ground–surface deformation intensity that was defined using the normalized ground deformation data obtained from DInSAR during the monitored time; the weight values of both were set at 0.5 after analyzing the geological environment and inducing conditions of landslides in the study area, respectively. Finally, the hazard indexes of slope units were calculated by summing the product of assessment factors and corresponding weight values. Four levels of hazard, i.e., very high (>0.594), high (0.594–0.416), moderate (0.416–0.269), and low (<0.269), were categorized based on the natural breaks classification, and the LOS displacement in different hazard levels were counted, which are illustrated in Figures 11 and 12, respectively.

The number and LOS displacement values of slope units in different hazard levels are illustrated in Figure 11. The distribution histogram shows the maximum and mean displacement values, as well as the number of slope units in different hazard levels. It shows that the displacement values of slope units are distributed in a normal curve and that the higher hazard of slope units presents a higher displacement value than the lower hazard on the whole. The results show that 40% (732) of the slope units in the study area are in the high- and very high-hazard areas for landslides, accounting for 24% and 16% of the total, respectively, with a total area of 11.1 km^2 (Figure 12). There was a small increase in the number and distribution of the hazard zones in the urban areas compared with the susceptibility map. About 34% (254) of the landslides and the unstable slopes are located in the very high-hazard areas, accounting for 14% of the total number of slopes; 32% (234) of the landslides and the unstable slopes are located in the high-hazard areas, accounting for 13% of the total number of slopes; 23% (170) of the landslides and the unstable slopes are located in the moderate-hazard areas, accounting for 9% of the total number of slopes; and 11% (83) of the landslides and the unstable slopes are located in the low-hazard areas, which are 5% of the total number of slopes. The results show that the spatial distribution of landslide hazard areas was consistent with the field investigations.

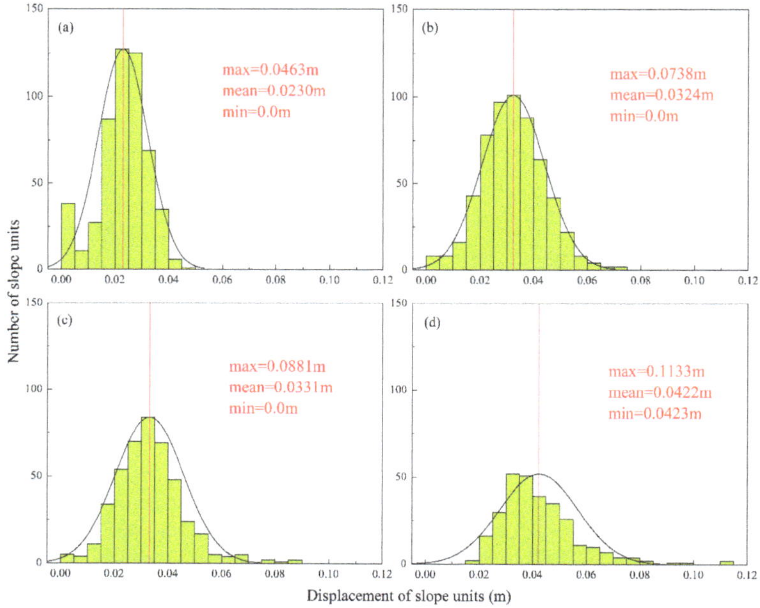

Figure 11. Distribution histogram of the number and LOS displacement values of slope units in different hazard levels: (**a**) low hazard; (**b**) moderate hazard; (**c**) high hazard; and (**d**) very high hazard. The red lines represent the average displacement values in different hazard levels.

4.4. Landslide Risk Mapping

The JTC-1 Joint Technical Committee on Landslides and Engineered Slopes noted that landslide risk is a measure of the probability and severity of the adverse effects of landslides on health or property, which must consider the hazard mapping and vulnerability of landslides [5]. Vulnerability assessment is a fundamental element in the evaluation of landslide risks [66]. Vulnerability to landslides is expressed in economic (monetary, quantitative) and heuristic (qualitative) scales. When using economic measurements, vulnerability is commonly expressed in the element values, such as monetary, intrinsic, and utilitarian values [67]. Due to a lack of information about properties and population distribution data, the Kriging interpolation of building distribution and building density was used for the vulnerability assessment in this study. The location and spatial distribution of buildings reflect the distance between buildings and slope units, which indirectly indicates the extent to which buildings and populations are threatened by landslides. Additionally, the building density can also indicate the properties and populations. Of course, this assumes that the sizes and values of the buildings are similar and that the differences in the populations attached to the different buildings are slight. The equation for landslide risk calculation is expressed as follows:

$$R = H_L \times V_L \tag{3}$$

where H_L and V_L represent the landslide hazard and vulnerability, respectively. The landslide risk index obtained from Equation (3) is divided into four levels according to the natural breaks method after normalization, namely, very high-risk (>0.406), high-risk (0.406–0.223), middle-risk (0.223–0.101), and low-risk (<0.101). The results of the risk assessment zones and statistics are shown in Figure 13 and Table 5.

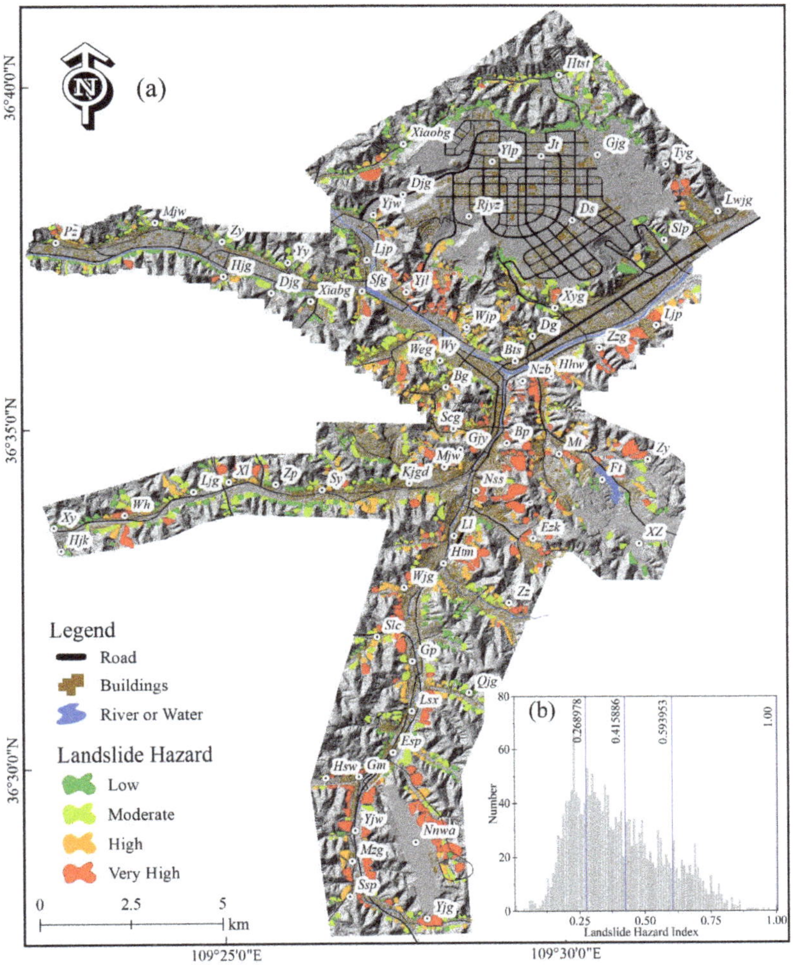

Figure 12. Landslide hazard map of the study area in (**a**), and the distribution of landslide hazard index using the natural breaks in (**b**).

The landslide risk assessment map shows that the risk in the urban center is higher than that in the suburban areas, where the risk decreases with increasing distance from the urban center (Figure 13). Additionally, a total of 20 and 167 extra slopes are in very high and high-hazard zones besides landslides and unstable slopes (Table 5). About 6% (116) of the slope units are located in very high-risk zones with a total area of about 2 km^2; 20% (377) of the slope units are located in high-risk zones with a total area of about 4.4 km^2, which are mainly distributed in concentrated areas (i.e., Yjl-Sy-Slc-Hhw) (Figure 13 and Table 5). The building and population densities in these areas are high, which may lead to significant economic losses and casualties, so it is necessary to pay more attention and conduct landslide risk management to mitigate the landslide risks. Compared with the landslide susceptibility and hazard maps, Nnwa and its surrounding areas are classified into moderate- and low-risk areas because of the low population density in the areas. In addition, many engineering solutions, including slope geometry modification, underground drainage systems, gravity retaining walls, and anti-slide piles, have been applied to stabilize the slopes. Therefore, the slopes, which are originally very highly prone to landslides, are classified as low-risk zones for landslides.

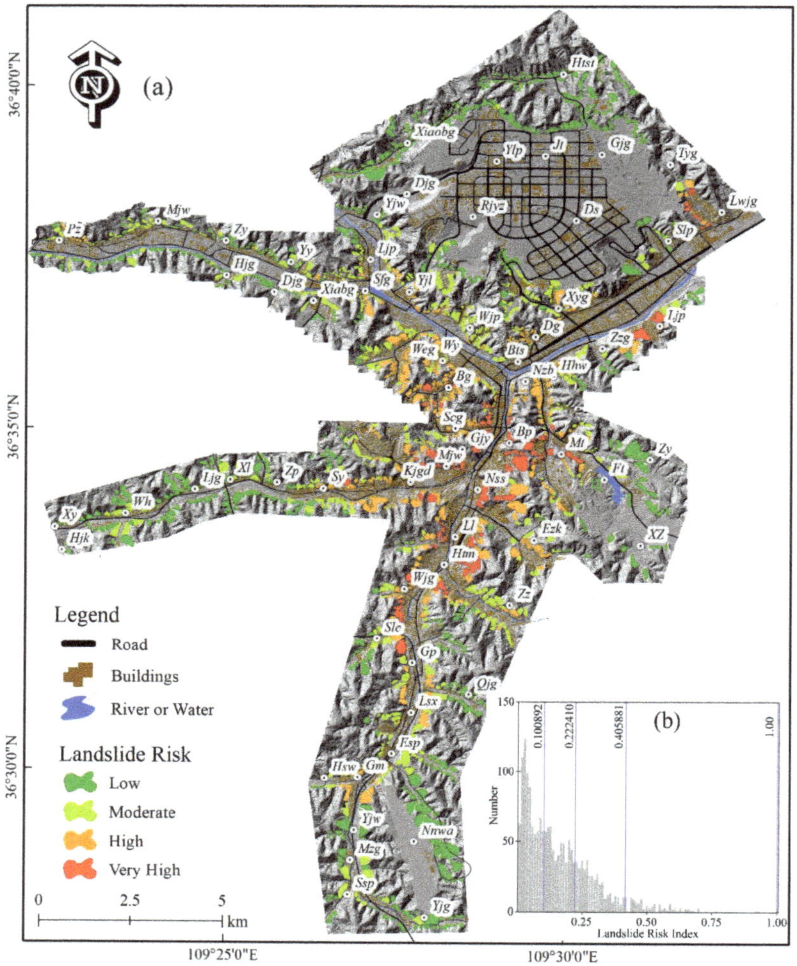

Figure 13. Landslide risk map of the study area in (**a**), and the distribution of the landslide risk index using the natural breaks in (**b**).

5. Discussion

Landslide risk assessment has been attracting the attention of researchers and governments in order to effectively deal with landslides in the study area. For this purpose, machine learning and DInSAR technology were used to evaluate the landslide susceptibility in the main urban parts of Yan'an City. Currently, the susceptibility, hazard, and risk of landslides in the whole of Yan'an City have been determined in existing studies. In terms of the methodology, they can be divided into the heuristic model and generalized objective functions based on experts' knowledge scoring [44,68,69], a quantitative model of evidence weight [42], and a physically deterministic model [43]. In terms of map units and scopes, they can be divided into grid cells (25 m or 30 m), catchment basin units [42–44,68,69], administrative boundaries [42–44], and watershed boundaries [68,69]. However, those results are not only subjective but also can only meet the needs of a wide range of risk management options and cannot truly reflect the geomorphic characteristics of the slope in the study area, which can be useful for risk management in a large administrative area. The research reviewed that an inventory including detailed landslide information and a reasonable mapping unit as well as model type is a prerequisite for obtaining highly accurate assessment results [20]. Firstly, a landslide inventory with more detailed information can provide more input data to the model to analyze the relationship between landslides and geological environment factors to obtain a comprehensive landslide susceptibility map [70]. Some previous studies in the study area may not have shown this. Therefore, a complete landslide inventory, mapping all landslides in the study area, was determined through the interpretation of UAV images and site-by-site investigations. Secondly, field surveys and risk assessment of slope units on a large scale in small areas can provide planners with an adequate and applicable landslide risk map, especially in areas of critical concern such as urban centers. Research shows that grid cells or pixels are still the most commonly used map units in current papers on landslide assessment, and only a few papers have used slope units [20,21]. To reflect the geomorphic characteristics of slopes, slope units were used as map units in this study. Thirdly, the selection of a model is also an important factor affecting the accuracy of landslide susceptibility assessment. There are more and more models and methods developed for landslide susceptibility assessment, among which machine learning with good performance can be used to solve the nonlinear relationship between landslides and geological environment factors [16,34]. For this reason, random forest was selected to predict landslide susceptibility in the study area, with good proven performance [34,71]. Therefore, the accuracy of the landslide susceptibility assessment in this study was improved by using a machine learning model and slope units.

In addition, DInSAR technology was introduced in the process of hazard assessment to calculate slope deformation, and the hazard was calculated by giving the same weight coefficient of susceptibility and slope deformation. InSAR technology was used to perform a time-effective analysis, and the results can present the active state of slopes directly to predict the failure time and assess the hazard class of landslides [72,73]. At present, combining the ground deformation products from the InSAR technology with a landslide risk assessment map has become a concern in the relevant research [63,74]. However, many existing studies focus on the early identification and long-term monitoring of temporal and spatial evolution using InSAR technology [75,76] but there is still insufficient attention paid to risk assessment. The application of InSAR data in landslide risk assessment can improve the reliability of landslide predictions and make a reliable landslide risk map [45].

Finally, landslide risk was calculated by multiplying the hazard with the vulnerability composed of the spatial distribution and density of the buildings. The susceptibility, hazard, and risk assessment in this paper have a similar trend to the previous paper on Yan'an City in the area of different levels and spatial distributions [42–44], such as the area and percentage of low-susceptibility or hazard zones being greater than that of the higher-susceptibility or hazard zones, and the high-risk zones in the spatial distribution patterns are similar, which can also imply the accuracy of this work to a certain extent. Moreover, the risk assessment in this study can provide more specific guidance for risk management and prevention in practice.

There are some limitations that need to be considered in future research. Firstly, due to the lack of detailed population and property data, only the spatial location information of buildings was used for the vulnerability. The precondition for this question is to assume that the values of the buildings in the study area are the same, which would lead to certain information loss for vulnerability. Secondly, influenced by the observation mode of the radar satellites, the deformations obtained by the SAR images are ultimately along the LOS. However, the deformation rate along the slope (Vslope) can more intuitively reflect the real motion of the slopes, which can be transformed through the spatial geometric relationship between the radar LOS and slope. Due to the limitations of the image numbers in the SAR dataset from the study area, we have to use the ascending SAR images for InSAR processing, with which it is difficult to form an effective complement for the descending data. Therefore, we will also try to transform the LOS displacement into the slope direction displacement in future research. PSI or SBAS algorithms can be selected to obtain long-term ground deformation products if the SAR datasets have sufficient and long-term images, which can reflect the long-term movement status and trend of slopes [63,74]. It is expected that more accurate risk assessment maps for the study or elsewhere could be produced by improving the above limitations.

6. Conclusions

Quantitative risk assessment is very effective for landslide risk management and urban areas need more detailed investigations and assessments. In this study, the quantitative landslide risk assessment was based on susceptibility and hazard assessments. The random forest classifier and eight environmental factors influencing landslides, including slope, profile curvature, relief, NDVI, landslides density, building density, the thickness of loess and the thickness of exposed bedrock were used to examine landslide susceptibility in Baota District, Yan'an City. Combined with DInSAR technology, landslide hazard mapping was developed to reflect the hazards quantitatively. Surface deformation, which can be caused by many factors (e.g., precipitation, slope groundwater, and engineering), can be detected by DInSAR technology with centimeter precision. Finally, the landslide risk map was obtained by being combined with the landslide susceptibility and hazard assessment and divided into very high-risk, high-risk, middle-risk, and low-risk areas according to the natural breaks method.

In this study, a total of 1841 slope units were mapped in the study area, including 334 landslides and 411 unstable slopes determined by field investigations, in which the main material of landslides and unstable slopes is loess and only a few of them contain bedrock. The length and width of landslides and unstable slopes are mainly between 50 m and 150 m, the slope angles are mainly between 20° and 50°, and the heights are predominantly between 30 and 90 m, where the slope angles and heights of most of the unstable slopes are larger than those of the landslides. The areas are usually less than 20×10^3 m^2. Reliable risk assessment was achieved using 1841 slope units, which were divided based on the terrain, optical images, and DEM. Remote sensing InSAR technology was applied to determine the quantitative landslide hazard zones. The classification results of the random forest classifier were evaluated with the receiver operating characteristics (ROC) curve and confusion matrix. The confusion matrix shows that the overall accuracy of the random forest classifier is 0.903 and that the AUC value is 0.96, with good prediction accuracy and classified ability of landslide susceptibility. The results of the landslide risk assessment indicate the risk level and the corresponding quantity of the slope units and total areas. Approximately 6% of the slope units located in the very high-risk zones and 20% of the slope units located in the high-risk zones must receive more attention to monitor the dynamics.

The present research has significant implications for landslide risk mitigation in Baota District, Yan'an City. Our scientific landslide risk map is expected to promote landslide prevention based on a zoning strategy and provide a valuable decision to support the local and regional government for disaster prevention, mitigation, and management, which eventually can effectively reduce the impacts of geohazards.

Author Contributions: Conceptualization, Y.Z.; methodology, W.L., Y.Z. and Y.L. (Yiwen Liang); software, W.L. and Y.L. (Yiwen Liang); validation, P.S., W.L., Y.Z. and Y.L. (Yiwen Liang); formal analysis, W.L. and Y.L. (Yiwen Liang); investigation, Y.L. (Yuanxi Li), X.S. and A.W.; resources, Y.Z.; data curation, Y.Z.; writing—original draft preparation, W.L. and Y.L. (Yiwen Liang); writing—review and editing, Y.Z. and X.M.; visualization, W.L. and Y.L. (Yiwen Liang); supervision, Y.Z. and X.M.; project administration, Y.Z.; funding acquisition, Y.Z. and X.M. All authors have read and agreed to the published version of the manuscript.

Funding: This study was funded by the National Natural Science Foundation of China (Grant No. 42007232, 42130709), the Natural Science Foundation for Young Scientists of Gansu Province (Grant No. 20JR5RA223), the Science and Technology Project of Gansu Province (Grant No. 18JR2JA006), and the Fundamental Research Funds for the Central Universities (Grant No. lzujbky-2021-ey05), Geological Survey Project of China (Grant No. DD20189270).

Data Availability Statement: Not applicable.

Acknowledgments: The ALOS-2 images were provided by the Japan Aerospace Exploration Agency (JAXA). We are grateful to the editor and four anonymous reviewers for their constructive suggestions and comments to improve the paper.

Conflicts of Interest: The authors declare no conflict of interest.

References

1. Froude, M.J.; Petley, D.N. Global fatal landslide occurrence from 2004 to 2016. *Nat. Hazards Earth Syst. Sci.* **2018**, *18*, 2161–2181. [CrossRef]
2. Xu, L.; Yan, D.; Zhao, T. Probabilistic evaluation of loess landslide impact using multivariate model. *Landslides* **2021**, *18*, 1011–1023. [CrossRef]
3. Peng, J.; Fan, Z.; Wu, D.; Zhuang, J.; Dai, F.; Chen, W.; Zhao, C. Heavy rainfall triggered loess–mudstone landslide and subsequent debris flow in Tianshui, China. *Eng. Geol.* **2015**, *186*, 79–90. [CrossRef]
4. Li, Y.; Wang, X.; Mao, H. Influence of human activity on landslide susceptibility development in the Three Gorges area. *Nat. Hazards* **2020**, *104*, 2115–2151. [CrossRef]
5. Fell, R.; Corominas, J.; Bonnard, C.; Cascini, L.; Leroi, E.; Savage, W.Z. Guidelines for landslide susceptibility, hazard and risk zoning for land use planning. *Eng. Geol.* **2008**, *102*, 85–98. [CrossRef]
6. Dai, F.C.; Lee, C.F.; Ngai, Y.Y. Landslide risk assessment and management: An overview. *Eng. Geol.* **2002**, *64*, 65–87. [CrossRef]
7. Huang, J.; Lyamin, A.V.; Griffiths, D.V.; Krabbenhoft, K.; Sloan, S.W. Quantitative risk assessment of landslide by limit analysis and random fields. *Comput. Geotech.* **2013**, *53*, 60–67. [CrossRef]
8. Tan, Q.; Bai, M.; Zhou, P.; Hu, J.; Qin, X. Geological hazard risk assessment of line landslide based on remotely sensed data and GIS. *Measurement* **2021**, *169*, 108370. [CrossRef]
9. Zhao, M.; Liu, X. Regional risk assessment for urban major hazards based on GIS geoprocessing to improve public safety. *Saf. Sci.* **2016**, *87*, 18–24. [CrossRef]
10. Trigila, A.; Iadanza, C.; Esposito, C.; Scarascia-Mugnozza, G. Comparison of Logistic Regression and Random Forests techniques for shallow landslide susceptibility assessment in Giampilieri (NE Sicily, Italy). *Geomorphology* **2015**, *249*, 119–136. [CrossRef]
11. Pourghasemi, H.R.; Rahmati, O. Prediction of the landslide susceptibility: Which algorithm, which precision? *Catena* **2018**, *162*, 177–192. [CrossRef]
12. Huang, F.; Yin, K.; Huang, J.; Gui, L.; Wang, P. Landslide susceptibility mapping based on self-organizing-map network and extreme learning machine. *Eng. Geol.* **2017**, *223*, 11–22. [CrossRef]
13. Godt, J.W.; Baum, R.L.; Savage, W.Z.; Salciarini, D.; Schulz, W.H.; Harp, E.L. Transient deterministic shallow landslide modeling: Requirements for susceptibility and hazard assessments in a GIS framework. *Eng. Geol.* **2008**, *102*, 214–226. [CrossRef]
14. Frattini, P.; Crosta, G.B.; Fusi, N.; Dal Negro, P. Shallow landslides in pyroclastic soils: A distributed modelling approach for hazard assessment. *Eng. Geol.* **2004**, *73*, 277–295. [CrossRef]
15. Pradhan, A.M.S.; Kim, Y.T. Evaluation of a combined spatial multi-criteria evaluation model and deterministic model for landslide susceptibility mapping. *Catena* **2016**, *140*, 125–139. [CrossRef]
16. Huang, F.; Cao, Z.; Guo, J.; Jiang, S.H.; Guo, Z. Comparisons of heuristic, general statistical and machine learning models for landslide susceptibility prediction and mapping. *Catena* **2020**, *191*, 104580. [CrossRef]
17. Yalcin, A. GIS-based landslide susceptibility mapping using analytical hierarchy process and bivariate statistics in Ardesen (Turkey): Comparisons of results and confirmations. *Catena* **2008**, *72*, 1–12. [CrossRef]
18. Zhu, A.; Miao, Y.; Wang, R.; Zhu, T.; Deng, Y.; Liu, J.; Yang, L.; Qin, C.; Hong, H. A comparative study of an expert knowledge-based model and two data-driven models for landslide susceptibility mapping. *Catena* **2018**, *166*, 317–327. [CrossRef]
19. Shu, H.; Hürlimann, M.; Molowny-Horas, R.; González, M.; Pinyol, J.; Abancó, C.; Ma, J. Relation between land cover and landslide susceptibility in Val d'Aran, Pyrenees (Spain): Historical aspects, present situation and forward prediction. *Sci. Total Environ.* **2019**, *693*, 133557. [CrossRef]
20. Reichenbach, P.; Rossi, M.; Malamud, B.; Mihir, M.; Guzzetti, F. A review of statistically-based landslide susceptibility models. *Earth Sci. Rev.* **2018**, *180*, 60–91. [CrossRef]
21. Pourghasemi, H.R.; Yansari, Z.T.; Panagos, P.; Pradhan, B. Analysis and evaluation of landslide susceptibility: A review on articles published during 2005–2016 (periods of 2005–2012 and 2013–2016). *Arab. J. Geosci.* **2018**, *11*, 193. [CrossRef]
22. Ramani, S.E.; Pitchaimani, K.; Gnanamanickam, V.R. GIS based landslide susceptibility mapping of Tevankarai Ar sub-watershed, Kodaikkanal, India using binary logistic regression analysis. *J. Mt. Sci. Engl.* **2011**, *8*, 505–517. [CrossRef]
23. Budimir, M.E.A.; Atkinson, P.M.; Lewis, H.G. A systematic review of landslide probability mapping using logistic regression. *Landslides* **2015**, *12*, 419–436. [CrossRef]
24. Khan, H.; Shafique, M.; Khan, M.A.; Bacha, M.A.; Shah, S.U.; Calligaris, C. Landslide susceptibility assessment using Frequency Ratio, a case study of northern Pakistan. *Egypt. J. Remote Sens. Space Sci.* **2019**, *22*, 11–24. [CrossRef]
25. Rabby, Y.W.; Li, Y. Landslide Susceptibility Mapping Using Integrated Methods: A Case Study in the Chittagong Hilly Areas, Bangladesh. *Geosciences* **2020**, *10*, 483. [CrossRef]

26. Neuhäuser, B.; Damm, B.; Terhorst, B. GIS-based assessment of landslide susceptibility on the base of the Weights-of-Evidence model. *Landslides* **2012**, *9*, 511–528. [CrossRef]
27. Kavzoglu, T.; Colkesen, I.; Sahin, E.K. Machine Learning Techniques in Landslide Susceptibility Mapping: A Survey and a Case Study. In *Landslides: Theory, Practice and Modelling*; Pradhan, S.P., Vishal, V., Singh, T.N., Eds.; Springer International Publishing: Cham, Switzerland, 2019; pp. 283–301. [CrossRef]
28. Ma, Z.; Mei, G.; Piccialli, F. Machine learning for landslides prevention: A survey. *Neural Comput. Appl.* **2021**, *33*, 10881–10907. [CrossRef]
29. Chen, X.; Chen, W. GIS-based landslide susceptibility assessment using optimized hybrid machine learning methods. *Catena* **2021**, *196*, 104833. [CrossRef]
30. Behnia, P.; Blais-Stevens, A. Landslide susceptibility modelling using the quantitative random forest method along the northern portion of the Yukon Alaska Highway Corridor, Canada. *Nat. Hazards* **2018**, *90*, 1407–1426. [CrossRef]
31. Hong, H.; Liu, J.; Bui, D.T.; Pradhan, B.; Acharya, T.D.; Pham, B.T.; Zhu, A.; Chen, W.; Ahmad, B.B. Landslide susceptibility mapping using J48 Decision Tree with AdaBoost, Bagging and Rotation Forest ensembles in the Guangchang area (China). *Catena* **2018**, *163*, 399–413. [CrossRef]
32. Goetz, J.N.; Brenning, A.; Petschko, H.; Leopold, P. Evaluating machine learning and statistical prediction techniques for landslide susceptibility modeling. *Comput. Geosci.* **2015**, *81*, 1–11. [CrossRef]
33. Xie, Z.; Chen, G.; Meng, X.; Zhang, Y.; Qiao, L.; Tan, L. A comparative study of landslide susceptibility mapping using weight of evidence, logistic regression and support vector machine and evaluated by SBAS-InSAR monitoring: Zhouqu to Wudu segment in Bailong River Basin, China. *Environ. Earth Sci.* **2017**, *76*, 313. [CrossRef]
34. Chen, W.; Xie, X.; Wang, J.; Pradhan, B.; Hong, H.; Bui, D.T.; Duan, Z.; Ma, J. A comparative study of logistic model tree, random forest, and classification and regression tree models for spatial prediction of landslide susceptibility. *Catena* **2017**, *151*, 147–160. [CrossRef]
35. Huang, F.; Ye, Z.; Jiang, S.; Huang, J.; Chang, Z.; Chen, J. Uncertainty study of landslide susceptibility prediction considering the different attribute interval numbers of environmental factors and different data-based models. *Catena* **2021**, *202*, 105250. [CrossRef]
36. Guzzetti, F.; Carrara, A.; Cardinali, M.; Reichenbach, P. Landslide hazard evaluation: A review of current techniques and their application in a multi-scale study, Central Italy. *Geomorphology* **1999**, *31*, 181–216. [CrossRef]
37. Erener, A.; Düzgün, H.S.B. Landslide susceptibility assessment: What are the effects of mapping unit and mapping method? *Environ. Earth Sci.* **2012**, *66*, 859–877. [CrossRef]
38. Jacobs, L.; Kervyn, M.; Poesen, J.; Reichenbach, P.; Rossi, M.; Marchesini, I.; Alvioli, M.; Dewitte, O. Dealing with heterogeneous landslide information for landslide susceptibility assessment: Comparing a pixel-based and slope unit-based approach. In Proceedings of the EGU General Assembly Conference, Vienna, Austria, 23–28 April 2017; p. 6260.
39. Ba, Q.; Chen, Y.; Deng, S.; Yang, J.; Li, H. A comparison of slope units and grid cells as mapping units for landslide susceptibility assessment. *Earth Sci. Inform.* **2018**, *11*, 373–388. [CrossRef]
40. Jacobs, L.; Kervyn, M.; Reichenbach, P.; Rossi, M.; Marchesini, I.; Alvioli, M.; Dewitte, O. Regional susceptibility assessments with heterogeneous landslide information: Slope unit- vs. pixel-based approach. *Geomorphology* **2020**, *356*, 107084. [CrossRef]
41. Peng, J.; Wang, Q.; Zhuang, J.; Leng, Y.; Fan, Z.; Wang, S. Dynamic formation mechanism of landslide disaster on the Loess Plateau. *J. Geomech.* **2020**, *26*, 714–730. [CrossRef]
42. Xue, Q.; Zhang, M.; Li, L. Loess landslide susceptibility evaluation based on slope unit and information value method in Baota District, Yan' an. *Geol. Bull. China* **2015**, *34*, 118–125.
43. Gao, B.; Wang, X. Risk Zoning of Landslide Based on SINMAP Model in Yan'an City. *Bull. Soil Water Conserv.* **2019**, *39*, 211–216. (In Chinese) [CrossRef]
44. Yang, L.; Mou, X.; Li, C.; Zheng, X.; Yue, D. Risk assessment of geological hazards in Baota District, Yan' an City, Shanxi, China. *Mt. Res.* **2020**, *38*, 679–690. [CrossRef]
45. Ciampalini, A.; Raspini, F.; Lagomarsino, D.; Catani, F.; Casagli, N. Landslide susceptibility map refinement using PSInSAR data. *Remote Sens. Environ.* **2016**, *184*, 302–315. [CrossRef]
46. Ma, R. Analysis of trend and mutation characteristics of precipitation in yan'an city during past 45 years. *J. Yanan Univ. Nat. Sci. Ed.* **2016**, *35*, 95–99. [CrossRef]
47. Zhu, J.; Chen, Z.; Zhu, Y. Distribution Regularity and Development Characteristics of Landslides in Yan'an. *Geol. Sci. Technol. Inf.* **2017**, *36*, 236–243. [CrossRef]
48. Van Westen, C.J.; Castellanos, E.; Kuriakose, S.L. Spatial data for landslide susceptibility, hazard, and vulnerability assessment: An overview. *Eng. Geol.* **2008**, *102*, 112–131. [CrossRef]
49. Sharma, R.H. Evaluating the effect of slope curvature on slope stability by a numerical analysis. *Aust. J. Earth Sci.* **2013**, *60*, 283–290. [CrossRef]
50. Tang, Y.; Feng, F.; Guo, Z.; Feng, W.; Li, Z.; Wang, J.; Sun, Q.; Ma, H.; Li, Y. Integrating principal component analysis with statistically-based models for analysis of causal factors and landslide susceptibility mapping: A comparative study from the loess plateau area in Shanxi (China). *J. Clean. Prod.* **2020**, *277*, 124159. [CrossRef]
51. Ho, J.Y.; Lee, K.T.; Chang, T.C.; Wang, Z.Y.; Liao, Y.H. Influences of spatial distribution of soil thickness on shallow landslide prediction. *Eng. Geol.* **2012**, *124*, 38–46. [CrossRef]

52. Zhang, M.; Tan, X.; Dong, Y.; Sun, P.; Dong, P.; Lu, N. Initial analysis on environmental effect of cutting hills to backfill ditch project on Loess Plateau—Take Yan' an New District as an example. *Geol. Rev.* **2019**, *65*, 1409–1421. [CrossRef]
53. Aksu, G.; Güzeller, C.; Eser, T. The Effect of the Normalization Method Used in Different Sample Sizes on the Success of Artificial Neural Network Model. *Int. J. Assess. Tools Educ.* **2019**, *6*, 170–192. [CrossRef]
54. Colesanti, C.; Wasowski, J. Investigating landslides with space-borne Synthetic Aperture Radar (SAR) interferometry. *Eng. Geol.* **2006**, *88*, 173–199. [CrossRef]
55. Wegmüller, U.; Strozzi, T.; Tosi, L. Differential SAR interferometry for land subsidence monitoring: Methodology and examples. In Proceedings of the Sixth International Symposium on Land Subsidence, Ravenna, Italy, 24–29 September 2000.
56. Pasquali, P.; Pellegrini, R.; Prati, C.; Rocca, F. Combination of interferograms from ascending and descending orbits. In Proceedings of the IGARSS '94—1994 IEEE International Geoscience and Remote Sensing Symposium, Pasadena, CA, USA, 8–12 August 1994; Volume 2, pp. 733–735.
57. Yu, X.; Xue, D.; Chen, F. Analysis of Influence of Vegetation Coverage and Slope on SAR Interferometric Coherence. *Mt. Res.* **2020**, *38*, 926–934. [CrossRef]
58. Breiman, L. Random Forests. *Mach. Learn.* **2001**, *45*, 5–32. [CrossRef]
59. Wu, Q.; Jia, C.; Chen, S.; Li, H. SBAS-InSAR Based Deformation Detection of Urban Land, Created from Mega-Scale Mountain Excavating and Valley Filling in the Loess Plateau: The Case Study of Yan'an City. *Remote Sens.* **2019**, *11*, 1673. [CrossRef]
60. Liao, M.; Zhang, R.; Lv, J.; Yu, B.; Pang, J.; Li, R.; Xiang, W.; Tao, W. Subsidence Monitoring of Fill Area in Yan'an New District Based on Sentinel-1A Time Series Imagery. *Remote Sens.* **2021**, *13*, 3044. [CrossRef]
61. Chen, W.; Pourghasemi, H.R.; Panahi, M.; Kornejady, A.; Wang, J.; Xie, X.; Cao, S. Spatial prediction of landslide susceptibility using an adaptive neuro-fuzzy inference system combined with frequency ratio, generalized additive model, and support vector machine techniques. *Geomorphology* **2017**, *297*, 69–85. [CrossRef]
62. Zhao, F.; Meng, X.; Zhang, Y.; Chen, G.; Su, X.; Yue, D. Landslide Susceptibility Mapping of Karakorum Highway Combined with the Application of SBAS-InSAR Technology. *Sensors* **2019**, *19*, 2685. [CrossRef]
63. Lu, P.; Catani, F.; Tofani, V.; Casagli, N. Quantitative hazard and risk assessment for slow-moving landslides from Persistent Scatterer Interferometry. *Landslides* **2014**, *11*, 685–696. [CrossRef]
64. Hammad, M.; Van Leeuwen, B.; Mucsi, L. Integration of gis and advanced remote sensing techniques for landslide hazard assessment: A case study of northwest syria. *ISPRS Ann. Photogramm. Remote Sens. Spat. Inf. Sci.* **2020**, *VI-3/W1-2020*, 27–34. [CrossRef]
65. Saaty, T.L. Multicriteria decision making: The analytic hierarchy process: Planning, priority setting resource allocation. *Resour. Alloc.* **1990**, *2*, 1–20.
66. Frédéric, L.; Asté, J.; Leroi, E. Vulnerability assessment of elements exposed to mass-movement: Working toward a better risk perception. In *Landslides-Glissements de Terrain*; Balkema: Rotterdam, The Netherlands, 1996; pp. 263–270.
67. Akgun, A.; Kıncal, C.; Pradhan, B. Application of remote sensing data and GIS for landslide risk assessment as an environmental threat to Izmir city (west Turkey). *Environ. Monit. Assess.* **2012**, *184*, 5453–5470. [CrossRef] [PubMed]
68. Tang, Y.; Xue, Q.; Li, Q.; Sun, P. Slope classification system for loess collapse risk assessment. *J. Eng. Geol.* **2012**, *20*, 378–386.
69. Tang, Y.; Zhang, M.; Xue, Q. Landslide risk assessment methods and flow on a large scale—A case study of loess landslides risk assessment in Yan'an urban districts, Shaanxi, China. *Geol. Bull. China* **2011**, *30*, 166–172.
70. Galli, M.; Ardizzone, F.; Cardinali, M.; Guzzetti, F.; Reichenbach, P. Comparing landslide inventory maps. *Geomorphology* **2008**, *94*, 268–289. [CrossRef]
71. Youssef, A.M.; Pourghasemi, H.R. Landslide susceptibility mapping using machine learning algorithms and comparison of their performance at Abha Basin, Asir Region, Saudi Arabia. *Geosci. Front.* **2021**, *12*, 639–655. [CrossRef]
72. Zhang, Y.; Meng, X.; Novellino, A.; Dijkstra, T.; Chen, G.; Jordan, C.; Li, Y.; Su, X. Characterization of pre-failure deformation and evolution of a large earthflow using InSAR monitoring and optical image interpretation. *Landslides* **2021**, *19*, 35–50. [CrossRef]
73. Confuorto, P.; Medici, C.; Bianchini, S.; Del Soldato, M.; Rosi, A.; Segoni, S.; Casagli, N. Machine Learning for Defining the Probability of Sentinel-1 Based Deformation Trend Changes Occurrence. *Remote Sens.* **2022**, *14*, 1748. [CrossRef]
74. Ambrosi, C.; Strozzi, T.; Scapozza, C.; Wegmüller, U. Landslide hazard assessment in the Himalayas (Nepal and Bhutan) based on Earth-Observation data. *Eng. Geol.* **2018**, *237*, 217–228. [CrossRef]
75. Mondini, A.C.; Guzzetti, F.; Chang, K.; Monserrat, O.; Martha, T.R.; Manconi, A. Landslide failures detection and mapping using Synthetic Aperture Radar: Past, present and future. *Earth Sci. Rev.* **2021**, *216*, 103574. [CrossRef]
76. Kang, Y.; Lu, Z.; Zhao, C.; Xu, Y.; Kim, J.; Gallegos, A.J. InSAR monitoring of creeping landslides in mountainous regions: A case study in Eldorado National Forest, California. *Remote Sens. Environ.* **2021**, *258*, 112400. [CrossRef]

remote sensing

MDPI

Article

Performance Testing of Optical Flow Time Series Analyses Based on a Fast, High-Alpine Landslide

Doris Hermle [1,*], Michele Gaeta [2], Michael Krautblatter [1], Paolo Mazzanti [2,3] and Markus Keuschnig [4]

1 Landslide Research Group, TU Munich, 80333 Munich, Germany; m.krautblatter@tum.de
2 NHAZCA S.r.l., Spin-Off from La SAPIENZA University of Rome, 00185 Rome, Italy; michele.gaeta@nhazca.com (M.G.); paolo.mazzanti@uniroma1.it (P.M.)
3 Department of Earth Sciences & CERI Research Center, "Sapienza" University of Rome, 00185 Rome, Italy
4 GEORESEARCH Forschungsgesellschaft mbH, 5412 Puch, Austria; markus.keuschnig@georesearch.ac.at
* Correspondence: doris.hermle@tum.de; Tel.: +49-(0)89-289-25896

Abstract: Accurate remote analyses of high-alpine landslides are a key requirement for future alpine safety. In critical stages of alpine landslide evolution, UAS (unmanned aerial system) data can be employed using image registration to derive ground motion with high temporal and spatial resolution. However, classical area-based algorithms suffer from dynamic surface alterations and their limited velocity range restricts detection, resulting in noise from decorrelation and hindering their application to fast landslides. Here, to reduce these limitations we apply for the first time the optical flow-time series to landslides for the analysis of one of the fastest and most critical debris flow source zones in Austria. The benchmark site Sattelkar (2130–2730 m asl), a steep, high-alpine cirque in Austria, is highly sensitive to rainfall and melt-water events, which led to a 70,000 m^3 debris slide event after two days of heavy precipitation in summer 2014. We use a UAS data set of five acquisitions (2018–2020) over a temporal range of three years with 0.16 m spatial resolution. Our new methodology is to employ optical flow for landslide monitoring, which, along with phase correlation, is incorporated into the software IRIS. For performance testing, we compared the two algorithms by applying them to the UAS image stacks to calculate time-series displacement curves and ground motion maps. These maps allow the exact identification of compartments of the complex landslide body and reveal different displacement patterns, with displacement curves reflecting an increased acceleration. Visually traceable boulders in the UAS orthophotos provide independent validation of the methodology applied. Here, we demonstrate that UAS optical flow time series analysis generates a better signal extraction, and thus less noise and a wider observable velocity range—highlighting its applicability for the acceleration of a fast, high-alpine landslide.

Keywords: digital image correlation; phase correlation; optical flow; time series image stack; landslides; ground motion identification; displacement mapping; UAS

Citation: Hermle, D.; Gaeta, M.; Krautblatter, M.; Mazzanti, P.; Keuschnig, M. Performance Testing of Optical Flow Time Series Analyses Based on a Fast, High-Alpine Landslide. *Remote Sens.* **2022**, *14*, 455. https://doi.org/10.3390/rs14030455

Academic Editor: Andrea Ciampalini

Received: 30 November 2021
Accepted: 14 January 2022
Published: 18 January 2022

1. Introduction

Landslides have a causal link to climate change, thus pose an increasing risk in magnitude and frequency for people and their livestock [1]. In particular, investigations of high-alpine landslide areas are often difficult and dangerous; hence, remote sensing techniques have to be employed to generate sufficient spatial and temporal coverage. Here, optical space and airborne remote sensing offers two key advantages: (i) Optical images with their close-to-nadir viewing geometry, with the image plane orthogonal to the sensor's line-of-sight (LOS), allow scientists to directly monitor and interpret geomorphic processes of steep slopes [2] without using derived products. (ii) Optical remote sensing for the calculation of ground motion by image registration is often the only feasible way to quantify horizontal surface displacements of both shallow and complex slope instabilities [3], where geomorphic processes are moving at rates too high for radar remote sensing techniques [4].

Motion analysis, applied to optical satellite and UAS imagery—including (terrestrial) LiDAR data [5–7]—to measure horizontal surface displacement is a well-established method. Image registration, also known as image matching and image correlation, geometrically aligns images and allows tracking for accurate 2D change measurements in optical images. It has been used on a local to regional scale to assess motion fields for glaciological [8–10], earthquake [11–13], dune migration [14,15], rock glacier [16] and landslide studies [5,17–25].

Image registration is among the most widely utilised techniques in computer vision. It is used to overlay two or more images of the same scene acquired at different times, from various viewpoints, by the same or different sensors [26]. Another application is stereo matching, using an image pair from the same scene taken at the same time from multiple opposing view angles [27]. Several approaches exist to estimate the relative translative offset, such as digital image correlation, optical flow and feature matching, which are often applied to measure displacement and strains. Two major approaches exist to estimate displacements to sub-pixel precision. The first is the area-based approach (also known as the correlation-like method), which recognises uniform amplitude patterns in both the reference and secondary image (Figure 1). Various algorithms calculate the correlation of these patterns in order to quantify the final displacement. The second is the feature-based method, less sensitive to illumination changes and image distortions; it depends on the existence of well-spread, salient features—detectable in both images—which are then extracted to estimate displacement vectors. This feature recognition approach is suitable for multisensor analyses and is computationally less expensive. However, its success is conditioned on surface structures, which therefore restrict its general application [28].

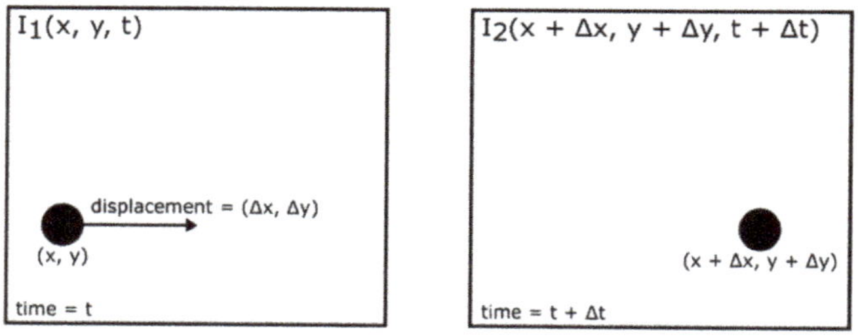

Figure 1. Displacement detection of an object in the reference (I_1) and secondary (I_2) image for t and t + Δt.

Classical area-based algorithms are cross-correlation, normalised cross-correlation (NCC) and minimum distance criteria [29], utilising the intensity without any structural analysis information to match areas or regions [26]. The Fourier shift theorem was proposed as a method for registering translated images [30] and was used to modify the original phase correlation (PC) algorithm (using the phase information) [31]. Thus, by introducing the theorem, the PC works within the frequency domain, utilising rotational and translational properties to calculate the relative transformation parameters based on a translational or similarity model [32,33]. Working with sub-pixel accuracy, PC is highly computationally efficient, and thus can handle large matching templates. Furthermore, it overcomes intensity contrasts, which are frequency-dependent noise, as well as non-uniform, temporal variations such as disturbing illumination influences [26,31]. Nonetheless, the constraints to applying Fast Fourier Transform (FFT) for measurements include both the picket-fence effect and spectral leakage [32]. Computational efficiency can be further enhanced, as can matching if an (inverse) pyramid approach (i.e., hierarchical cross-correlation) is applied [34]. Starting with a coarser image resolution on a high pyramid level, matched patches, i.e., measured displacement, and areas with few matching errors are propagated to a finer resolution and can be used to guide matching on finer levels down to the original resolution [35].

Several area-based applications for motion studies exist in the geoscience community. One of the first tools was IMCORR (NCC), implemented in SAGA GIS, and thus freely available [36,37]; another well-known and often-used tool is COSI-Corr (PC and NCC-based matching) [11,12]. The add-on is free but implemented within the commercial ENVI Classic. Other open source options are the library MicMac, which employs hierarchical image correlation, a combination of NCC-based matching and spatial regularisation [38], and the DIC-FFT (FFT), whose code runs in MATLAB [20]. Two other freely available tools are CIAS, performing an NCC procedure [9,39], and EMT [40], combining cross correlation and least square matching.

Several limitations nevertheless still exist for the abovementioned area-based methods. Geometrical inaccuracies arise from co-registration, with additional problems resulting from vegetation changes and dynamically significant surface processes, leading to mismatches [11,18]. While less sensitive to illumination differences, which are particularly important and challenging for high-alpine environments (arising from low contrast, moving cast shadows and cloud shadows), these factors can still result in erroneous displacement results [11,18,25]. The most important problems for their application to fast landslides are velocities exceeding the search window-related matching limit, which causes decorrelation and ambiguous signals [5,41,42]. Finally, external problems influence the calculation results from orthorectification and sensor model errors [11].

In contrast, the intensity-based approach analyses motions using the differential matching technique of optical flow, i.e., the determination of the dense deformation field of two dynamic images by computing motion vectors at every pixel (Figure S1). For more than 40 years, optical flow, also known as motion analysis, has been one of the classic research problems in computer vision [43]. Following Horn's taxonomy, a motion field is the 2D representation of a 3D surface based on the brightness patterns of an apparent motion. Thus, it is the dense information of a dynamic motion field between two consecutive images. Based on the assumption of a globally smooth motion field [44], or if the dynamic motion field is constant within a certain interrogation window [34], the brightness constancy term is valid, so changes in illumination are resolved in motion [43,45]. As a function of space (x, y) and time (t), the first image I_1 (x, y, t) moved by Δx, Δy will correspond to the intensity of the second image, with an offset I_2 (x + Δx, y + Δy, t + Δt), which can be expressed as the optical flow problem: $(u,\ v) = \left(\frac{\Delta x}{\Delta t}, \frac{\Delta y}{\Delta t}\right)$.

Fundamental works by Horn and Schunck [44], as well as Lucas and Kanade, [34] accelerated computation times, e.g., with the coarse-to-fine search strategy for an inverse image pyramid approach, which decreases computational costs. Today, due to this computational effectiveness while handling large displacements at sub-pixel resolutions, this strategy is widely employed in medical image registration, automotive driver assistance, human motion analysis, and has been applied in geosciences to determine glacier flow [46,47].

Nevertheless, based on the brightness consistency assumption, limitations arise due to considerable changes in illumination induced by shadows, seasonal effects such as non-uniform glacial crevasse patterns, surface feature changes of tumbling rocks, and large textureless regions due to snow cover and shade [43,46]. Accordingly, optical flow was suggested for images of low noise and brightness variance to estimate small displacements only [42]. A recent work by Kroeger et al. [45,48] utilises the fast and noise-robust inverse compositional image alignment approach [49,50] and proposes a fast dense inverse search method to capture matches quickly in order to deal with large displacements, deformations, appearance alterations such as illumination, chromaticity and blur, as well as motion discontinuities or outliers [45,51].

In general, the application of optical image registration methods can encompass a large velocity range as these methods are less sensitive to large displacements and long measurement intervals, leading to decorrelation when using Differential Interferometric Synthetic Aperture Radar (DInSAR). Its application is restricted to relatively slow motions ($\leq$1 m/yr), i.e., remaining below a quarter of the SAR sensors' wavelength λ ($\geq\lambda/4$), with some exceptions of $\lambda/2$ [4]. Although active radar sensors are relatively independent of atmospheric constraints and of shadows, further limiting high-alpine factors include snow cover, slope exposition, layover effects and foreshadowing [52,53]. Nevertheless, using optical remote sensing for research on high-alpine sites is often difficult due to meteorological constraints such as snow cover, clouds or cloud shadows and mountain ridge topographic shadowing effects for certain seasons and times of day. Even though satellite revisit rates have significantly increased (e.g., Sentinel-2, five days, PlanetScope, daily), their actual net revisit rate for high-alpine sites is restricted to a few images per month [25,54]. Therefore, UAS campaigns offer the highest temporal flexibility to overcome these obstacles [17], but good quality data depends on GCPs (ground control points) with reproducible flight plans [55], which requires significantly more time, and hence, higher costs.

While experience for landslide displacement calculations on area-based image matching exists [14,56], no previous research has applied intensity-based optical flow to derive ground motion for landslide behaviour assessment. In this study, we test the performance of the intensity-based fast optical flow method using dense inverse search and compare the results with the widely known area-based phase correlation algorithm, both of which are implemented into the commercial software IRIS. We employed these algorithms to compute the deformation fields for the Sattelkar, a complex landslide in a steep high-alpine cirque (2130–2739 m asl) exhibiting varying displacement rates from slow (few meters) to moderate velocities (<30 m/yr) [57]. We conduct a single interval analysis, together with a time-series approach, on five UAS orthophotos acquired during a three-year period (2018–2020) of both high spatial accuracy and resolution (0.08 m) [58]. We evaluate the results based on trajectories of large boulder blocks (<10 m) which are traceable in the UAS orthophotos and stable bedrock. Accordingly, we seek to answer the following research questions:

1. Is the dense inverse search (DIS) method applicable to the large displacements of landslides with complex behaviours?
2. Can the method investigate both slow and moderate velocities between repeated observation intervals?
3. Is the DIS method robust enough to cope with the changing and unfavourable illumination of a high-alpine steep study site?
4. How does the DIS method perform in comparison to the well-established phase correlation algorithm?

2. A Complex Landslide

The Sattelkar is a ~30° steep high-alpine deglaciated cirque in the Obersulzbach valley, Großvenedigergruppe, Austria (Figure 2). It is located at an altitude between 2130–2730 m asl and is west-oriented. The cirque is surrounded by a headwall of central granitic gneiss and is filled with an abundant volume of deposits from past and current rockfalls, glacial and periglacial debris, moraine walls and relicts of a dissolving rock glacier [59,60]. The cirque infill is characterised by a wide grain size distribution, with boulders up to 10 m.

Since 2003, surface alterations have taken place: the vegetation cover has degraded, and been replaced with loose mobile rock material [59]. The deep-seated, retrogressive movement is sensitive to rainfall and meltwater, causing high water (over)saturation and leading to a spreading and sliding behaviour on the glacially smoothed bedrock, developing into a flow-like behaviour while huge blocks tumble and turn—all of which can be classified as a complex landslide [57]. These characteristics make this a challenging benchmark object and suitable site for landslide displacement analyses using optical remote sensing.

Based on aerial orthophotos, damage documentation, and witness reports, during the last decade a continuous intensification of mass wasting and debris flow activities has taken place. Heavy precipitation on 30 and 31 July 2014 led to a debris flow of about 70,000 m^3 from the catchment area above the cirque threshold (at ~2000 m asl), and a further 100,000 m^3 of mobilized material was entrained from within the channel [59].

Initial investigations estimated an unstable area of 130,000 m^2 where 1 mio. m^3 of debris showed high activity; displacement rates up to 10 m a^{-1} between 2003–2015 were obtained from aerial orthophotos and repeated field measurements [59,60]. As the debris consisted of boulders up to 10 m, continuous visual block tracking could be employed to estimate the displacement within the active area based on aerial orthophotos. Recent studies conducted with COSI–Corr confirmed this ongoing increase of mass wasting processes, with variable displacement rates ranging from 1–14 m in only 42 days [25].

Today long-term monitoring is conducted with high-accuracy UAS; nine differential GPS-measured GCPs provide continuous stable and precise conditions. Additional monitoring instruments provide geomorphologic insights into rockfall behaviour via one autarkic seismograph. Thirteen near-surface temperature loggers at a 0.1 m depth also record mean annual ground surface temperatures, indicating potential, sporadic permafrost conditions [61]. Recent empirical statistical permafrost modelling in the Hohe Tauern Range supports permafrost occurrence at our study site [62].

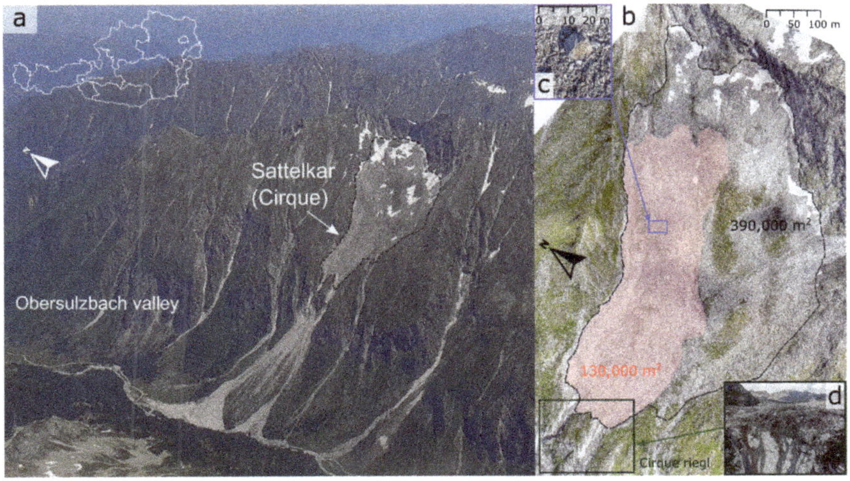

Figure 2. (**a**) Sattelkar, 30 June 2019, with the debris cone of the 2014 debris flow event; overview map of Austria in the top right corner (white) (Österreichischer Bundesverlag Schulbuch GmbH & Co. KG and Freytag–Berndt & Artaria KG, Wien, Austria), (**b**) UAS orthophoto (4 September 2019) with the landslide in transparent red and the entire cirque indicated with a dashed grey outline, (**c**) boulder size of 5–10 m used for manual motion tracking and (**d**) view on the front of the cirque threshold.

3. Materials and Methods

There are several reasons to utilise UAS orthoimages for our study. We have a stable set-up for the study site which guarantees a high quality, reliable data set. Due to their time flexibility, UAS flights are conducted under best illumination conditions, with similar time periods. Additionally, we have full control of the image acquisition format (same UAS and flight plan, management over spatial resolution, extent, acquisition altitude and snow coverage) [55] and the subsequent manual post-processing (georeferencing and spatial accuracy, orthorectification) [58].

We applied two image matching algorithms to optical multi-temporal UAS orthophotos to identify and quantify horizontal displacements. The first is phase correlation [33], and the second is optical flow, where we make use of a dense inverse search method (DIS) [48] which applies the inverse search approach [49,50] based on the fundamental work of the coarse-to-fine Lucas–Kanade algorithm [34]. The DIS code is freely available online (OpenCV, https://docs.opencv.org/3.4/d4/dee/tutorial_optical_flow.html, last access 11 October 2021; the script is based on Kroeger et al. [48]). Both algorithms are incorporated into the commercial software IRIS, developed by NHAZCA S.r.l., in which the displacements were calculated (https://www.photomonitoring.com/iris, last access 12 January 2022). Furthermore, we used ArcGIS (calculations, statistics and map creation), ArcGIS Pro (multihillshade calculations), open source QGIS (data handling and management) and SAGA (displacement vector calculations).

3.1. UAS Image Acquisition and Processing

Five UAS (unmanned aerial system) acquisition campaigns took place between 2018 and 2020 (Table 1). All UAS flights were conducted at approximately the same time of day in order to have best similarity of illumination conditions regarding shadows. Additionally, the flight campaigns were set up around the same dates to have generally equal time intervals. Thus, for our study approach, the four time intervals of our data set allowed us—with a sufficient number and an adequate approach—to interpret the processes of both two one-year-long as well as two summer season intervals, with flights in mid-July and the beginning of September (Table 1). The flights were planned with UgCS (identical flight plans with four flights at different elevations of high overlap for front: 80% and side: 70%) and carried out with a DJI Phantom 4. The ground sampling distance was 7 cm for the area of ~3.4 km^2 and with a flight speed of ~8 m/s, with a total flight time of ~3.5 h (Table 2). Images were taken in RAW format, improved for contrast, highlights, shadows and clarity using Adobe Exposer, then exported as JPGs (95% compression), and finally processed with Pix4Dmapper to 0.08 m resolution. Based on nine permanent ground control points on bare rock (GCPs, 30 × 30 cm), the pictures were georeferenced and orthorectified. GCPs were repeatedly registered (1000 measurements/position) with the TRIMBLE R5 dGPS (differential GPS). We post-processed the data using the baseline data of the Austrian Positioning Service (APOS) provided by the Bundesamt für Eich und Vermessungswesen (BEV). The horizontal RMSE was ~0.05 m and the vertical RMSE was ~0.10 m, and they were used to rectify all UAS campaigns. Lastly, the data (orthophotos and DEMs) was clipped to a consistent area of interest (AOI), projected to UTM 33N (EPSG 32633) and downsampled to 0.16 m (bilinear interpolation) with GDAL to enhance processing time. In addition, to better understand surface processes, hillshades (ArcGIS) and multi-hillshades (ArcGIS Pro) were calculated and visualised as GIFs and combined with total displacement results from PC and DIS (see online supplementary material, OSM).

Table 1. UAS acquisition dates and time interval overview for single (I–IV) and multimaster analysis (1.-3.-2.-5., 001–004) (see Section 3.2).

Acquisition Dates	Intervals Single Analysis		Intervals Multimaster Analysis									
13 July 2018 24 July 2019 4 September 2019 9 July 2020 11 September 2020	(I): 376 (III): 309	(II): 42 (IV): 64	(I): 376	(1.-3.): 418	(1.-4.): 727	(1.-5.): 791	(2.-4.): 351	(2.-5.): 415	(0001): 376	(0002): 418	(0003): 727	(0004): 791

Table 2. UAS flight plans.

Flight Plan Parts	Length of Flightpath [km]	Flight Time [min]	Passes	No. of Images	GSD	Altitude Startpoint [m]	Highest Flight Position [m]	Lowest Terrain Point [m]
Top	6.8	17	6	121	7	2630	3120	2365
Middle	7.5	19	6	135	7	2200	2682	1820
Low 1	7.3	17	6	130	7	1768	2115	1620
Low 2	5.6	14	6	81	7	1768	2110	1620
Total	27.2	67	24	467	7		3120	1620

3.2. Displacement Calculation and Derivation of Displacement Curves

We systematically tested parameter sensitivities in order to determine the best settings for both algorithms, PC and DIS. To do so, we conducted horizontal displacement analyses for all possible time interval combinations on both single analyses (i.e., $t_n - t_{n+1}$) as well as multimaster analyses (Table 1). Multispectral (rgb) UAS orthophotos were used as the input, which were transformed into grayscale images.

The parameter settings of the phase correlation in IRIS are (i) step size, which is the pixel size in x, y between the two sliding windows. This defines the final output resolution, and thus, the final information density (the smaller the step, the denser the coverage), and significantly affects computation time. Similarly, the size of the moving window is a critical parameter, as it defines the physical resolution of the result and represents a compromise between noisy or homogeneous data, hence producing stable output results. Enlarging the window size, the correlation can be increased and homogeneity enhanced; however, computation time will increase. For our input data, we achieved the best results (ii) using a matching window size of 256 pixels. We set (iii) the subpixel resolution to 0.25, which resulted in an upsampled cross-correlation by a factor of 4 and a final subpixel resolution of 0.04 m. Although we had (iv) the option to use the coarse-to-fine pyramid approach, working with one pyramid level was sufficient, and thus the original resolution was kept. In the post-processing the results were (v) resampled using nearest neighbour. In order to identify the matching limitations, i.e., decorrelations, we did (vi) not apply a correlation coefficient threshold, thus keeping the results raw. For final total displacement visualisation, a minimum threshold for values below 0.5 m was applied to eliminate noise.

Thereafter, a multitemporal analysis was performed and the same image matching parameter settings (i–vi) were applied as used for the single analysis. For each reference image we calculated every secondary image, i.e., retaining all possible image combinations (Figure 3). Prior to each single displacement analysis, image pairs were co-registered using the stable area around the landslide. Then, a weighted average using the correlation coefficient was applied to all single analysis results in order to calculate the final multimaster outcome. Each single analysis resultant map was saved to facilitate the final analysis.

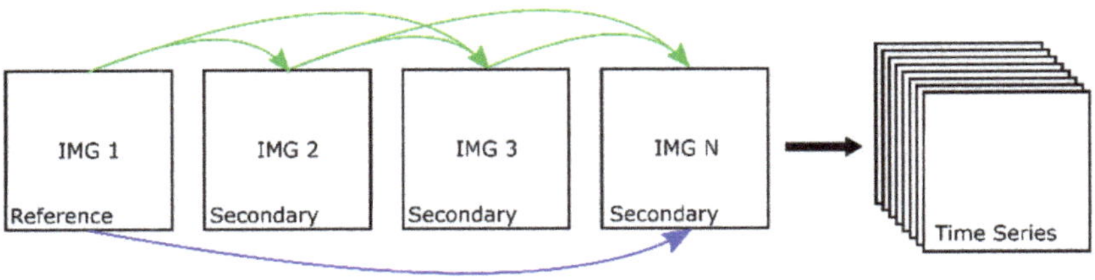

Figure 3. Multimaster approach. For each co-registered image of the single analysis (green arrow), and thus each secondary image pair, the average was weighted with the correlation coefficient of the first to the last image result (blue arrow).

Using the OpenCV [63] algorithm, which implemented a dense inverse search optical flow [48], we analysed the UAS data set and tested different parameter combinations for each single time interval in order to find the most suitable one (Table A1). The matching patch sizes employed 8×8 or 16×16 pixels. A mean normalisation for the patches was applied, which improves the robustness of illumination changes as it calculates every band's mean.

Once we determined the best single-interval combinations, we used these results for the final multimaster analysis of DIS. We used the same multimaster settings as those applied for the PC multimaster calculation.

In order to calculate the time series (TS) displacement curves based on the multimaster analysis, in QGIS we created six large rectangles (TS AOI 1–6; see Figure 8). On the basis of preliminary results during sensitivity tests for both PC and DIS, the AOIs were placed beginning in the centre of the landslide process and moving uphill towards the rear (Table 3) to prevent the AOIs from being disturbed by any noise. Afterwards, they were imported into IRIS as *.kml. For both TS results (PC and DIS), we then derived for each TS AOI (1–6) a displacement curve (Figure 8).

Table 3. Areas of rectangles for TS calculation of displacement curves (TS AOI 1–6) and bedrock identification number (1–5) for the assessment of modelled bedrock displacement.

TS AOI No./Bedrock No.	1	2	3	4	5	6
Time Series AOI [m^2]	337.94	457.09	988.15	974.03	969.85	893.25
Bedrock [m^2]	162.095	50.31	23.95	13.79	167.50	-

3.3. Accuracy Assessment and Result Validation

We demonstrated the validity of our algorithms in two ways: by (a) checking the displacement for stable bedrock areas and (b) using the trajectories of manually measured boulders detectable in the orthophotos. In QGIS we selected five different stable bedrock areas of sufficient spatial extent outside the landslide process area (Table 3). These areas were unaffected by cast shadow and remnants of debris from past rainfall events. The exported *.kml were used in IRIS to derive displacement curves for north–south (NS) and east–west (EW) displacements.

As image registration is based on the matching of pixel patches, we assumed that adjacent pixels represented a similar displacement magnitude. To estimate the accuracy of fit resulting from the total displacement calculations (PC and DIS), we calculated a spatial mean total displacement of the boulder trajectories with a buffer of 0.1 m to plot against the manually measured travel distances of the boulder trajectories (Figure 4). We ran a regression model to see if boulder motion significantly predicted data distance with and without outliers. In order to detect outliers, we used Mahalanobis distance, a reliable outlier detection measure, as it focuses on multivariate distributions of more than one variable, and hence is suitable for our data [64]. Accordingly, we show regression lines without outliers.

After exporting the *.raw results of the total, NS and EW displacement into ArcGIS, we used the former to visualise the ground motion above 0.5 m displacement for the known process area, with no further filter applied. This threshold was selected due to the NS and EW components of the bedrock times series calculations indicating the accuracy of our calculations (see Section 4.1). With the SAGA tool "Gradient vectors from Directional Components", displacement arrows (*.raw NS and EW displacement results, mean value with a step of 100 and range of 50–250) were calculated. As some arrows were slightly outside the total displacement results, we filtered and cleaned up the vectors (extent of total displacement) to avoid ambiguity (see left column, Figure 5).

3.4. Atmospheric and Hydrological Conditions

In order to interpret the calculated displacement results of this hydrologically sensitive complex landslide, we considered precipitation data. An automatic weather station at the Kürsinger cabin in close vicinity to the Sattelkar measures rainfall during the opening season of the cabin (spring to autumn; Tables 4 and 5).

Table 4. Number of days with daily amount of precipitation greater than 20, 30, 40 and 50 mm (meteorological station Kürsinger cabin).

	>20 [mm]	>30 [mm]	>40 [mm]	>50 [mm]
2009	12	5	3	2
2010	8	4	2	1
2011	10	3	1	0
2012	16	5	2	1
2013	6	1	0	0
2014	9	5	3	2
2015	18	6	1	0
2016	10	5	0	0
2017	10	4	0	0
2018	10	5	2	1
2019	11	1	1	1
2020	16	8	6	2

Table 5. Ten highest days of total precipitation for the observation period 2009–2020 (meteorological station Kürsinger cabin, descending order).

Date	Precipitation [mm]
29 August 2020	82.9
30 July 2014	76.1
27 April 2009	70.1
3 October 2020	62.8
11 June 2014	60.4
19 August 2017	57.2
6 June 2009	52.5
17 July 2010	52.3
28 July 2019	51.2
3 August 2020	50.0

4. Results

This section outlines the results of the studies of the total displacement for the single and the multimaster analysis; for the latter only we present displacement curves. We further evaluate the findings based on manually measured boulder tracks and stable bedrock areas.

4.1. Accuracy Assessment: Stable Areas and Ground Truth Comparison

In order to estimate the quality of the algorithms, displacement curves of the stable bedrock areas 1–5 (Table 3) for all intervals are presented in Figure 4. The maximum displacement for DIS NS (c) and EW components (d) was the only outlier of −0.5 m for the last interval IV, Bedrock 2 (Figure 8b). Apart from that, the displacements for the stable bedrock ranged by ±0.3 m. By contrast, PC returned a smaller distribution around zero, with lower values of ±0.2 m for NS components (a) and ±0.3 m for EW components (b).

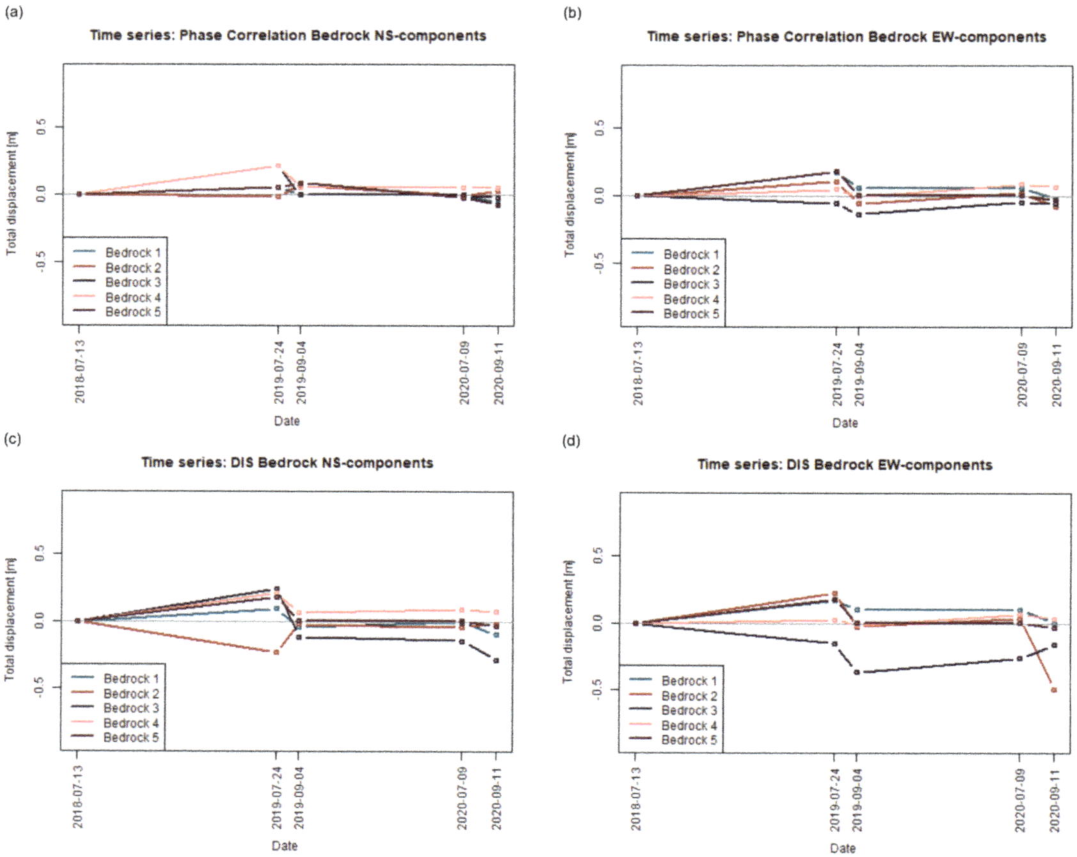

Figure 4. TS calculated for stable bedrock areas 'Bedrock 1 to 5' for NS and EW components for DIS (**a**,**b**) and Phase Correlation (**c**,**d**). Locations of bedrock areas are displayed in the orthophoto map of Figure 8.

We assessed the accuracy of the total displacement calculations for both algorithms by comparing them to a mean buffer around the manually measured boulder paths for the corresponding time interval. The Mahalanobis outlier detection [64] yielded two outliers for both DIS and PC (marked in red Figure 5). We found a significant positional relationship between boulder motion and the modelled DIS mean (b = 0.49, t = 2.72, $p < 0.01$) and PC mean (b = 0.55, t = 2.89, $p < 0.01$), and determined that boulder motion accounts for 16% and 17% of variance in the data distance, respectively. After outlier removal, the variance in the data distance yielded 56% for DIS ($R^2 = 0.5564$) and 65% for PC ($R^2 = 0.6471$). The plots present results for time interval II for PC (a) and DIS (b) with the regression line after outlier removal (Figure 5). Figure 6 shows boulder trajectories for PC and DIS.

Displacement vectors indicate a smooth downslope flow direction. There are minor patches of chaotic directions for a heterogeneous displacement patch in the PC results (central north and towards the northern end) and for displacements between 0 and 0.5 m at the landslide head flowing downslope. For the same area, DIS vectors point in different directions as well as towards the northern rim of the landslide.

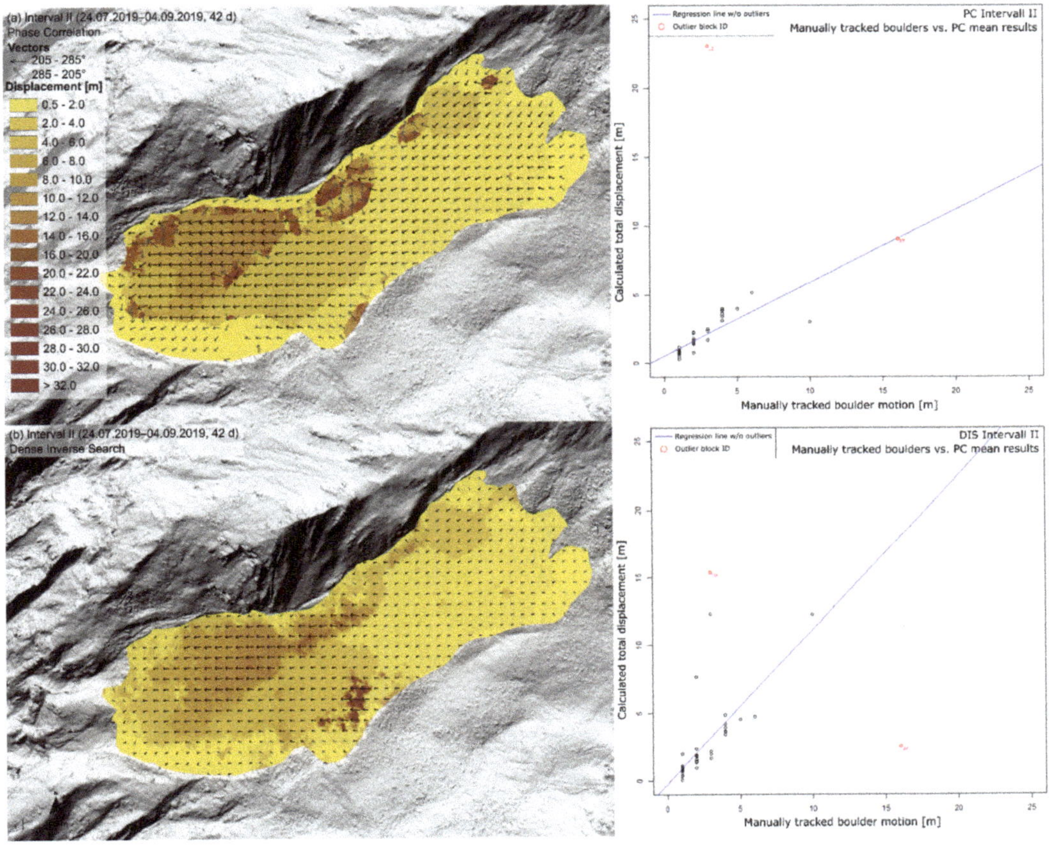

Figure 5. The left top row represents the results from the PC (**a**), the left bottom row the results from DIS (**b**)—both for time interval II: 24 July 2019–4 September 2019, 42 d. Arrows in black represent a realistic downslope range between 205–285°, while putting unrealistic directions (285–205°) in grey. Background: hillshade of UAS DEM, 0.16 m resolution. On the right column, displacements of manually traced boulder trajectories (x-axis) are plotted against the mean total displacement for the corresponding trajectory with a buffer of 0.1 m (y-axis) for PC (**top right**) and DIS (**bottom right**). Outliers are marked in red (with block ID), and the regression line in blue after outlier detection (Mahalanobis distance) and removal. Boulder trajectories are displayed in Figure 6.

4.2. Total Displacement for Single Analysis

This section outlines the results based on single intervals, i.e., $t_n - t_{n+1}$, twice covering an approximately one-year period (I, III) and a summer season (II, IV; see Table 1).

The total displacement from PC results, for all intervals I–IV (Figure 6a–d), yielded a clearly demarcated landslide body for values above 0.5 m, limited to values of about 20 m. Homogeneous areas (landslide's rear body in the east to the centre) are replaced by a patchy, inhomogeneous area for I, III and IV. The homogeneous displacement is lowest (0.5–2 m) in II (42 d), slightly higher for I (376 d), and returns highest values (12–14 m) for both the longer and shorter intervals III (308 d) and IV (62 d). By contrast, for I, III and IV, the landslide head is less noise-affected for II, with more homogeneous patches. Comparing displacements to boulder trajectories, the values are consistent apart from the landslide's head, with ambiguous signals as well as some 1 m trajectories not reflected in the landslide's rear area (Figure 6b,c).

Total displacements (I–IV) derived from DIS (Figure 6e–g) lie within the process area, with patches of particularly high motion (>32 m and 22–25 m) in the northern frontal half (I) and centre (IV), respectively. Interval II reveals the lowest overall displacement, increasing towards the front. Displacements in the rear are higher for III and IV (1–8 m) than I and II. Boulder trajectories match well from the centre to the rear (I–IV), except for the foremost front of the landslide's head (I–IV).

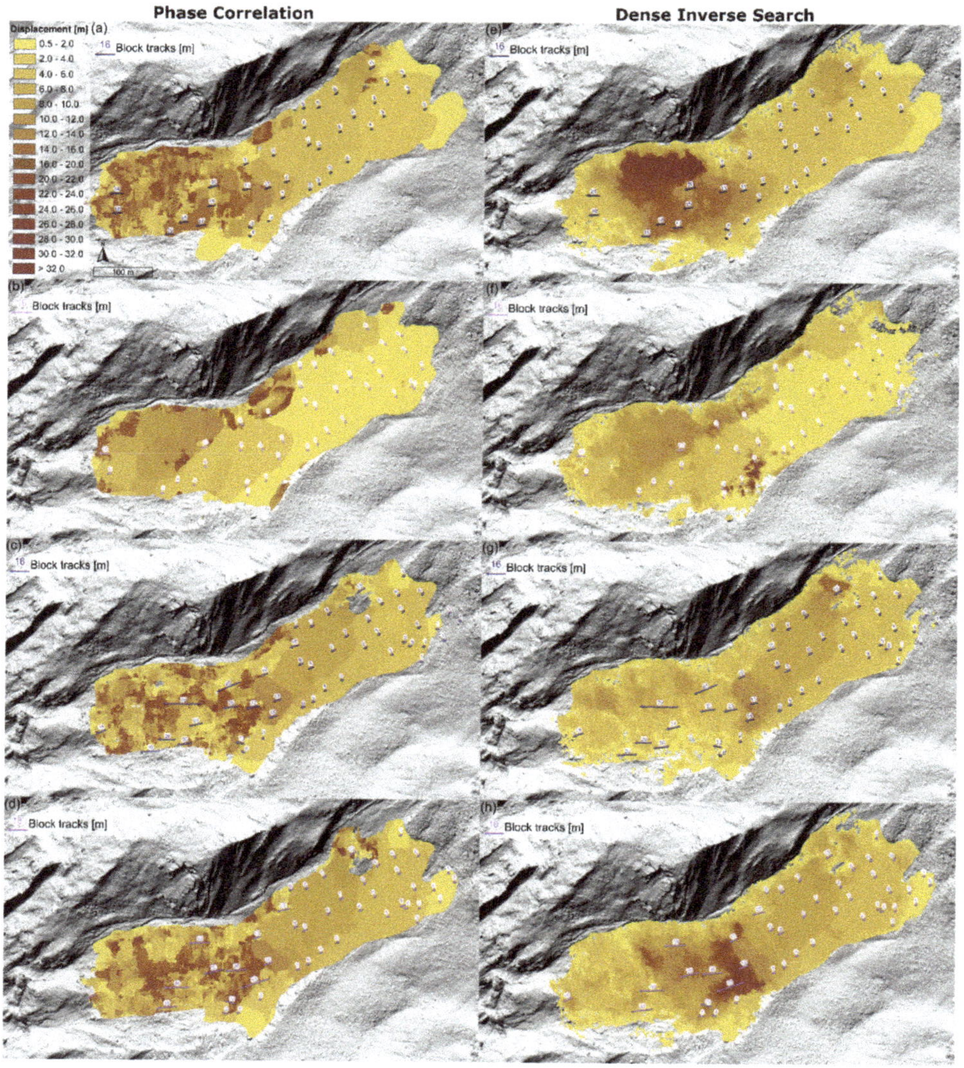

Figure 6. Results of single total displacement calculations of UAS orthoimages at 0.16 m resolution for (**a–d**) using PC for the left column and (**e–h**) using DIS for the right column. The time intervals I–IV (I: 13 July 2018–24 July 2019, 376 d; II: 24 July 2019–4 September 2019, 42 d; III: 4 September 2019–8 July 2020, 308 d; IV: 8 July 2020–8 September 2020, 62 d) follow from top to bottom and are to be read in rows to compare the two algorithms. The arrows represent manually measured boulder trajectories for the corresponding time interval, with displacements in meters indicated on top (ref. Section Material and Methods). Background: hillshade of UAS DEM, 0.16 m resolution.

Comparing the two algorithms on the basis of boulder trajectories, ambiguous areas for PC return significantly lower displacements for DIS, as indicated by trajectories (23–54 m) in the centre for I, III and IV (a, c, d and e, f, h) and at the front, where DIS returns particularly low motion (3–4 m)—although boulder trajectories appeared to indicate higher displacements (6–70 m). For I, in the south of the frontal half, some minor heterogeneous patches exist for PC (a), with similar values for DIS (b), which are confirmed by boulder trajectories (15, 16 and 17 m). Except for the landslide's head, both algorithms indicate consistent displacement values for II, with larger homogeneous areas by PC (b).

4.3. *Total Displacement for Multimaster Analysis and Displacement Curves*

Here, the results of the multimaster approach focus on interval 002 as well as the longest final time interval 004 for algorithms PC and DIS (Figure 7). In the previous section, the first master interval 001 was described, as it is identical to the single analysis interval I. An overview of the individual image combinations for the multimaster analysis is provided in Tables 1 and A1.

The PC multimaster displacement for 002 (a) is characterised by a large ambiguous and heterogeneous area at the landslide's head, transitioning into a homogeneous area for the last two-thirds. Here, values increase from 0.5 in the rear to 10 m. The accumulated displacement for 004 returns values up to 23 m from the centre decreasing towards the rear. The results derived for DIS 002 reveal values up to 7 m at the landslide's head and increase to the highest values (15–32 m) in the centre. There, for 004, both algorithms show a similar boundary, with displacements decreasing rearwards (PC, 9–2 m, DIS 12–22 m), with the highest values towards the front (>32 m) for DIS; similar to DIS single analysis results, values at the foremost landslide's head do not exceed 10 m.

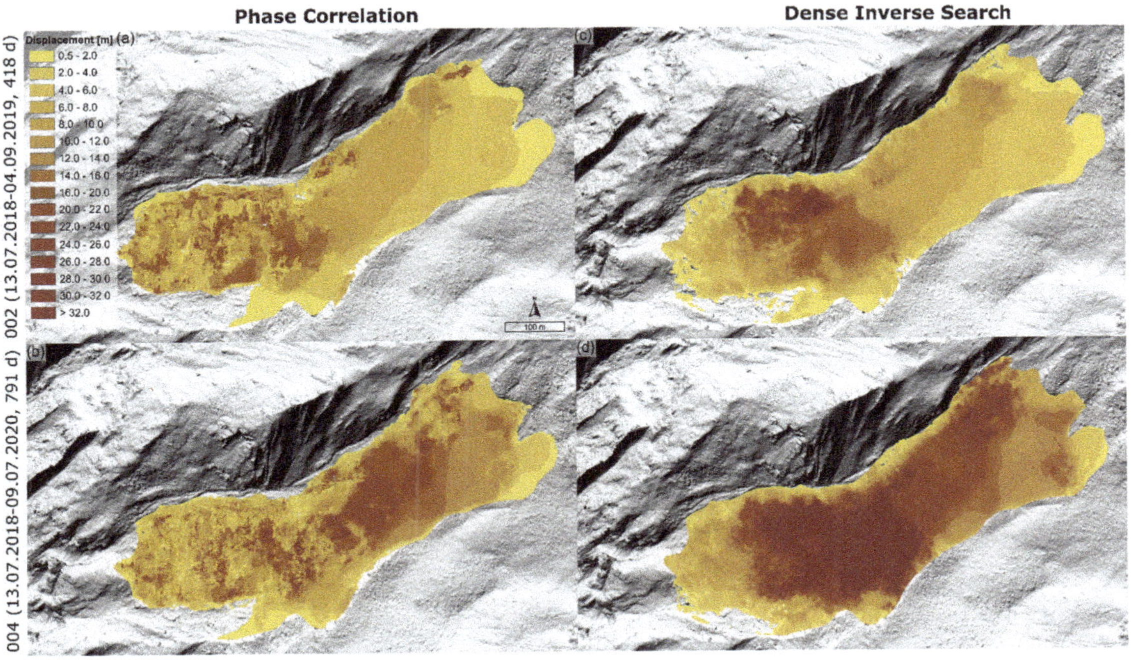

Figure 7. Results of multimaster TS analysis for total accumulated displacement calculations of UAS orthoimages at 0.16 m resolution for top row 002 (13 July 2018–4 September 2019, 418 d) and bottom row 004 (13 July 2018–11 September 2020, 791 d). Left column with PC algorithm (**a**,**b**) and right column DIS (**c**,**d**). Background: hillshade of UAS DEM, 0.16 m resolution.

Our calculations of TS displacement curves (Figure 8) for PC (a) and DIS (b) indicate a continuous increase in displacement. PC exhibited a smooth, linear increase with similar rates for all five AOIs, apart from AOI 6, and a clear limit at approximately 20 m. For DIS, by contrast, the maximum accumulated displacement exceeds 42 m for the foremost AOI 1, followed by AOI 2 with ~40 m—both of which show the strongest increase for an additional 17–20 m (IV), with higher accumulated displacement values than PC. Generally, all AOIs increase steadily, similar to PC for I and II, with the rear AOIs (AOI 3–6) more or less identical for both algorithms (8–20 m).

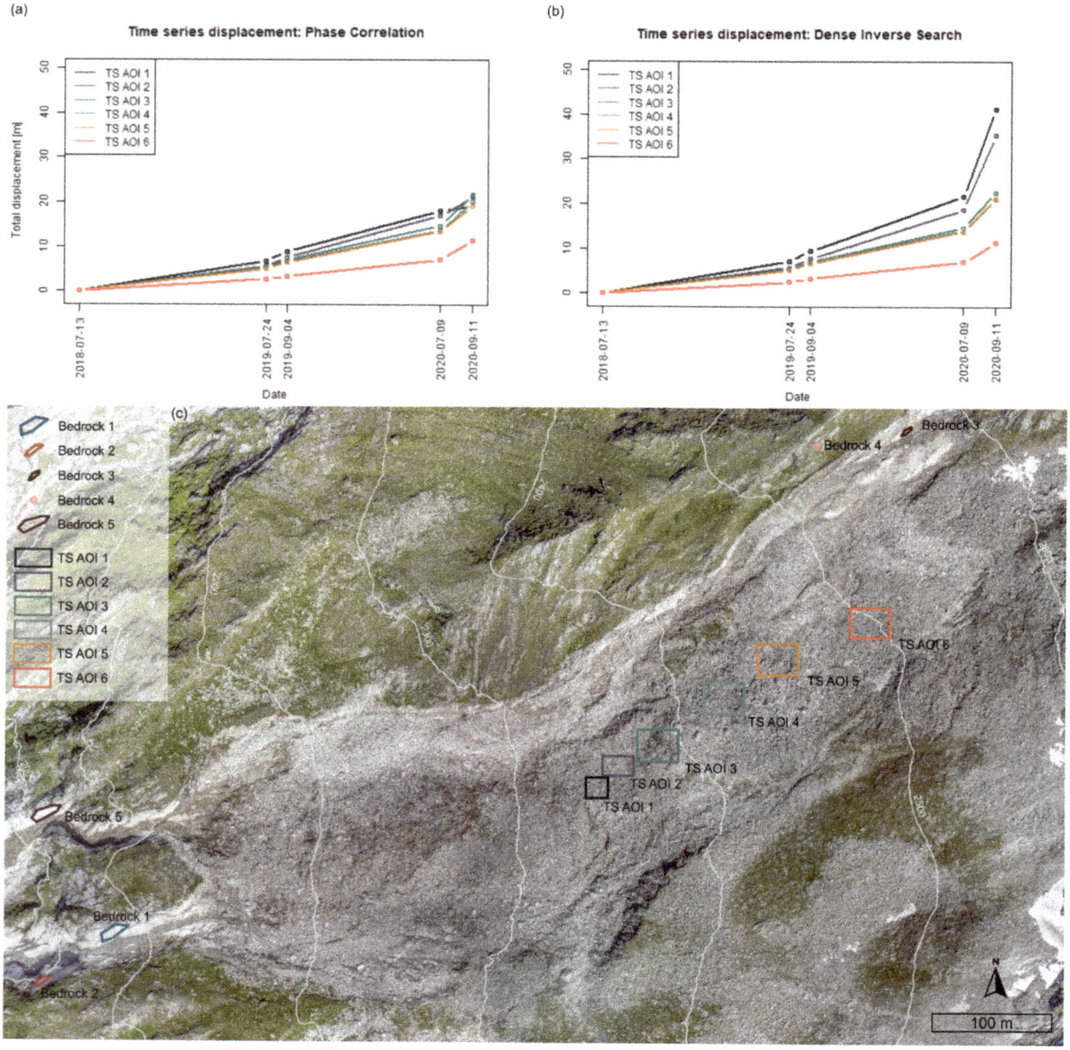

Figure 8. Top row represents the accumulated total displacement resulting from the multimaster analysis for the TS AOI 1–6 from 13 July 2018–8 September 2020 for PC (**a**) and DIS (**b**). The orthophoto map (8 September 2020, UAS 0.16 m resolution) in the middle (**c**) represents the locations of the TS AOI 1–6 within the landslide and the stable bedrock areas 'Bedrock 1 to 5'.

4.4. Meteorological Data

Precipitation measurements during the UAS observation period 2018–2020 indicate that 2020 had the highest number of days with a daily amount of precipitation greater than 20, 30, 40 and 50 mm (Table 4). The comparison for the highest days of total precipitation shows the highest and 10th-highest amount in 2020 at 82.0 and 50.0 mm, respectively, although the number of days at 50.0 mm are lower than the highest records (Table 5). The year 2020, with highest amount of total precipitation, is followed by the second-highest, which was the 2014 debris flow event-year of 76.1 mm.

5. Discussion

We have calculated time series displacements for complementary image-matching algorithms PC and DIS for a three-year long set of UAS orthophotos. We derived displacements for single time intervals (Figure 6) as well as a combined reference-secondary image approach (Figure 7), aka multimaster analysis, in order to investigate the complex behaviour of the Sattelkar, with its heterogeneous motion fields with low to medium velocities [57,59,60]. In so doing, we were able to derive displacement curves (Figure 8), interpret our observations in relation to meteorological data (Tables 4 and 5) and confirm the results in our accuracy assessment on the basis of traceable boulder trajectories (Figures 5 and 6), as well as stable bedrock displacement curves (Tables 4 and 5).

In answer to research question (1) we determined that, for the most part, the displacement results show that DIS is an applicable method to capture large displacements—even for a landslide with complex behaviour.

For research question (2), given a very heterogeneous landslide behaviour, the method allows us to investigate both slow and moderate velocities, which we can in large part support by manually measuring boulder trajectories and statistical observations (Figure 6). For a single interval II, apart from two outliers, DIS returns valid ground motion values (Figure 5). Though derived displacements for the process area from the centre to the rear are well represented and confirmed by the boulder trajectories, the foremost area of the landslide's head is significantly underestimated by a factor of four (I, 18–24 m boulder trajectories) and a factor of five to ten (III, IV). This area leads directly over the cirque threshold into the steep channel, and from field observations we know that surface processes are particularly dynamic and complex, including tumbling boulders changing their surface appearance by turning. Nevertheless, DIS shows its capability to reveal the ongoing process of the dissolving rock glacier, visible in IV (Figure 6h), and less pronounced in III (g) due to the sharp difference in ground motion. For optical flow to work reliably, the brightness consistency has to be valid, with illumination changes resulting in motion. Until now, the use of optical flow has been restricted to ground motion observations of small displacements, little noise and difference in illumination [47].

However, our results confirm research question (3) that optical flow, using the improved approach of DIS [48] is robust enough for the orthophoto dataset of our high-alpine steep study site to cope with changing and unfavourable illumination conditions. This is supported by displacement vectors calculated from NS and EW displacements, which to a large degree reveal a correct flow direction.

With regards to research question (4), comparing the results of DIS to those of PC, DIS has been shown to overcome the correlation limits of the PC algorithm. Ambiguous signals of PC results come from noise resulting from decorrelation as the detection limit; hence, the maximum possible correlation for the amount of displacement and/or number of surface changes is reached. The results show that for our high resolution in the UAS dataset of 0.16 m with temporal baselines the limitation is reached (I, III, IV) [14,16]. In essence, matching failed due to massive changes in pixel values. Our field observations provide evidence of the deformation of rock masses with strong surface alterations due to rotational block behaviour and the high mobility of rock material, which are thus likely to be responsible for this decorrelation. Similar observations on the matching limitations and other reasons leading to decorrelation have been confirmed by others [14,42,52]. We decided to keep these noise-affected areas to be able to differentiate the results. Consequently, while towards the front PC decorrelates and DIS seems to be underrepresented, from the distinctive boundary in the landslide's centre towards the rear, both algorithms return similar displacement values, as confirmed by the boulder paths (I, III and IV).

These boulder trajectories are reliable and support the correctly derived high displacement values for interval III (308 d), and in particular, the higher ones for interval IV—although shorter in duration covering the summer period (62 d). This displacement increase can be explained as a sensitive reaction of the debris due to the heavy precipitation events in 2020. The year 2020 had by far the highest number of days with daily precipitation, followed by the 2014 debris flow event-year. In the summer interval IV (8 July 2020–8 September 2020), the two highest days of total precipitation were measured (see Tables 4 and 5: highest, 82.9 mm, 29 August 2020; 10th, 50.0 mm, 3 August 2020). We assume that this heavy rainfall event at the end of August indicates how the debris reacts to hydrological influences with a massive acceleration [65,66]. It seems unlikely that the 50.0 mm event significantly contributed to this acceleration, as in the previous year the ninth-highest measure was recorded and showed no signs of acceleration.

Apart from single interval calculations, we further performed a multimaster approach and compared the results of both PC and DIS algorithms for intervals 002 and 004 (Figure 7). The previously discussed total displacement distribution for the single intervals is more pronounced due to the summing up of all possible image interval combinations. While 002 reveals a very similar pattern for the rear of the landslide mass, the front is again partitioned beginning at the centre, leading to decorrelation for PC (Figure 7a) and high displacements between 20 and 32 m for DIS. However, these values becomes unrealistically low towards the landslide's head (Figure 7). Displacement values reach their maximum for PC of 20 m (c), whereas DIS exceeds 32 m (d), in these areas. Single analysis interval I (Figure 6b) and multimaster interval 002 (Figure 7b) reveal an area of particularly high motion at the northern rim towards the landslide's head. This high motion can be confirmed by field observations of severe, high dynamic surface changes—a behaviour visible in the GIFs (Figures S2–S4).

The limitations of ground motion are confirmed by calculated displacement curves for PC and DIS (Figure 8). Where PC has a definite limit of 20 m for total displacement detection (a), there is no upper limit for DIS. Generally, displacement curves for both PC and DIS indicate a clear acceleration behaviour—in particular for the heavy rainfall season 2020. The explanatory power of DIS to derive ground motion and displacement curves is high.

We applied the approach of displacement curve calculations to areas outside of the landslide process in order to estimate the quality of the algorithm, as there should be zero to limited displacement. The NS and EW components for PC show values very close to zero (± 0.2 and ± 0.3 m), and DIS never exceeds $\pm$ 0.5 m, indicating the high accuracy and reliability of our results.

Our results demonstrated the possibilities and limitations of the optical flow dense inverse search algorithm. Backed by the comparison to the well-known and robust phase correlation algorithm, we find that DIS is a more sensitive, less rigid and more flexible algorithm. While computationally very efficient, both small and large displacements can be detected without upper limitations. We can confirm the results from the DIS algorithm and for the first time, image registration methods reflect motions we know through our field observations (area of high motion at the northern rim towards the front, see Figures 6e, 7c and S3; and at the front of the dissolving rock glacier, see Figure 6g,h).

However, the results must be interpreted with caution as there is a clear underrepresentation for the landslide's head. Areas of too-high surface dynamics and/or displacements lead to a drastic change in pixel values. Therefore, PC with an upper detection limit of 20 m fails, returning areas of decorrelated noise, whereas DIS still returns some displacement values, but they are too low. These returned values imply a correct signal, but based on the comparison to PC we know that DIS could lead to underrepresented displacements, which is why the results have to be carefully interpreted. Thus, areas of decorrelation can be interpreted as a valid limit of PC with no over- or underestimation, as with DIS, and a higher robustness towards illumination changes.

6. Conclusions

This study evaluated the potential of the dense inverse search (DIS) algorithm to derive ground motions to assess a high-alpine, complex landslide. Our research has made a substantial contribution, as for the first-time, optical flow was applied to study landslide behaviour. We tested the algorithm on time interval combinations of single intervals as well as multimaster pairs of reference-secondary images based on five high accuracy UAS orthophotos of 0.16 m acquired between 2018 and 2020.

We compared total displacement results of DIS to those of the well-known phase correlation algorithm (PC), both of which are implemented in the software IRIS, with regard to trajectories of traceable boulders. These results were contrasted to trajectories and confirm a high goodness of fit. In an accuracy assessment we evaluated our results by deriving NS and EW displacements for five stable bedrock areas, ranging between ±0.2 and ±0.3 m. Our findings show that DIS is applicable to determine ground motions of both slow and moderate velocities, as it detected displacements from 0.5 to 42 m for our observation intervals. This was supported by boulder trajectories and correlated, heterogeneous displacements derived from PC. DIS further overcame the correlation limits of PC, which occurred at about 20 m, and we obtained decorrelation even with a larger template. It is likely that for both severe surface changes and very high displacements, DIS underestimated values at the landslide's head, while PC decorrelated due to excessive surface changes. The findings are based on our experiments and are confirmed by our own field observations, as well as published descriptions of geomorphological processes [59–61]. In addition, we calculated displacement curves, which indicated acceleration and high ground motions—thus confirming the displacement increase in summer 2020, which can be explained by a high rainfall event.

Apart from the complex, high-alpine study site investigated here, DIS could be profitably employed for landslide types from pre-alpine to alpine sites. DIS could also be of high value for earthquake and glacier studies, as it is able capture displacement rates exceeding the detection capability of DInSAR. Nevertheless, future studies should focus on the applicability of complementary optical data from other sensors, improving the accuracy as well as the robustness for real world illumination conditions to confirm the detection capability of DIS for landslide displacement. Further research is also needed to exploit the potential of image-matching techniques for an improved understanding of landslide kinematics ranging from single block sliding to complex flow-like behaviour, as well as early warnings for landslides.

Supplementary Materials: The following supporting information can be downloaded at: https://www.mdpi.com/article/10.3390/rs14030455/s1, Figure S1: An example of optical flow for estimating mouth motion. Two consecutive images show regions of a mouth in motion (a,b) and the estimated flow field using dense optical flow method (c) adapted from reference [67]. Figure S2: GIF showing UAS multi-hillshades [0.16 m] of 13 July 2018; 24 July 2019; 4 September 2019; 8 July 2020; 8 September 2020. Figure S3: GIF showing DIS derived total displacement (transparent) superimposed on UAS multi-hillshades [0.16 m] for the corresponding time interval (I: 13 July 2018–24 July 2019, 376 d; II: 24 July 2019–4 September 2019, 42 d; III: 4 September 2019–8 July 2020, 308 d; IV: 8 July 2020–8 September 2020, 62 d). Figure S4: GIF showing PC derived total displacement (transparent) superimposed on UAS multi-hillshades [0.16 m] for the corresponding time interval (I: 13 July 2018–24 July 2019, 376 d; II: 24 July 2019–4 September 2019, 42 d; III: 4 September 2019–8 July 2020, 308 d; IV: 8 July 2020–8 September 2020, 62 d).

Author Contributions: This study was conceptualised by D.H. together with M.K. (Markus Keuschnig), M.G. and P.M., who supervised this work. The two algorithms were implemented within the software IRIS by M.G.; D.H. analysed the data, explored the possibilities and limitations of the method and the software, and discussed first stages with M.G.; M.G. and M.K. (Markus Keuschnig) discussed the outcomes of final results and their validation with D.H.; D.H. drafted the manuscript, which was revised by all authors, M.K. (Michael Krautblatter), M.G. and P.M. The software was provided by NHAZCA S.r.l., Rome. All authors have read and agreed to the published version of the manuscript.

Funding: This research was funded by a scholarship of the Hanns–Seidel Foundation and the AlpSenseRely project, which is funded by the Bavarian State Ministry of the Environment and Consumer Protection (StMUV).

Informed Consent Statement: Not applicable.

Acknowledgments: The authors are grateful for the software support from NHAZCA S.r.l., a spin-off company of La Sapienza, University of Rome and GeoResearch, Puch, Austria. We thank Robert Delleske for contacting UAS flight campaigns and deriving the data, as well as Riccardo Scandroglio and Stephen Starck for proof-reading.

Conflicts of Interest: The authors declare no conflict of interest.

Appendix A

In order to test IRIS, please visit https://www.photomonitoring.com/iris/ to request a free trial of the software via the contact form. Last access 12 January 2022.

Table A1. Settings for DIS single analysis.

	Finest Scale	Patch Size	Patch Stride	Gradient Descending Iterations	Use Mean Normalisation	Use Spatial Propagation	Var. Refinement Alpha	Var. Refinement Delta	Var. Refinement Gamma	Var. Refinement Iterations
Interval I	0	8	3	25	x	x	20	5	10	5
Interval II	0	8	3	25	x	x	20	5	10	5
Interval III	0	16	2	30	x	x	15	5	10	10
Interval IV	0	8	3	25	x	x	15	5	10	10
Interval 1–3	0	8	3	25	x	x	15	5	10	10
Interval 1–4	0	8	3	25	x	x	20	5	10	5
Interval 2–4	0	8	3	25	x	x	20	5	10	5
Interval 2–5	0	8	3	25	x	x	20	5	10	5

References

1. Huggel, C.; Clague, J.; Korup, O. Is climate change responsible for changing landslide activity in high mountains? *Earth Surf. Processes Landf.* **2012**, *37*, 77–91. [CrossRef]
2. Lacroix, P.; Araujo, G.; Hollingsworth, J.; Taipe, E. Self-Entrainment Motion of a Slow-Moving Landslide Inferred from Landsat-8 Time Series. *J. Geophys. Res. Earth Surf.* **2019**, *124*, 1201–1216. [CrossRef]
3. Scaioni, M.; Longoni, L.; Melillo, V.; Papini, M. Remote Sensing for Landslide Investigations: An Overview of Recent Achievements and Perspectives. *Remote Sens.* **2014**, *6*, 9600–9652. [CrossRef]
4. Moretto, S.; Bozzano, F.; Esposito, C.; Mazzanti, P.; Rocca, A. Assessment of Landslide Pre-Failure Monitoring and Forecasting Using Satellite SAR Interferometry. *Geosciences* **2017**, *7*, 36. [CrossRef]
5. Travelletti, J.; Delacourt, C.; Allemand, P.; Malet, J.-P.; Schmittbuhl, J.; Toussaint, R.; Bastard, M. Correlation of multi-temporal ground-based optical images for landslide monitoring: Application, potential and limitations. *J. Photogramm. Remote Sens.* **2012**, *70*, 39–55. [CrossRef]
6. Romeo, S.; Cosentino, A.; Giani, F.; Mastrantoni, G.; Mazzanti, P. Combining Ground Based Remote Sensing Tools for Rockfalls Assessment and Monitoring: The Poggio Baldi Landslide Natural Laboratory. *Sensors* **2021**, *21*, 2632. [CrossRef] [PubMed]
7. Voordendag, A.; Goger, B.; Klug, C.; Prinz, R.; Rutzinger, M.; Kaser, G. Automated and permanent long-range terrestrial laser scanning in a high mountain environment: Setup and first results. *ISPRS Ann. Photogramm. Remote Sens. Spat. Inf. Sci.* **2021**, *2*, 153–160. [CrossRef]
8. Scherler, D.; Leprince, S.; Strecker, M. Glacier-surface velocities in alpine terrain from optical satellite imagery—Accuracy improvement and quality assessment. *Remote Sens. Environ.* **2008**, *112*, 3806–3819. [CrossRef]
9. Heid, T.; Kääb, A. Evaluation of existing image matching methods for deriving glacier surface displacements globally from optical satellite imagery. *Remote Sens. Environ.* **2012**, *118*, 339–355. [CrossRef]
10. Nagy, T.; Andreassen, L.M.; Duller, R.A.; Gonzalez, P.J. SenDiT: The Sentinel-2 Displacement Toolbox with Application to Glacier Surface Velocities. *Remote Sens.* **2019**, *11*, 1151. [CrossRef]
11. Leprince, S.; Barbot, S.; Ayoub, F.; Avouac, J.-P. Automatic and Precise Orthorectification, Coregistration, and Subpixel Correlation of Satellite Images, Application to Ground Deformation Measurements. *IEEE Trans. Geosci. Remote Sens.* **2007**, *45*, 1529–1558. [CrossRef]

12. Ayoub, F.; Leprince, S.; Avouac, J.-P. Co-registration and correlation of aerial photographs for ground deformation measurements. *ISPRS J. Photogramm. Remote Sens.* **2009**, *64*, 551–560. [CrossRef]
13. Kääb, A.; Altena, B.; Mascaro, J. Coseismic displacements of the 14 November 2016 Mw 7.8 Kaikoura, New Zealand, earthquake using the Planet optical cubesat constellation. *Nat. Hazards Earth Syst. Sci.* **2017**, *17*, 627–639. [CrossRef]
14. Leprince, S.; Berthier, E.; Ayoub, F.; Delacourt, C.; Avouac, J.-P. Monitoring Earth Surface Dynamics with Optical Imagery. *Eos* **2008**, *89*, 1–12. [CrossRef]
15. Michel, S.; Avouac, J.-P.; Ayoub, F.; Ewing, R.C.; Vriend, N.; Heggy, E. Comparing dune migration measured from remote sensing with sand flux prediction based on weather data and model, a test case in Qatar. *Earth Planet. Sci. Lett.* **2018**, *497*, 12–21. [CrossRef]
16. Debella-Gilo, M.; Kääb, A. Sub-pixel precision image matching for measuring surface displacements on mass movements using normalized cross-correlation. *Remote Sens. Environ.* **2011**, *115*, 130–142. [CrossRef]
17. Lucieer, A.; de Jong, S.; Turner, D. Mapping landslide displacements using Structure from Motion (SfM) and image correlation of multi-temporal UAV photography. *Prog. Phys. Geogr.* **2014**, *38*, 97–116. [CrossRef]
18. Stumpf, A.; Malet, J.-P.; Allemand, P.; Ulrich, P. Surface reconstruction and landslide displacement measurements with Pléiades satellite images. *ISPRS J. Photogramm. Remote Sens.* **2014**, *95*, 1–12. [CrossRef]
19. Turner, D.; Lucieer, A.; de Jong, S. Time Series Analysis of Landslide Dynamics Using an Unmanned Aerial Vehicle (UAV). *Remote Sens.* **2015**, *7*, 1736–1757. [CrossRef]
20. Bickel, V.; Manconi, A.; Amann, F. Quantitative Assessment of Digital Image Correlation Methods to Detect and Monitor Surface Displacements of Large Slope Instabilities. *Remote Sens.* **2018**, *10*, 865. [CrossRef]
21. Caporossi, P.; Mazzanti, P.; Bozzano, F. Digital Image Correlation (DIC) Analysis of the 3 December 2013 Montescaglioso Landslide (Basilicata, Southern Italy): Results from a Multi-Dataset Investigation. *ISPRS Int. J. Geo-Inf.* **2018**, *7*, 372. [CrossRef]
22. Lacroix, P.; Bièvre, G.; Pathier, E.; Kniess, U.; Jongmans, D. Use of Sentinel-2 images for the detection of precursory motions before landslide failures. *Remote Sens. Environ.* **2018**, *215*, 507–516. [CrossRef]
23. Mazzanti, P.; Caporossi, P.; Muzi, R. Sliding Time Master Digital Image Correlation Analyses of CubeSat Images for landslide Monitoring: The Rattlesnake Hills Landslide (USA). *Remote Sens.* **2020**, *12*, 592. [CrossRef]
24. Mulas, M.; Ciccarese, G.; Truffelli, G.; Corsini, A. Integration of Digital Image Correlation of Sentinel-2 Data and Continuous GNSS for Long-Term Slope Movements Monitoring in Moderately Rapid Landslides. *Remote Sens.* **2020**, *12*, 2605. [CrossRef]
25. Hermle, D.; Keuschnig, M.; Hartmeyer, I.; Delleske, R.; Krautblatter, M. Timely prediction potential of landslide early warning systems with multispectral remote sensing: A conceptual approach tested in the Sattelkar, Austria. *Nat. Hazards Earth Syst. Sci.* **2021**, *21*, 2753–2772. [CrossRef]
26. Zitová, B.; Flusser, J. Image registration methods: A survey. *Image Vis. Comput.* **2003**, *21*, 977–1000. [CrossRef]
27. Smith, M.W.; Carrivick, J.L.; Quincey, D.J. Structure from motion photogrammetry in physical geography. *Prog. Phys. Geogr. Earth Environ.* **2016**, *40*, 247–275. [CrossRef]
28. Scheffler, D.; Hollstein, A.; Diedrich, H.; Segl, K.; Hostert, P. AROSICS: An Automated and Robust Open-Source Image Co-Registration Software for Multi-Sensor Satellite Data. *Remote Sens.* **2017**, *9*, 676. [CrossRef]
29. Dawn, S.; Vikas, S.; Sharma, B. Remote Sensing Image Registration Techniques: A Survey. In *Image and Signal Processing, Proceedings of the 4th International Conference, ICISP 2010, Trois-Rivières, QC, Canada, 30 June—2 July 2010*; Elmoataz, A., Lezoray, O., Nouboud, F., Mammass, D., Meunier, J., Eds.; Springer: Berlin, Germany, 2010; pp. 103–112. ISBN 3-642-13680-X.
30. Bracewell, R.N. *The Fourier Transform and Its Applications*; McGraw-Hill: New York, NY, USA, 1965.
31. Foroosh, H.; Zerubia, J.B.; Berthod, M. Extension of phase correlation to subpixel registration. *IEEE Trans. Image Process.* **2002**, *11*, 188–200. [CrossRef] [PubMed]
32. Mohammed, A.S.; Helmi, A.K.; Fkirin, M.A.; Badwai, S.M. Subpixel Accuracy Analysis of Phase Correlation Shift Measurement Methods Applied to Satellite Imagery. *Int. J. Adv. Comput. Sci. Appl.* **2012**, *3*, 202–206. [CrossRef]
33. Tong, X.; Ye, Z.; Xu, Y.; Gao, S.; Xie, H.; Du, Q.; Liu, S.; Xu, X.; Liu, S.; Luan, K.; et al. Image Registration with Fourier-Based Image Correlation: A Comprehensive Review of Developments and Applications. *IEEE J. Sel. Top. Appl. Earth Obs. Remote Sens.* **2019**, *12*, 4062–4081. [CrossRef]
34. Lucas, B.; Kanade, T. An iterative image registration technique with an application to stereo vision. In Proceedings of the Seventh International Joint Conference on Artificial Intelligence, Vancouver, BC, Canada, 24–28 August 1981; pp. 674–679.
35. Rosenfeld, A.; Vanderbrug, G.J. Coarse-fine template matching. *IEEE Trans. Syst. Man Cybern.* **1977**, *7*, 104–107.
36. Scambos, T.A.; Dutkiewicz, M.J.; Wilson, J.C.; Bindschadler, R.A. Application of image cross-correlation to the measurement of glacier velocity using satellite image data. *Remote Sens. Environ.* **1992**, *42*, 177–186. [CrossRef]
37. Fahnestock, M.; Scambos, T.; Moon, T.; Gardner, A.; Haran, T.M.; Klinger, M. Rapid large-area mapping of ice flow using Landsat 8. *Remote Sens. Environ.* **2016**, *185*, 84–94. [CrossRef]
38. Rosu, A.-M.; Pierrot-Deseilligny, M.; Delorme, A.; Binet, R.; Klinger, Y. Measurement of ground displacement from optical satellite image correlation using the free open-source software MicMac. *ISPRS J. Photogramm. Remote Sens.* **2015**, *100*, 48–59. [CrossRef]
39. Kääb, A.; Vollmer, M. Surface Geometry, Thickness Changes and Flow Fields on Creeping Mountain Permafrost: Automatic Extraction by Digital Image Analysis. *Permafr. Periglac. Processes* **2000**, *11*, 315–326. [CrossRef]
40. Schwalbe, E.; Maas, H.-G. The determination of high-resolution spatio-temporal glacier motion fields from time-lapse sequences. *Earth Surf. Dynam.* **2017**, *5*, 861–879. [CrossRef]

41. Delacourt, C.; Allemand, P.; Berthier, E.; Raucoules, D.; Casson, B.; Grandjean, P.; Pambrun, C.; Varel, E. Remote-sensing techniques for analysing landslide kinematics: A review. *Bull. Soc. Geol. Fr.* **2007**, *178*, 89–100. [CrossRef]
42. Stumpf, A.; Malet, J.-P.; Puissant, A.; Travelletti, J. Monitoring of Earth Surface Motion and Geomorphologic Processes by Optical Image Correlation: Chapter 5. In *Land Surface Remote Sensing: Environment and Risks*; Baghdadi, N., Zribi, F.M., Eds.; ISTE Press—Elsevier: Amsterdam, The Netherlands, 2016; pp. 147–190. ISBN 9781785481055.
43. Fortun, D.; Bouthemy, P.; Kervrann, C. Optical flow modeling and computation: A survey. *Comput. Vis. Image Underst.* **2015**, *134*, 1–26. [CrossRef]
44. Horn, B.K.; Schunck, B.G. Determining optical flow. *Artif. Intell.* **1981**, *17*, 185–203. [CrossRef]
45. Baker, S.; Scharstein, D.; Lewis, J.P.; Roth, S.; Black, M.J.; Szeliski, R. A Database and Evaluation Methodology for Optical Flow. *Int. J. Comput. Vis.* **2011**, *92*, 1–31. [CrossRef]
46. Vogel, C.; Bauder, A.; Schindler, K. Optical flow for glacier motion estimation. *ISPRS Ann. Photogramm. Remote Sens. Spat. Inf. Sci.* **2012**, *I-3*, 359–365. [CrossRef]
47. Altena, B.; Kääb, A. Weekly Glacier Flow Estimation from Dense Satellite Time Series Using Adapted Optical Flow Technology. *Front. Earth Sci.* **2017**, *5*, 53. [CrossRef]
48. Kroeger, T.; Timofte, R.; Dai, D.; van Gool, L. Fast Optical Flow Using Dense Inverse Search. In *Computer Vision—ECCV 2016, Lecture Notes in Computer Science, Proceedings of the European Conference on Computer Vision, Amsterdam, The Netherlands, 11–14 October 2016*; Leibe, B., Matas, J., Sebe, N., Welling, M., Eds.; Springer: Cham, Switzerland, 2016; pp. 471–488. ISBN 978-3-319-46492-3.
49. Baker, S.; Matthews, I. Equivalence and efficiency of image alignment algorithms. In Proceedings of the 2001 IEEE Computer Society Conference on Computer Vision and Pattern Recognition, CVPR, Kauai, HI, USA, 8–14 December 2001; pp. 1–8.
50. Baker, S.; Matthews, I. Lucas-Kanade 20 Years On: A Unifying Framework. *Int. J. Comput. Vis.* **2004**, *56*, 221–255. [CrossRef]
51. Brox, T.; Bregler, C.; Malik, J. Large displacement optical flow. In Proceedings of the IEEE Conference on Computer Vision, Miami, FL, USA, 20–25 June 2009; pp. 41–48. ISBN 1063-6919.
52. Plank, S.; Krautblatter, M.; Thuro, K. Feasibility Assessment of Landslide Monitoring by Means of SAR Interferometry: A Case Study in the Ötztal Alps, Austria. In *Engineering Geology for Society and Territory*; Lollino, G., Giordan, D., Crosta, G.B., Corominas, J., Azzam, R., Wasowski, J., Sciarra, N., Eds.; Springer International Publishing: Cham, Switzerland, 2015; Volume 2, pp. 375–378. ISBN 978-3-319-09056-6.
53. Moretto, S.; Bozzano, F.; Esposito, C.; Mazzanti, P. Lesson learned from the pre-collapse time series of displacement of the Preonzo landslide (Switzerland). *Rend. Online Soc. Geol. Ital.* **2016**, *41*, 247–250. [CrossRef]
54. Sudmanns, M.; Tiede, D.; Augustin, H.; Lang, S. Assessing global Sentinel-2 coverage dynamics and data availability for operational Earth observation (EO) applications using the EO-Compass. *Int. J. Digit. Earth* **2019**, 768–784. [CrossRef]
55. Clapuyt, F.; Vanacker, V.; van Oost, K. Reproducibility of UAV-based earth topography reconstructions based on Structure-from-Motion algorithms. *Geomorphology* **2016**, *260*, 4–15. [CrossRef]
56. Peppa, M.V.; Mills, J.P.; Moore, P.; Miller, P.E.; Chambers, J.E. Brief communication: Landslide motion from cross correlation of UAV-derived morphological attributes. *Nat. Hazards Earth Syst. Sci.* **2017**, *17*, 2143–2150. [CrossRef]
57. Hungr, O.; Leroueil, S.; Picarelli, L. The Varnes classification of landslide types, an update. *Landslides* **2014**, *11*, 167–194. [CrossRef]
58. Barsi, Á.; Kugler, Z.; László, I.; Szabó, G.; Abdulmutalib, H.M. Accuracy dimensions in remote sensing. *Int. Arch. Photogramm. Remote Sens. Spat. Inf. Sci.* **2018**, *XLII-3*, 61–67. [CrossRef]
59. Anker, F.; Fegerl, L.; Hübl, J.; Kaitna, R.; Neumayer, F.; Keuschnig, M. Geschiebetransport in Gletscherbächen der Hohen Tauern: Beispiel Obersulzbach. *Wildbach Lawinenverbauung* **2016**, *80*, 86–96.
60. GeoResearch. Projekt Sattelkar: Zwischenbericht 1: 2018. *Klimasensitive, Hochalpine Kare.* 2018. Available online: http://www.parcs.at/nphts/mmd_fullentry.php?docu_id=37640 (accessed on 29 November 2021).
61. GeoResearch. Projekt Sattelkar: Zwischenbericht 3: 2020. *Klimasensitive, Hochalpine Kare.* 2020. Available online: http://www.parcs.at/nphts/mmd_fullentry.php?docu_id=41348 (accessed on 29 November 2021).
62. Schrott, L.; Otto, J.-C.; Keller, F. Modelling alpine permafrost distribution in the Hohe Tauern region, Austria. *Austrian J. Earth Sci.* **2012**, *105*, 169–183.
63. OpenCV: Dense Optical Flow. Available online: https://docs.opencv.org/3.4/d4/dee/tutorial_optical_flow.html (accessed on 6 November 2021).
64. Leys, C.; Delacre, M.; Mora, Y.L.; Lakens, D.; Ley, C. How to Classify, Detect, and Manage Univariate and Multivariate Outliers, With Emphasis on Pre-Registration. *Int. Rev. Soc. Psychol.* **2019**, *32*, 5. [CrossRef]
65. Segoni, S.; Battistini, A.; Rossi, G.; Rosi, A.; Lagomarsino, D.; Catani, F.; Moretti, S.; Casagli, N. Preface: An operational landslide early warning system at regional scale based on space–time-variable rainfall thresholds. *Nat. Hazards Earth Syst. Sci.* **2015**, *15*, 853–861. [CrossRef]
66. Corominas, J.; Remondo, J.; Farias, P.; Zézere, J.; Estevao, M.; Díaz de Teràn, J.; Schrott, L.; Dikau, R.; Moya, J.; González, A. Debris flow. In *Landslide Recognition: Identification, Moevement and Courses*; Dikau, R., Brundsen, D., Schrott, L., Ibsen, M.-L., Eds.; John Wiley & Sons: New York, NY, USA, 1996; p. 161. ISBN 0471964778.
67. Fleet, D.J.; Black, M.J.; Yacoob, Y.; Jepson, A.D. Design and Use of Linear Models for Image Motion Analysis. *Int. J. Comput. Vis.* **2000**, *36*, 171–193. [CrossRef]

remote sensing

MDPI

Article

A Zoning Earthquake Casualty Prediction Model Based on Machine Learning

Boyi Li [1,2,3], Adu Gong [1,2,3,*], Tingting Zeng [1,2,3], Wenxuan Bao [1,2,3], Can Xu [1,2,3] and Zhiqing Huang [1,2,3]

1 State Key Laboratory of Remote Sensing Science, Beijing Normal University, Beijing 100875, China; 202121051190@mail.bnu.edu.cn (B.L.); zengtt@mail.bnu.edu.cn (T.Z.); 202021051176@mail.bnu.edu.cn (W.B.); 202021051188@mail.bnu.edu.cn (C.X.); 202121051188@mail.bnu.edu.cn (Z.H.)

2 Beijing Key Laboratory of Environmental Remote Sensing and Digital City, Beijing Normal University, Beijing 100875, China

3 Faculty of Geographical Science, Beijing Normal University, Beijing 100875, China

* Correspondence: gad@bnu.edu.cn

Abstract: The evaluation of mortality in earthquake-stricken areas is vital for the emergency response during rescue operations. Hence, an effective and universal approach for accurately predicting the number of casualties due to an earthquake is needed. To obtain a precise casualty prediction method that can be applied to regions with different geographical environments, a spatial division method based on regional differences and a zoning casualty prediction method based on support vector regression (SVR) are proposed in this study. This study comprises three parts: (1) evaluating the importance of influential features on seismic fatality based on random forest to select indicators for the prediction model; (2) dividing the study area into different grades of risk zones with a strata fault line dataset and WorldPop population dataset; and (3) developing a zoning support vector regression model (Z-SVR) with optimal parameters that is suitable for different risk areas. We selected 30 historical earthquakes that occurred in China's mainland from 1950 to 2017 to examine the prediction performance of Z-SVR and compared its performance with those of other widely used machine learning methods. The results show that Z-SVR outperformed the other machine learning methods and can further enhance the accuracy of casualty prediction.

Keywords: earthquake; casualty prediction; importance assessment; spatial division; support vector regression

Citation: Li, B.; Gong, A.; Zeng, T.; Bao, W.; Xu, C.; Huang, Z. A Zoning Earthquake Casualty Prediction Model Based on Machine Learning. *Remote Sens.* **2022**, *14*, 30. https://doi.org/10.3390/rs14010030

Academic Editors: Paolo Mazzanti and Saverio Romeo

Received: 27 November 2021
Accepted: 18 December 2021
Published: 22 December 2021

1. Introduction

Earthquakes are among the most unpredictable and destructive natural hazards around the world and have caused extremely heavy damage to human life and possessions [1–4]. China is located at the intersection of the Alpine-Himalayan and Circum-Pacific seismic zones, and is subjected to the collision and compression of the Eurasian Plate, Philippine Plate and Indian Plate [5,6]; hence, it has always been prone to earthquakes [7,8]. To date, there have been nine catastrophic earthquakes with more than 200,000 casualties in the world, of which three occurred in China. Since 1949, more than 100 destructive earthquakes have occurred in 22 provinces of China, which have caused 270,000 casualties in total, thereby accounting for 54% of all deaths from natural disasters in this country [5]. Considering the heavy destruction of earthquakes in China's mainland, this study selected it as the study area.

After an earthquake, it is necessary to promptly and efficiently conduct emergency rescue to reduce damage and prevent further increases in the damage degree. An early prediction of the death toll that is caused by the earthquake is an essential reference for the government to determine which grade of emergency response [9] to be launched and what amount of relief supplies to be mobilized to the affected areas [10]. Therefore, rapid and accurate prediction of the number of earthquake casualties is a focus of disaster assessment research.

Related studies on seismic casualty prediction focus mainly on two aspects. One aspect is the relationships between relevant factors and the number of earthquake casualties; these studies can be broadly classified into three categories. The studies in the first category explore the impact of seismic parameters on earthquake fatality. Xiao [11] analyzed the relationship of seismic intensity and population density with the mean mortality rate, and proposed an empirical formula for rapidly assessing the death toll, which has been recommended as an effective method for evaluating the mortality rate by *Assessment of Earthquake Disaster Situation in Emergency Period* (a China's national standard). Jaiswal and Wald [12] analyzed the mortality rates of earthquakes with various shaking intensity levels all around the world and proposed a country/region-specific empirical model by using an optimization method to evaluate seismic mortality. The studies in the second category seek to identify the relationship between building vulnerability and earthquake fatality. In the 1980s, commissioned by the Federal Emergency Management Agency (FEMA), the Applied Technology Council (ATC) [13] surveyed and classified buildings in California and proposed the ATC-13 earthquake damage matrix for systematically studying and forecasting possible earthquake losses in this region. Ceferino et al. [14] proposed a probabilistic model for evaluating the number and spatial distribution of casualties due to earthquakes, which improved methods that focused only on a single-building by taking multiple buildings into consideration. The studies in the third category consider the impact of other factors, such as secondary disasters or demographic characteristics, on human loss. Bai et al. [15] scientifically assessed the possible casualties that were caused by secondary disasters and developed a logical regression model for predicting the death toll caused by landslides in the 2014 Yunnan Ludian M_S 6.5 earthquake. Shapira et al. [16] integrated risk factors that are related to population characteristics (age, gender, physical disability and socioeconomic status) and proposed a model on the basis of the widely used loss estimation model HAZUS.

Other studies focus on enhancing the accuracy of prediction models by improving models or proposing new methods [17,18]. Karimzadeh et al. [19] presented a GIS-oriented procedure in combination with geo-related parameters for identifying the destruction in earthquake-stricken areas and evaluated the seismic loss based on damage functions and relational analyses. Feng et al. [20] regarded building damage as a major cause of earthquake deaths, and used high-resolution satellite imagery to detect building damage in disaster areas. They developed a model for estimating the mortality rate due to an earthquake based on remote sensing and a geographical information system. To solve the problems in the evaluation systems (low precision, long time consumption and poor stability), Zhang [21] proposed a seismic disaster casualty assessment system based on mobile communication big data. Considering that seismic data has the characteristics of small scale, nonlinearity and high dimensionality, many scholars have applied machine learning methods, such as support vector machine (SVM), artificial neural network (ANN), and random forest (RF), to earthquake casualty prediction models in recent years. Xing et al. [22] improved SVM with a robust loss function and used it to construct a robust wavelet earthquake casualty prediction model. Gul and Guneri [23] used earthquake magnitude, occurrence time, and population density as input parameters and built a model for earthquake casualty prediction based on the theory of ANN. Jia et al. [24] used the RF model to compare the importance of features affecting the number of earthquake casualties and proposed a deep learning model for casualty prediction.

According to the literature review above, relatively complete earthquake casualty prediction methodologies have been presented by researchers from various aspects, which provide references for feature selection and model construction in our study. However, an analysis of the previous studies on earthquake casualty prediction reveals the following shortcomings: (1) many prediction methods, especially those that utilize empirical functions, can only be implemented with abundant historical seismic data, which makes it difficult to obtain reliable prediction results when a limited quantity of data are available; (2) some scholars simply considered one earthquake as the case and used a small number of samples to predict the death toll, whose achievements may be difficult to apply and deploy due to the under-representativeness of predictors and methods; and (3) most studies simply focused on the statistical relations between influential features and earthquake casualties, which led to inadequate representativeness and lack of a theoretical basis for the generality of such prediction models.

Based on the above observations, this study aimed to (1) evaluate the importance of influential features on seismic fatality, study the regional variations in natural and human geographical environments, and propose a spatial division approach for dividing the study area into three degrees of risk zones; (2) improve the support vector regression (SVR) model with reasonable input factors and the best model parameters for all risk zones; and (3) evaluate the performance of the proposed zoning model through experiments.

The remainder of this paper is structured as follows. Section 2 introduces the geographical and seismic background of the study area and describes the data and methodology that are used in this study. Section 3 presents the process and result of importance assessment and proposes the approach of spatial division. Section 4 derives the SVR algorithm in detail and presents the flow of the data processing and model construction. Section 5 presents the experimental results of the proposed method. Section 6 discusses the results and compares them with those of other models. The conclusions of this study are contained in Section 7.

2. Materials and Methods

2.1. Study Area

China's mainland is located at the intersection of the Alpine-Himalayan and Circum-Pacific seismic zones, where destructive earthquakes occur frequently [25]. Seismicity in China's mainland is characterized by high frequency, wide distribution, great intensity, shallow seismic focus, and clear regional differences. Most earthquakes in this area are shallow focus earthquakes that occurred within the continental crust, whose principal type are strike-slip type [26]. Based on statistical data from the Earthquake Science Knowledge Service System (http://earthquake.ckcest.cn/earthquake_n/dzml/ch5.html, accessed on 15 July 2021), we developed a chart of the spatial distribution of historical earthquakes in China's mainland. Figure 1 shows the positions of plates and all earthquakes over M_S 4.0 that have occurred in China's mainland since 1950. These earthquakes are widely distributed in China's mainland and the spatial pattern of seismic activities in this area is featured by strong activities in the west and weak activities in the east.

With an area of 9.6 million square kilometers (including Taiwan Province), China has diverse natural and human geographical environments that differ in terms of climates, landforms and geological conditions in China; hence, it is difficult to build a single seismic casualty prediction model that is suitable for the whole area. Seismic destructive effects in this vast area are obviously regional. Figure 2 shows the distribution of population and historical earthquakes in China's mainland. The frequency of earthquakes and life losses caused by these disasters are roughly bounded by a population dividing line called the Hu Line [27]. To the east of the Hu Line, earthquakes have caused lager death tolls than those to the west of this boundary, although high seismicity has been observed in the west. Since 1949, 19 provinces in China's mainland suffered deaths due to earthquakes, among which Hebei, Sichuan, and Yunnan Provinces suffered the most life loss events, accounting for more than 90% of all casualties [28].

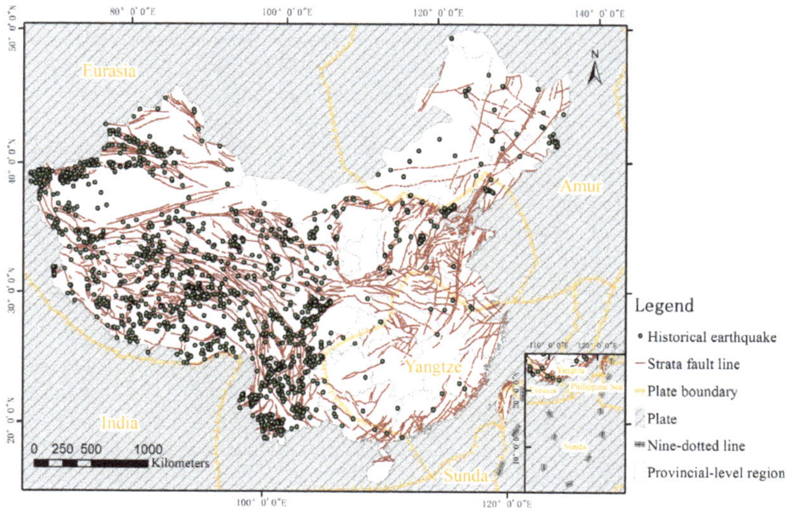

Figure 1. Historical earthquakes and plate distribution in China's mainland; nine-dotted line is the boundary of China's territory in the South China sea.

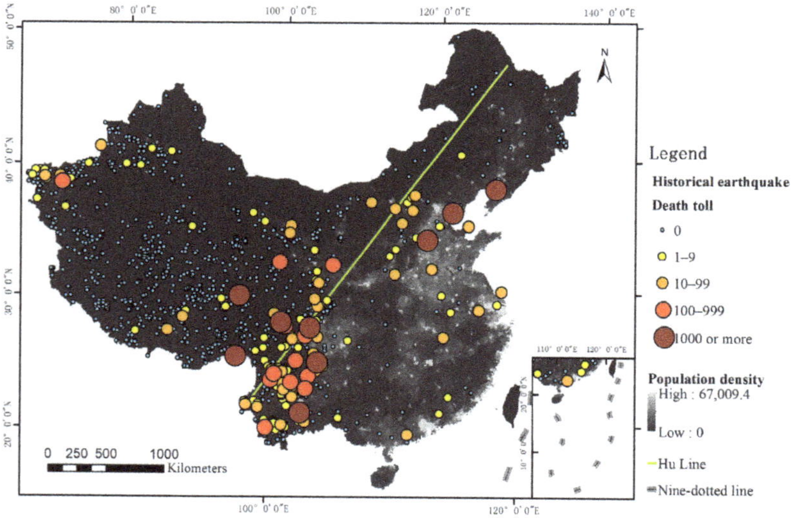

Figure 2. Historical earthquakes and population distribution in China's mainland.

2.2. Materials

The data that were used in this study included a geological fault dataset, a population dataset and an earthquake case dataset. This study trained and verified the proposed prediction model using the earthquake case dataset, which was also used to evaluate the importance of factors affecting seismic fatality. Geological fault and population datasets were used to divide the study area into defined risk zones based on regional differences.

2.2.1. Earthquake Case Dataset

The majority of the earthquake cases were collected from the Earthquake Science Knowledge Service System (http://earthquake.ckcest.cn/featured_resources/disaster_show.html, accessed on 20 July 2021), which includes 479 records of earthquakes over M_S

4.0 that have occurred in China's mainland since 1950. We deleted cases without deaths, corrected and supplemented the dataset with relevant literature and reports [29–32], and finally selected a total of 152 seismic cases with death registers in China's mainland. The original earthquake case dataset only had attributes such as location, occurrence date, magnitude, focal depth and death toll. Because information about historical earthquakes is very limited and difficult to acquire, a large part of the data mining process was devoted to collecting and supplementing relevant attributes. We complemented the attributes of earthquakes, including epicenter intensity, aftershock, landform, climatic condition, secondary disaster, collapsed buildings and rescue capability, from their disaster situation evaluation reports and relevant literature [24]. The attributes of occurrence time and day were converted from the occurrence date. We calculated the linear density of strata faults in ArcGIS software, and used the statistical analysis tool in ArcGIS to acquire the earthquake attribute of geological fault density. The attributes of population density and the Gross domestic product (GDP) were collected from statistical yearbooks of provinces where earthquakes occurred. GDP is a monetary measure of the market value of all the final goods and services produced in a specific time period. The data we collected is per capital GDP, which is the ratio of GDP to the total population of the earth-quake-stricken region. Detailed information about each attribute in the earthquake case dataset is provided in Table 1.

Table 1. Specification of attributes in the earthquake case dataset.

No.	Attribute	Description & Qualification
1	Occurrence day	There are 7 categories where 1~7 correspond to Monday to Sunday, respectively.
2	Occurrence time	The time when the earthquake occurred, which is defined as the minutes after 0:00 on the day.
3	Location	The province and city where the earthquake occurred, including longitude and latitude.
4	Magnitude	Defined as the surface wave magnitude.
5	Focal depth	The vertical distance from the hypocenter to the surface of the earth (km).
6	Epicenter intensity	Measured according to *The China Seismic Intensity Scale* (China's national standard).
7	Aftershock	The number of shocks of magnitude greater than M_S 5.0 after the occurrence of the main shock.
8	Geological fault density	The average density of strata faults in the earthquake-stricken area.
9	Landform	There are five categories, which are labelled 1 to 5, and represent plain, basin, hill, mountain and plateau, respectively.
10	Climatic condition	There are two levels where 0 indicates normal and 1 indicates abnormal.
11	Secondary disaster	There are two categories, where 0 indicates no secondary disaster and 1 indicates the occurrence of a secondary disaster.
12	Population density	The number of people who live in the earthquake-stricken area per square kilometer.
13	Collapsed buildings	The number of collapsed houses.
14	Rescue capability	There are three levels where 1 indicates lacking assignment, 2 indicates general assignment and 3 indicates improved assignment.
15	GDP	The ratio of GDP to the total population of the earthquake-stricken region.
16	Death toll	The number of casualties due to the earthquake.

To describe the data distribution characteristics of earthquake cases, we divided their numbers of casualties into 6 categories: 0–9, 10–99, 100–999, 1000–9999, 10,000–99,999, and ≥100,000. Then, we calculated the piecewise frequency statistics for each category and plotted a statistical chart, which is shown in Figure 3. As shown in this graph, the death tolls of most earthquakes in the dataset were within the ranges of less than 10, 10–99 and 100–999. Strong earthquakes with many casualties occurred with lower frequency; hence, this study focuses on accurately predicting the death toll for earthquakes with less than 1000 casualties.

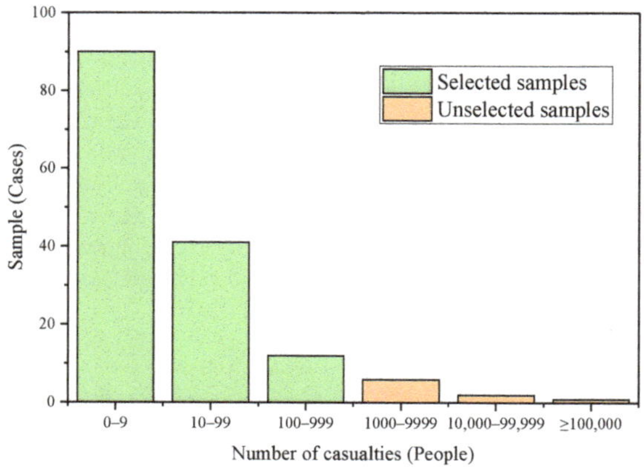

Figure 3. Piecewise frequency statistics of earthquake casualties.

In the construction process of the machine learning model, earthquake samples with many casualties will exert a significant impact on the performance of the prediction model. To evaluate the influence of samples with great values, we conducted an experiment to compare the prediction performance between two data groups: Group A and Group B. Group A was the dataset including all the 152 seismic cases with 1000 casualties or more. Group B was the dataset excluding samples whose numbers of casualties were more than 1000. We took Group A as the training dataset and input it into SVR model, and used the 10-fold cross-validation method to evaluate its prediction performance. The evaluation indicators employed in this experiment were root mean square error (RMSE) and mean absolute error (MeaAE), which are described in detail in Section 6.1. The same experiment was also conducted in Group B. We calculated the average RMSE and MeaAE values for the two groups. The result showed that the RMSE and MeaAE of Group A were 6579.29 and 2346.96, respectively. By contrast, the RMSE and MeaAE of Group B were 48.27 and 40.41 respectively, which means Group B shows significantly better prediction performance due to the exclusion of extreme value samples.

Considering that the devastating earthquakes with more than 1000 casualties occur extremely unfrequently, and their disaster mechanisms are much more complicated, the study focuses on accurately predicting the death toll for earthquakes with less than 1000 casualties. Therefore, we removed cases with more than 1000 casualties in order to avoid the influence of great values. A total of 143 seismic cases with death registers were finally selected. The procedure of dataset division is as follows. (1) In Section 3, we proposes a spatial division method and divides the study area into three groups: high, moderate and low risk zones. Based on the result of spatial division, those selected cases were divided into three parts, including 49 cases in low risk areas, 13 in moderate risk areas, and 81 in high risk areas. (2) To evaluate the prediction accuracy of the Z-SVR model for three degrees of risk zones, we divided the dataset into training and testing datasets. For earthquake cases in each degree of risk zones, we randomly extracted 1/5 of them as the testing dataset, and the remainder was divided into the training dataset. We finally extracted 10 cases in low risk zones, 3 in moderate risk zones, and 17 in high risk zones as the testing dataset to evaluate the performance of the seismic casualty prediction model. The remainder was used as training dataset for building Z-SVR model. Table 2 presents the division of sample dataset. Figure 4 shows the spatial distributions of historical cases.

Table 2. Numbers of training and testing samples in the defined risk zones.

Zone	Training Sample (Cases)	Testing Sample (Cases)	Total (Cases)
Low risk	39	10	49
Moderate risk	10	3	13
High risk	64	17	81
Total	113	30	143

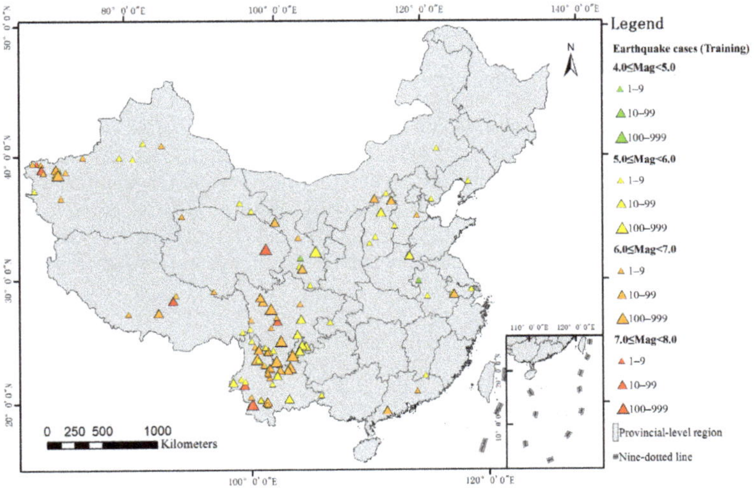

(**a**)

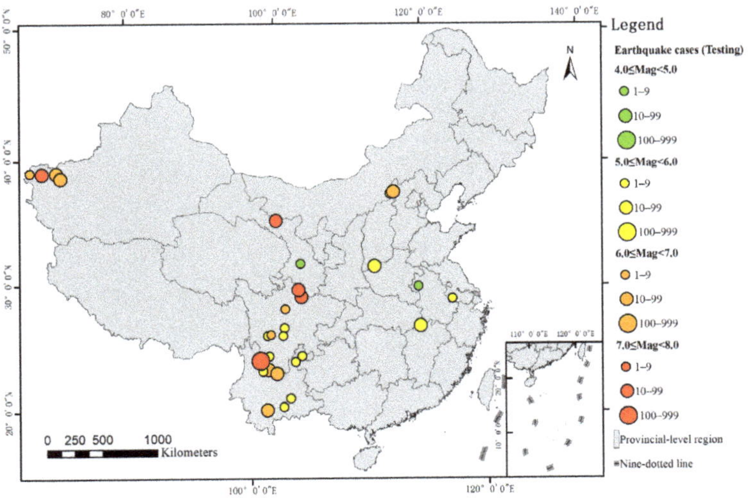

(**b**)

Figure 4. Spatial distributions of the earthquake case dataset: (**a**) Training samples; (**b**) testing samples.

2.2.2. Geological Fault Dataset

We collected the geological fault dataset from the China Earthquake Data Center (http://datashare.igl.earthquake.cn/map/ActiveFault/introFault.html, accessed on 24 July 2021). It provides the spatial distribution of strata faults in China; the data are in vector format and can be used for spatial analysis in ArcGIS software. This dataset includes 1966 fault segments. For 456 of these segments, detailed parameters such as age, orientation and sliding rate are provided; for 664, only the name and number are specified; for 846, only graphical features are provided, without any attributes. Since the coordinate system of the dataset is the Krassovsky ellipsoid with the Albers projection, we used the projection raster tool in ArcGIS to convert it into the WGS 1984 to ensure the consistency of the spatial reference.

2.2.3. Population Dataset

The population dataset was collected from WorldPop (https://www.worldpop.org/, accessed on 28 July 2021). It details the spatial distribution of the population with a spatial resolution of 100 m. Its units are number of people per pixel with country totals adjusted to match United Nations national population estimates. The format of this dataset is raster, where the digital value of every pixel reflects the total population within this grid. Considering that the samples in the earthquake case dataset have a long time series while population data of a single year have difficulty reflecting demographic changes, we collected population records in China's mainland every five years from 2000 to 2020 (2000, 2005, 2010, 2015 and 2020) to explore the change in population in a long time series.

2.3. *Methods*

A methodological flowchart of the investigation is shown in Figure 5.

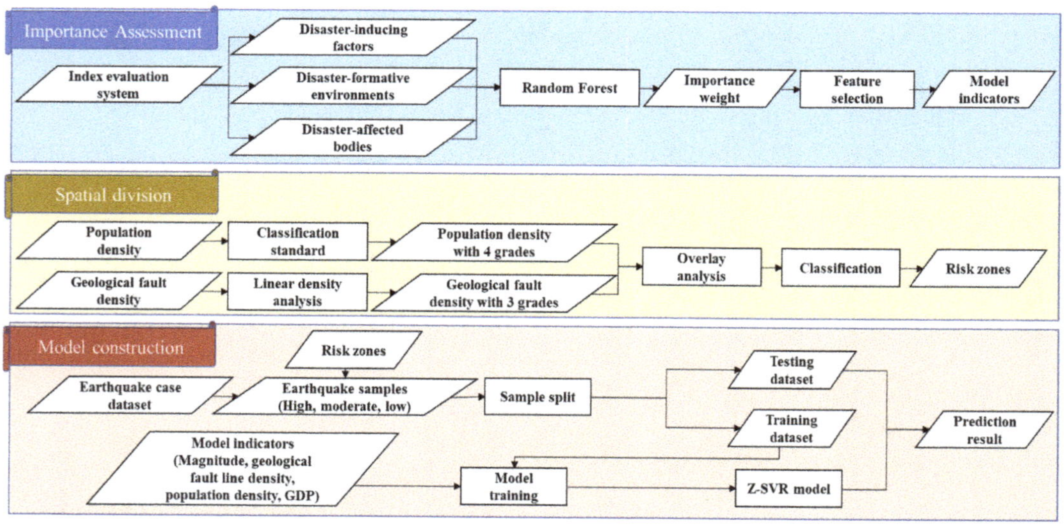

Figure 5. Framework of the Z-SVR model.

Seismic fatality is a comprehensive result that is influenced by diverse factors, and whether a factor has a decisive impact on earthquake casualties is an essential question for feature selection of prediction models [33]. Therefore, before constructing a prediction model for earthquake casualties, it is crucial to establish a reasonable index system and analyze the importance of relevant indicators, which will serve as a reference for the prediction model to select more important features. Based on regional disaster system theory, this study established an evaluation index system for 14 major features that affect earthquake fatality. We used the earthquake case dataset and the random forest model to assess the importance weights of features, of which the ranking served as an important reference for feature selection of the prediction model.

Because of the variations among regions, there will be different numbers of casualties due to earthquakes with the same ground motion parameter. Therefore, in earthquake cases with the same seismicity, the diversity of disaster-formative environments and disaster-affected bodies reflects the difference among regions [34]. Due to the vast area of China's mainland, it is difficult to build a universal prediction model that is suitable for all regions. To enhance the accuracy of earthquake disaster assessment in emergency periods, it is effective to divide the study area into risk zones based on regional differences and construct a model that performs well for each risk zone. Based on the results of the importance assessment and feature selection, geological fault density and population density are the most important features of disaster-formative environments and disaster-affected bodies, respectively. Therefore, we chose these two features with relatively high importance weights as representative factors for developing a partition standard and dividing the study area into the defined grades of risk zones. The accuracy and applicability of the earthquake casualty prediction approach can be improved by building different submodels for areas with different regional characteristics.

As an extension of support vector machine (SVM) for solving regression problems support vector regression (SVR) has attracted much attention in the field of machine learning and displayed strong predictive ability in mortality evaluation. Compared with other machine learning algorithms, SVR can achieve the optimal solution with a small number of samples and avoid problems such as overfitting and local extremum as much as possible, which makes its generalization ability and performance stand out [35]. However, as a machine learning method that is based on historical statistics, it may be difficult for the SVR model to accurately predict casualties due to earthquakes occurring in different regions of the study area, especially those with vast acreage and diverse environments. Therefore, based on the characteristics of SVR and regional differences in the study area, we constructed a zoning SVR model (Z-SVR) for various regions in the study area; for which the optimal model parameters for all risk zones were identified using training samples from the earthquake case dataset.

3. Spatial Division

3.1. Importance Assessment

According to regional disaster system theory, a seismic disaster is a complex mechanism that is a comprehensive result of interactions between disaster-inducing factors, disaster-affected bodies and disaster-formative environments [36]. Among them, disaster-inducing factors, such as seismic magnitude and focal depth, are the sufficient conditions for disaster occurrence; disaster-affected bodies, such as population distribution and building destruction, represent the necessary conditions for disaster resilience; and disaster-formative environments, such as climatic condition and secondary disaster, provide a natural and human geological background that affects disaster-inducing factors and disaster-affected bodies [17]. The loss due to a disaster is attributed to the combined effects of these three factors; therefore, for screening the prediction indicators, we constructed an evaluation index system on the basis of regional disaster system theory, which is presented in Table 3.

Table 3. Evaluation index system of features that influence earthquake fatality.

Target Level	Rule Level	Index Level
Seismic fatality	Disaster-inducing factors	Magnitude Epicenter intensity Focal depth Geological fault density Occurrence time Occurrence day Aftershock
	Disaster-affected bodies	Collapsed buildings Rescue capability Population density GDP
	Disaster-formative environments	Climatic condition Landform Secondary disaster

Determining the importance weights of all features in the evaluation index system is a quantitative task in importance assessment. Although traditional linear models show good performance in the importance assessment of factors that affect earthquake fatality, the result can be easily disturbed by the uncertainty and fuzziness of input data [37]. An integrated ensemble model is an effective approach for mitigating the above problem and improving the accuracy and generalization performance of the evaluation method [38], which was demonstrated by previous studies [39]. Random forest (RF) is an effective integrated ensemble model with random binary decision trees for classification or regression [39]. As an expansion of the bagging method, this algorithm constructs multiple independent estimators that determine the output result by average or majority voting. This approach enhances the precision and stability of the prediction model, reduces the sensitivity of the model to noise and outliers, and avoids problems such as overfitting [40]. In contrast to other machine learning methods, the RF model can provide the quantified importance of prediction indicators by calculating their increases in predictive error by randomly permuting the values of a variable through out-of-bag observations of each tree.

We chose 7 indicators of disaster-inducing factors, 4 of disaster-affected bodies and 3 of disaster-formative environments as the input parameters of the RF model to evaluate their importance to earthquake fatality. The values of the input parameters were extracted from the earthquake case dataset. We utilized the machine learning package scikit-learn of the Python programming language to construct the RF model. The "feature_importances_" is an attribute of the RF model in the scikit-learn package. The importance of a feature is computed as the normalized total reduction of the criterion brought by that feature. The procedure is summarized as follows:

- Inputs: Disaster-inducing factors (7 variables), disaster-affected bodies (4 variables) and disaster-formative environments (3 variables).
- Parameters: Number of estimators = 150, criterion = 'squared_error', max depth = 6, min samples split = 2, min samples leaf = 1, min weight fraction leaf = 0.0, max features = 'auto', max leaf nodes = None, min impurity decrease = 0.0, bootstrap = Frue, oob score = False, number of jobs = None, random state = None, verbose = 0, warm start = False, ccp_alpha = 0.0, max samples = None.
- Step 1: Use bootstrap sampling to extract subtraining sets from the training set.
- Step 2: Generate the feature subsets by randomly selecting features before node splitting.
- Step 3: Establish decision trees.
- Step 4: Obtain the results for the sample to be tested.
- Step 5: Calculate the importance of the input parameters.
- Output: Importance weight of the prediction indicators.

The ranking of all factors according to the importance weights from low to high is shown in Figure 6.

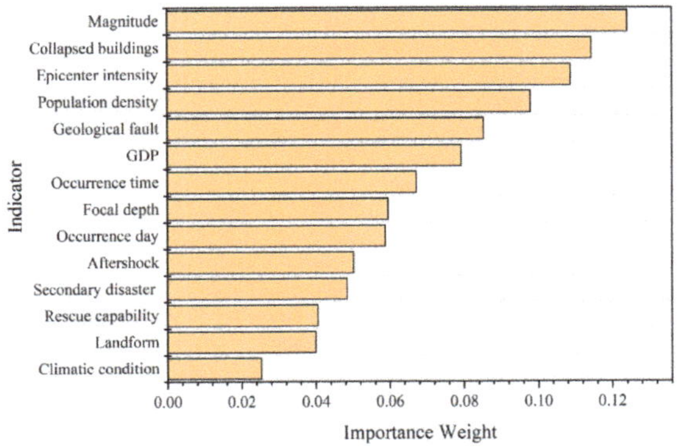

Figure 6. Importance weights of indicators on the index levels.

Based on the results of the importance assessment of influential features, magnitude, collapsed buildings, epicenter intensity, population density, geological fault density and GDP are major factors that affect seismic fatality. Magnitude and epicenter intensity are the two most important parameters to depict the severity of an earthquake and exert substantial influence on the seismic fatality; however, there is a strong correlation between these two features. To avoid information redundancy, we selected magnitude, which has greater importance weight, as the input parameter of the Z-SVR model. Building destruction is the direct cause of earthquake injuries and deaths [41], and the primary task of emergency rescue is to search for people who are buried in collapsed constructions. However, the aim of the proposed model in this study is to rapidly predict the possible casualties of an instantly occurring earthquake, which requires an extremely fast response speed. It will take some time to identify the situation of building destruction and count the number of collapsed buildings. Population density is the most important feature among the disaster-affecting bodies; since human beings are the major victims of earthquakes, it is significant to choose this feature as one of the prediction indicators. Geological fault is the most important factor under the level of disaster-formative environments, where the density of strata fault lines can be used to quantitively analyze regional differentiation and merits consideration. GDP is a comprehensive indicator that is mutually restricted with population density in terms of earthquake casualties; therefore, it is significant to introduce this factor as an input parameter and consider its comprehensive effect with population density to ensure the stability and accuracy of the prediction results. In conclusion, based on the result of the importance assessment and the principles of rapid evaluation and avoiding information redundancy, we finally selected magnitude, population density, geological fault density and GDP as the input parameters for the construction of the Z-SVR model, among which geological fault line density and population density were also applied to divide the study area into risk zones.

3.2. Population Density

Disaster-affected bodies reflect the necessary conditions for disaster resilience, of which population density has a major influence on the number of earthquake casualties and the degree of destruction. High population density provides a vital motivation for the increase in earthquake casualties [42]. In this study, the population dataset that was collected from WorldPop includes raster data on the population distribution of China's mainland every five years from 2000 to 2020 (2000, 2005, 2010, 2015 and 2020). For those five raster datasets, we converted the population count value to population density and calculated the average density, which was implemented using the raster calculator tool in ArcGIS software.

The general classification standard of population density was used to divide different population densities into four categroies: extremely sparsely (less than 1 people/km^2), sparsely (from 1 to 25 people/km^2), moderately (from 25 to 100 people/km^2), and densely populated (greater than 100). Through this standard, we divided China's population distribution dataset into four parts, as shown in Figure 7.

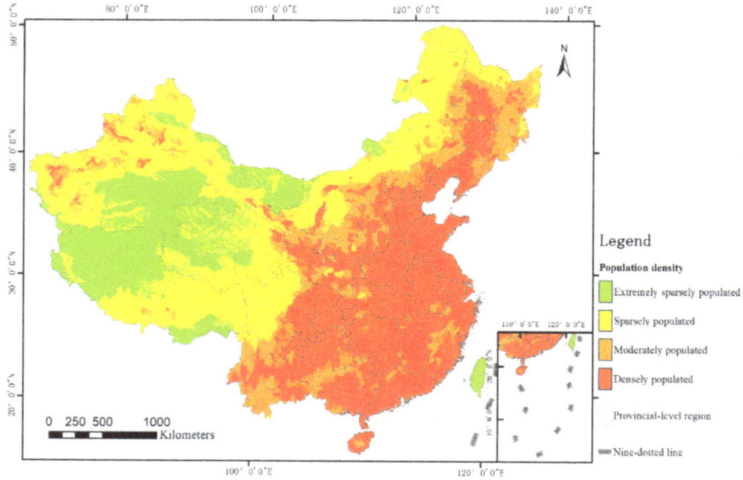

Figure 7. Distribution of classified population density in China's mainland.

3.3. *Geological Fault Density*

Disaster-formative environments refer to the natural and human geological background that affects disaster-inducing factors and disaster-affected bodies [17], among which geological faults are the zone blocks that bump into each other and generate shakes. Previous work [28] has demonstrated that the distance from a geological fault is correlated with the number of casualties that are caused by an earthquake. Therefore, we calculated the linear densities of strata faults in China using ArcGIS software. The linear densities were divided into three grades (high, moderate and low) by natural breaks. Figure 8 shows the spatial distribution of the classified geological fault densities in the study area.

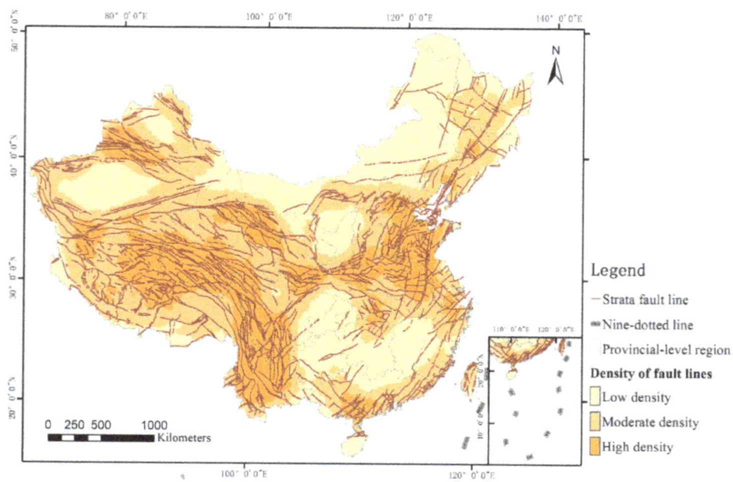

Figure 8. Distribution of the classified strata fault densities in China's mainland.

3.4. Overlay Analysis

Overlay analysis is a frequently used geographic computing operation and a significant spatial analysis tool in GIS software, which is widely used in applications that are related to spatial computing [43]. This operation integrates different data layers and their corresponding attributes in the study area, which connects multiple spatial objects from multiple data sources and quantitatively analyzes the spatial range and characteristics of the interactions among different forms of spatial objects. Based on the feature selection results, geological faults are the birthplace of an earthquake, and humans are the victims of seismic disasters. In earthquakes with similar seismicity, denser strata fault lines and higher population density will lead to a greater risk to personnel safety [28]. For the above reasons, this study divided the study area into parts according to the variations in population density and strata fault density and established a corresponding partition standard. We developed a comprehensive partition standard that was used to overlay the classification results. Then, we divided the study area into risk areas of three grades: low risk, moderate risk, and high risk zones. The theory and procedure of the proposed spatial division method are illustrated in Figure 9.

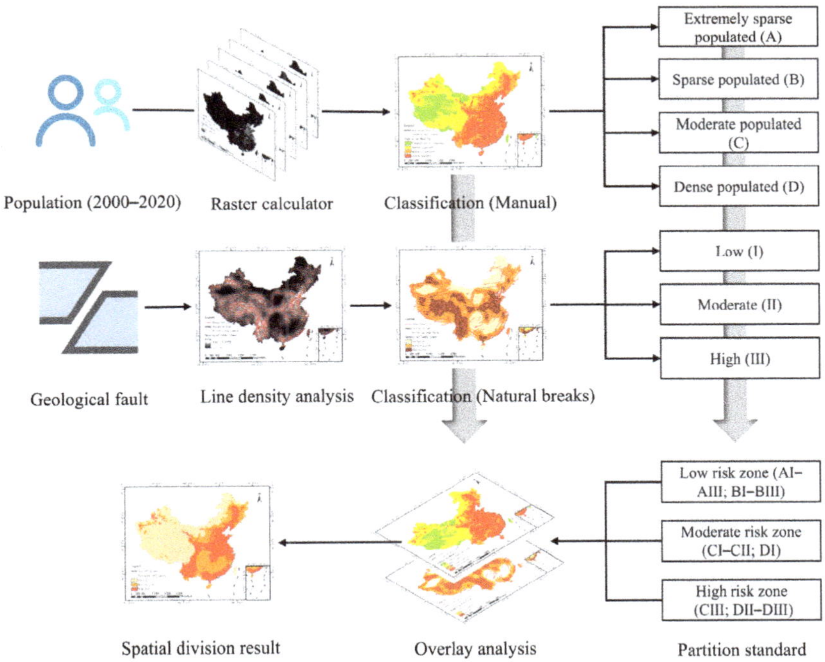

Figure 9. Spatial division process.

4. Prediction Model

4.1. Algorithm

Support vector machine (SVM) is a kind of machine learning method that is based on statistical learning theory and is a supervised learning model [44]. SVM implements the structural risk minimization principle rather than the empirical risk minimization principle [45], which gives it unique advantages in solving small-sample, nonlinear and high-dimensional pattern recognition problems. Although SVM was initially applied to classification problems, it has been gradually used to solve regression problems due to its good performance in function fitting [46]. SVR is an extension of SVM for solving regression problems. Compared with other machine learning algorithms, SVR can obtain the optimal solution with a small number of samples and avoid problems such as overfitting and partial extreme values as much as possible [28], and its generalization ability and performance have been well demonstrated.

The SVR algorithm is explained as follows. Consider a given training sample set $D = \{(x_1, y_1), (x_2, y_2), \ldots\ldots, (x_m, y_m)\}$, where $x_i = (x_{i1}, x_{i2}, \ldots, x_{id})^T \in \mathbb{R}^d$, $y_i \in \mathbb{R}$, $i = 1, 2, \ldots, \mathrm{m}$, x_i is the ith sample and has feature dimensionality d, x_{ij} is the value of the jth feature, $y_i \in R$ is the corresponding target value of the ith sample, and m is the number of samples. The goal of SVR is to find a regression model $f(x) = \omega^T x + b$ such that $f(x)$ is close to its corresponding target value y, where ω and are parameters to be calculated. In the traditional regression model, the function loss is calculated based on the difference between $f(x)$ and y, which is too strict and will eventually lead to overfitting [47]. To overcome this disadvantage, SVR sets a maximum deviation ϵ between $f(x)$ and y, and the function loss is counted only when the difference between $f(x_i)$ and y_i is greater than ϵ (Figure 10). This is equivalent to constructing a spacer band of width 2ε with $f(x)$ as the center; when the training sample is within the spacer band, the prediction result will be designated as correct [48]. Therefore, the SVR problem can be formulated as

$$\min_{\omega,b} \frac{1}{2}||\omega||^2 + C\sum_{i=1}^{m} \ell_\epsilon(f(x_i) - y_i) \tag{1}$$

where $C > 0$ is a regularization constant and ℓ_ϵ is an ϵ-insensitive loss function (Figure 11), which is expressed as

$$\ell_\epsilon(z) = \begin{cases} 0, & if\ |z| \leq \epsilon; \\ |z| - \epsilon, & otherwise. \end{cases} \tag{2}$$

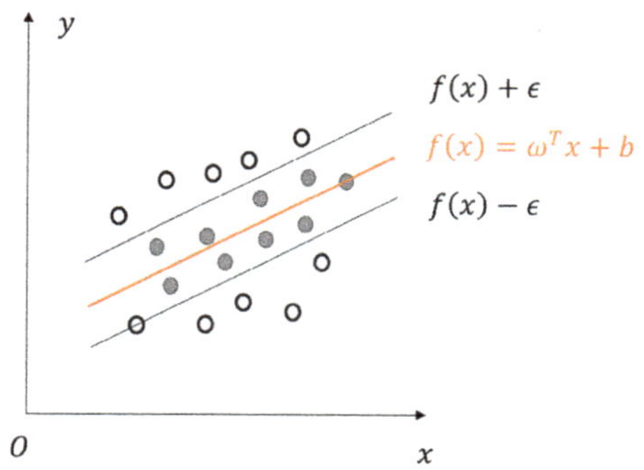

Figure 10. Sketch diagram for SVR.

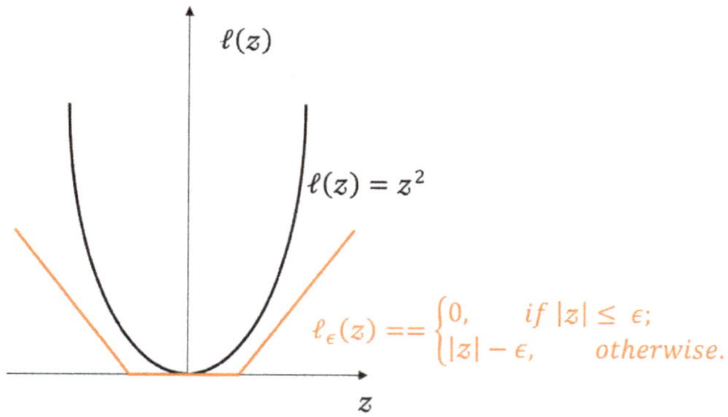

Figure 11. Sketch diagram for the ϵ-insensitive loss function.

The first term of Equation (1) represents the flatness of the function, which is also called the structural risk, and the second term of the equation, namely, $\sum_{i=1}^{m} \ell_\epsilon(f(x_i) - y_i)$, represents the fitness between $f(x)$ and its corresponding target values, which is also called the empirical risk [48]. The regularization constant C is a compromise between the structural risk and empirical risk. The constant $C > 0$ determines the trade-off between the flatness of $f(x)$ and the amount up to which deviations larger than ϵ are tolerated [49]. To describe the real deviation, two slack variables, namely, ξ_i and $\hat{\xi}_i$, are introduced, and Equation (1) can be reformulated as

$$\min_{\omega,b,\xi,\xi^*} \frac{1}{2}||\omega||^2 + C\sum_{i=1}^{m}(\xi_i + \hat{\xi}_i); s.t. \begin{cases} f(x_i) - y_i \leq \epsilon + \xi_i; \\ y_i - f(x_i) \leq \epsilon + \hat{\xi}_i; \\ \xi_i, \hat{\xi}_i \geq 0,\ i = 1,2,\ldots,m. \end{cases} \tag{3}$$

To efficiently solve the above optimization problem with inequality constraints, multipliers $\mu_i \geq 0$, $\hat{\mu}_i \geq 0$, $\alpha_i \geq 0$, and $\hat{\alpha}_i \geq 0$ are introduced. Based on the Lagrange multiplier method, the following function can be deduced from Equation (3):

$$\begin{aligned} L(\omega, b, \alpha, \hat{\alpha}, \xi, \hat{\xi}, \mu, \hat{\mu}) \\ &= \tfrac{1}{2}||\omega||^2 + C\sum_{i=1}^{n}(\xi_i + \hat{\xi}_i) - \sum_{i=1}^{m}\mu_i\xi_i - \sum_{i=1}^{m}\hat{\mu}_i\hat{\xi}_i \\ &+ \sum_{i=1}^{m}\alpha_i(f(x_i) - y_i - \epsilon - \xi_i) + \sum_{i=1}^{m}\hat{\alpha}_i(y_i - f(x_i) - \epsilon - \hat{\xi}_i). \end{aligned} \tag{4}$$

$f(x) = \omega^T x + b$ is substituted into Equation (4), the partial derivatives of $L(\omega, b, \alpha, \hat{\alpha}, \xi, \hat{\xi}, \mu, \hat{\mu})$ with respect to ω, b ξ_i and $\hat{\xi}_i$ are calculated, and these partial derivatives are set equal to 0. The following system of equations is obtained:

$$\omega = \sum_{i=1}^{m}(\hat{\alpha}_i - \alpha_i)x_i, \tag{5}$$

$$0 = \sum_{i=1}^{m}(\hat{\alpha}_i - \alpha_i), \tag{6}$$

$$C = \alpha_i + \mu_i, \tag{7}$$

$$C = \hat{\alpha}_i + \hat{\mu}_i. \tag{8}$$

After solving the above system of equations, the dual problem of SVR can be formulated as

$$\begin{aligned} &\max_{\alpha,\hat{\alpha}} \sum_{i=1}^{m}(y_i(\hat{\alpha}_i - \alpha_i) - \epsilon(\hat{\alpha}_i + \alpha_i)) - \tfrac{1}{2}\sum_{i=1}^{m}\sum_{j=1}^{m}(\hat{\alpha}_i - \alpha_i)(\hat{\alpha}_j - \alpha_j)x_i^T x_j; \\ &\quad s.t.\ \sum_{i=1}^{m}(\hat{\alpha}_i - \alpha_i) = 0,\ 0 \leq \alpha_i, \hat{\alpha}_i \leq C. \end{aligned} \tag{9}$$

To solve the above quadratic programming problem, the Karush-Kuhn–Tucker (KKT) conditions [50] are used:

$$\begin{cases} \alpha_i(f(x_i) - y_i - \epsilon - \xi_i) = 0, \\ \hat{\alpha}_i(y_i - f(x_i) - \epsilon - \hat{\xi}_i) = 0, \\ \alpha_i\hat{\alpha}_i = 0,\ \xi_i\hat{\xi}_i = 0, \\ (C - \alpha_i)\,\xi_i = 0,\ (C - \hat{\alpha}_i)\hat{\xi}_i = 0. \end{cases} \tag{10}$$

Substituting Equation (5) into $f(x) = \omega^T x + b$ yields the following solution of the SVR:

$$f(x) = \sum_{i=1}^{m}(\hat{\alpha}_i - \alpha_i)x_i^T x + b. \tag{11}$$

If the term $(\hat{\alpha}_i - \alpha_i)$ of Equation (11) is not equal to 0, the corresponding sample is a support vector of SVR that is located outside the spacer band. Based on the KKT conditions, it is found that in Equation (10), every sample $(x_i,\ y_i)$ satisfies the conditions $(C - \alpha_i)\ \xi_i = 0$ and $\alpha_i(f(x_i) - y_i - \epsilon - \xi_i) = 0$; therefore, ξ_i is equal to 0 when $0 < \alpha_i < C$. Then, the value of b can be deduced from Equation (11) as

$$b = y_i + \epsilon - \sum_{i=1}^{m}(\hat{\alpha}_i - \alpha_i)x_i^T x. \tag{12}$$

However, Equation (11) is merely a solution for linear SVR. For real-world problems with high feature dimensionality, it is impossible to find a hyperplane that satisfies both fitness and flatness simultaneously [47]. An efficient approach is to map samples from the original space to a higher-dimensional feature space where the samples are linearly separable [48], and Equation (5) can be reformulated as

$$\omega = \sum_{i=1}^{m}(\hat{\alpha}_i - \alpha_i)\phi(x_i) \tag{13}$$

where $\phi(x_i)$ is the feature vector after mapping to a higher-dimensional feature space.

With the utilization of the kernel function method, the following solution for nonlinear SVR is obtained:

$$f(x) = \sum_{i=1}^{m}(\hat{\alpha}_i - \alpha_i)\kappa(x, x_i) + b \tag{14}$$

where $\kappa(x, x_i) = \phi(x)^T\phi(x_i)$ is the kernel function. Table 4 presents various widely used kernel functions.

Table 4. Specification of kernel functions.

Type	Expression [1]
Linear kernel	$K(u,v) = u^T v$
Gaussian kernel	$K(u,v) = \mathrm{e}^{-\gamma\|\|u-v\|\|^2},\ \gamma > 0$
Polynomial kernel	$K(u,v) = (\gamma u^T v + r)^d$
Sigmoid kernel	$K(u,v) = \tanh(\gamma u^T v + r)$

[1] u and v are multivariate vectors, and $d \geq 1$ is the degree of the polynomial.

4.2. Model Construction

Based on the results of the importance assessment and feature selection, we selected the magnitude, population density, geological fault density and GDP as the input variables and selected the number of earthquake casualties as the output variable. Considering that different prediction indicators have different units of measurements, it is necessary to normalize the sample dataset to enhance the convergence speed in finding the optimal solution and to improve the accuracy of the Z-SVR model. The normalization method that was used in this study was z-score normalization, which can be formulated as

$$z_i = \frac{x_i - \overline{x}}{\sqrt{\frac{1}{n}\sum_{i=1}^{n}(x_i - \overline{x})^2}} \tag{15}$$

where n is the number of samples in the dataset, x_i. is the initial value of the i th sample, z is its corresponding normalized value, and $\overline{x} = \sum_{i=1}^{n} x_i$ is the average initial value of all samples.

Previous studies [51,52] have shown that the type of kernel function and corresponding parameters have substantial impacts on the prediction performance of the SVR model. To construct a fine-tuned Z-SVR model, parameter C for the linear kernel, parameters (C, gamma) for the Gaussian kernel and sigmoid kernel, and parameters (*C*, gamma, degree) for the polynomial kernel should be selected [47]. *C* is the regularization parameter; gamma and degree are equivalent to γ and *d*. in Table 4, respectively. Grid search is a general and effective method for parameter optimization, which is usually combined with cross-validation [17]. To find the best SVR model for each risk zone, this study invoked the GridSearchCV module in the scikit-learn package to search for optimal kernel functions and their corresponding model parameters in a specified range based on grid search. The selected parameters of the Z-SVR model are presented in Table 5.

Table 5. Model parameters of Z-SVR.

Zone	Kernel Function	Parameters
Lowisk	Gaussian kernel	C = 100, gamma = 0.1
Moderate risk	Gaussian kernel	C = 100, gamma = 1
High risk	Gaussian kernel	C = 1000, gamma = 0.1

This study obtained the Z-SVR model using the Python programming language and machine learning package scikit-learn. The procedure of model establishment is summarized as follows: (1) Select suitable features as input parameters. (2) Preprocess the sample dataset by normalizing and dividing samples into training data and testing data. (3) Establishing a scoring rule for comparing the predicted results with the actual number of death casualties; if these two values are of the same order of magnitude, the prediction will be considered correct. (4) Invoke the SVR module in the scikit-learn package to build a model for each risk zone. (5) Invoke the GridSearchCV module in the scikit-learn package, and obtain parameters and search ranges; then, use the 10-fold cross-validation method to test the robustness of the model. (6) Input the training dataset into the SVR model for each risk zone to obtain optimal kernel functions and their corresponding model parameters for the Z-SVR model. (7) Input the testing dataset into Z-SVR model and predict the earthquake death tolls. (8) Since the number of earthquake casualties should not be negative, revise negative prediction results by setting them to 0. (9) Assess the performance of the Z-SVR model on the testing dataset.

5. Results

5.1. Spatial Division of the Study Area

Considering the vast area and diverse environments of China's mainland, to build an earthquake casualty prediction model with better applicability, it is helpful to propose a machine learning approach with submodels that are applied to different regions. Using the strata fault dataset and population dataset, we divided the study area into risk zones using the raster calculator tool in ArcGIS software according to the proposed partition standard. We plotted the spatial division results and overlaid historical earthquakes with various magnitudes and numbers of casualties onto it, as shown in Figure 12.

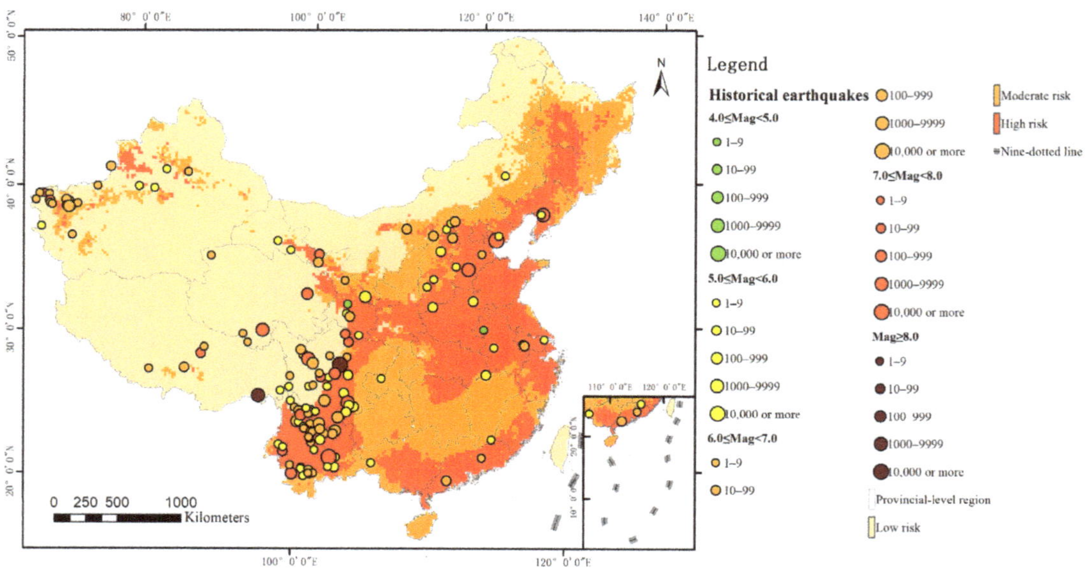

Figure 12. Distribution of risk zones and historical earthquake in China's mainland.

As shown in Figure 12, low risk zones were the most extensive, which accounted for 51.94% of China's mainland, followed by high risk zones, which accounted for 25.59%. The area of moderate risk zones was the smallest, which accounted for 22.47% of the study area. According to the distribution of historical earthquakes, the majority of destructive earthquakes occurred in high risk areas, which indicates the validity of the proposed spatial division method. Fewer destructive earthquakes occurred in some provinces of Northern China (Heilongjiang, Jilin, Beijing and Shanxi), Southern China (Hubei, Hunan and Guizhou) and Eastern China (Zhejiang and Fujian), while these regions were divided into high or moderate risk zones. This can be explained by the presence of dense strata fault lines or high population density in these provinces. Considering that regions with fewer earthquakes usually encounter more casualties due to failure to take necessary precautions for disasters, it is significant for people in high and moderate risk zones to be trained with anti-seismic knowledge and to engage in evacuation practices. Interestingly, although earthquakes occurred in Xizang, Qinghai and Xinjiang Provinces of Western China, most parts of these regions were divided into low risk zones. This inconsistency is due to the low population densities of these provinces, which contain vast depopulated zones; this is supported by the observation that most earthquakes with high seismicity caused minor casualties in low risk zones.

5.2. Prediction Result of Z-SVR Model

This study improved the SVR model and proposed the Z-SVR model with optimal parameters for different risk areas. We randomly selected 10 samples in low risk zones (L1~L10), 3 in moderate risk zones (M1~M3) and 17 in high risk zones (H1~H17) to predict the numbers of casualties and compare them with corresponding true values, which are presented in Figure 13 and Table 6. Although the number of casualties varied over a large range in the risk zones, the differences between the majority of the predicted values by Z-SVR and the true values were acceptable. However, there were three samples with noticeable error. Among these three earthquake cases, 2 occurred in Puer (H7 and H14), and 1 occurred in Lijiang (H17); both cities are located in Yunnan Province. Considering that Yunnan is a region with significant variation of the geological environment and a huge economic gap between cities and villages, further research should be conducted to develop a specific approach for predicting earthquakes in this region.

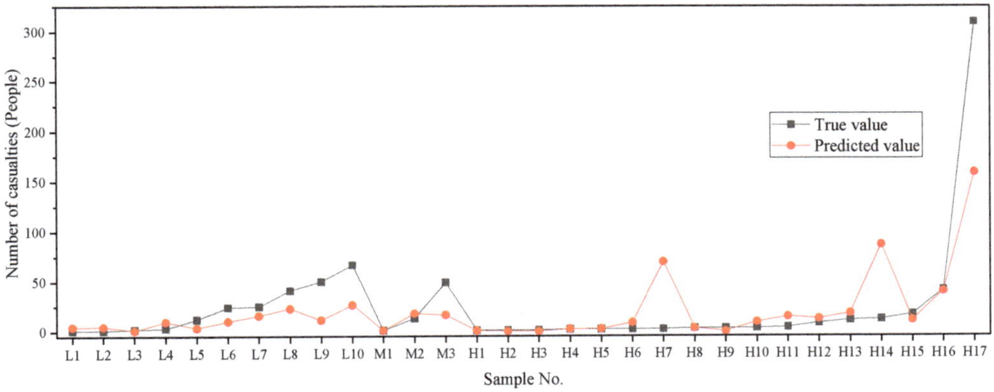

Figure 13. Prediction result of Z-SVR compared with the corresponding true values.

Table 6. Representative earthquakes in testing samples.

Sample No.	Time	Place	True Value	Predicted Value
L1	1989/9/22	Xiaojin	1	4.6
L3	1986/8/7	Litang	2	1.2
L7	2017/8/8	Jiuzhaigou	25	15.8
M1	1991/3/26	Datong-Yanggao	1	1.1
M2	2005/11/26	Jiujiang-Ruichang	13	17.8
H8	1953/5/4	Mile	3	3
H13	1965/1/13	Yuanqu	11	17.9
H16	2008/8/30	Renhe-Huili	41	39.6

6. Discussion

6.1. Comparison between Z-SVR and Other Models

To evaluate the effectiveness of the proposed model, this study selected training samples and used a cross-validation method to evaluate the robustness of the Z-SVR model. The regression and classification performances of the proposed model were also assessed by predicting the numbers of casualties in testing samples and comparing the results in terms of numerical difference and order of magnitude. Similar experiments were also implemented on other widely used machine learning methods, including random forest (RF), back propagation neural network (BP) and logistic regression (LR). This was followed by a series of experiments and detailed analyses.

Several commonly used regression model evaluation indicators were employed in this study, including root mean square error (RMSE) and mean absolute error (MeaAE), which are defined as follows:

$$\text{RMSE} = \sqrt{\frac{1}{n}\sum_{i=1}^{n}(y_i - \hat{y}_i)^2} \quad (16)$$

$$\text{MeaAE} = \frac{1}{n}\sum_{i=1}^{n}|y_i - \hat{y}_i| \quad (17)$$

where $\hat{y}_i$ is the predicted death toll of the ith sample, y_i is the corresponding true death toll, and n is the number of samples.

The classification model evaluation indicators that were applied in this study were *Precision*, *Recall* and $F1$, which are defined as follows:

$$Precision = \frac{TP}{TP + FP} \tag{18}$$

$$Recall = \frac{TP}{TP + FN} \tag{19}$$

$$\frac{1}{F1} = \frac{1}{2}\left(\frac{1}{Precision} + \frac{1}{Recall}\right) \tag{20}$$

where TP is the number of true-positive samples, FP is the number of false-positive samples, TN is the number of true-negative samples, and FN is the number of false-negative samples.

6.1.1. Cross-Validation

The robustness of each model was evaluated using the cross-validation method. As discussed in Section 2.2.1, 113 seismic cases were selected as the training dataset, among which 49 cases were in low risk areas, 13 in moderate risk areas, and 81 in high risk areas. We randomly divided the cases in low and high risk zones into ten groups, respectively; considering the limited number of samples, we randomly divided the cases in moderate risk zones into five groups. The sample data in each group were not repeated. We used RMSE and MeaAE to compare the regression precision between the Z-SVR model and other machine learning models using the spatial division method. RMSE and MeaAE were calculated for three degrees of risk zones (L, M and H) and the average values (RMSE(A) and MeaAE(A)) were also given. The comparison result of all models is shown in Figure 14.

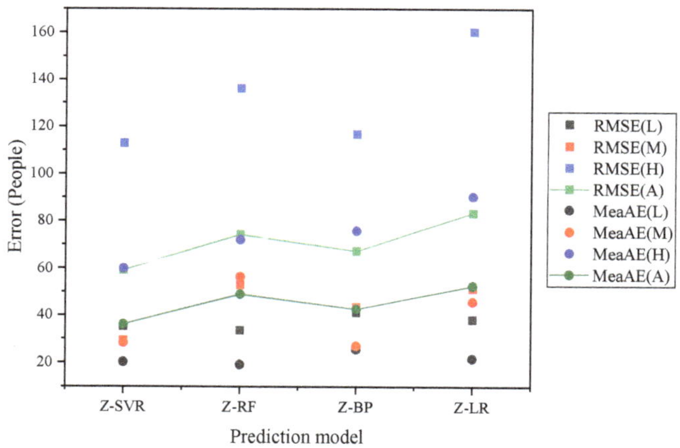

Figure 14. Model performance evaluated by the cross-validation method.

Judging from the stability of the prediction results on the training samples, all models performed relatively better in low and moderate risk zones than in high risk zones. A possible explanation is that there are 64 training samples in high risk zones, much more than in low and moderate risk zones. In addition, the true numbers of casualties in these 64 samples vary from 1 to 748, which is a huge range and increases the difficulty for machine learning models to achieve accurate prediction. Among all prediction models, Z-LR performed the worst, as its RMSE and MeaAE were 83.37 and 52.72, respectively, which ranked last in the two evaluation indicators. Z-BP and Z-RF outperformed the Z-LR model, with RMSEs of 67.30 and 74.27, respectively, and MeaAEs of 42.80 and 49.17, respectively. In contrast to the above prediction methods, Z-SVR showed higher overall accuracy in cross-validation experiments for all risk zones. Its RMSE was 59.15, and its

MeaAE was 36.16, which were significantly lower than those of the compared models; this indicates that the proposed Z-SVR model had the smallest dispersion and the highest stability.

6.1.2. Regression Accuracy Evaluation

For samples in low, moderate and high risk zones, this study used Z-SVR and other models to predict their death tolls. Evaluation indicators of RMSE (L, M and H) and MeaAE (L, M and H) were calculated for the risk zones, and the overall regression performances (RMSE(A) and MeaAE(A)) of all models were also calculated, which are plotted in Figure 15. For samples in low and moderate risk zones, the majority of models showed relatively high regression accuracy, while for those in high risk zones, the Z-SVR and Z-BP models showed good regression performance. Among all prediction models, in terms of overall MeaAE, the Z-BP model showed the best regression accuracy with the lowest value of 16.73, and the Z-SVR model also performed well with MeaAE(A) of 17.39. In terms of the overall RMSE, the average value of Z-SVR was 35.61, which was the lowest value, followed by 35.89 for Z-BP. The precision evaluation results from Figure 15 further prove that the proposed spatial division method has the advantages of enhancing prediction accuracy and stability. For example, the RMSE of the Z-SVR model was the lowest, namely, nearly half that of the SVR model; a similar result was obtained between the Z-BP and BP models. In addition, the best fitting results were obtained by the Z-SVR and Z-BP models, while the worst results were obtained by the RF, SVR and LR models, among which the SVR and BP algorithms showed obviously improved performance with the utilization of the spatial division method. The above analysis demonstrates that spatial division is an effective method for improving the performance of machine learning algorithms in predicting earthquake casualties and that the proposed Z-SVR model showed good and stable performance in casualty prediction.

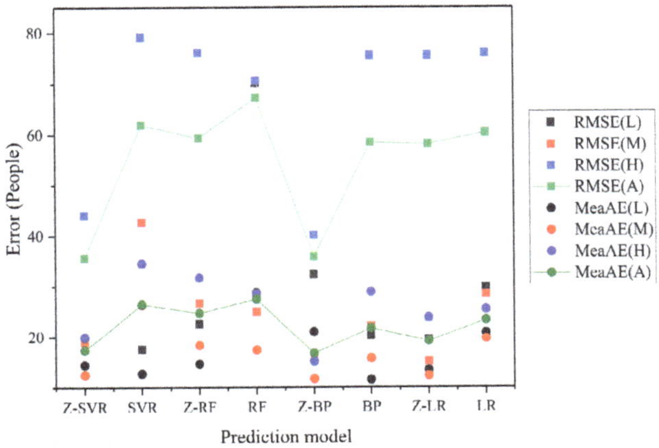

Figure 15. Regression performances of Z-SVR and other models.

6.1.3. Classification Accuracy Evaluation

The prediction results of Z-SVR, Z-RF, Z-BP, Z-LR and their initial models were also compared with the corresponding true values in terms of classification performance, where pairs of prediction and true values with the same order of magnitude were considered correct. Based on this criterion, we calculated the evaluation indicators of *Precision*, *Recall*, and *F*1 for all prediction models for the risk zones, which are presented in Table 7. In low and moderate risk zones, although the *Precision* of the LR model was 1, its *Recall* performance was unsatisfactory, which led to a low *F*1 value; compared with LR and other models, Z-SVR showed better classification performance in low and moderate risk areas with relatively high *Precision* values and the highest *Recall* and *F*1 values. With regard to samples in high risk zones, Z-BP was the model with the best prediction performance, with an *F*1 value of 0.87. However, the classification result of Z-SVR in high risk zones was also excellent, with the highest *Recall*, the second-highest*Precision* and the third-highest *F*1 values. In general, the Z-SVR model showed significant stability in classification prediction, with the highest values of *Recall* and *F*1 and a relatively high value of *Precision*. The *F*1 order of Z-SVR in all risk areas from high to low is moderate, low, and high risk zones. However, only a few earthquakes with casualties occurred in moderate risk areas; hence, we obtained a limited number of historical cases for training prediction models and verifying their performances, which made it difficult to evaluate the difference in classification performance order between the two models.

Table 7. Comparison of classification performance between Z-SVR and other models for three degrees of risk zones.

Indicator	Model	Low Risk Zones	Moderate Risk Zones	High Risk Zones	Total
Precision	Z-SVR	0.92	1	0.87	0.87
	SVR	0.92	0.5	0.47	0.63
	Z-RF	0.85	1	0.52	0.64
	RF	0.77	1	0.5	0.51
	Z-BP	0.72	0.83	1	0.94
	BP	0.87	0.83	0.71	0.67
	Z-LR	0.87	0.83	1	0.93
	LR	1	1	0.86	0.91
Recall	Z-SVR	0.9	1	0.82	0.87
	SVR	0.9	0.67	0.47	0.63
	Z-RF	0.7	0.33	0.53	0.57
	RF	0.6	0.33	0.47	0.5
	Z-BP	0.5	0.67	0.76	0.67
	BP	0.6	0.67	0.65	0.63
	Z-LR	0.6	0.67	0.71	0.67
	LR	0.4	0.33	0.65	0.53
*F*1	Z-SVR	0.9	1	0.81	0.87
	SVR	0.9	0.56	0.46	0.63
	Z-RF	0.71	0.5	0.52	0.59
	RF	0.61	0.5	0.45	0.5
	Z-BP	0.54	0.67	0.87	0.74
	BP	0.63	0.67	0.64	0.65
	Z-LR	0.63	0.67	0.83	0.74
	LR	0.57	0.5	0.74	0.67

We also divided the testing samples into three groups according to the number of casualties, where the division criterion was order of magnitude (1 to 9, 10 to 99, 100 and greater). We compared the classification performances of Z-SVR and other models in the groups and calculated the evaluation indicators of *Precision*, *Recall*, and *F*1 for all prediction models. Figure 16 presents the comparison results of classification performance between Z-SVR and other models on samples with various numbers of casualties. Z-SVR provided the most balanced and accurate classification into the three groups. Although models such as Z-BP and Z-LR showed better classification performance in terms of *Precision* or *Recall* in some groups, the *Precision* and *Recall* values of the Z-SVR model in the three groups were high, balanced and stable; thus, Z-SVR had the highest *F*1 values in each group. In general, the Z-SVR model was the most precise and stable model, which provided accurate classification results for earthquakes with various numbers of casualties.

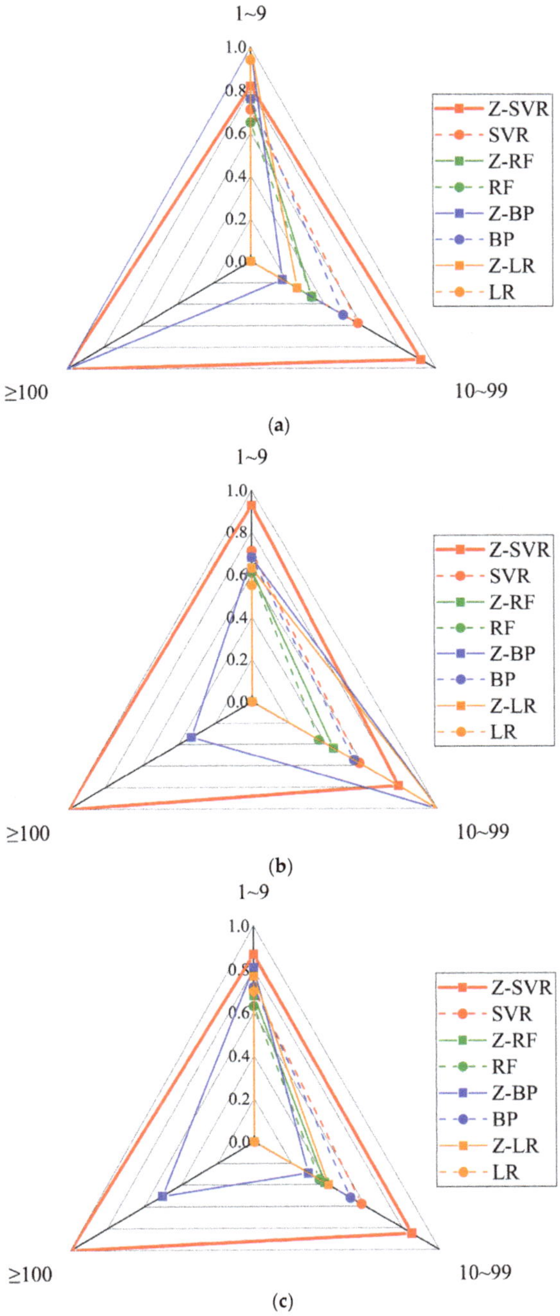

Figure 16. Classification results of Z-SVR and other models for earthquakes with casualties of different orders of magnitude: (**a**) Comparison of *Precision*; (**b**) comparison of *Recall*; and (**c**) comparison of *F*1.

6.2. Future Work

Further extensive studies are needed, and recommendations for future research are discussed as follows. First, this study analyzes the importance of features that affect seismic mortality, which simply collects 14 features and classifies them into disaster-inducing factors, disaster-affected bodies and disaster-formative environments. Future studies can extend the research by refining the classification standard and increasing the number of factors. Second, this study divides the study area into risk zones of three grades based on regional differences, where the partition standard exerts a potential influence on the accuracy and applicability of the proposed model. Future studies can explore more reasonable criteria for different study areas. Third, the proposed prediction approach is a regression model that is based on SVR, which is essentially a data-driven model. Future studies can build models based on deeper seismic mechanisms to predict deaths that are caused by earthquakes.

7. Conclusions

This study evaluated the importance of 14 features that affect seismic fatality based on the RF model. On the basis of the importance assessment, we selected magnitude, population density, geological fault density and GDP as the input parameters of the prediction model, among which the densities of population and geological faults were also integrated for spatial division. This study also proposed a spatial division method based on the theory of regional difference. We studied the regional diversity of geological fault density and population in China's mainland using the WorldPop population dataset (100 m resolution) every five years from 2000 to 2020 and the strata fault line dataset and, finally, divided the study area into zones of various risk grades by overlay analysis. Based on the results of feature selection and spatial division, this study proposed a zoning prediction model based on SVR. Using 113 samples in the earthquake case dataset, we implemented model training and obtained the optimal model parameters for each risk zone to enhance the prediction accuracy of earthquake death tolls. The following conclusions were drawn from the results that were obtained in this study:

1. Among all selected features from the evaluation index system, the order of importance from high to low is as follows: magnitude, collapsed buildings, epicenter intensity, population density, geological fault density, GDP, occurrence time, focal depth, occurrence day, aftershock, secondary disaster, rescue capability, landform, and climatic condition.
2. The proposed method of spatial division based on regional diversity could be used as an effective tool to refine complex study areas. Using this method, we divided China's mainland into high, moderate, and low risk zones, which laid the foundation for the construction of a prediction model with submodels that are suitable for different risk zones. The verification results demonstrated that the proposed division method is feasible for classifying study regions, especially those with vast area and complex environments.
3. The proposed Z-SVR model realizes accurate prediction and good generalization performance. We collected 143 historical earthquake cases, of which 113 cases were selected as the training dataset and 30 for examining the prediction performance of the model. The best model parameters were selected for each risk zone, which led to precise prediction results in risk zones of various grades. The proposed model also showed accurate regression and classification accuracy in the various risk zones compared with other machine learning models, including RF, BP and LR. Moreover, the proposed Z-SVR model was compared to the initial SVR model using the same database. Similar experiments were also implemented on comparative machine learning models, and we found that the prediction performances of all models with spatial division significantly improved. The above results prove the advantages and significance of the proposed model and spatial division method.

Author Contributions: B.L. and T.Z. implemented the research and wrote the original manuscript. A.G. provided the original idea for the study and supervised the research. W.B., C.X. and Z.H. aided with the manuscript revision. All authors have read and agreed to the published version of the manuscript.

Funding: This research was jointly supported by the National Key Research and Development Program of China (Grant No. 2019YFE01277002, No. 2017YFB0504102 and No. 2017YFC1502704) and the National Natural Science Foundation of China (41671412).

Institutional Review Board Statement: Not applicable.

Informed Consent Statement: Not applicable.

Data Availability Statement: Our research data are from relevant open data websites, which can be obtained according to the links listed in our references.

Acknowledgments: The authors would like to express deep gratitude to Jianghong Zhao from Beijing University of Civil Engineering and Architecture for her guidance on the framework design of the paper.

Conflicts of Interest: The authors declare no conflict of interest.

References

1. Zhang, Y.; Weng, W.G.; Huang, Z.L. A scenario-based model for earthquake emergency management effectiveness evaluation. *Technol. Forecast. Soc.* **2018**, *128*, 197–207. [CrossRef]
2. Alizadeh, M.; Zabihi, H.; Rezaie, F.; Asadzadeh, A.; Wolf, I.D.; Langat, P.K.; Khosravi, I.; Beiranvand Pour, A.; Mohammad Nataj, M.; Pradhan, B. Earthquake Vulnerability Assessment for Urban Areas Using an ANN and Hybrid SWOT-QSPM Model. *Remote Sens.* **2021**, *13*, 4519. [CrossRef]
3. Schilling, J.; Hertig, E.; Tramblay, Y.; Scheffran, J. Climate change vulnerability, water resources and social implications in North Africa. *Reg. Environ. Chang.* **2020**, *20*, 15. [CrossRef]
4. Yariyan, P.; Zabihi, H.; Wolf, I.D.; Karami, M.; Amiriyan, S. Earthquake risk assessment using an integrated Fuzzy Analytic Hierarchy Process with Artificial Neural Networks based on GIS: A case study of Sanandaj in Iran. *Int. J. Disaster Risk Re.* **2020**, *50*, 101705. [CrossRef]
5. Jian, W. The Research of Earthquake Information Extraction and Assessment Based on Object-Oriented Technology with Remotely-Sensed Data. Doctor's Thesis, Wuhan University, Wuhan, China, 5 June 2010.
6. Zhu, Y.; Diao, F.; Fu, Y.; Liu, C.; Xiong, X. Slip rate of the seismogenic fault of the 2021 Maduo earthquake in western China inferred from GPS observations. *Sci. China Earth Sci.* **2021**, *64*, 1363–1370. [CrossRef]
7. Chen, L.; Huang, Y.; Bai, R.; Chen, A. Regional disaster risk evaluation of China based on the universal risk model. *Nat. Hazards* **2017**, *89*, 647–660. [CrossRef]
8. Zhou, W.; Guo, S.; Deng, X.; Xu, D. Livelihood resilience and strategies of rural residents of earthquake-threatened areas in Sichuan Province, China. *Nat. Hazards* **2021**, *106*, 255–275. [CrossRef]
9. National Earthquake Emergency Plan. Available online: http://www.gov.cn/yjgl/2012-09/21/content_2230337.htm (accessed on 17 May 2021).
10. Maqsood, S.T.; Schwarz, J. Estimation of Human casualties from earthquakes in Pakistan—An engineering approach. *Seismol. Res. Lett.* **2011**, *82*, 32–41. [CrossRef]
11. Guangxian, X. Rapid assessment of disaster losses in post-earthquake. *J. Catastrophology* **1991**, *4*, 12–17.
12. Jaiswal, K.; Wald, D. An empirical model for global earthquake fatality estimation. *Earthq. Spectra* **2010**, *26*, 1017–1037. [CrossRef]
13. ATC. *Earthquake Damage Evaluation Data for California (ATC-13)*; Applied Technology Commission: Redwood City, CA, USA, 1985.
14. Ceferino, L.; Kiremidjian, A.; Deierlein, G. Probabilistic model for regional multiseverity casualty estimation due to building damage following an earthquake. *ASCE-ASME J. Risk Uncertain. Eng. Syst. Part A Civil. Eng.* **2018**, *4*, 4018023. [CrossRef]
15. Xianfu, B.; Gaozhong, N.; Yuqian, D.; Qingkun, Y.; Weidong, L.; Liaoyuan, Y. Modeling and Testing Earthquake-induced Landslide Casualty Rate Based on a Grid in a Kilometer Scale: Taking the 2014 Yunnan Ludian MS6. 5 Earthquake as a Case. *J. Seismol. Res.* **2021**, *44*, 87–95.
16. Stav, S.; Lena, N.; Yaron, B.D.; Limor, A.D.; Asim, Z. An Integrated and Interdisciplinary Model for Predicting the Risk of Injury and Death in Future Earthquakes. *PLoS ONE* **2016**, *11*, e151111.
17. Cui, S.; Yin, Y.; Wang, D.; Li, Z.; Wang, Y. A stacking-based ensemble learning method for earthquake casualty prediction. *Appl. Soft Comput.* **2020**, *101*, 107038. [CrossRef]
18. Gao, Z.; Li, Y.; Shan, X.; Zhu, C. Earthquake Magnitude Estimation from High-Rate GNSS Data: A Case Study of the 2021 Mw 7.3 Maduo Earthquake. *Remote Sens.* **2021**, *13*, 4478. [CrossRef]
19. Karimzadeh, S.; Miyajima, M.; Hassanzadeh, R.; Amiraslanzadeh, R.; Kamel, B. A GIS-based seismic hazard, building vulnerability and human loss assessment for the earthquake scenario in Tabriz. *Soil. Dyn. Earthq. Eng.* **2014**, *66*, 263–280. [CrossRef]

20. Feng, T.; Hong, Z.; Fu, Q.; Ma, S.; Jie, X.; Wu, H.; Jiang, C.; Tong, X. Application and prospect of a high-resolution remote sensing and geo-information system in estimating earthquake casualties. *Nat. Hazards Earth Syst. Sci.* **2014**, *1*, 7137–7166. [CrossRef]
21. Wenjuan, Z. Design of the Population Casualty Acquisition and Evaluation System in Earthquake Disaster Areas Based on Mobile Communication Big Data. *China Earthq. Eng. J.* **2019**, *41*, 1066–1071.
22. Huang, X.; Zhou, Z.; Wang, S. The prediction model of earthquake casuailty based on robust wavelet v-SVM. *Nat. Hazards* **2015**, *77*, 717–732.
23. Gul, M.; Guneri, A.F. An artificial neural network-based earthquake casualty estimation model for Istanbul city. *Nat. Hazards* **2016**, *84*, 2163–2178. [CrossRef]
24. Jia, H.; Lin, J.; Liu, J. An Earthquake Fatalities Assessment Method Based on Feature Importance with Deep Learning and Random Forest Models. *Sustainability* **2019**, *11*, 2727. [CrossRef]
25. Sousa, J.J.; Liu, G.; Fan, J.; Perski, Z.; Steger, S.; Bai, S.; Wei, L.; Salvi, S.; Wang, Q.; Tu, J. Geohazards Monitoring and Assessment Using Multi-Source Earth Observation Techniques. *Remote Sens.* **2021**, *13*, 4269. [CrossRef]
26. Shi, P. *Natural Disasters in China*; Springer: Berlin/Heidelberg, Germany, 2016.
27. Wen, L.; Wenkai, C.; Zhonghong, Z. Assessing the applicability of life vulnerability models for earthquake disasters in typical regions of China. *J. Beijing Norm. Univ.* **2019**, *55*, 284–290.
28. Tingting, Z. Assessment of Earthquake Fatality and Disaster Degree Based on Spatio-Temporal Method. Bachelor's Thesis, Beijing Normal University, Beijing, China, 3 June 2020.
29. China Earthquake Administration. *Compilation of Earthquake Disaster Loss Assessment in China's Mainland*; Seismological Press: Beijing, China, 1996.
30. Monitoring and Forecasting Department of China Earthquake Administration. *Compilation of Earthquake Disaster Loss Assessment in China's Mainland*; Seismological Press: Beijing, China, 2001.
31. Earthquake Emergency Rescue Department of China Earthquake Administration. *Compilation of Earthquake Disaster Loss Assessment in China's Mainland from 2001 to 2005*; Seismological Press: Beijing, China, 2010.
32. Earthquake Emergency Rescue Department of China Earthquake Administration. *Compilation of Earthquake Disaster Loss Assessment in China's Mainland from 2006 to 2010*; Seismological Press: Beijing, China, 2015.
33. Liang, S.; Chen, D.; Li, D.; Qi, Y.; Zhao, Z. Spatial and Temporal Distribution of Geologic Hazards in Shaanxi Province. *Remote Sens.* **2021**, *13*, 4259. [CrossRef]
34. Hoffmann, S.; Beierkuhnlein, C. Climate change exposure and vulnerability of the global protected area estate from an international perspective. *Divers. Distrib.* **2020**, *26*, 1496–1509. [CrossRef]
35. Xiong, K.; Adhikari, B.R.; Stamatopoulos, C.A.; Zhan, Y.; Wu, S.; Dong, Z.; Di, B. Comparison of different machine learning methods for debris flow susceptibility mapping: A case study in the Sichuan Province, China. *Remote Sens.* **2020**, *12*, 295. [CrossRef]
36. Peijun, S. Theory on Disaster Science and Disaster Dynamics. *J. Nat. Disasters* **2002**, *11*, 1–9.
37. Chen, W.; Shirzadi, A.; Shahabi, H.; Ahmad, B.B.; Zhang, S.; Hong, H.; Zhang, N. A novel hybrid artificial intelligence approach based on the rotation forest ensemble and naïve Bayes tree classifiers for a landslide susceptibility assessment in Langao County, China. *Geomat. Nat. Hazards Risk* **2017**, *8*, 1955–1977. [CrossRef]
38. Altmann, A.; Toloşi, L.; Sander, O.; Lengauer, T. Permutation importance: A corrected feature importance measure. *Bioinformatics* **2010**, *26*, 1340–1347. [CrossRef] [PubMed]
39. Chen, W.; Sun, Z.; Han, J. Landslide susceptibility modeling using integrated ensemble weights of evidence with logistic regression and random forest models. *Appl. Sci.* **2019**, *9*, 171. [CrossRef]
40. Stumpf, A.; Kerle, N. Object-oriented mapping of landslides using Random Forests. *Remote Sens. Environ.* **2011**, *115*, 2564–2577. [CrossRef]
41. Yuanyuan, L.; Guofeng, S.; Wenguo, W. A Review of Researches on Seismic Casualty Estimation. *J. Catastrophology* **2014**, *29*, 223–227.
42. Fan, Y.; Baozhu, Z.; Liangliang, Y. System of Earthquake Casualty Assessment Based on BP Neural Network. *Technol. Earthq. Disaster Prev.* **2009**, *4*, 428–435.
43. Zhao, K.; Jin, B.; Fan, H.; Song, W.; Zhou, S.; Jiang, Y. High-Performance Overlay Analysis of Massive Geographic Polygons That Considers Shape Complexity in a Cloud Environment. *Int. J. Geo-Inf.* **2019**, *8*, 290. [CrossRef]
44. Thomas, S.; Pillai, G.N.; Pal, K. Prediction of peak ground acceleration using ϵ-SVR, v-SVR and Ls-SVR algorithm. *Geomat. Nat. Hazards Risk* **2017**, *8*, 177–193. [CrossRef]
45. Lin, J.Y.; Cheng, C.T.; Chau, K.W. Using support vector machines for long-term discharge prediction. *Hydrol. Sci. J.* **2006**, *51*, 599–612. [CrossRef]
46. Guirong, W.; Juan, Y.; Lixia, X. *Machine Learning and Its Application*; China Machine Press: Beijing, China, 2019.
47. Tao, D.; Ma, Q.; Li, S.; Xie, Z.; Lin, D.; Li, S. Support Vector Regression for the Relationships between Ground Motion Parameters and Macroseismic Intensity in the Sichuan—Yunnan Region. *Appl. Sci.* **2020**, *10*, 3086. [CrossRef]
48. Zhihua, Z. *Machine Learning*; Tsinghua University Press: Beijing, China, 2016.
49. Smola, A.J.; Schölkopf, B. A tutorial on support vector regression. *Stat. Comput.* **2004**, *14*, 199–222. [CrossRef]
50. Chih-Chung, C.; Chih-Jen, L. LIBSVM: A library for support vector machines. *ACM Trans. Intell. Syst. Technol.* **2011**, *2*, 1–39. [CrossRef]

51. Ghorbani, M.; Zargar, G.; Jazayeri-Rad, H. Prediction of asphaltene precipitation using support vector regression tuned with genetic algorithms. *Petroleum* **2016**, *2*, 301–306. [CrossRef]
52. Bamakan, S.; Wang, H.; Ravasan, A.Z. Parameters Optimization for Nonparallel Support Vector Machine by Particle Swarm Optimization. *Procedia Comput. Sci.* **2016**, *91*, 482–491. [CrossRef]

remote sensing

MDPI

Article

DSMNN-Net: A Deep Siamese Morphological Neural Network Model for Burned Area Mapping Using Multispectral Sentinel-2 and Hyperspectral PRISMA Images

Seyd Teymoor Seydi [1], Mahdi Hasanlou [1,*] and Jocelyn Chanussot [2,3]

[1] School of Surveying and Geospatial Engineering, College of Engineering, University of Tehran, Tehran 14174-66191, Iran; seydi.teymoor@ut.ac.ir

[2] Aerospace Information Research Institute, Chinese Academy of Sciences, Beijing 100094, China; jocelyn.chanussot@grenoble-inp.fr

[3] CNRS, Grenoble INP, GIPSA-Lab, Université Grenoble Alpes, 38000 Grenoble, France

* Correspondence: hasanlou@ut.ac.ir; Tel.: +98-21-6111-4525

Citation: Seydi, S.T.; Hasanlou, M.; Chanussot, J. DSMNN-Net: A Deep Siamese Morphological Neural Network Model for Burned Area Mapping Using Multispectral Sentinel-2 and Hyperspectral PRISMA Images. *Remote Sens.* **2021**, *13*, 5138. https://doi.org/10.3390/rs13245138

Academic Editors: Saverio Romeo and Paolo Mazzanti

Received: 14 November 2021
Accepted: 15 December 2021
Published: 17 December 2021

Abstract: Wildfires are one of the most destructive natural disasters that can affect our environment, with significant effects also on wildlife. Recently, climate change and human activities have resulted in higher frequencies of wildfires throughout the world. Timely and accurate detection of the burned areas can help to make decisions for their management. Remote sensing satellite imagery can have a key role in mapping burned areas due to its wide coverage, high-resolution data collection, and low capture times. However, although many studies have reported on burned area mapping based on remote sensing imagery in recent decades, accurate burned area mapping remains a major challenge due to the complexity of the background and the diversity of the burned areas. This paper presents a novel framework for burned area mapping based on Deep Siamese Morphological Neural Network (DSMNN-Net) and heterogeneous datasets. The DSMNN-Net framework is based on change detection through proposing a pre/post-fire method that is compatible with heterogeneous remote sensing datasets. The proposed network combines multiscale convolution layers and morphological layers (erosion and dilation) to generate deep features. To evaluate the performance of the method proposed here, two case study areas in Australian forests were selected. The framework used can better detect burned areas compared to other state-of-the-art burned area mapping procedures, with a performance of >98% for overall accuracy index, and a kappa coefficient of >0.9, using multispectral Sentinel-2 and hyperspectral PRISMA image datasets. The analyses of the two datasets illustrate that the DSMNN-Net is sufficiently valid and robust for burned area mapping, and especially for complex areas.

Keywords: deep learning; PRISMA; burned area; Sentinel-2; morphological operator; convolutional neural network

1. Introduction

As a natural hazard, wildfires represent one of the most important reasons for the evolution of ecosystems in the Earth's system on a global scale [1–3]. Recently, the frequency of occurrence of wildfires has increased significantly due to climate change and human activities around the world [4,5]. Wildfires can be influenced by the environment from different aspects, such as soil erosion, increasing flood risk, and habitat degradation for wildlife [6,7]. Furthermore, wildfires generate a wide range of pollutants, including greenhouse gases (i.e., methane and carbon dioxide) [8].

Burned area mapping (BAM) can be useful to predict the behavior of a fire, to define the burning biomass, for compensation from insurance companies, and for estimation of greenhouse gases emitted [9,10]. As result, the generation of reliable and accurate burned area maps is necessary for their management and planning in the support of decision making. BAM by traditional methods (e.g., field surveys) is a major challenge, and these methods have some limitations, such as the wide areas to be covered and the lack of direct access to the region of interest, which leads to large time and financial costs [10].

The Earth observation satellite fleet has steadily grown over the last few decades [11]. The diversity of Earth observation datasets means that remote sensing (RS) is now known as a key tool in the provision of valuable information about the Earth that is available at low cost and time needs on a global scale [12]. Currently, the upcoming new series of RS sensors (e.g., Landsat-9, *PRecursore IperSpettrale della Missione Applicativa* (PRISMA), Sentinel-5) provides improvements in terms of spatial, temporal, and spectral detail, with RS now becoming a routine tool with an extensive range of applications [13,14]. The most common applications of RS include classification [15,16] and detection of targets [17,18] and changes [19,20].

The diversity of RS Earth observation imagery and its free availability has meant that monitoring of changes following disasters has turned into a hot topic for research [21–28]. Indeed, we are witnessing many BAM products on a global scale that differ in terms of spatial resolution and reliability of the burned areas mapped. Based on spatial resolution, the recent BAM methods can be categorized into two main groups: (1) coarse spatial resolution satellite sensors and (2) fine spatial resolution sensors.

Burned area mapping based on the low and medium resolution of satellite imagery is common in the RS community. In recent years, many studies have used BAM based on Moderate Resolution Imaging Spectroradiometer (MODIS), Sentinel-3, Medium Resolution Imaging Spectrometer (MERIS), and Visible/Infrared Imager Radiometer Suite (VIIRS) [29–31]. However, while these sensors have a high temporal resolution, they suffer from low spatial resolution. Accurate BAM for small areas is a major challenge due to the mixing of pixels. Furthermore, the complex diversity of scenes can result in spectrum gains in one burned pixel to be mixed with some other material. Furthermore, these are based on ruleset classification and manual feature extraction such that the extraction of suitable features and the finding of optimum threshold values are time consuming.

Recently, with the arrival of a new series of cloud computing platforms (e.g., Google Earth Engine, Microsoft Azure), BAM using fine-resolution datasets has been considered by researchers. The capacity of cloud computing platforms has created a great opportunity for BAM based on high-resolution datasets and advanced machine-learning-based methods for accurate mapping. Based on the structure of the algorithm, we can categorize these methods into two main categories: (1) BAM by conventional machine-learning methods and (2) BAM via deep-learning-based frameworks.

Burned area mapping based on conventional machine-learning-based methods can be used to extract spectral and spatial features, and then to define the burned areas according to a classifier [10]. For instance, Donezar, et al. [32] designed a BAM framework based on the multitemporal change-detection method and time series synthetic aperture radar (SAR) imagery. They used an object-based image analysis method for classification of the SAR imagery. They also used the Shuttle Radar Topography Mission (SRTM) for digital elevation models to enhance their BAM results. Additionally, Xulu, et al. [33] considered a BAM method based on differenced normalized burned ratios and Sentinel-2 imagery in the cloud-based Google Earth engine. A random forest classifier method was used for the BAM. They reported an overall accuracy close to 97% for detection of burned areas. Moreover, Seydi, Akhoondzadeh, Amani and Mahdavi [10] evaluated the performance of a statistical machine-learning method for BAM using the Google Earth Engine and pre/post-fire Sentinel-2 imagery. Furthermore, they evaluated the potential spectral and spatial texture features using a Harris hawks optimization algorithm for the BAM. They reported an accuracy of 92% by the random forest classifier on the validation dataset. Liu, et al. [34] proposed a new index for BAM for bi-temporal Landsat-8 imagery and an automatic thresholding method. They evaluated the efficiency of their proposed method in different areas. Their BAM results showed that their presented method had high efficiency.

Recently, deep-learning-based approaches have been applied increasingly for mapping RS imagery, with promising results obtained. These methods can extract high-level features from the raw data automatically, by convolution layers. This advantage of deep-learning-based methods has resulted in their use for BAM. BAM based on deep-learning-based methods has become a hot topic of research, with many methods being proposed. For instance, Nolde, et al. [35] designed large-scale burned area monitoring in near-real-time based on the morphological active contour approach. This framework was applied through several steps: (1) generation of a normalized difference vegetation index (NDVI) for pre/post-fire; (2) determination of the region of interest based on active fires, anomaly detection, and region-growing methodologies; (3) accurate shape of the burned area perimeter extraction based on morphological snakes; (4) confidence evaluation based on a burn area index; and (5) tracking. The result was accuracy of 76% by evaluation with reference data. Knopp, et al. [36] carried out BAM by deep-learning-based semantic segmentation based on mono-temporal Sentinel-2 imagery. They used the U-Net architecture for their BAM. A binary change map was obtained based on the thresholding of the U-Net probability map. They reported the difficulty of the segmentation model in some areas, such as agriculture fields, rocky coastlines, and lake shores. de Bem, et al. [37] investigated the effects of patch sizes on the result of burned area classification with deep-learning methods using Landsat-8 OLI. Here, three different deep-learning methods were investigated: simple convolutional neural network (CNN), U-Net, and Res-U-Net. Their results showed that Res-U-Net had high efficiency, with a patch-size of 256 $\times$ 266. Hu, Ban and Nascetti [26] evaluated the potential of deep learning methods for BAM based on the unitemporal multispectral Sentinel-2 and Landsat-8 datasets. Their study showed that deep-learning methods have a high potential for BAM in comparison to machine-learning methods. Ban, et al. [38] experimented with the capacity of time series SAR imagery for BAM by a deep-learning method. To this end, their deep-learning framework was based on CNN and was developed to automatically detect burned areas by investigating backscatter variations in the time series of Sentinel-1 SAR imagery. They reported accuracy of <95% for BAM. Zhang, et al. [39] proposed a deep-learning framework for mapping burned areas based on fusion Sentinel-1 and Sentinel-2 imagery. Furthermore, they investigated two scenarios for training the deep-learning method: (1) continuous joint training with all historical data and (2) learning-without-forgetting based on new incoming data alone. They reported that the second scenario for BAM showed accuracy close to 90%, in terms of overall accuracy. Zhang, Ban and Nascetti [39] presented a deep-learning-based BAM framework by fusion of optical and radar datasets. They proposed a deep-learning framework based on CNN, with two convolution layers, max-pooling, and two fully connected layers. They showed an increase in the complexity of the network that resulted in rising computing needs, while the results for the burned area detection were not enhanced. Farasin, et al. [40] presented an automatic framework for evaluation of the damage severity level based on a supervised deep-learning method and post-fire Sentinel-2 satellite imagery. They used double-step U-Net architecture for two tasks (classification and regression). The classification generated binary damage maps and the regression was used to generate damage severity levels. Lestari, et al. [41] increased the efficiency of statistical machine-learning methods and a CNN classifier for BAM using optical and SAR imagery. Their BAM results showed that the CNN method has high efficiency in comparison with other machine-learning methods with texture features. Furthermore, the fusion of optical and SAR imagery can enhance the results of BAM. Belenguer-Plomer, et al. [42] developed a CNN-based BAM method by combining active and passive datasets. Sentinel-1 and Sentinel-2 were used by the CNN algorithm to generate burned areas. Their proposed CNN architecture included two convolution layers, a max-pooling layer, and two fully connected layers. The results of BAM have shown that combining Sentinel-1 and Sentinel-2 imagery can provide improvements.

Although many research efforts have proposed several algorithms for BAM and applied them to fine-resolution optical and SAR RS imagery, many limitations remain: (1) Semantic segmentation based methods (e.g., U-Net DeeplabV3+ and Seg-Net) have provided promising results, but they need large numbers of labeled datasets, and finding large amounts of sample data with specific sizes (i.e., 512 × 512 or 1024 × 1024) for small areas is a major challenge. (2) The performance of statistical classification methods such as random forest or support vector machine classifiers depend on the setting of the input features, while the selection and extraction of the informative manual features can be a time-consuming process. (3) Some studies have focused on only spectral features for BAM, while the efficiency of spatial features in BAM has been shown in many studies; furthermore, an unsupervised thresholding manner on the spectral index is not always effective due to the complexity of the background and ecosystem characteristics, and to the topographic effects on the surface reflectance [43,44]. (4) Shallow feature representation methods have been shown not to be applicable in complex areas, especially for BAM tasks. (5) More methods have focused on time series Sentinel-1 imagery; however, preprocessing and processing of SAR imagery is very difficult due to noise conditions.

To overcome these problems, the present study presents a novel framework for BAM with heterogeneous datasets that has many advantages compared to other state-of-the-art methods. The method proposed here is applied in an end-to-end manner without additional processes, based on a deep morphological network. This method is based on change detection that uses pre/post-fire datasets based on deep Siamese morphological operators. Additionally, the efficiency of the hyperspectral dataset in comparison with the multispectral dataset shows that this study takes advantage of the hyperspectral dataset. The proposed framework is additive with the type of datasets, whereby the pre-event dataset is Sentinel-2 imagery while the post-event dataset can be either Sentinel-2 or hyperspectral PRISMA datasets.

The main contributions of this study follow: (1) BAM is based on deep morphological layers for the first time; (2) it takes advantage of the hyperspectral PRISMA sensor dataset for accurate BAM for the first time; and (3) it includes evaluation of the performance of the multispectral and hyperspectral dataset in BAM and comparison of the results with state-of-the-art methods.

This paper is outlined as follows: Section 2 provides the details of the DSMNN-Net for BAM. Section 3 introduces the study areas and the datasets. The evaluation results of this study area are provided in Section 4, and the experimentation results are discussed in Section 5.

2. Methodology

The proposed framework is conducted in three steps, according to the flowchart in Figure 1. The first step is image preparation, and in this step, some preprocessing (i.e., registration) is applied. The second step is the training of the proposed network to tune the network parameters based on reference sample data. The training and validation datasets are exploited in the training process to optimize the model parameters, while the testing dataset is used to evaluate model hyperparameters. The third step is burned area map generation and accuracy assessment of the result of the BAM.

2.1. Proposed Deep Learning Architecture

The proposed DSMNN-Net architecture for the detection of burned areas is illustrated in Figure 2. Accordingly, the framework has two parts: (1) two streams of deep-feature-extraction models and (2) classification. The deep-feature-extraction task is conducted in a double-stream manner, such that these streams are for post-event and pre-event datasets, respectively. Then, the deep features are transformed to the next task, which is the classification. The classification task included two fully connected layers and a soft-max layer for making a decision. More details of the DSMNN-Net are explained in the next subsection.

2.2. Deep-Feature Extraction

Feature extraction can be defined as an image processing technique to determine the identity of the mutual importance of imaged areas. There are many procedures for feature extraction in the field of image processing and RS, such as the texture Haralick feature [45,46], spectral features (e.g., the NDVI) [10], and transformation-based (e.g., principal component analysis) and deep-feature [15,47,48] extraction. Among the types of feature-extraction methods, the deep-feature-extraction methods have found a specific place in RS communication because they have great potential for the extraction of complex features from an image [49]. Deep-learning methods can automatically extract high-level spatial and spectral features simultaneously [50]. This advantage of deep-learning methods means that they have been used for many applications in RS, such as change detection [51], classification [52], anomaly detection [53], and damage mapping [54].

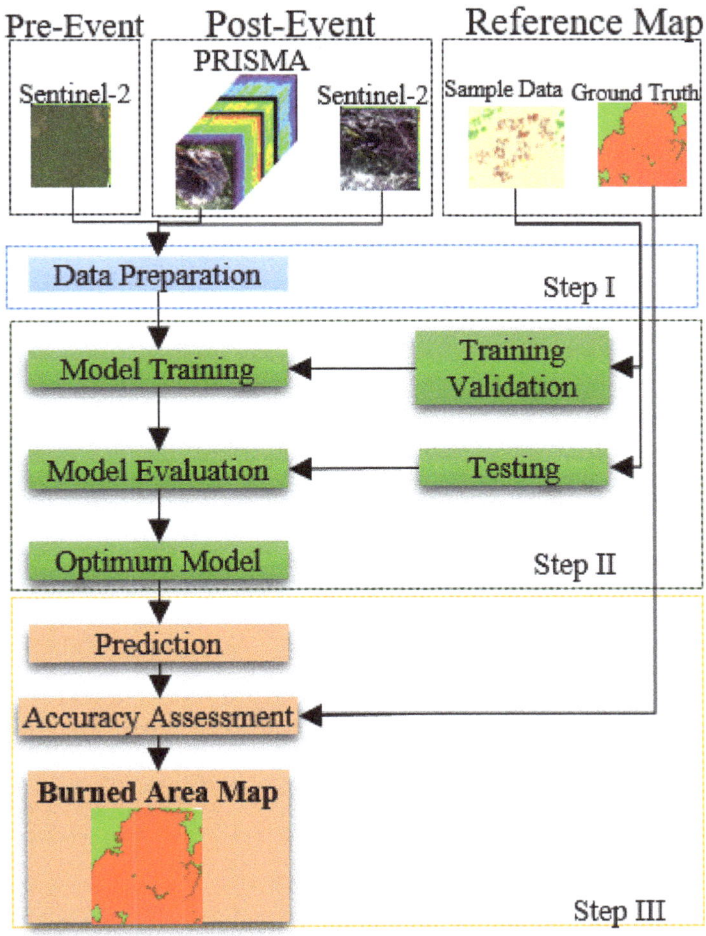

Figure 1. Overview of the general framework for the burned area mapping.

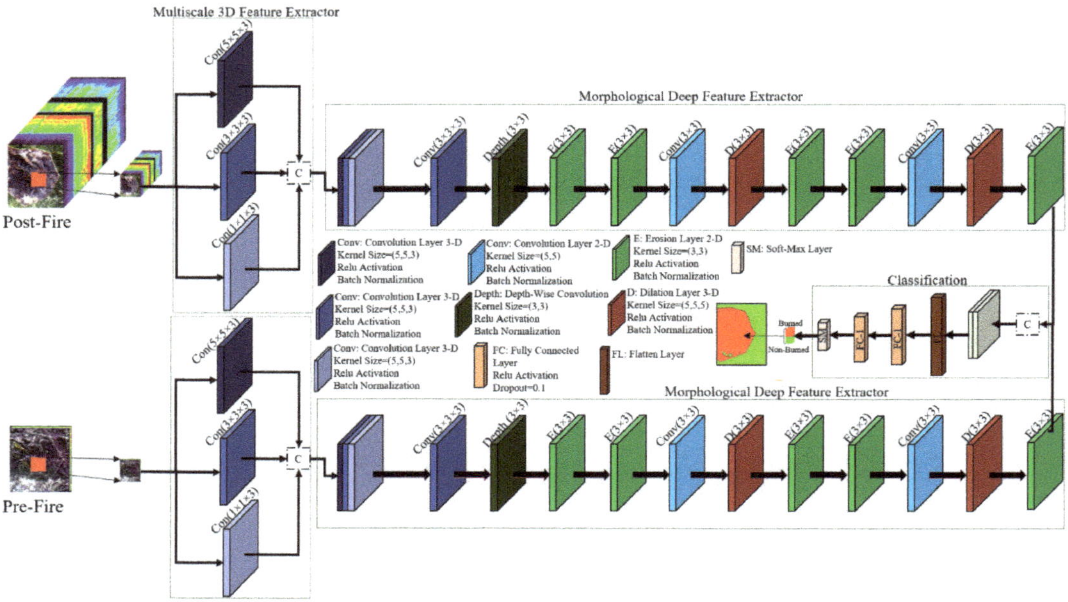

Figure 2. Overview of the proposed DSMNN-Net architecture for burned area mapping.

The deep features are extracted by convolution layers, and the arrangement of the convolution layers and their diversity has caused many deep-learning-based methods to be proposed [55–57]. Presenting the informative structure of convolution layers can be a major challenge. In this regard, the present study presents a novel framework based on standard 3D, 2D, and depthwise convolution layers, with their combination with morphological layers. As illustrated in Figure 2, the method proposed here has two deep-feature extractor streams. The first stream investigates the pre-fire dataset, and the second stream explores the deep features from the post-event dataset. Each stream includes 2D-depthwise, 3D/2D standard convolution layers, and morphological layers based on erosion and dilation operators. Initially, the deep-features extraction is based on 3D multiscale convolution layers, and then the extracted features are fed into the 3D convolution layer. The main advantage of 3D convolution layers is to take the full content of the spectral information of the input dataset by considering the relation among all of the spectral bands. Furthermore, the multiscale block enhances the robustness of the DSMNN-Net against variations in the object size [12]. The multiscale block uses a type kernel size of convolution layers that increase the efficiency of the network. The expected features are reshaped and converted to 2D feature maps, and then the 2D-depthwise convolution layers are used. Next, the hybrid morphological layers based on 2D dilation and erosion combine with 2D convolution layers to explore more high-level features. For this, first, we use two erosion layers, and then the 2D convolution layer and dilation layers are used (see Figure 2). Finally, the 2D convolution, erosion, and dilation layers have been used in the last part of the morphological deep-feature extractor. The extracted deep features are concatenated for two streams and then they are flattened and transferred to two fully connected layers, and finally, the soft-max layer is entitled to decide the input data. The main differences between the proposed architecture and other CNN frameworks are:

(1) We take advantage of multiscale convolution layers that increase the robustness of the network against the scale of variations.
(2) We use the trainable morphological layers, which can increase the efficiency of the network for the extraction of nonlinear features.

(3) We use 3D convolution layers to make use of the full content of the spectral information in the hyperspectral and multispectral datasets.
(4) We use depthwise convolution layers that are computationally cheaper and can help to reduce the number of parameters and to prevent overfitting.

2.3. Convolution Layer

The convolution layers are the core building block of deep-learning methods that can learn feature representations of input data. A convolution layer builds several convolution kernels to extract the type of meaningful features. This study used 3D/2D convolution layers for deep-feature extraction [58–60]. Mathematically, the feature value (Ψ) in the lth layer is expressed according to Equation (1) [61]:

$$\nu^l = g\left(w^l x^{l-1}\right) + b^l, \tag{1}$$

where x is the input data, g is the activation function, b is the bias vector for the current layer, and w is the weighted vector. The value (ν) at position (x,y,z) on the jth feature ith layer for the 3D convolution layer is given by Equation (2) [62]:

$$\nu_{i,j}^{xyz} = g(b_{i,j} + \sum_{\chi} \sum_{\omega=0}^{\Omega_i-1} \sum_{\varphi=0}^{\Phi_i-1} \sum_{\lambda=0}^{\Lambda_i-1} W_{i,j,\chi}^{\omega,\varphi,\lambda} v_{i-1,\chi}^{(x+\omega)(y+\varphi)(z+\lambda)}) \tag{2}$$

where χ is the feature cube connected to the current feature cube in the $(i-1)$th layer, and Ω, Φ, and Λ are the length, width, and depth of the convolution kernel size, respectively. In 2D convolution, the output of the jth feature map in the ith layer at the spatial location of (x,y) can be computed using Equation (3):

$$\nu_{i,j}^{xy} = g\left(b_{i,j} + \sum_{\chi} \sum_{\omega=0}^{\Omega_i-1} \sum_{\varphi=0}^{\Phi_i-1} W_{i,j,\chi}^{\omega,\varphi} v_{i-1,\chi}^{(x+\omega)(y+\varphi)}\right) \tag{3}$$

2.4. Morphological Operation Layers

Topological operators are applied to images by morphological operators to recover or filter out specific structures [63,64]. Mathematical morphology operators are nonlinear image operators that are based on the image spatial structure [65–67]. *Dilation* and *Erosion* are shape-sensitive operations that can be relatively helpful to extract discriminative spatial-contextual information during the training stage [67–69]. *Erosion*($\ominus$) and *Dilation*($\oplus$) are two basic operations in morphology operators that can be defined for a grayscale image X with size $M \times N$ and W structuring elements, as follows in Equation (4) [65,66]:

$$\begin{gathered}(X \oplus W)(x,y) = \max_{(i,m)\in S}(X(x-l,y-m) + W_d(l,m)) \\ S = \{(l,m) | l \in \{1,2,3,\ldots,a\}; m \in \{1,2,3,\ldots,b\};\}\end{gathered} \tag{4}$$

where W_d is the structuring element of dilation that can be defined on domain S. Accordingly, the erosion operator with structuring element W_d can be defined as follows in Equation (5):

$$(X \ominus W)(x,y) = \min_{(i,m)\in S}(X(x+l,y+m) - W_e(l,m)) \tag{5}$$

The structure element is initialized based on random values in the training process. The back-propagation algorithm is used to update the structure elements in the morphological layers. The propagation of the gradient through the network is very similar to that of a neural network.

2.5. Classification

After deep-feature extraction by convolution and morphological layers, the deep features are transformed for the flattening layer to reshape as 1D vectors. Then, these vectors are fed to the first fully connected layer and the second fully connected layer. The latest layer is soft-max, which assigns probabilities to each class for input pixels. Figure 1 presents the classification procedure for this framework.

2.6. Training Process

The network parameters are initialized based on the initial values and then are tuned iteratively based on optimizers, such as stochastic gradient descent. The DSMNN-Net is trained based on the training data, and the error of the network is obtained based on the calculation of the loss value on the validation dataset. The error of the training model is fed to the optimizer and is used to update the parameters. Due to back-propagation, the parameters are updated at each step to decrease the error of comparing the results obtained from the network with the validation dataset. The Tversky loss function is used to calculate the network error in the training process, which is a generalization of the dice score [70]. The Tversky index (*TI*) between $\hat{\Psi}$ (predicted value) and Ψ (truth value) is defined as in Equation (6):

$$TI(\hat{\Psi},\Psi,\alpha,\beta) = \frac{|\hat{\Psi}\Psi|}{|\hat{\Psi}\Psi| + \alpha|{}^{\hat{\Psi}}/{}_{\Psi}| + \beta|{}^{\Psi}/{}_{\hat{\Psi}}|} \tag{6}$$

where α and β control the magnitude of penalties for false positive and false negative pixels, respectively. These parameters are often chosen based on trial and error.

2.7. Accuracy Assessment

We assessed the results of the BAM based on visual and numerical analysis. The numerical analysis was applied as the standard measurement indices. To this end, the five most common quantitative assessment metrics were selected to evaluate the results. These indices are the overall accuracy (OA), the kappa coefficient (KC), and the F1-score, Recall, and intersection over union (IOU).

To compare the performance of the method proposed here, two state-of-the-art deep-learning methods were selected for this study. The first method was the deep Siamese network, which has been proposed in many studies for change detection purposes [71–73]. This method has three convolution layers in each stream, and then fully connected was used for classification. Then, the second method was CNN, based on a framework designed by Belenguer-Plomer, Tanase, Chuvieco and Bovolo [42] for mapping of burned areas. This method has two convolution layers and a max-pooling layer, then two fully connected layers were used. More details of this method can be found in [42].

3. Case Study and Satellite Images

This section investigates the case study area and the satellite data in more detail.

3.1. Study Area

Both study areas in this research were located in the Australian continent. The main reason for choosing the areas was the availability of the PRISMA hyperspectral datasets for these areas. Reference is the most important factor in the evaluation of BAM results. Thus, the reference data were obtained based on visual analysis and the interpretation of the results of BAM in previous papers. Figure 3 presents the locations of two study areas, in the southern Australian continent.

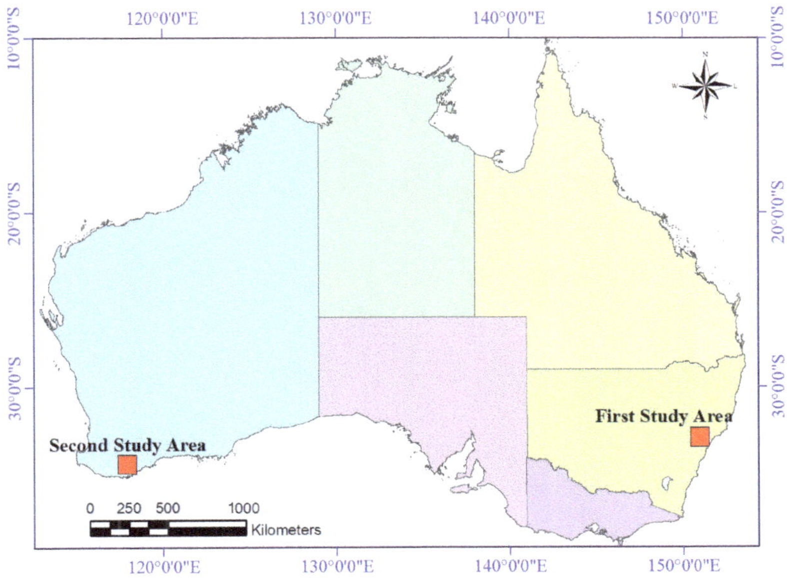

Figure 3. The locations of the two study areas for burned area mapping.

Figure 4 shows the incorporated burned area datasets for the first study area. Figure 5 illustrates the original incorporated dataset for the BAM for the second study area. The details for the incorporated datasets for both of the study areas are given in Table 1.

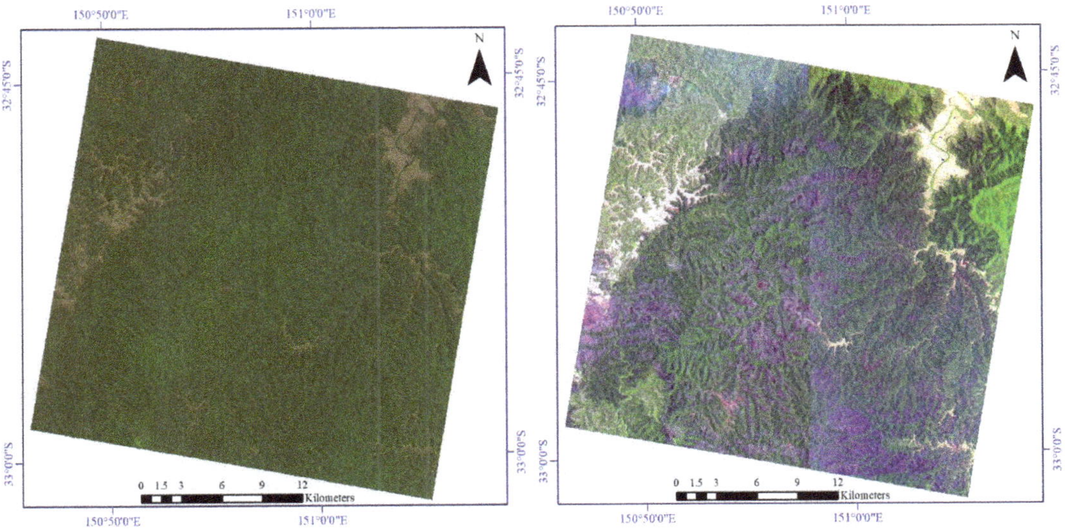

(a) (b)

Figure 4. *Cont.*

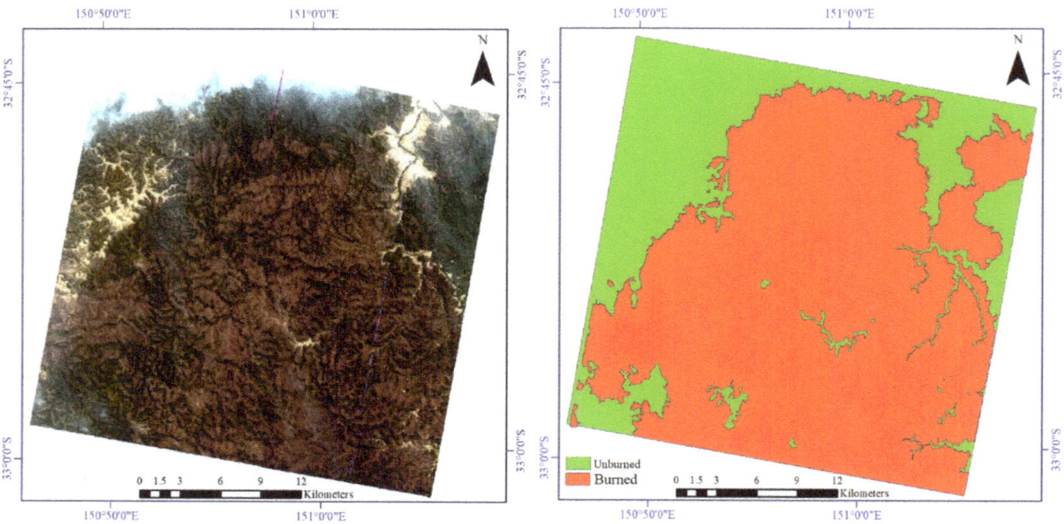

(c) (d)

Figure 4. The dataset used for the burned area mapping for the first study area. (**a**) Pre-event Sentinel-2 imagery. (**b**) Post-event Sentinel-2 dataset. (**c**) Post-event PRISMA hyperspectral imagery. (**d**) Ground truth.

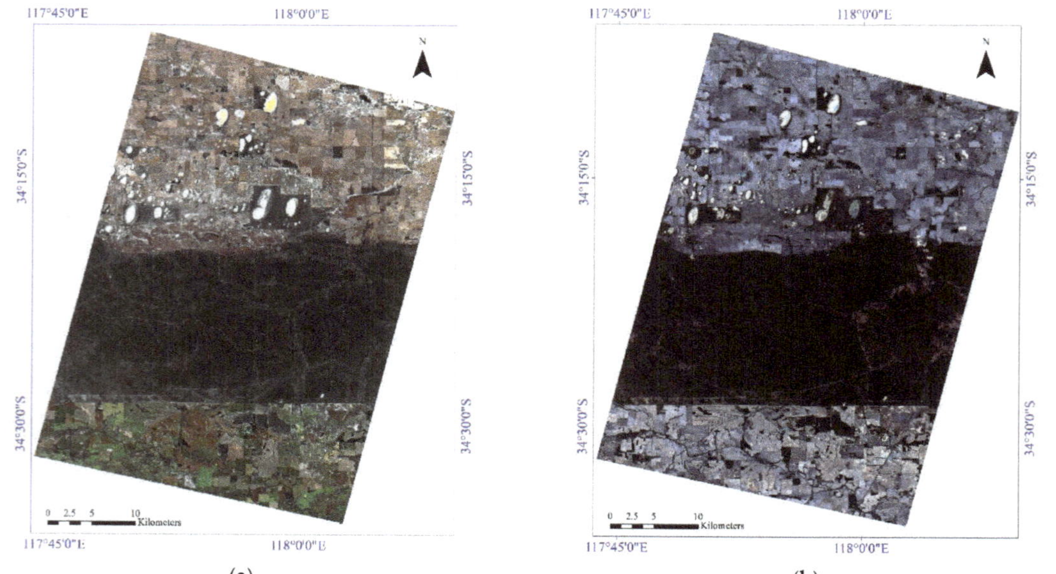

(a) (b)

Figure 5. *Cont.*

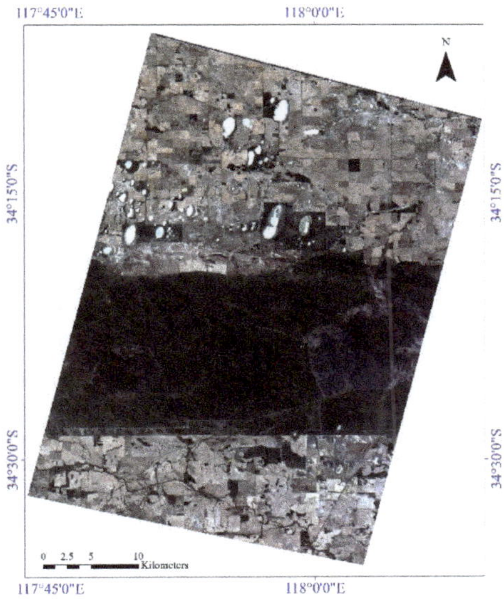

(c)

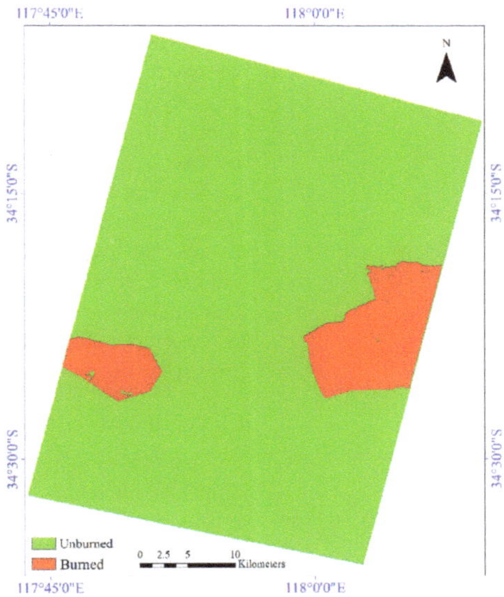

(d)

Figure 5. Illustration of the various incorporated datasets for the burned area mapping for the second study area. (**a**) Pre-event Sentinel-2 imagery. (**b**) Post-event Sentinel-2 dataset. (**c**) Post-event PRISMA hyperspectral imagery. (**d**) Ground truth.

Table 1. The main characteristics of the incorporated datasets for both case studies.

Sensor	Properties	First Study Area	Second Study Area
Sentinel-2	Spectral bands	13	13
	Spatial resolution (m)	10	10
	Resampled spatial resolution (m)	30	30
	Data size (pixel)	1168 × 1168	1159 × 1853
	Pre-event acquired date	December 2019	October 2019
	Post-event acquired date	November 2020	January 2020
PRISMA	Spectral bands	169	169
	Spatial resolution (m)	30	30
	Data size (pixel)	1168 × 1168	1159 × 1853
	Post-event acquired date	December 2019	January 2020

3.2. *Sentinel-2 Images*

Sentinel-2 is a European Space Agency Earth observation project that provides continuity to services dependent on multispectral high-spatial-resolution observations over the whole land surface of the Earth. This mission consists of two satellites, Sentinel-2-A and Sentinel-2-B, which have completed the existing Landsat and Spot missions and have enhanced data availability for RS communications. One satellite has a temporal resolution of 10 days, while the two satellites have a temporal resolution of 5 days [10]. The Sentinel-2 main sensor, the multispectral instrument, is based on the push-broom principle. Sentinel-2 has 13 spectral bands and broad spectral coverage.

This study used the Level-2A product as input data for BAM, which are surface reflectance data. Furthermore, it was necessary to convert the spatial resolution of the Sentinel-2 dataset into the spatial resolution of the hyperspectral dataset (30 m).

3.3. PRISMA Images

PRISMA is a medium-resolution hyperspectral imaging mission of the Italian Space Agency that was launched in March 2019 [74]. The PRISMA sensor is a spaceborne system that acquires hyperspectral datasets continuously, with a repeat orbital cycle of approximately 29 days [75]. PRISMA images the Earth surface in 240 contiguous spectral bands (66 visible to near-infrared, plus 174 short-wave infrared) with a push-broom scanning mode, covering the wavelengths between 400 and 2500 nm, at a spatial resolution of 30 m. The high dimensional spectral bands provide the possibility to analyze complex land-cover objects [14]. We chose the level-2-D product for the BAM, which was preprocessed (i.e., atmospheric correction and geolocation, orthorectification) [14]. The PRISMA hyperspectral dataset is freely available on this website: http://prisma.asi.it/missionselect/, accessed on 16 November 2021. After removing the noisy and no-data bands, 169 spectral bands were chosen for the next analysis.

4. Experiments and Results

Gathering of sample data is required to estimate the burned area due to using a supervised learning method. The quality and quantity of the sample data have a key role in BAM. In this study, the numbers of the sample data were kept at a sufficient level for the two classes (i.e., burned areas, unburned areas). Figure 6 shows the spatial distribution of the sample data.

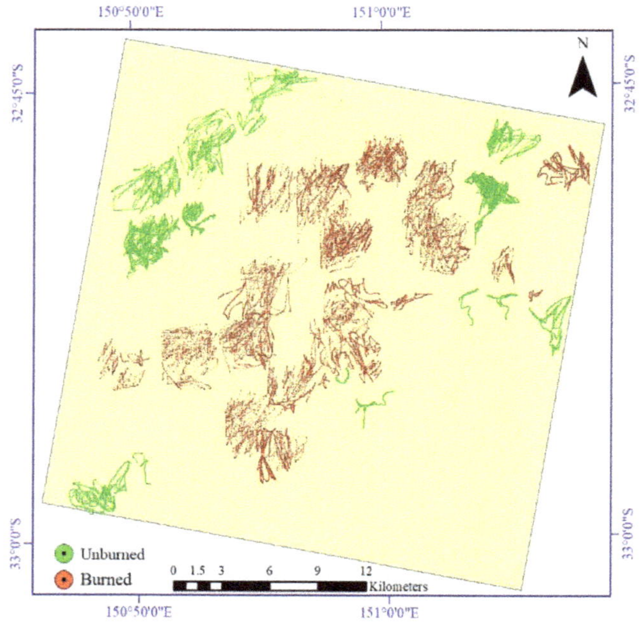

(a)

Figure 6. *Cont.*

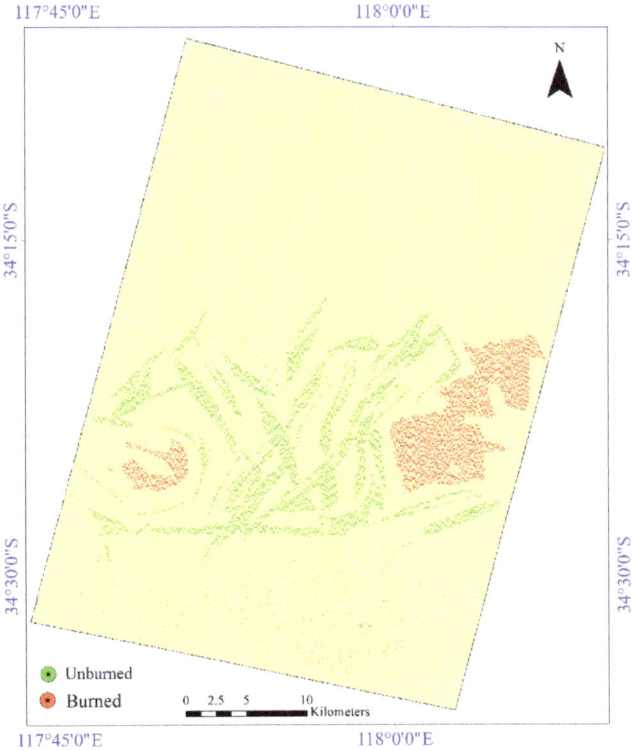

(b)

Figure 6. The spatial distributions of the two study areas for the burned area mapping. (**a**) Sample data for the first study area. (**b**) Sample data for the second study area.

In addition, Table 2 shows the sizes of sample data for two classes in the study areas.

Table 2. The number of samples used for mapping of burned area in the two study areas.

Case Study	Number of Pixels in the Study Area	Class	Number of Samples	Training	Validation	Testing
First study area	989,764	Unburned	15,318	9803	2450	3065
		Burned	21,387	13,687	3421	4459
Second study area	1,955,898	Unburned	6590	4217	1054	1318
		Burned	3206	2051	513	642

4.1. Parameter Setting

The DSMNN-Net has hyperparameters that need to be set. These hyperparameters were set manually based on trial and error. The optimum values of these parameters were set as follows: the input patch-size for Sentinel-2 and PRISMA sensors were $11 \times 11 \times 13$ and $11 \times 11 \times 169$, respectively, with 500 epochs; the weight initializer was set as He-normal-Initializer [76] for convolution layers; the random value for initializing of the morphological layers, number of neurons at the fully-connected layer was 900; the initial learning rate was 10^{-4}; and the minibatch size was 550. It is worth noting that all of the hyperparameters were constant during the process for all of the CNN methods. Similarly, the two other methods set such values. Additionally, the selection of some of these parameters was related to hardware (e.g., increasing minibatch size quickly filled the RAM of the system). Moreover, the weight initializer by the He-normal-initializer increased the speed of network convergence compared to the random initializer.

4.2. Results

The results for the BAM for the two study areas are considered in this section. For the two main scenarios, these were investigated according to Table 3.

Table 3. Different scenarios for mapping of burned areas in two case-study areas.

Scenario	Pre-Event Dataset	Post-Event Dataset
S#1	Sentinel-2	Sentinel-2
S#2	Sentinel-2	PRISMA

4.2.1. First Study Area

Figure 7 shows the results of the BAM based on the post/pre-event Sentinel-2 imagery. Based on these results, the DSMNN-Net differed from the BAM. Most methods detected the burned areas, with differences seen in the detail. For example, there are some missed detection areas in the results of the two CNN-based methods (center of scene) while the method proposed here detected these well.

Figure 8 shows the mapping results for the heterogeneous dataset provided by various methods. As shown in Figure 8, all of the methods provided better performance in comparison with the first scenario (S#1) as a result of the small missed detection area that was significantly decreased. The results of the DSMNN-Net fit better with ground truth while results of other methods have many false pixels; in particular, for the Siamese network (Figure 8a). The main differences among these results are obvious at the edges of the burned areas.

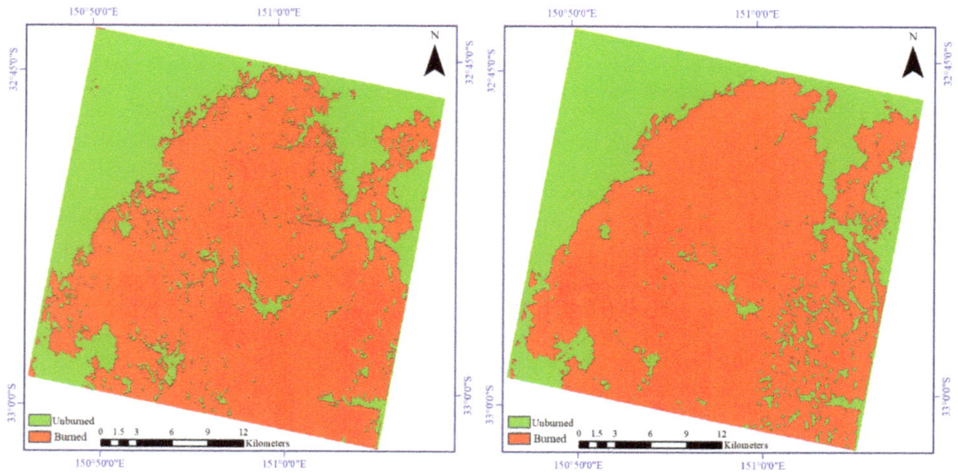

(a) (b)

Figure 7. *Cont.*

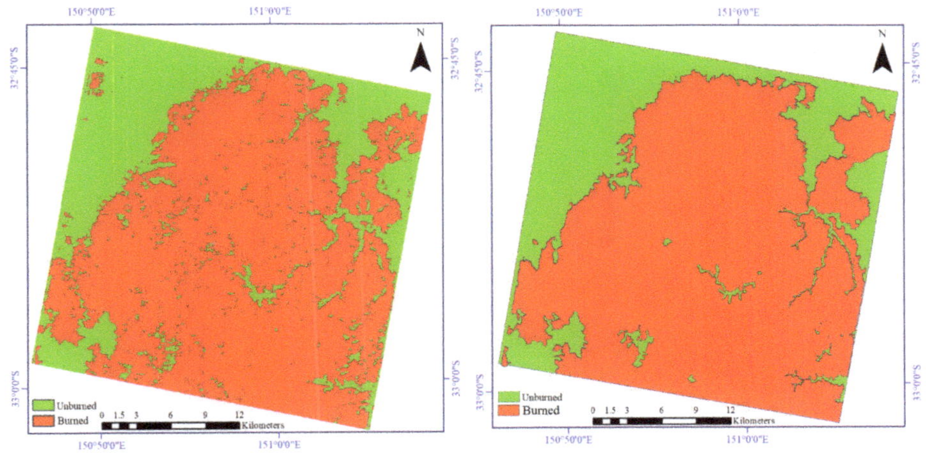

(c) (d)

Figure 7. Visual comparisons of the results for the burned area mapping based on the post/pre-event Sentinel-2 imagery for the first study area. (**a**) Deep Siamese network. (**b**) Using the CNN method proposed in [42]. (**c**) Using the method proposed in the present study. (**d**) Ground truth map.

The numerical results for the BAM for the first study area are given in Table 4. Based on these data, the accuracies of all of the methods in both scenarios were >87% in all terms. The accuracy of the algorithms in combining the hyperspectral datasets with the multispectral dataset was significantly better than only the multispectral datasets. The accuracy of the BAM results based on the fusion of the PRISMA imagery and Sentinel-2 imagery was >94% by OA index. However, the results of BAM based on only the Sentinel-2 imagery were very close together, but these were considerably different in the second scenario (S#2). The method proposed here provided an accuracy of >97% in terms of the OA, Recall, and F1-score indices. Furthermore, this provided the highest score by KC index for the second scenario.

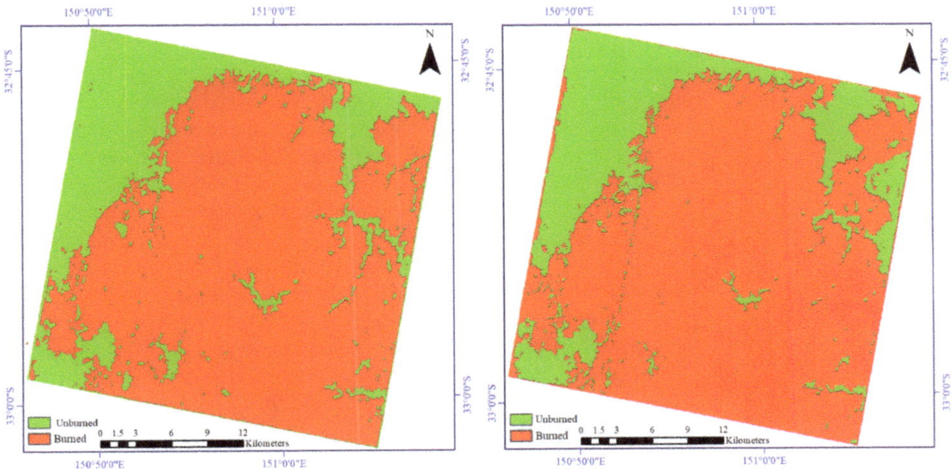

(a) (b)

Figure 8. *Cont.*

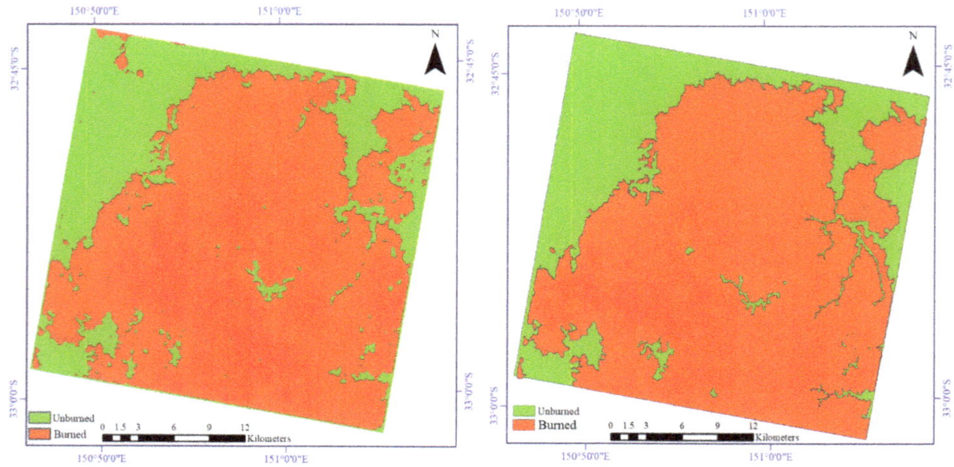

(c) (d)

Figure 8. Visual comparisons of the results for the burned area mapping based on pre-event Sentinel-2 and post-event PRISMA imagery for the first study area. (**a**) Deep Siamese network. (**b**) Using the CNN method proposed in [42]. (**c**) Using the DSMNN-Net. (**d**) Ground truth map.

Table 4. Accuracy assessment for the burned area mapping for the first study area. S#1, pre/post-event Sentinel-2 imagery; S#2, pre-event Sentinel-2 imagery and post-event PRISMA imagery.

Method	Scenario	OA (%)	Recall (%)	F1-Score (%)	IOU	KC
Siamese network	*S#1*	87.94	87.10	91.34	0.740	0.716
	S#2	94.79	96.19	96.43	0.786	0.868
CNN method proposed by [42]	*S#1*	89.35	89.40	92.46	0.842	0.744
	S#2	94.35	97.13	96.17	0.851	0.853
DSMNN-Net	*S#1*	**90.24**	**92.51**	**93.26**	**0.864**	**0.755**
	S#2	**97.46**	**97.99**	**98.25**	**0.901**	**0.936**

OA, overall accuracy; IOU, intersection over union; KC, kappa coefficient.

4.2.2. Second Study Area

Figure 9 illustrates the results of the BAM based on the bi-temporal multispectral/hyperspectral datasets for the second study area. Based on these results, there are some differences among the algorithms seen for the details. Figure 9a shows the performance of the deep Siamese network, in that it has low false pixels, although many missed detection pixels can be seen in the result. However, the lowest missed pixels can be seen in the BAM for the method proposed by [42] in Figure 9b, although it shows high false pixels in the results presented. The result of BAM by the DSMNN-Net can be seen in Figure 9c, which shows the lowest false pixels and missed pixels in the mapping.

The results of the BAM based on the Sentinel-2 and PRISMA sensors for the second area are presented in Figure 10. Based on the comparisons of the results presented with the multispectral dataset, there are some improvements in the details of the mapping. These improvements are more evident in the results of the DSMNN-Net, as some false pixels were classified correctly.

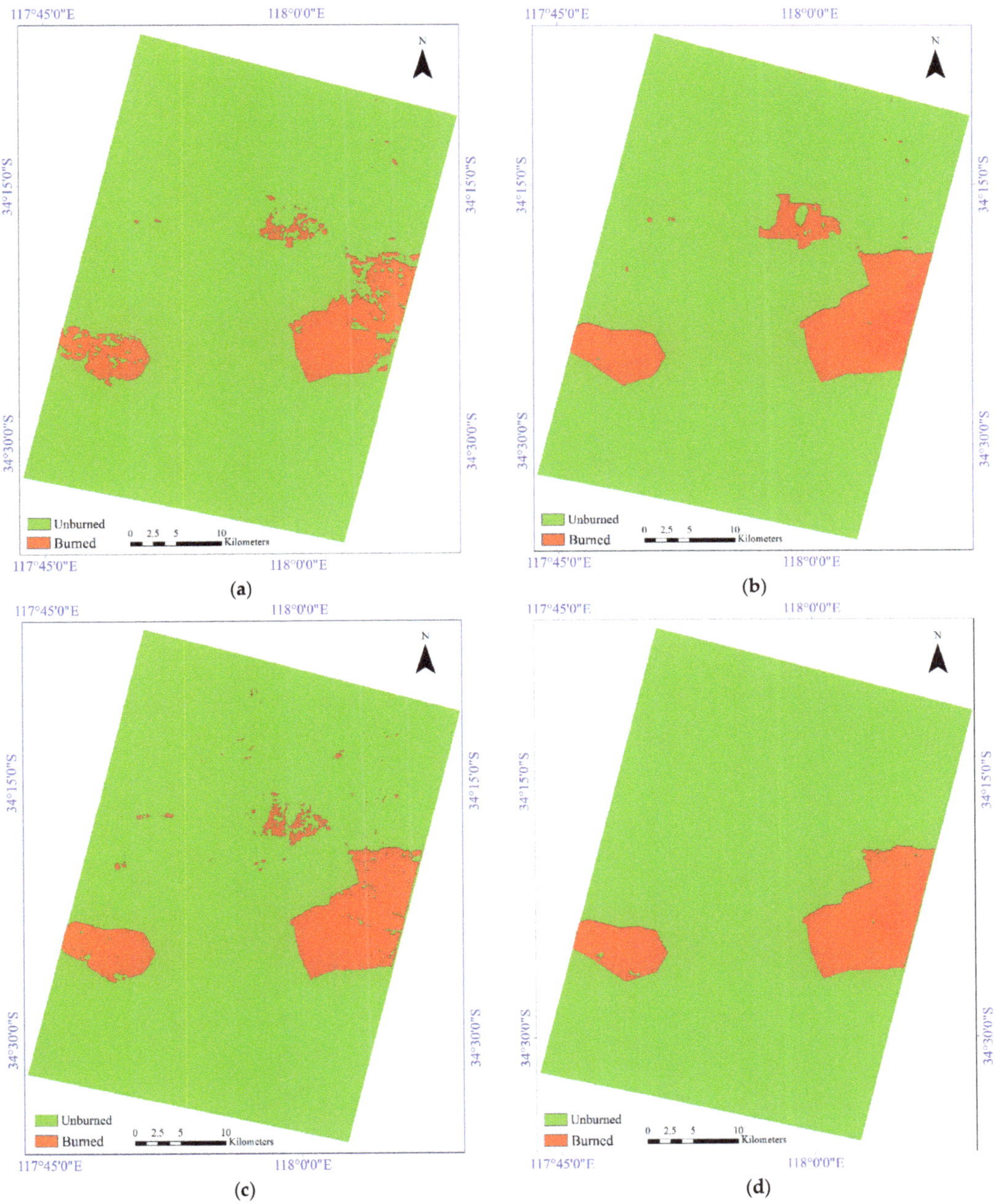

Figure 9. Visual comparisons of the results for the burned area mapping based on the post/pre-event Sentinel-2 imagery for the second study area. (**a**) Deep Siamese network. (**b**) Using the CNN method proposed in [42]. (**c**) Using the method proposed in the present study. (**d**) Ground truth map.

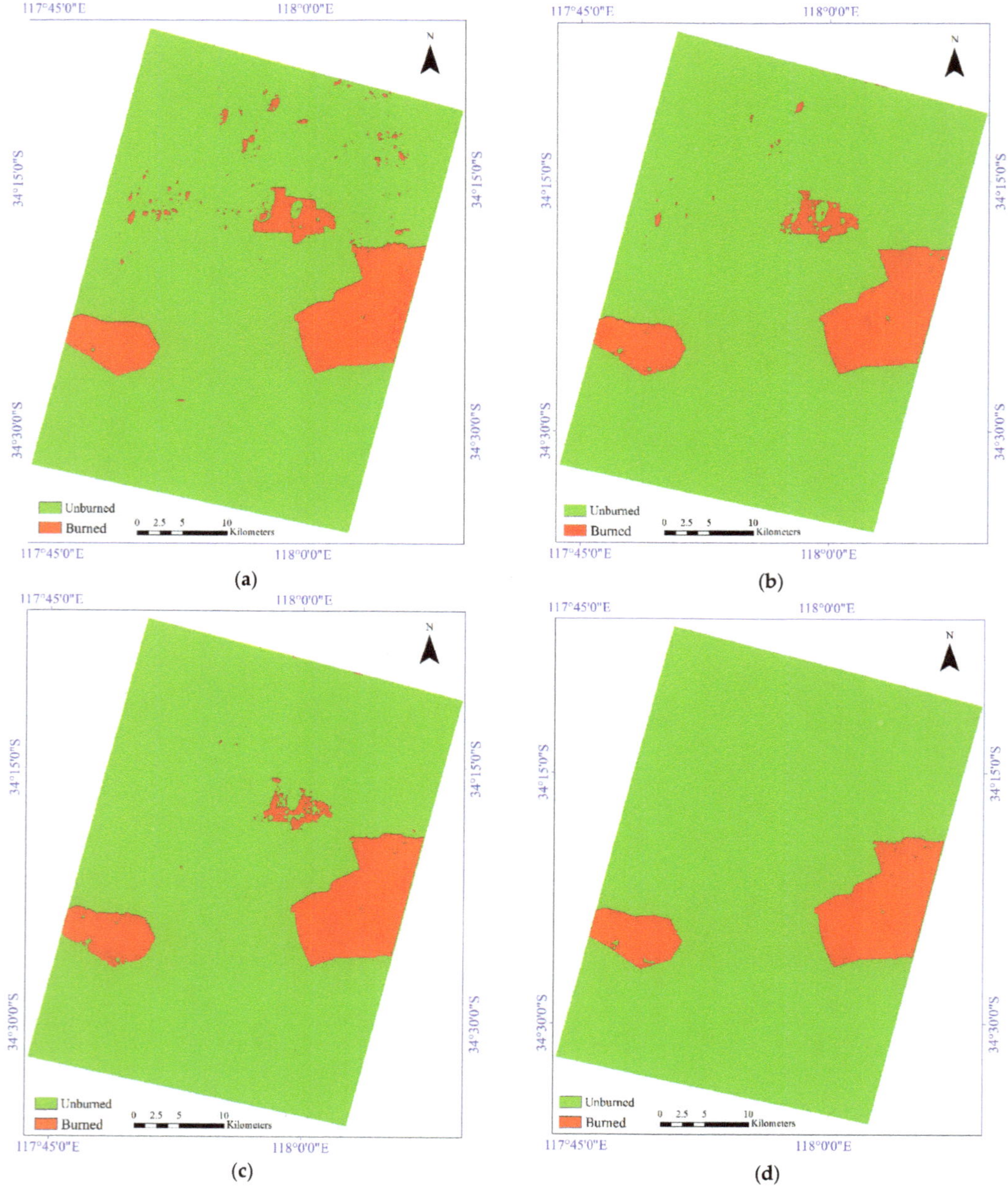

Figure 10. Visual comparisons of the results for the burned area mapping based on the pre-event Sentinel-2 and post-event PRISMA imagery for the second study area. (**a**) Deep Siamese network. (**b**) Using the CNN method proposed in [42]. (**c**) Using the DSMNN-Net. (**d**) Ground truth map.

Table 5 presents the quantitative performance comparison of the methods for the second study area for the BAM. Based on these data, there are some enhancements in the results of the BAM in all terms for the case study area. The enhancement of the methods is more evident for the Recall, F1-Score, IOU, and KC indices. For example, the difference between Recall of the Siamese network in the first and second scenarios is >20%; moreover, in other terms, this improvement is >3%. The improvement of the BAM in the CNN method proposed by [42] was slight. The DSMNN-Net has some improvements that are more evident by terms KC, IOU, F1-Score, and Recall. For example, these terms show greater improvement for the method proposed here in the detection of burned pixels, while there is some improvement in the detection of nonburned pixels.

Table 5. Accuracy assessment for the burned area mapping for the second study area. S#1, pre/post-event Sentinel-2 imagery; S#2, pre-event Sentinel-2 imagery and post-event PRISMA imagery.

Method	Scenario	OA (%)	Recall (%)	F1-Score (%)	IOU	KC
Siamese network	*S#1*	97.32	78.90	85.03	0.739	0.835
	S#2	97.41	98.79	88.05	0.786	0.866
CNN method proposed by [42]	*S#1*	98.21	98.94	91.44	0.842	0.904
	S#2	98.35	97.75	91.94	0.851	0.910
DSMNN-Net	*S#1*	**98.56**	**95.13**	**92.75**	**0.864**	**0.919**
	S#2	**98.95**	**98.90**	**94.80**	**0.901**	**0.942**

OA, overall accuracy; IOU, intersection over union; KC, kappa coefficient.

5. Discussion

This study focused on BAM based on deep-learning methods based on bi-temporal multispectral and hyperspectral imagery. BAM has mainly been applied based on low-resolution satellite imagery (e.g., MODIS, VIIRS, and Sentinel-3). However, BAM based on these sensors has provided promising results, although mapping of small burned areas is the most important challenge. These methods support the high coverage areas but do not provide suitable results for small areas. Furthermore, there are some burned area products on a global scale based on the MODIS satellite imagery. Many studies have evaluated the accuracy obtained by BAM based on the MODIS collection, where this has been reported as <80%, while for the BAM for both study areas, the DSMNN-Net provided an accuracy of >98% by the OA index (Tables 4 and 5).

Most BAM is mainly based on high-resolution imagery (e.g., Landsat-8, Sentinel-2) for the normalized burned ratio index. Although this index has provided some promising results for BAM, due to the dependency of burned areas on the environmental features and the behavior of the fire, it is hard to discriminate burned areas from the background. This issue has reduced the efficiency of the BAM methods by the need to threshold the normalized burned ratio indices. Furthermore, some high-resolution burned area products on a global scale have been obtained based on this. Thus, these products do not support accurate BAM in practical real-world burned area estimation. Similarly, some unsupervised thresholding methods have been used, but due to the complexity of the background and noise conditions, the selection of suitable thresholds is another limitation of these methods.

Many BAM methods have been proposed based on machine-learning methods, such as random forest, K-nearest neighbor, and support machine vector. While these methods have provided acceptable results for BAM, they use handcrafted feature extraction. Manual feature extraction and then selection of suitable features is a time-consuming process. This issue needs to be considered when the study area is very large scale and the number of features is high. Additionally, these methods mainly focus on spectral features and ignore the potential of spatial features. The potential of spatial features has been shown in many studies on BAM based on machine-learning methods. The deep-learning-based methods can automatically extract deep features that are a combination of spectral and spatial features. This study has used the deep-feature extraction manner for BAM based on convolution and morphological layers.

To show further the effectiveness of the DSMNN-Net process, we visualized the feature maps in the different layers to look inside their internal operation and behavior. Figure 11 illustrates the visualization of the feature maps extracted from some layers in the proposed DSMNN-Net for random pixels. The first layers show the shallow features, while the middle layers focus on the general structure around the central pixel.

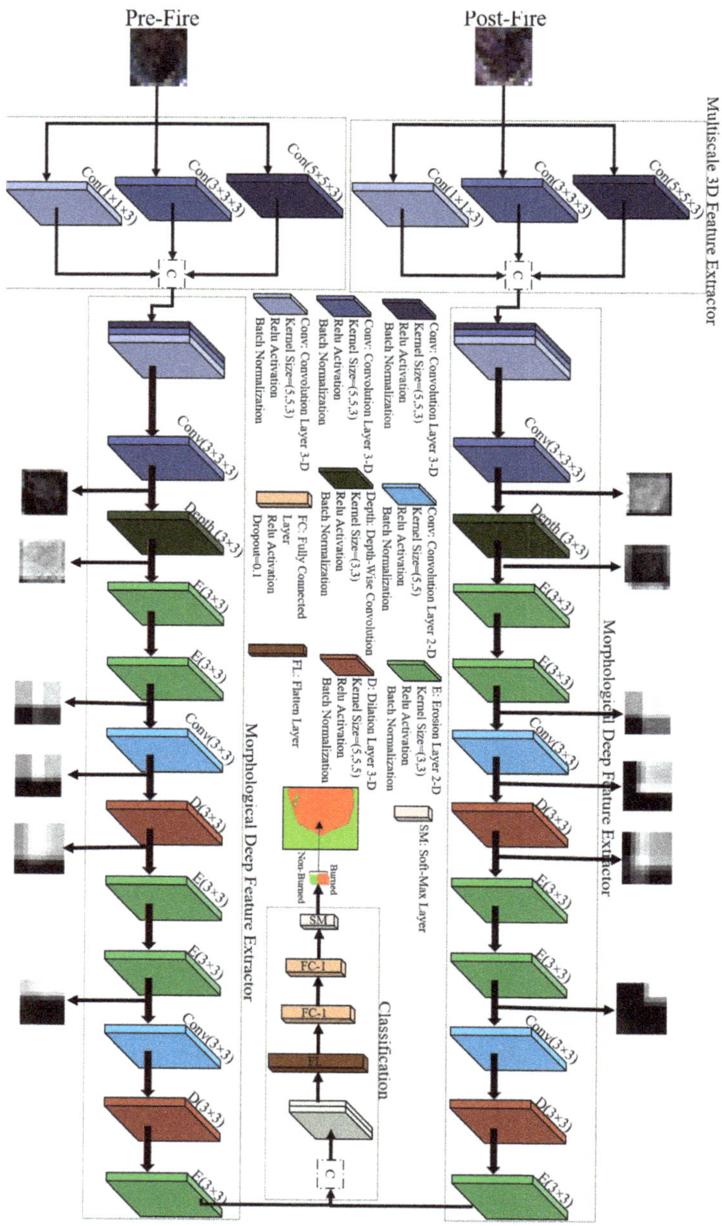

Figure 11. Visualization of feature maps in the DSMNN-Net for a random pixel.

The efficiency of deep features and handcrafted features can be seen in Table 6. Based on these data, the DSMNN-Net provided greater accuracy compared to other state-of-the-art methods for BAM using the Sentinel-2 imagery.

Table 6. Comparison of performance of the DSMNN-Net with other burned area mapping methods.

Reference	Accuracy	Method	Dataset
Grivei, et al. [77]	(F1-Score: 0.873)	Support vector machine algorithm and spectral indices, factor analysis	Sentinel-2
Barboza Castillo, et al. [78]	94.4	Thresholding on the spectral index	Sentinel-2
Syifa, et al. [79]	92	Support vector machine and imperialist competitive algorithm	Sentinel-2
Quintano, et al. [80]	84	Spectral index and thresholding	Combination of Landsat-8 and Sentinel-2
Ngadze, et al. [81]	92	Random forest	Sentinel-2
Roy, et al. [82]	92	Random forest change regression, and a region growing manner	Combination of Landsat-8 and Sentinel-2
Lima, et al. [83]	96	Thresholding on the spectral index	Sentinel-2
Seydi, Akhoondzadeh, Amani and Mahdavi [10]	91	Spectral and spatial features and random forest	Sentinel-2
DSMNN-Net	98	Deep-learning based	Sentinel-2

OA, overall accuracy.

Hyperspectral imagery has a high content of spectral information in comparison with multispectral imagery. This advantage of hyperspectral imagery helps to detect burned areas with a highly complex background. Thus, the main reason behind the robust results provided by the method proposed here is the use of the hyperspectral dataset for the BAM. The burned pixels have a high similarity to some unburned pixels, and to clarify this subject, we presented some spectral signatures of burned and unburned pixels in the different areas. Figure 12 illustrates the similarities of the spectral signatures for the two main classes. Based on these data, the burned and unburned pixels have similar behaviors in the 0.45 to 0.8 μm range, while for other areas there are some differences in the reflectance. Therefore, hyperspectral imagery and combining pre-event datasets can be useful for BAM.

Additionally, the proposed suitable deep-feature extraction framework is very important in deep-learning-based methods. Among these three deep-learning-based methods, the method proposed here provided the best performance in all of the scenarios and for both study areas. This issue originated from the architecture of the deep-learning methods in the extraction of deep features. The DSMNN-Net extracts the deep features based on the type of kernel convolution and morphological layers. Initially, the DSMNN-Net uses multiscale 3D kernel convolution that investigates the relation among the spectral bands in deep-feature extraction. Based on Figure 12, there are some differences among the spectral signatures for the same classes, and these differences are greater for some spectral bands. Furthermore, there is some overlap between the two classes in the spectral bands. Therefore, using 3D convolution can enhance the efficiency of the network, because this can consider the relations between the spectral bands and the relations between the central pixel and the neighboring pixels. This advantage is the most important factor in taking the full content of spectral information for the BAM. Then, the morphological layers are used to explore nonlinear characterizers of the input dataset in the mapping. Thus, the DSMNN-Net can extract high-level and informative deep features based on proposed architectures; as a result, accurate BAM is possible using this proposed method. Additionally, the diversity of objects and the complexity of unburned areas mean that the BAM changes, which is a challenge. Solving this challenge mainly requires increasing the depth of the network, which results in an increasing number of parameters, and the need for greater training data and time. Here, the DSMNN-Net uses morphological operation cases to investigate the complexity of the background. The morphological operators use nonlinear operations that increase the efficiency of the network for the BAM.

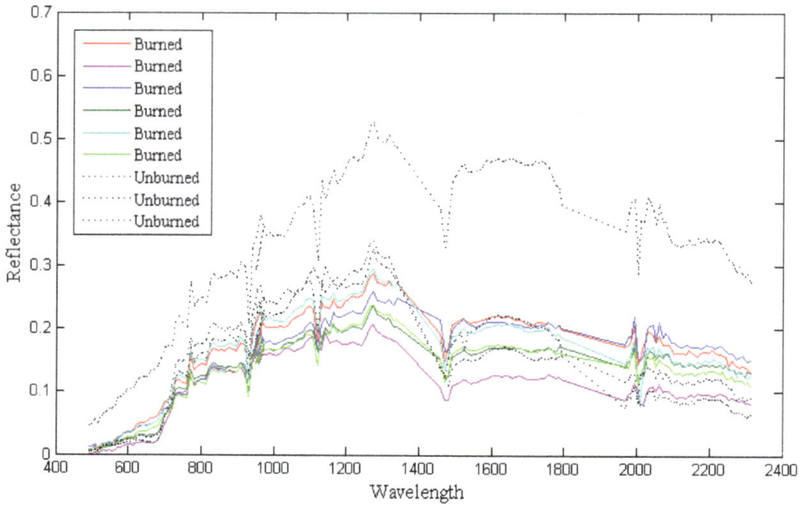

Figure 12. Spectral signatures among the burned and unburned pixels.

Recently, some semantic-segmentation-based (U-Net architecture) methods have been proposed. While these methods can provide considerably improved results for BAM, it needs to be noted that they require large amounts of labeled data of a specific size (i.e., 512 × 512 or 256 × 256). Obtaining a large amount of labeled datasets for such an application is very difficult and time consuming. Furthermore, these models are more complex due to the higher number of parameters and the need for more time for training the network. The DSMNN-Net used close to 45,000 pixels for the BAM, and obtaining this amount of sample data is easy according to the extent of the study areas.

One of the most common challenges for BAM based on changes in detection methods is the detection of nontarget changes. For example, the second study area has some nontarget change areas where their changes originated from changes in the water level of the lakes. This issue meant that the methods considered these as burned areas, while they are nonburned areas. This challenge is more evident in the BAM by the CNN method proposed in [42]. The sample data should cover more areas in the background, although the method proposed here controlled this issue in the BAM.

The method proposed here uses adaptive heterogeneous datasets in the mapping of burned areas. However, pixel-based change detection methods can be applied for bi-temporal multispectral/hyperspectral datasets easily, while they are difficult to apply for a heterogeneous dataset. In other words, some methods (i.e., image differencing algorithm) compare the pixel-to-pixel of the first and second time of bi-temporal dataset for BAM, while for heterogeneous datasets this is very difficult due to difference in a number of spectral bands, and content of datasets. The proposed DSMNN-Net can be applied to heterogeneous datasets without any additional processing (e.g., dimensional reduction). These advantages will also help in BAM when applied in a near real-time manner. It is worth noting that the proposed DSMNN-Net applied based on pre-event multispectral and post-event hyperspectral datasets while the bi-temporal pre-event and post-event hyperspectral datasets can improve the result of the BAM.

6. Conclusions

Accurate and timely BAM is the most important factor in wildfire damage assessment and management. In this study, a novel framework based on a deep-learning method (DSMNN-Net) and the use of bi-temporal multispectral and hyperspectral datasets was proposed. We evaluated the performance of the DSMNN-Net for two study areas in two scenarios: (1) BAM based on bi-temporal Sentinel-2 datasets and (2) BAM based on pre-event Sentinel-2 and post-event PRISMA datasets. Furthermore, the results for the BAM are compared with other state-of-the-art methods, both visually and numerically. The results of the BAM show that the method proposed here has high efficiency in comparison with the other methods for BAM. Additionally, the use of hyperspectral datasets can improve the performance of BAM based on deep-learning-based methods. The experimental results of this study illustrate that the DSMNN-Net has some advantages: (1) it provides high accuracy for BAM; (2) it has a high sensitivity for BAM for complex background areas; (3) it is adaptive, with heterogeneous datasets for BAM (multispectral and hyperspectral); and (4) it can be applied in an end-to-end framework without any additional processing.

Author Contributions: Conceptualization, S.T.S. and M.H.; methodology, S.T.S.; writing—original draft preparation, S.T.S.; writing—review and editing, S.T.S., M.H. and J.C.; visualization, S.T.S.; supervision, M.H. and J.C.; funding acquisition, J.C. All authors have read and agreed to the published version of the manuscript.

Funding: This study received no external funding.

Institutional Review Board Statement: Not applicable.

Informed Consent Statement: Not applicable.

Data Availability Statement: Publicly available datasets were analyzed in this study. These datasets can be found at https://scihub.copernicus.eu/, accessed on 16 November 2021 and http://prisma.asi.it/missionselect/, accessed on 16 November 2021.

Acknowledgments: The authors would like to thank the European Space Agency and the Italian Space Agency for providing the datasets. We thank the anonymous reviewers for their valuable comments on our manuscript. This research was partially supported by the AXA fund for research and by MIAI @ Grenoble Alpes, (ANR-19-P3IA-0003).

Conflicts of Interest: The authors declare that they have no conflict of interest.

References

1. Llorens, R.; Sobrino, J.A.; Fernández, C.; Fernández-Alonso, J.M.; Vega, J.A. A methodology to estimate forest fires burned areas and burn severity degrees using Sentinel-2 data. Application to the October 2017 fires in the Iberian Peninsula. *Int. J. Appl. Earth Obs. Geoinf.* **2021**, *95*, 102243. [CrossRef]
2. Nohrstedt, D.; Mazzoleni, M.; Parker, C.F.; Di Baldassarre, G. Exposure to natural hazard events unassociated with policy change for improved disaster risk reduction. *Nat. Commun.* **2021**, *12*, 193. [CrossRef]
3. Roy, D.P.; Li, Z.; Giglio, L.; Boschetti, L.; Huang, H. Spectral and diurnal temporal suitability of GOES Advanced Baseline Imager (ABI) reflectance for burned area mapping. *Int. J. Appl. Earth Obs. Geoinf.* **2021**, *96*, 102271. [CrossRef]
4. de Luca, G.; Silva, J.M.; Modica, G. A workflow based on Sentinel-1 SAR data and open-source algorithms for unsupervised burned area detection in Mediterranean ecosystems. *GIScience Remote Sens.* **2021**, *58*, 516–541. [CrossRef]
5. Duane, A.; Castellnou, M.; Brotons, L. Towards a comprehensive look at global drivers of novel extreme wildfire events. *Clim. Chang.* **2021**, *165*, 43. [CrossRef]
6. Palaiologou, P.; Kalabokidis, K.; Troumbis, A.; Day, M.A.; Nielsen-Pincus, M.; Ager, A.A. Socio-Ecological Perceptions of Wildfire Management and Effects in Greece. *Fire* **2021**, *4*, 18. [CrossRef]
7. Serra-Burriel, F.; Delicado, P.; Prata, A.T.; Cucchietti, F.M. Estimating heterogeneous wildfire effects using synthetic controls and satellite remote sensing. *Remote Sens. Environ.* **2021**, *265*, 112649. [CrossRef]
8. Chowdhury, S.; Zhu, K.; Zhang, Y. Mitigating Greenhouse Gas Emissions Through Generative Adversarial Networks Based Wildfire Prediction. *arXiv* **2021**, arXiv:2108.08952.
9. Haque, M.K.; Azad, M.A.K.; Hossain, M.Y.; Ahmed, T.; Uddin, M.; Hossain, M.M. Wildfire in Australia during 2019–2020, Its Impact on Health, Biodiversity and Environment with Some Proposals for Risk Management: A Review. *J. Environ. Prot.* **2021**, *12*, 391–414. [CrossRef]

10. Seydi, S.T.; Akhoondzadeh, M.; Amani, M.; Mahdavi, S. Wildfire damage assessment over Australia using sentinel-2 imagery and MODIS land cover product within the google earth engine cloud platform. *Remote Sens.* **2021**, *13*, 220. [CrossRef]
11. Pandey, P.C.; Koutsias, N.; Petropoulos, G.P.; Srivastava, P.K.; Ben Dor, E. Land use/land cover in view of earth observation: Data sources, input dimensions, and classifiers—A review of the state of the art. *Geocarto Int.* **2021**, *36*, 957–988. [CrossRef]
12. Seydi, S.T.; Hasanlou, M.; Amani, M.; Huang, W. Oil Spill Detection Based on Multi-Scale Multi-Dimensional Residual CNN for Optical Remote Sensing Imagery. *IEEE J. Sel. Top. Appl. Earth Obs. Remote Sens.* **2021**, *14*, 10941–10952. [CrossRef]
13. Masek, J.G.; Wulder, M.A.; Markham, B.; McCorkel, J.; Crawford, C.J.; Storey, J.; Jenstrom, D.T. Landsat 9: Empowering open science and applications through continuity. *Remote Sens. Environ.* **2020**, *248*, 111968. [CrossRef]
14. Vangi, E.; D'Amico, G.; Francini, S.; Giannetti, F.; Lasserre, B.; Marchetti, M.; Chirici, G. The new hyperspectral satellite PRISMA: Imagery for forest types discrimination. *Sensors* **2021**, *21*, 1182. [CrossRef]
15. Wambugu, N.; Chen, Y.; Xiao, Z.; Wei, M.; Bello, S.A.; Junior, J.M.; Li, J. A hybrid deep convolutional neural network for accurate land cover classification. *Int. J. Appl. Earth Obs. Geoinf.* **2021**, *103*, 102515. [CrossRef]
16. Xu, L.; Li, J.; Brenning, A. A comparative study of different classification techniques for marine oil spill identification using RADARSAT-1 imagery. *Remote Sens. Environ.* **2014**, *141*, 14–23. [CrossRef]
17. Gu, Y.; Liu, T.; Gao, G.; Ren, G.; Ma, Y.; Chanussot, J.; Jia, X. Multimodal hyperspectral remote sensing: An overview and perspective. *Sci. China Inf. Sci.* **2021**, *64*, 1–24. [CrossRef]
18. Hong, D.; Chanussot, J.; Zhu, X.X. An Overview of Multimodal Remote Sensing Data Fusion: From Image to Feature, from Shallow to Deep. In Proceedings of the 2021 IEEE International Geoscience and Remote Sensing Symposium IGARSS, Brussels, Belgium, 11–16 July 2021; pp. 1245–1248.
19. Mardian, J.; Berg, A.; Daneshfar, B. Evaluating the temporal accuracy of grassland to cropland change detection using multitemporal image analysis. *Remote Sens. Environ.* **2021**, *255*, 112292. [CrossRef]
20. Sun, Y.; Lei, L.; Li, X.; Sun, H.; Kuang, G. Nonlocal patch similarity based heterogeneous remote sensing change detection. *Pattern Recognit.* **2021**, *109*, 107598. [CrossRef]
21. ElGharbawi, T.; Zarzoura, F. Damage detection using SAR coherence statistical analysis, application to Beirut, Lebanon. *ISPRS J. Photogramm. Remote Sens.* **2021**, *173*, 1–9. [CrossRef]
22. Hashemi-Beni, L.; Gebrehiwot, A.A. Flood extent mapping: An integrated method using deep learning and region growing using UAV optical data. *IEEE J. Sel. Top. Appl. Earth Obs. Remote Sens.* **2021**, *14*, 2127–2135. [CrossRef]
23. Liu, S.; Chi, M.; Zou, Y.; Samat, A.; Benediktsson, J.A.; Plaza, A. Oil spill detection via multitemporal optical remote sensing images: A change detection perspective. *IEEE Geosci. Remote Sens. Lett.* **2017**, *14*, 324–328. [CrossRef]
24. Moya, L.; Geiß, C.; Hashimoto, M.; Mas, E.; Koshimura, S.; Strunz, G. Disaster Intensity-Based Selection of Training Samples for Remote Sensing Building Damage Classification. *IEEE Trans. Geosci. Remote Sens.* **2021**, *59*, 8288–8304. [CrossRef]
25. Zheng, Z.; Zhong, Y.; Wang, J.; Ma, A.; Zhang, L. Building damage assessment for rapid disaster response with a deep object-based semantic change detection framework: From natural disasters to man-made disasters. *Remote Sens. Environ.* **2021**, *265*, 112636. [CrossRef]
26. Hu, X.; Ban, Y.; Nascetti, A. Uni-Temporal Multispectral Imagery for Burned Area Mapping with Deep Learning. *Remote Sens.* **2021**, *13*, 1509. [CrossRef]
27. Muñoz, D.F.; Muñoz, P.; Moftakhari, H.; Moradkhani, H. From local to regional compound flood mapping with deep learning and data fusion techniques. *Sci. Total Environ.* **2021**, *782*, 146927. [CrossRef]
28. Zhang, Q.; Ge, L.; Zhang, R.; Metternicht, G.I.; Du, Z.; Kuang, J.; Xu, M. Deep-learning-based burned area mapping using the synergy of Sentinel-1&2 data. *Remote Sens. Environ.* **2021**, *264*, 112575.
29. Chiang, S.-H.; Ulloa, N.I. Mapping and Tracking Forest Burnt Areas in the Indio Maiz Biological Reserve Using Sentinel-3 SLSTR and VIIRS-DNB Imagery. *Sensors* **2019**, *19*, 5423. [CrossRef]
30. Lizundia-Loiola, J.; Franquesa, M.; Boettcher, M.; Kirches, G.; Pettinari, M.L.; Chuvieco, E. Implementation of the Burned Area Component of the Copernicus Climate Change Service: From MODIS to OLCI Data. *Remote Sens.* **2021**, *13*, 4295. [CrossRef]
31. Pinto, M.M.; Trigo, R.M.; Trigo, I.F.; DaCamara, C.C. A Practical Method for High-Resolution Burned Area Monitoring Using Sentinel-2 and VIIRS. *Remote Sens.* **2021**, *13*, 1608. [CrossRef]
32. Donezar, U.; De Blas, T.; Larrañaga, A.; Ros, F.; Albizua, L.; Steel, A.; Broglia, M. Applicability of the multitemporal coherence approach to sentinel-1 for the detection and delineation of burnt areas in the context of the copernicus emergency management service. *Remote Sens.* **2019**, *11*, 2607. [CrossRef]
33. Xulu, S.; Mbatha, N.; Peerbhay, K. Burned Area Mapping over the Southern Cape Forestry Region, South Africa Using Sentinel Data within GEE Cloud Platform. *ISPRS Int. J. Geo-Inf.* **2021**, *10*, 511. [CrossRef]
34. Liu, S.; Zheng, Y.; Dalponte, M.; Tong, X. A novel fire index-based burned area change detection approach using Landsat-8 OLI data. *Eur. J. Remote Sens.* **2020**, *53*, 104–112. [CrossRef]
35. Nolde, M.; Plank, S.; Riedlinger, T. An Adaptive and Extensible System for Satellite-Based, Large Scale Burnt Area Monitoring in Near-Real Time. *Remote Sens.* **2020**, *12*, 2162. [CrossRef]
36. Knopp, L.; Wieland, M.; Rättich, M.; Martinis, S. A deep learning approach for burned area segmentation with Sentinel-2 data. *Remote Sens.* **2020**, *12*, 2422. [CrossRef]

37. de Bem, P.P.; de Carvalho Júnior, O.A.; de Carvalho, O.L.F.; Gomes, R.A.T.; Fontes Guimarães, R. Performance analysis of deep convolutional autoencoders with different patch sizes for change detection from burnt areas. *Remote Sens.* **2020**, *12*, 2576. [CrossRef]
38. Ban, Y.; Zhang, P.; Nascetti, A.; Bevington, A.R.; Wulder, M.A. Near real-time wildfire progression monitoring with Sentinel-1 SAR time series and deep learning. *Sci. Rep.* **2020**, *10*, 1322. [CrossRef]
39. Zhang, P.; Ban, Y.; Nascetti, A. Learning U-Net without forgetting for near real-time wildfire monitoring by the fusion of SAR and optical time series. *Remote Sens. Environ.* **2021**, *261*, 112467. [CrossRef]
40. Farasin, A.; Colomba, L.; Garza, P. Double-step u-net: A deep learning-based approach for the estimation of wildfire damage severity through sentinel-2 satellite data. *Appl. Sci.* **2020**, *10*, 4332. [CrossRef]
41. Lestari, A.I.; Rizkinia, M.; Sudiana, D. Evaluation of Combining Optical and SAR Imagery for Burned Area Mapping using Machine Learning. In Proceedings of the 2021 IEEE 11th Annual Computing and Communication Workshop and Conference (CCWC), Virtual, 27–30 January 2021; pp. 0052–0059.
42. Belenguer-Plomer, M.A.; Tanase, M.A.; Chuvieco, E.; Bovolo, F. CNN-based burned area mapping using radar and optical data. *Remote Sens. Environ.* **2021**, *260*, 112468. [CrossRef]
43. Proy, C.; Tanre, D.; Deschamps, P. Evaluation of topographic effects in remotely sensed data. *Remote Sens. Environ.* **1989**, *30*, 21–32. [CrossRef]
44. Hao, D.; Wen, J.; Xiao, Q.; Wu, S.; Lin, X.; You, D.; Tang, Y. Modeling anisotropic reflectance over composite sloping terrain. *IEEE Trans. Geosci. Remote Sens.* **2018**, *56*, 3903–3923. [CrossRef]
45. Pouyap, M.; Bitjoka, L.; Mfoumou, E.; Toko, D. Improved Bearing Fault Diagnosis by Feature Extraction Based on GLCM, Fusion of Selection Methods, and Multiclass-Naïve Bayes Classification. *J. Signal Inf. Process.* **2021**, *12*, 71–85.
46. Raja, G.; Dev, K.; Philips, N.D.; Suhaib, S.M.; Deepakraj, M.; Ramasamy, R.K. DA-WDGN: Drone-Assisted Weed Detection using GLCM-M features and NDIRT indices. In Proceedings of the IEEE INFOCOM 2021-IEEE Conference on Computer Communications Workshops (INFOCOM WKSHPS), Vancouver, BC, Canada, 10–13 May 2021; pp. 1–6.
47. Liu, W.; Wang, C.; Chen, S.; Bian, X.; Lai, B.; Shen, X.; Cheng, M.; Lai, S.-H.; Weng, D.; Li, J. Y-Net: Learning Domain Robust Feature Representation for Ground Camera Image and Large-scale Image-based Point Cloud Registration. *Inf. Sci.* **2021**, *581*, 655–677. [CrossRef]
48. Yu, Y.; Wang, J.; Qiang, H.; Jiang, M.; Tang, E.; Yu, C.; Zhang, Y.; Li, J. Sparse anchoring guided high-resolution capsule network for geospatial object detection from remote sensing imagery. *Int. J. Appl. Earth Obs. Geoinf.* **2021**, *104*, 102548. [CrossRef]
49. Chen, Y.; Jiang, H.; Li, C.; Jia, X.; Ghamisi, P. Deep feature extraction and classification of hyperspectral images based on convolutional neural networks. *IEEE Trans. Geosci. Remote Sens.* **2016**, *54*, 6232–6251. [CrossRef]
50. Li, Y.; Cui, P.; Ye, C.; Junior, J.M.; Zhang, Z.; Guo, J.; Li, J. Accurate Prediction of Earthquake-Induced Landslides Based on Deep Learning Considering Landslide Source Area. *Remote Sens.* **2021**, *13*, 3436. [CrossRef]
51. Seydi, S.T.; Hasanlou, M. A New Structure for Binary and Multiple Hyperspectral Change Detection Based on Spectral Unmixing and Convolutional Neural Network. *Measurement* **2021**, *186*, 110137. [CrossRef]
52. Roy, S.K.; Kar, P.; Hong, D.; Wu, X.; Plaza, A.; Chanussot, J. Revisiting Deep Hyperspectral Feature Extraction Networks via Gradient Centralized Convolution. *IEEE Trans. Geosci. Remote Sens.* **2021**. [CrossRef]
53. Yu, Q.; Wang, S.; He, H.; Yang, K.; Ma, L.; Li, J. Reconstructing GRACE-like TWS anomalies for the Canadian landmass using deep learning and land surface model. *Int. J. Appl. Earth Obs. Geoinf.* **2021**, *102*, 102404. [CrossRef]
54. Wu, C.; Zhang, F.; Xia, J.; Xu, Y.; Li, G.; Xie, J.; Du, Z.; Liu, R. Building Damage Detection Using U-Net with Attention Mechanism from Pre-and Post-Disaster Remote Sensing Datasets. *Remote Sens.* **2021**, *13*, 905. [CrossRef]
55. Dhaka, V.S.; Meena, S.V.; Rani, G.; Sinwar, D.; Ijaz, M.F.; Woźniak, M. A survey of deep convolutional neural networks applied for prediction of plant leaf diseases. *Sensors* **2021**, *21*, 4749. [CrossRef] [PubMed]
56. Li, Z.; Liu, F.; Yang, W.; Peng, S.; Zhou, J. A survey of convolutional neural networks: Analysis, applications, and prospects. *IEEE Trans. Neural Netw. Learn. Syst.* **2021**, 1–21. [CrossRef] [PubMed]
57. Rawat, W.; Wang, Z. Deep convolutional neural networks for image classification: A comprehensive review. *Neural Comput.* **2017**, *29*, 2352–2449. [CrossRef]
58. Goodfellow, I.; Bengio, Y.; Courville, A. *Deep Learning*; MIT Press: Cambridge, MA, USA, 2016.
59. Lu, L.; Wang, X.; Carneiro, G.; Yang, L. *Deep Learning and Convolutional Neural Networks for Medical Imaging and Clinical Informatics*; Springer: Cham, Switzerland, 2019.
60. DeLancey, E.R.; Simms, J.F.; Mahdianpari, M.; Brisco, B.; Mahoney, C.; Kariyeva, J. Comparing deep learning and shallow learning for large-scale wetland classification in Alberta, Canada. *Remote Sens.* **2020**, *12*, 2. [CrossRef]
61. Seydi, S.; Rastiveis, H. A Deep Learning Framework for Roads Network Damage Assessment Using Post-Earthquake Lidar Data. *Int. Arch. Photogramm. Remote Sens. Spat. Inf. Sci.* **2019**. [CrossRef]
62. Yu, C.; Han, R.; Song, M.; Liu, C.; Chang, C.-I. A simplified 2D-3D CNN architecture for hyperspectral image classification based on spatial–spectral fusion. *IEEE J. Sel. Top. Appl. Earth Obs. Remote Sens.* **2020**, *13*, 2485–2501. [CrossRef]
63. Nogueira, K.; Chanussot, J.; Dalla Mura, M.; Dos Santos, J.A. An introduction to deep morphological networks. *IEEE Access* **2021**, *9*, 114308–114324. [CrossRef]
64. Limonova, E.E.; Alfonso, D.M.; Nikolaev, D.P.; Arlazarov, V.V. Bipolar Morphological Neural Networks: Gate-Efficient Architecture for Computer Vision. *IEEE Access* **2021**, *9*, 97569–97581. [CrossRef]

65. Shen, Y.; Zhong, X.; Shih, F.Y. Deep morphological neural networks. *arXiv Prepr.* **2019**, arXiv:1909.01532.
66. Franchi, G.; Fehri, A.; Yao, A. Deep morphological networks. *Pattern Recognit.* **2020**, *102*, 107246. [CrossRef]
67. Islam, M.A.; Murray, B.; Buck, A.; Anderson, D.T.; Scott, G.J.; Popescu, M.; Keller, J. Extending the morphological hit-or-miss transform to deep neural networks. *IEEE Trans. Neural Netw. Learn. Syst.* **2020**, *32*, 4826–4838. [CrossRef] [PubMed]
68. Klemen, M.; Krsnik, L.; Robnik-Šikonja, M. Enhancing deep neural networks with morphological information. *arXiv Prepr.* **2020**, arXiv:2011.12432.
69. Mondal, R. *Morphological Network: Network with Morphological Neurons*; Indian Statistical Institute: Kolkata, India, 2021.
70. Salehi, S.S.M.; Erdogmus, D.; Gholipour, A. Tversky loss function for image segmentation using 3D fully convolutional deep networks. In Proceedings of the International Workshop on Machine Learning in Medical Imaging, Quebec City, QC, Canada, 10 September 2017; pp. 379–387.
71. Zhan, Y.; Fu, K.; Yan, M.; Sun, X.; Wang, H.; Qiu, X. Change detection based on deep siamese convolutional network for optical aerial images. *IEEE Geosci. Remote Sens. Lett.* **2017**, *14*, 1845–1849. [CrossRef]
72. Arabi, M.E.A.; Karoui, M.S.; Djerriri, K. Optical remote sensing change detection through deep siamese network. In Proceedings of the IGARSS 2018—2018 IEEE International Geoscience and Remote Sensing Symposium, Valencia, Spain, 22–27 July 2018; pp. 5041–5044.
73. Yang, L.; Chen, Y.; Song, S.; Li, F.; Huang, G. Deep Siamese networks based change detection with remote sensing images. *Remote Sens.* **2021**, *13*, 3394. [CrossRef]
74. Guarini, R.; Loizzo, R.; Facchinetti, C.; Longo, F.; Ponticelli, B.; Faraci, M.; Dami, M.; Cosi, M.; Amoruso, L.; De Pasquale, V. PRISMA Hyperspectral Mission Products. In Proceedings of the IGARSS 2018—2018 IEEE International Geoscience and Remote Sensing Symposium, Valencia, Spain, 22–27 July 2018; pp. 179–182.
75. Loizzo, R.; Guarini, R.; Longo, F.; Scopa, T.; Formaro, R.; Facchinetti, C.; Varacalli, G. PRISMA: The Italian Hyperspectral Mission. In Proceedings of the IGARSS 2018—2018 IEEE International Geoscience and Remote Sensing Symposium, Valencia, Spain, 22–27 July 2018; pp. 175–178.
76. He, K.; Zhang, X.; Ren, S.; Sun, J. Delving deep into rectifiers: Surpassing human-level performance on imagenet classification. In Proceedings of the IEEE International Conference on Computer Vision, Santiago, Chile, 7–13 December 2015; pp. 1026–1034.
77. Grivei, A.-C.; Văduva, C.; Datcu, M. Assessment of Burned Area Mapping Methods for Smoke Covered Sentinel-2 Data. In Proceedings of the 2020 13th International Conference on Communications (COMM), Bucharest, Romania, 18–20 June 2020; pp. 189–192.
78. Barboza Castillo, E.; Turpo Cayo, E.Y.; de Almeida, C.M.; Salas López, R.; Rojas Briceño, N.B.; Silva López, J.O.; Barrena Gurbillón, M.Á.; Oliva, M.; Espinoza-Villar, R. Monitoring wildfires in the northeastern peruvian amazon using landsat-8 and sentinel-2 imagery in the GEE platform. *ISPRS Int. J. Geo-Inf.* **2020**, *9*, 564. [CrossRef]
79. Syifa, M.; Panahi, M.; Lee, C.-W. Mapping of post-wildfire burned area using a hybrid algorithm and satellite data: The case of the camp fire wildfire in California, USA. *Remote Sens.* **2020**, *12*, 623. [CrossRef]
80. Quintano, C.; Fernández-Manso, A.; Fernández-Manso, O. Combination of Landsat and Sentinel-2 MSI data for initial assessing of burn severity. *Int. J. Appl. Earth Obs. Geoinf.* **2018**, *64*, 221–225. [CrossRef]
81. Ngadze, F.; Mpakairi, K.S.; Kavhu, B.; Ndaimani, H.; Maremba, M.S. Exploring the utility of Sentinel-2 MSI and Landsat 8 OLI in burned area mapping for a heterogenous savannah landscape. *PLoS ONE* **2020**, *15*, e0232962. [CrossRef]
82. Roy, D.P.; Huang, H.; Boschetti, L.; Giglio, L.; Yan, L.; Zhang, H.H.; Li, Z. Landsat-8 and Sentinel-2 burned area mapping-A combined sensor multi-temporal change detection approach. *Remote Sens. Environ.* **2019**, *231*, 111254. [CrossRef]
83. Lima, T.A.; Beuchle, R.; Langner, A.; Grecchi, R.C.; Griess, V.C.; Achard, F. Comparing Sentinel-2 MSI and Landsat 8 OLI imagery for monitoring selective logging in the Brazilian Amazon. *Remote Sens.* **2019**, *11*, 961. [CrossRef]

remote sensing

MDPI

Article

Assessment of Wildfire Activity Development Trends for Eastern Australia Using Multi-Sensor Earth Observation Data

Michael Nolde [1,*], Norman Mueller [2], Günter Strunz [1] and Torsten Riedlinger [1]

1 German Aerospace Center (DLR), German Remote Sensing Data Center (DFD), Department for Geo-Risks and Civil Security, Oberpfaffenhofen, 82234 Wessling, Germany; guenter.strunz@dlr.de (G.S.); torsten.riedlinger@dlr.de (T.R.)

2 Geoscience Australia/Australian Government, Symonston, ACT 2609, Australia; norman.mueller@ga.gov.au

* Correspondence: michael.nolde@dlr.de

Abstract: Increased fire activity across the Amazon, Australia, and even the Arctic regions has received wide recognition in the global media in recent years. Large-scale, long-term analyses are required to postulate if these incidents are merely peaks within the natural oscillation, or rather the consequence of a linearly rising trend. While extensive datasets are available to facilitate the investigation of the extent and frequency of wildfires, no means has been available to also study the severity of the burnings on a comparable scale. This is now possible through a dataset recently published by the German Aerospace Center (DLR). This study exploits the possibilities of this new dataset by exemplarily analyzing fire severity trends on the Australian East coast for the past 20 years. The analyzed data is based on 3503 tiles of the ESA Sentinel-3 OLCI instrument, extended by 9612 granules of the NASA MODIS MOD09/MYD09 product. Rising trends in fire severity could be found for the states of New South Wales and Victoria, which could be attributed mainly to developments in the temperate climate zone featuring hot summers without a dry season (Cfa). Within this climate zone, the ecological units featuring needleleaf and evergreen forest are found to be mainly responsible for the increasing trend development. The results show a general, statistically significant shift of fire activity towards the affection of more woody, ecologically valuable vegetation.

Keywords: burnt area monitoring; Australia; Sydney; wildfire; earth observation; mid-resolution sensors; time series analysis; burn severity; climate zones

Citation: Nolde, M.; Mueller, N.; Strunz, G.; Riedlinger, T. Assessment of Wildfire Activity Development Trends for Eastern Australia Using Multi-Sensor Earth Observation Data. *Remote Sens.* **2021**, *13*, 4975. https://doi.org/10.3390/rs13244975

Academic Editor: Paolo Mazzanti

Received: 9 November 2021
Accepted: 3 December 2021
Published: 7 December 2021

1. Introduction

Wildfires have always played a significant role in the evolution of various ecosystems and are the predominant natural disturbance factor in many parts of the world. They significantly influence ecological patterns and processes on a global scale. This includes vegetation distribution and structure, as well as the carbon cycle [1]. While humans and wildfires have always coexisted, changes in wildfire patterns represent an increasing threat to human lives and property. Apart from the direct implications, wildfires have also been found to contribute to the greenhouse effect through CO_2 emissions, thus fostering atmospheric changes on a global level [2,3]. Research has shown that forest loss has increased substantially over the past two decades in many parts of the world, and that the underlying dynamics are strongly linked to fire activity [4]. Several studies have discovered changes in the frequency and size of wildfires and also in the length of the fire season, for example, regarding the Canadian boreal forest [5] and the Western United States [6].

In recent years, large wildfires have occurred in regions formerly unaffected by fire, such as the Arctic regions. Some regions regularly affected by fire have experienced unprecedented large-scale fire events, such as the Australian East coast, the Brazilian Amazonas region, or the state of California in the United States. Investigating the question if these recent events are part of a natural oscillation, or must instead be regarded as a result of a long-term trend, is a crucial task in fire science. Studies usually analyze parameters such as the frequency of occurrence, spatial extent, and burn severity of fire incidents to derive meaningful trends [7].

To obtain insights regarding shifts in global fire activity, studies have to incorporate multi-decadal time spans and continental-scale study areas. Global, long-term burnt area datasets are readily available for analyses, the most widely used ones being the National Aeronautics and Space Administration (NASA) MCD64A1 dataset [8] and the European Space Agency (ESA) Fire_cci BA 5.1 dataset [9,10]. A third global burnt area product from the Global Fire Emissions Database, version 4 (GFED4, [11]) is meanwhile discontinued and is only available until 2015 [12]. A semi-automatically generated product is also provided by the Joint Research Center of the European Commission (JRC) in the frame of the European Forest Fire Information System (EFFIS) [13]. However, these data are only available for Europe, Northern Africa, and the Middle East [14].

While the listed datasets all include the fire perimeter as well as the burning detection date, they do not feature information regarding the fire severity, and thus do not allow the derivation of trends in this regard. Yet, fire severity is a critical aspect of fire regimes, determining fire impacts on ecosystem attributes and associated post-fire recovery [15].

The German Aerospace Center (DLR) recently published a global, long-term burnt area dataset, which includes information regarding the burn severity, together with fire perimeter and the burning date. This dataset is closely linked to the burnt area monitoring service operated by DLR. It is maintained by the Department of Geo-Risks and Civil Security (GZS) of the German Remote Sensing Data Center (DFD). The service is based on mid-resolution Sentinel-3 Ocean and Land Color Instrument (OLCI) satellite imagery and provides burnt area information for the region of Europe twice a day in near-realtime. The service is fully automated and targeted at supporting rapid mapping activities and timely post fire damage assessment throughout Europe. A quality-optimized version called fusion product is generated after a time delay of 10 days, when additional post-event data is available.

The recently published, global dataset is build upon this fusion product. In addition to the data available for Europe, equivalent products are generated using the same methodology for North and South America, Africa, Oceania, and Asia. The methodology is briefly described consecutively, the complete description can be found in Nolde et al. [16]. As the dataset is based on Sentinel-3 data, it is only available since 2016, which is the year the first Sentinel 3 satellite was launched. For this study, the data was extended using data from the NASA MODIS MOD09/MYD09 product [17] in order to allow the derivation of trends on a longer time scale.

The complete input dataset comprises 9612 granules of the MODIS MOD09/MYD09 product in conjunction with 3503 tiles of the OLCI instrument onboard the Sentinel-3 satellite. The area of Eastern Australia has been selected as a study region, covering the states and territories of Queensland, New South Wales, the Australian Capital Territory (ACT), and Victoria, respectively.

This study region is chosen because it experienced destructive burnings in the 2019/2020 fire season, and because it also was regularly affected by wildfires in recent decades.

The year 2019 was Australia's warmest year on record so far, with significant heat waves occurring in January and December [18]. The national, average maximum temperature was as high as 43.6 °C, which is more than 1.8 °C above the long-term average [19]. In addition, 2019 has also been the driest year on record in Australia, caused by an extraordinary strong positive Indian Ocean Dipole [20]. The nationally-averaged rainfall was 40% below average, amounting to only 278 mm. As a consequence, the annual, cumulative Forest Fire Danger Index reached the highest values since the beginning of the national records in 1950 [18]. Three of the four investigated states and territories, namely, New South Wales, the ACT, and Victoria, reside in a temperate climate zone, with dryness conditions usually reaching moderate levels at most [21]. However, in 2019, these states experienced severe drought conditions, with New South Wales suffering the most severe conditions throughout Australia. These preconditions contributed significantly to the unprecedented fire activity in December 2019 and January 2020 [19].

This study analyses the existence of stable, wildfire related trends in this region, focusing on fire severity.

2. Materials and Methods

2.1. Area of Interest

The chosen study region comprises the states of Queensland, Victoria, and New South Wales together with the Australian Capital Territory (ACT). This region is vastly heterogeneous regarding climate and vegetation cover, with a pronounced inter-annual variability. Both Queensland and New South Wales include tropical, temperate, and arid climate zones. Figure 1 shows the area of interest, together with the respective climate zones and the burnt area from the 2019/2020 wildfire season. The highlighted climate zones are the ones found to feature increasing trends regarding fire severity in this study. The full names corresponding to the climate zone abbreviations can be found in the respective tables in the result section. This study is prepared in a hierarchical manner. Trends are analyzed on a state level, as well as regarding climate zones, and finally ecological units. These are consecutively set in relation to each other.

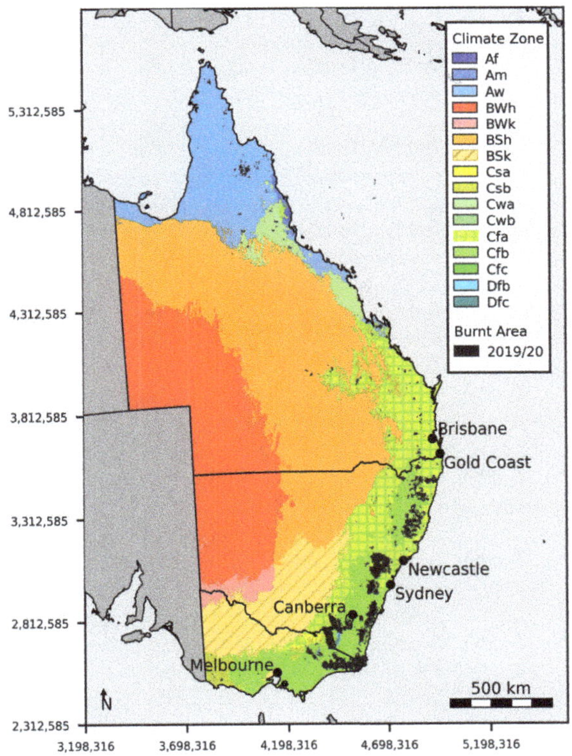

Figure 1. The area of interest, comprising Queensland, Victoria, and New South Wales together with the Australian Capital Territory (ACT). The climate zones are shown additionally, as well as the burnt area for the 2019/2020 fire season. The climate zones found to feature a significant increasing trend in this study are highlighted.

The utilized climate zone map, prepared by Beck et al. [22] following the methodology of Peel et al. [23], is derived using a long-term time series of weather station data regarding monthly precipitation and air temperature. The system classifies climatic regions into five main classes and 30 subtypes. The discrimination between classes is based on fixed thresholds addressing the seasonality of precipitation and temperature. Climate is recognized as the major driver of global vegetation. The classification is therefore regarded as an empirical mapping of biome distributions around the world. Although developed in the 19th century, it is widely used today, for example, in ecological modelling [22].

The Australian Forest Fire Danger Index [24] relies on four meteorological parameters, which have proven their reliability with regard to wildfire activity: temperature, wind speed, relative humidity, and the Drought Factor, a component representing fuel availability [25].

The latter parameter, which is strongly influenced by seasonal variations in rainfall, has been found to be pivotal for fire occurrence, together with the vegetation structure [26]. As described by Russel-Smith et al., fire frequency as well as fire extent throughout Australia are strongly influenced by rainfall seasonality. Fire occurrence is therefore most pronounced in the tropics of Northern Australia, which are intensely seasonal. Rainfall positively influences the dynamics of biomass growth, which provides the fuel for wildfire activity. Precipitation amounts above average could be attributed to large burnings in arid, central Australia [27].

Dryness, on the other hand, strongly increases the availability of the vegetation to burn [28]. It could be shown that drought conditions are associated with major fires in the forested areas of Southern Australia [29,30]. These ecosystems usually feature sufficient litter for propagation of fire at most times. The most influential factor for fire propagation is the availability of vegetation to burn, controlled by drought conditions and the weather at the time of ignition [28].

2.2. Utilized Data Sources

The study utilizes the DLR-GZS burnt area dataset, which is based on mid-resolution optical satellite data from two different sensors. First, the OLCI instrument onboard the Sentinel-3 A and B satellites of the European Copernicus Programme [31], and second, the MODIS instrument onboard the NASA Aqua/Terra satellites. Band information from the red and near-infrared (NIR) domain are utilized for the retrieval of burnt area perimeters and burn severity estimation. Imagery of the Sentinel-3 OLCI instrument can be retrieved via the Copernicus Open Access Hub [32] and the Copernicus online data access website [33]. However, the available time span for OLCI data is considerably shorter than the one regarding MODIS. Sentinel-3A was launched in 2016, with Sentinel-3B following in 2018. Therefore, the OLCI dataset is extended by imagery from the MODIS sensor for this study. This data is available for an extended time range of more than two decades, starting in late 1999 with the launch of the NASA Terra satellite. The sensor is designed to observe the ocean, atmosphere, land, and ice on the Earth's surface [34]. It features 36 discrete spectral bands with differing spatial resolutions from 250 m to 1 km. Terra's twin satellite, Aqua, was launched in 2002, carrying a second MODIS instrument. This study makes use of MODIS information provided through the MOD09A1/MYD09A1 surface reflectance product [17], which represents a cloud-free 8-day composite of bands in the visible spectrum. The data are freely available from the NASA Land Processes Distributed Active Archive Center (LP DAAC) [35]. Furthermore, thermal anomaly information derived from the MODIS MOD14A2/MYD14A2 [36] product is used as auxiliary data. Equivalent data is utilized from the Visible Infrared Imaging Radiometer Suite (VIIRS) instrument onboard the Suomi National Polar-orbiting Partnership (SUOMI-NPP) satellite [37]. Both products are available for download on the NASA Fire Information for Resource Management System (FIRMS) website [38].

In order to reduce the total data volume to a significant selection, only the Australian summer months from November to February have been analyzed in this study.

Subsequent to the derivation of burnt area perimeters and the burn severity, the results are combined with land use and land cover (LULC) data to gain insights regarding affected vegetation classes. For this purpose, the CCI-LC (Climate Change Initiative-Land Cover) product from ESA is used. It provides mid-resolution land cover information on a global scale. Moreover, land cover maps are available for each individual year, starting in 1992 [39]. To investigate the relationship between fire activity and climatic conditions, a climate zone map based on the Köppen-Geiger classification system is utilized, published by Beck et al., 2018 [22]. Furthermore, ecological units provided by the United States Geological Survey (USGS) are used [40].

Two further burnt area datasets have been incorporated within this study to validate the presented results. First, the NASA MCD64A1 dataset, featuring global burnt area information derived from MODIS imagery, with a spatial resolution of 500 m [8]. These data are made available by the University of Maryland [41] and has been widely utilized in academic research, for example, for the Brazilian Savannas [42,43]. Second, a high resolution burnt area map for the state of New South Wales is utilized, which was prepared by the Department of Planning, Industry and Environment of New South Wales/Australia. This dataset, named Google Earth Engine burnt area map (GEEBAM) [44], is based on Sentinel-2 data and makes use of manually derived thresholds from aerial photography [45]. The National Indicative Aggregated Fire Extent Datasets [46], which are published by the Australian Government, have also been taken into consideration as a reference data source. However, the GEEBAM dataset has been found to feature a higher thematic accuracy, and is therefore chosen as reference in this study.

Table 1 lists the complete set of available MOD09/MYD09 composites and Sentinel-3-OLCI scenes used for generating the DLR-GZS burnt area dataset, which this study is build upon. Besides the number of available scenes for each time range and state, the average number of cloud free observations per pixels is given, representing a measure of the interpretability of the data. As the MOD09/MYD09 data are available as an eight-day composite, the number of cloud-free observations is considerably lower when compared to Sentinel-3. In total, 9612 MODIS MOD09/MYD09 granules from both Terra and Aqua have been analyzed, together with 3503 OLCI scenes from Sentinel-3 A and B. This amounts to an entirety of 13,115 scenes for the complete study time span.

Table 1. Analyzed data sources, listed separately for Queensland (QLD), New South Wales (NSW), Australian Capital Territory (ACT), and Victoria (VIC).

Time Span	Sensor	Scenes	Avg. Cloud-Free Overpasses per Pixel			
			QLD	NSW	ACT	VIC
2000/11–2001/02	MODIS	240	10	13	13	14
2001/11–2002/02	MODIS	240	15	15	14	15
2002/11–2003/02	MODIS	480	30	30	27	30
2003/11–2004/02	MODIS	480	30	30	27	30
2004/11–2005/02	MODIS	480	28	29	26	28
2005/11–2006/02	MODIS	484	30	30	27	30
2006/11–2007/02	MODIS	480	30	30	27	30
2007/11–2008/02	MODIS	480	30	30	27	30
2008/11–2009/02	MODIS	480	28	29	26	28
2009/11–2010/02	MODIS	480	30	30	27	30
2010/11–2011/02	MODIS	480	30	30	27	30
2011/11–2012/02	MODIS	480	30	30	27	30
2012/11–2013/02	MODIS	480	28	29	26	28
2013/11–2014/02	MODIS	480	30	30	27	30
2014/11–2015/02	MODIS	488	30	30	27	30
2015/11–2016/02	MODIS	480	30	30	27	30
2016/11–2017/02	MODIS	480	28	29	26	28
	OLCI	620	33	45	36	34
2017/11–2018/02	MODIS	480	30	30	27	30
	OLCI	616	35	41	30	35
2018/11–2019/02	MODIS	480	30	30	27	30
	OLCI	1019	58	75	61	61
2019/11–2020/02	MODIS	480	28	29	26	28
	OLCI	1248	76	91	68	71

2.3. Burnt Area Derivation Methodology

The accurate, automatic monitoring of burnt area evolution from satellite imagery represents a demanding task for the scientific community. This is mostly due to the spatial-temporal variability of the state of the Earth's surface. Inaccuracies are also introduced by the utilized sensor and related geometrical resolution [47]. Furthermore, the presence of clouds disturbs the derivation of meaningful surface features at reflective wavelengths [48]. Adding to the complexity inherent in optical sensor data, burnt areas are highly heterogeneous regarding size, shape, and spectral reflectance on the ground surface, and can thus be difficult to differentiate from shadows cast by clouds and mountains. These circumstances significantly limit the possibilities for an automated approach, especially on a large geographical scale. Many recent research activities have assessed these challenges using Machine Learning technologies. Methodologies such as Random Forests [49], Support Vector Machines [50], and Deep Learning classifiers [51,52] have been applied in this regard. However, research has mainly been focused on feature detection using high resolution satellite imagery on a small geographic scale. Due to the spatial resolution however, these satellites inherently feature a low temporal frequency, and are thus unsuited for daily monitoring purposes. Approaches utilizing mid-resolution satellite data, thus allowing a high temporal coverage for an extended geographical region, usually use a time series approach in combination with a burn-sensitive vegetation index. Thermal anomaly detections are often used as auxiliary data (see [53]). However, these approaches feature a significant amount of uncertainty. Humber et al. analyzed four global burnt area products, and concluded that the estimates of burned area vary greatly between products in terms of total area affected, the location of burning, and the timing of the burning [54]. In a similar study, Padilla et al. found that the commission error ratio was above 40% and omission error ratio was above 65% for the analyzed products [55]. Oliva et al. conducted a study investigating if thermal anomaly data could be used as a replacement for burnt area datasets, but found high omission and commission errors especially for grasslands, savannas, and agricultural areas [56].

The methodology utilized here is based on an approach developed at DLR [16]. It is primarily designed for monitoring continental-scale regions in near-real time, but can also be invoked to perform retrospective time series analysis. The derived information comprises burnt area perimeters, the date of detection, and the burn severity by means of the differential Normalized Difference Vegetation Index ($NDVI_{diff}$ [57], see Equations (1) and (2). As auxiliary information, the number of detections for each burnt area pixel is available, as well as the number of cloud-free satellite overpasses for each pixel.

$$NDVI = \frac{NIR - Red}{NIR + Red} \tag{1}$$

$$NDVI_{diff} = NDVI_{pre} - NDVI_{post} \tag{2}$$

At its core, the method exploits the synergetic effects of data from the red/NIR and the thermal wavelengths in order to derive burnt area information. Substantial work in this regard has been performed by Fraser et al. and Li et al. for the boreal forest of Canada as early as the year of 2000 [58,59].

As a basis for the processing, mosaics of pre- and post-$NDVI$ information are generated. Consecutively, the concept of Morphological Active Contours without Edges (MorphACWE [60,61]) is used to derive accurate burnt area perimeters. The method is closely related to Geodesic Active Contour Level Sets [62], which have been used for burnt area derivation [63] as well as other domains, such as crop field size estimation [64].

The term Active Contour refers to a dynamical curve, which grows starting from a set of seed pixels and converges when an optimal segmentation result is reached. For the generation of the burnt area dataset, Active Fire locations are used as seed information. This proceeding was shown to yield results of high geometric accuracy when inter-compared with the JRC/EFFIS dataset [13] as well as the NASA MCD64A1 dataset [8]. The accuracy validation, together with the detailed description of the methodology, can be found in Nolde et al. (2020) [16]. The methodology is schematically visualized in Figure 2.

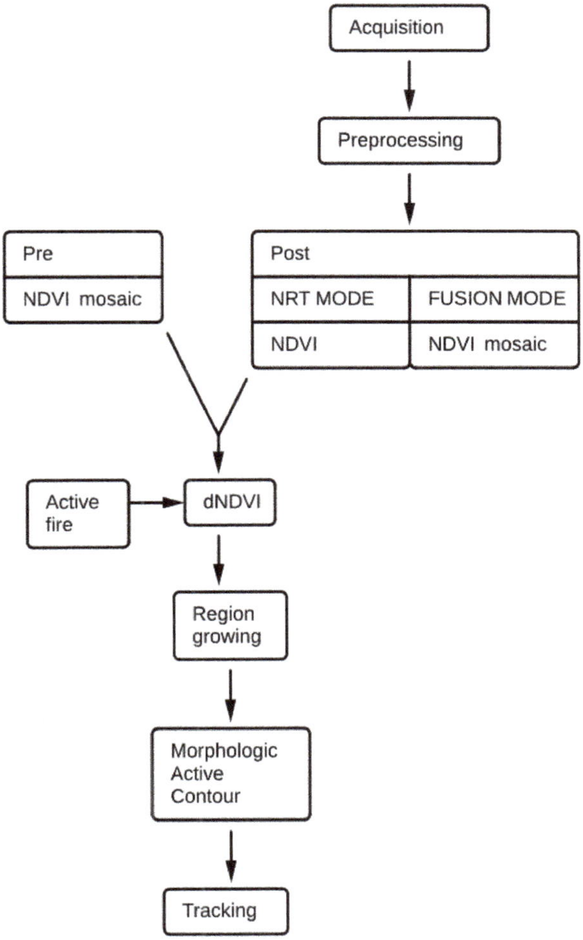

Figure 2. Scheme of the methodology for burnt area perimeter derivation, from Nolde et al. (2020) [16].

Figure 3 shows exemplary visualizations of the data used for the preparation of this study. Sub-figure (a) visualizes the maximally detected burn severity regarding the mega fire in the greater area of Sydney/New South Wales, which was active during November and December, 2019. Sub-figure (b) shows the temporal evolution of the burn activity, by means of detection date.

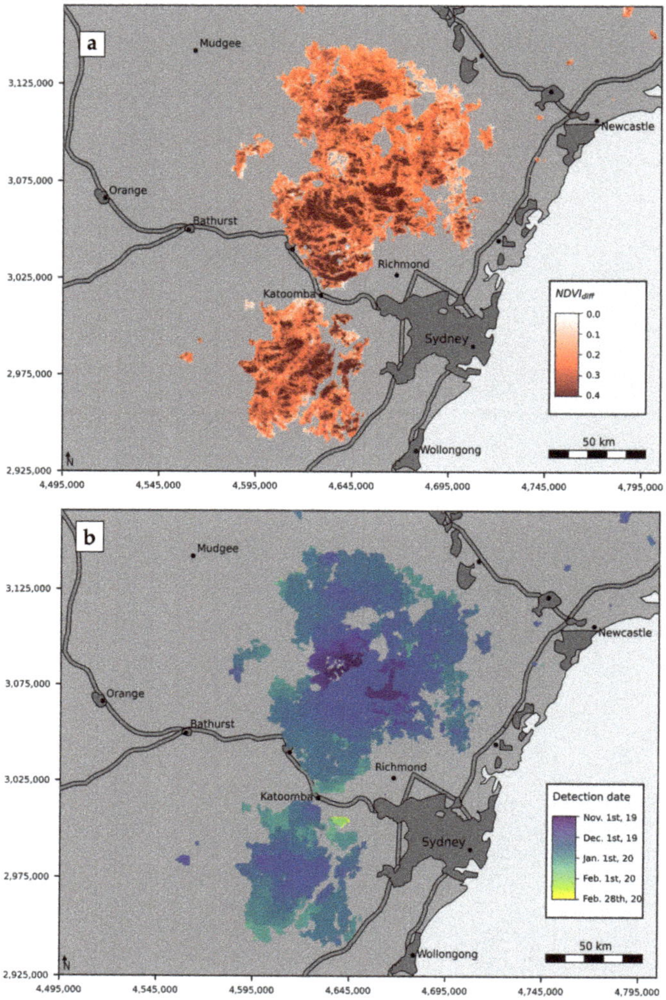

Figure 3. (**a**) $NDVI_{diff}$ values regarding the mega fire in the greater area of Sydney/New South Wales, 2019/2020. (**b**) Temporal evolution of the fire activity.

2.4. Validation of Burnt Area Data

The presented results are compared against the two reference datasets, NASA MCD64A1 and GEEBAM, regarding three criteria:

- True positives (TP): The total burnt area contained in the presented results as well as the reference data, in relation to the total burnt area of the reference data.
- False negatives (FN): The total burnt area not contained in the presented results, but contained in the reference, in relation to the total burnt area of the reference data.
- False positives (FP): The total burnt area contained in the presented results, but not contained in the reference area, in relation to the total burnt area of the reference data.

The fourth criterion, which represents the true negatives (TN), refers to the percentage of area neither contained in the presented results, nor in the reference data. However, as the total size of unburnt area greatly predominates the total size of burnt area, this true negative percentage is implicitly very close to 100 percent. This is, however, mostly due to the size of the study region, so this measure does not represent a meaningful value for this kind of study. The same applies to the overall accuracy [65]. These measures have therefore been omitted in Table 2, which shows the results of the evaluation. A more suited means of measure is to calculate the average of the true positive percentage and the inverted false positive percentage (false positives subtracted from 100%, named FP_{inv} in Table 2).

The inter-comparison of the burnt area of the presented results with the burnt area of the high resolution GEEBAM 2019/2020 data for New South Wales reveals a percentage of overlapping area of 77%, with an error of 9%. The Jaccard index, also known as Intersection over Union, yields a similarity of 70.8% [66]. To enable an evaluation of these numbers, the NASA MCD64A1 dataset is equivalently checked against the same reference, yielding 71% overlap and an error of 7%. The combination of true positives and inverted false negatives account for 84% for the presented results, and 82% for the MCD64A1 data. The results obtained in this study regarding the 2019/2020 fire season are in accordance with burnt area extent information published by Boer [67]. The data basis is therefore considered to be of satisfactory accuracy.

Table 2. Inter-comparison of burnt area extent with NASA MCD64A1 and GEEBAM reference datasets.

		Presented Results 4,577,850 ha	**MCD64A1 4,176,018 ha**	**GEEBAM 5,306,688 ha**
presented results	TP	x	83.8%	77.1%
	FN	x	16.2%	22.9%
	FP	x	25.8%	9.2%
	TP/FP_{inv}	x	79.0%	84.0%
MCD64A1	TP	76.4%	x	71.1%
	FN	23.6%	x	28.9%
	FP	14.8%	x	7.6%
	TP/FP_{inv}	80.8%	x	81.7%
GEEBAM	TP	89.4%	90.3%	x
	FN	10.6%	9.7%	x
	FP	26.5%	36.8%	x
	TP/FP_{inv}	81.4%	81.4%	x

2.5. *Trend Derivation Methodology*

The trend derivation in this study is based on linear regression, whereby the slope of the regression line represents the actual trend. The input values for the burnt area extent analysis are the accumulated burnt area amounts per unit of investigation (state, climate zone, or ecological unit) for each year of the analyzed time span. Regarding the burn severity analysis, the input values are derived by averaging all values within the unit of investigation for each respective year. The correlation coefficient indicates to what extent the actual values are in concordance with the calculated trend line, while the RMSE (Root Mean Square Error) illustrates the error.

Finally, the 5- and 95-percentiles (named "Perc 5" and "Perc 95" in the result tables) represent the error margins, indicating how robust the result actually is. They are derived through repeatedly and randomly altering the input values within the range of the RMSE, so that the results reflect the average of a set of possible outcomes. To eliminate outliers, the 5 and 95-percentiles are used as upper and lower limits of the yielded results.

3. Results

The consecutive sub-sections show the results of the analysis on a state-wide level, as well regarding climatic zones and finally ecological units. These results are then set in relation to each other. Each subsection contains a visualization of the burnt area extent for each year within the analyzed time span, followed by the actual burn severity analysis.

3.1. Fire Trends Regarding the States in the Study Area

Figure 4 shows the total, annual extent of burnt area for the four Australian states, and territories of interest, regarding the time period of 2000 to 2020. The results derived from Sentinel-3 OLCI data are depicted as a green line, while results regarding MODIS MOD09A1/MYD09A1 data are shown in blue. As the first one of the Sentinel-3 satellites was only launched in 2016, the analysis could only be carried out for this limited timespan. The red, dotted line represents the NASA MCD64A1 [8] burnt area dataset. The latter is included as a reference, to allow an estimation of the accuracy of the presented results. Finally, the black, dotted line is the regression line, corresponding to the MODIS MOD09/MYD09 based results.

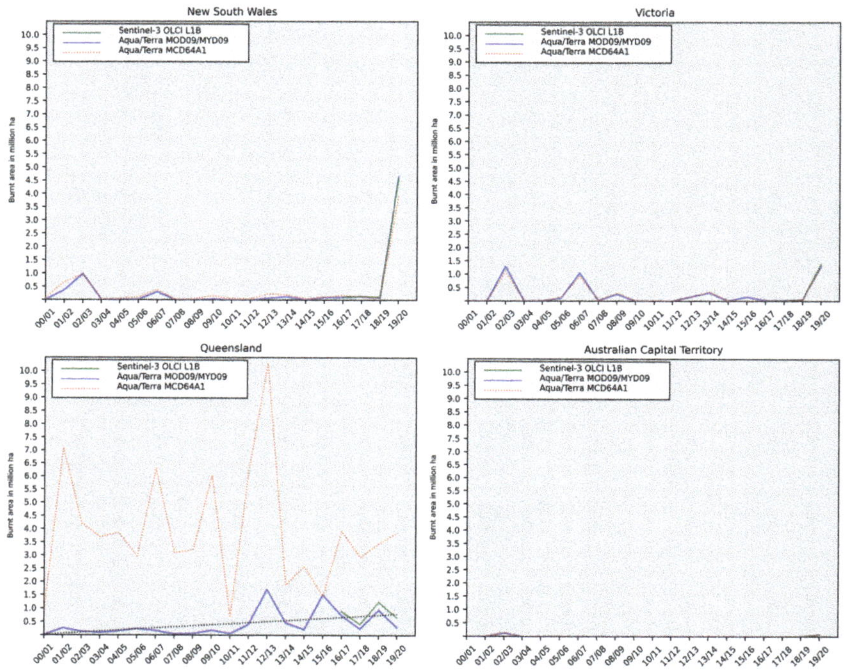

Figure 4. Total yearly burnt area amount in million hectares for New South Wales, Victoria, Queensland, and the Australian Capital Territory (ACT). MCD64A1 reference data is additionally visualized as dotted, red line. The black, dotted line in the subplot for Queensland represents the linear regression line.

It can be seen that the results show a high correlation of burnt area extent between the utilized DLR-GZS burnt area dataset and the MCD64A1 reference, regarding three of the four analyzed states. For Queensland, however, this is not the case. The discrepancies are due to differences in the methodologies: Unlike in the NASA dataset, burn sites where the vegetation has quickly recovered are excluded in the presented data. As stated before, wildfire is a natural phenomenon, and in many parts of Australia the regular burning of huge areas of bush and grassland vegetation is part of the natural cycle. The affected vegetation recovers quickly, some species do so even within a period of a few weeks. In the case of Queensland, these areas account for the majority of the overall burnt area. In order to confine the results to potentially harmful wildfire events, it was decided to consider burnt area in this study only when distinctive traces of burning activity could still be detected after a period of three weeks. This filtering is not performed for the generation of the MCD64A1 dataset, and as a consequence, the overall burnt area for Queensland differs considerably between the two products.

Furthermore, the figure shows that there is no linear trend in fire extent regarding New South Wales, Victoria, and the ACT. This is also reflected by Table 3, which lists the respective statistics for Figure 4. The *p*-value, symbolizing the statistical significance, illustrates that there is no linear development regarding the derived trends. Complying with the common standard, a *p*-value below 0.05 is regarded to represent statistical significance. Such a low value is found in the case of Queensland, which features a robust, increasing trend regarding the extent of wildfires over the last two decades.

Table 3. Trend statistics regarding the extent of burnt area from 2000 to 2020. Only Queensland shows a statistically significant upward trend (highlighted in dark gray).

State	Slope	Corr. Coef.	*p*-Value	RMSE
New South Wales	0.054	0.31	0.18	0.962
Queensland	0.04	0.495	0.026	0.402
Victoria	0.002	0.028	0.91	0.421
ACT	0.0	−0.072	0.764	0.032

A reason for the extraordinary extent of burnt area, which is especially pronounced for the state of New South Wales, is shown in Figure A1 in the Appendix A. The burnt area extent is depicted for the past 20 years, sub-divided by month of the fire season. It can be seen that the fire activity in New South Wales reached a significant level at the beginning of November, while the main activity usually only occurs towards the end of the year. The fire activity started several weeks earlier than usual in the 2019/2020 season.

While the figures above provide a general impression by showing the fire extents over the past 20 years, this study focuses on fire severity, which is analyzed consecutively. The term is defined as a measure of the degree of environmental change caused by fire [68]. It represents a critical aspect of fire regimes, indicating the impacts on ecosystems and associated post-fire recovery [15]. The respective impact ranges from the partial consumption of litter to the complete dieback of canopy trees [69].

Figure 5 shows the yearly trend of average burn severity for the four analyzed states and territories. The green line depicts the severity results regarding Sentinel-3 OLCI, while the results for the MODIS data are depicted in blue. For the latter data source, not only the burnt areas have been investigated, but also the complete complementary area, which has not been affected by fire. This represents a cross check, showing that the developments in severity are not actually caused by unrelated factors such as climate, soil or moisture related conditions. It can be seen that the states of New South Wales and Victoria feature a general upward trend, while the development is stable or even decreasing for the ACT and Queensland.

Table 4 shows that, even if the severity trend for New South Wales and Victoria is increasing, the *p*-values are too heterogeneous for these trends to be statistically significant. Only the slightly negative trend for Queensland can be considered robust. The results of the cross check are listed in Table 5. The mentioned heterogeneity is addressed in the Discussion section.

Table 4. Trend statistics regarding the mean severity of burnt area from 2000 to 2020. Queensland features a statistically significant trend (highlighted in dark gray), while Victoria shows a trend close to statistical significance (depicted in light gray).

State	Slope	Corr. Coef.	*p*-Value	RMSE
New South Wales	0.001	0.208	0.378	0.038
Queensland	−0.002	−0.664	0.001	0.010
Victoria	0.003	0.423	0.06	0.037
ACT	−0.002	−0.247	0.29	0.035

Table 5. Cross check trend statistics regarding the mean decline in vegetation fitness of area unaffected by fire, from 2000 to 2020.

State	Slope	Corr. Coef.	p-Value	RMSE
New South Wales	0.0	−0.148	0.53	0.018
Queensland	0.0	−0.044	0.85	0.038
Victoria	0.0	0.05	0.84	0.019
ACT	−0.002	−0.304	0.19	0.044

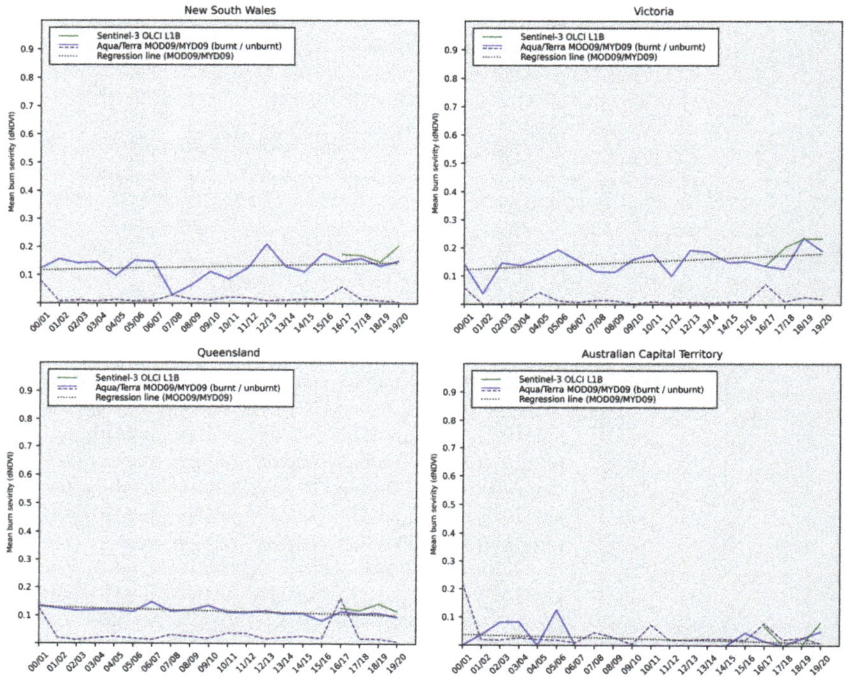

Figure 5. Yearly average burn severity for the period of 2000 to 2020. The line depicted in light blue represents the mean severity values for the burnt areas derived from MODIS MOD09/MYD09 data. The dashed, dark blue line shows the equivalent results for areas not affected by fire. Green color is used for Sentinel-3 OLCI burnt area results (available only since 2016). The black, dotted line is the regression line, with respect to the MODIS burnt area results.

In order to discriminate regions which account for the rising trend in New South Wales and Victoria, the study region is subdivided into regions which share similarities regarding available fuel and species composition. These are consecutively analyzed equivalent to the states. As a reasonable classification, climate zone mapping information is used. Parks et al. could show that climatic conditions, next to fuel and weather, represent a major driver of fire intensity [70]. As a next step, the climate zones featuring statistically relevant trends are further subdivided into ecological units, which are provided by the United States Geological Survey (USGS) [40]. This allows the fine-granular attribution of state-wide trends to small scale vegetation types.

The following section analyzes the climate zones contained in the study area regarding their burn severity trends. For the zones featuring developments with statistical significance, it is shown to which extent they overlap with the area of New South Wales and Victoria.

3.2. Fire Trends Regarding Climate Zones

Figure 6 shows, first and foremost, the exceptionality of the 2019/2020 wildfire events, regarding the total extent of the burnings. The calculated size reaches 6.5 million hectares, which is more than twice as high as in every other year in the analyzed time span. In addition to the total extent, a subdivision regarding the affected climate zones is shown. For this division, the Köppen–Geiger classification system published by Beck et al. (2018) [22] is used. The color scheme follows the one proposed by the authors.

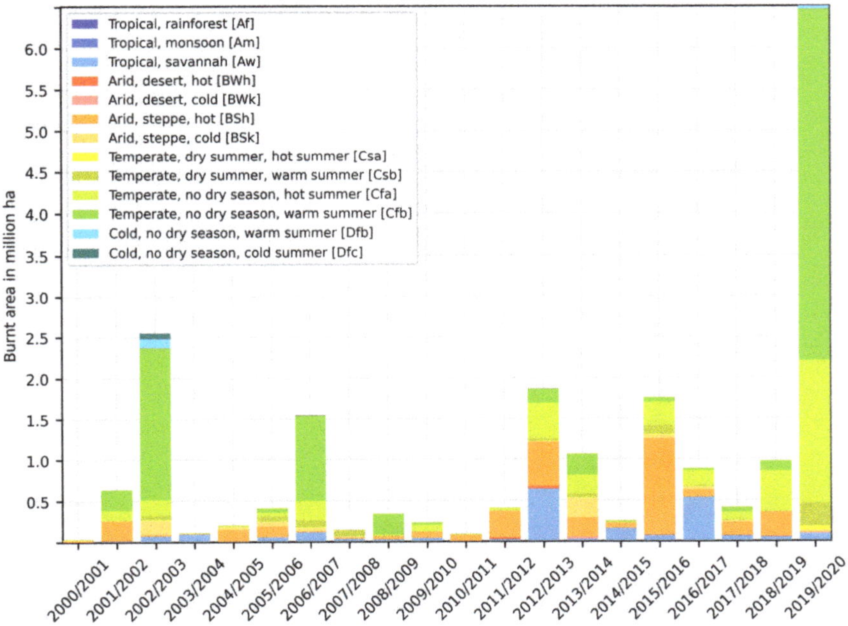

Figure 6. Total yearly burnt area amount in million hectares, subdivided by climate zones.

Table 6 represents a listing of all climate zones in the area of interest together with their p-value as a measure of statistical significance. A threshold of $p \leq 0.05$ is applied to indicate the significance, the respective table row is marked in dark gray. Climate zones featuring a p-value close to statistical significance ($0.05 < p \leq 0.1$) are highlighted with light gray color.

Table 6. Trends regarding the size of affected area for each climate zone in the area of interest. The climate zone featuring a statistically significant trend is marked in dark gray. The one showing a trend close to statistical significance is depicted in light gray.

Class	Slope (%)	Perc 5	Perc 95	Corr. Coef.	p-Value	RMSE	Label
Csa	0.070	0.057	0.080	0.385	0.09	0.009	Temperate, dry summer, hot summer
BSk	−0.053	−0.358	0.175	−0.051	0.83	0.060	Arid, steppe, cold
Cfb	3.519	0.443	7.165	0.206	0.38	0.962	Temperate, no dry season, warm summer
Am	0.004	−0.000	0.011	0.234	0.32	0.001	Tropical, monsoon
BWk	0.020	0.000	0.045	0.223	0.34	0.005	Arid, desert, cold
Csb	0.360	0.064	0.499	0.322	0.17	0.061	Temperate, dry summer, warm summer
Aw	0.883	0.675	1.449	0.307	0.19	0.158	Tropical, savannah
Cfa	3.422	3.006	4.467	0.519	0.019	0.325	Temperate, no dry season, hot summer
Dfb	−0.061	−0.117	0.066	−0.142	0.55	0.024	Cold, no dry season, warm summer
Dfc	−0.078	−0.123	−0.047	−0.295	0.21	0.014	Cold, no dry season, cold summer
BSh	1.494	0.741	2.058	0.325	0.16	0.250	Arid, steppe, hot
Af	−0.000	−0.001	0.001	−0.022	0.93	0.000	Tropical, rainforest
BWh	0.014	−0.024	0.038	0.108	0.65	0.007	Arid, desert, hot

Note that, apart from the temperate zones featuring dry winters and warm to hot summers (Cwa/Cwb), all climate zones existing in the Eastern part of Australia are affected by wildfire.

The table shows that only one climate zone satisfies the *p*-value condition for statistical significance (Cfa: Temperate, no dry season, hot summer). The respective correlation coefficient lies in the moderate range, even if the RMSE (Root Mean Square Error) shows a considerable oscillation around the regression line. The column "Slope (%)" shows the actual inclination of this line, given in percent. This value represents the actual trend. For this climate zone, the yearly rate is 3.4%, indicating a considerable increment in fire size over recent years.

As can be seen in Figure 7, robust trends regarding fire severity can be derived for two climate zones. First, the temperate zone featuring dry and hot summers (Csa), and second, the arid desert zone featuring cold conditions throughout the year (BWk). The first one shows an inclination of 0.42% per year on average, the second one features a value of 0.11%. The zone of arid steppe with year-round cold conditions (BSk) shows the second largest positive trend inclination, but features a *p*-value just above the threshold for statistical significance.

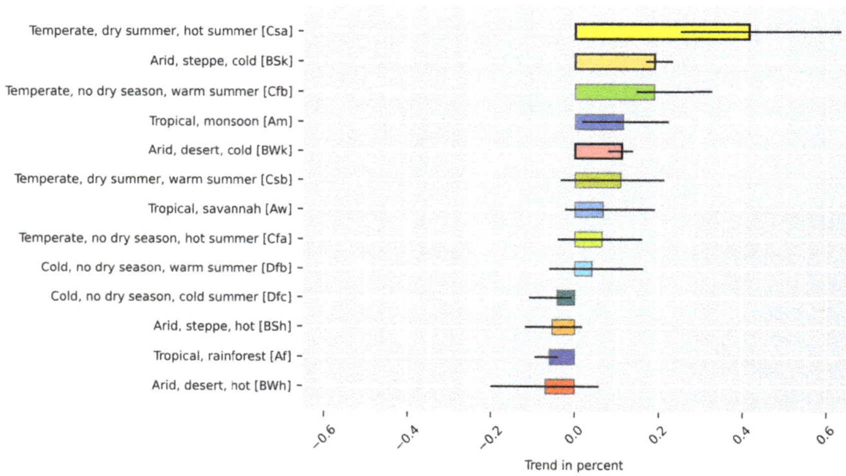

Figure 7. Trends regarding fire severity for each climate zone. The lengths of the bars visualize the strength of the trend. The bars for climate zones featuring a statistically significant trend are drawn with a thick, black border. A gray border is used to identify trends close to statistical significance. The horizontal, black lines represent the error margins.

Table 7 lists the severity trend for each climate zone.

Table 7. Trends regarding fire severity for each climate zone, given in percent of increase/decrease. Climate zones featuring a statistically significant trend are marked in dark gray. The one showing a trend close to statistical significance is depicted in light gray.

Class	Slope (%)	Perc 5	Perc 95	Corr. Coef.	p-Value	RMSE	Label
Csa	0.415	0.251	0.631	0.457	0.043	0.046	Temperate, dry summer, hot summer
BSk	0.188	0.169	0.229	0.418	0.07	0.023	Arid, steppe, cold
Cfb	0.187	0.147	0.325	0.317	0.17	0.032	Temperate, no dry season, warm summer
Am	0.113	0.018	0.219	0.209	0.38	0.030	Tropical, monsoon
BWk	0.111	0.080	0.135	0.566	0.009	0.009	Arid, desert, cold
Csb	0.108	−0.033	0.209	0.123	0.61	0.050	Temperate, dry summer, warm summer
Aw	0.066	−0.022	0.188	0.126	0.6	0.030	Tropical, savannah
Cfa	0.064	−0.039	0.157	0.108	0.65	0.034	Temperate, no dry season, hot summer
Dfb	0.040	−0.059	0.160	0.07	0.77	0.033	Cold, no dry season, warm summer
Dfc	−0.040	−0.107	−0.009	−0.117	0.62	0.020	Cold, no dry season, cold summer
BSh	−0.053	−0.116	0.016	−0.17	0.47	0.017	Arid, steppe, hot
Af	−0.059	−0.093	−0.041	−0.285	0.22	0.011	Tropical, rainforest
BWh	−0.069	−0.198	0.055	−0.109	0.65	0.036	Arid, desert, hot

In order to verify that the trends depicted in the above figures and tables are actually connected to fire occurrence, instead of being a general phenomenon or an effect by a cause not investigated, a cross-check has to be performed. Figure A2, which is located in the Appendix A, shows severity trends for each climate zone, where only areas are considered that have not been affected by fire. As can be expected, no general trend of increased fire severity is observable. In fact there is a generally negative development, indicating a general increase in vegetation fitness. The p-values and the correlation coefficients are generally low, meaning that there is no connection between vegetation fitness and the progression of years in these unaffected areas.

Table A1, which can be found in the Appendix A, shows the statistical information regarding the cross-check.

3.3. *Fire Trends Regarding Ecological Units*

In order to draw conclusions regarding the vegetation types causing the increasing severity trends in some of the climate zones, the analysis is also carried out on the basis of ecological units. These units feature a higher spatial and thematic resolution, and are thus better suited for analyses on a smaller scale.

For the incorporation of ecological units, the Global Ecological Land Units global dataset provided by the United States Geological Survey (USGS) [40] is used. The ecological units are a combination of bioclimate region, landform type, surficial lithology, and land cover information [71], and allow for a very high thematic resolution. This results in more than 3600 different units covering the area of interest. In order to reduce the number of units to be analyzed to an appropriate level, a subset is generated from the original data in a preceding step. This subset contains all ecological units which were affected by the 2019/2020 burnings and covered more than 1% of the burnt area. Furthermore, it comprises all ecological units covering more than 1% of the area of interest.

Figure 8 shows the analysis results of the total burnt area extent development, regarding the ecological units. It can be seen that the three most prominent units (1750, 1529, and 1664) feature either needleleaf or evergreen forest, which has not been the case in former years. Statistically significant results can, however, only be derived for the class of "Hot Wet Mountains on Non-Carbonate Sedimentary Rock with Mostly Needleleaf/Evergreen Forest" (2268). The statistical results are found in Table A2 in the Appendix A.

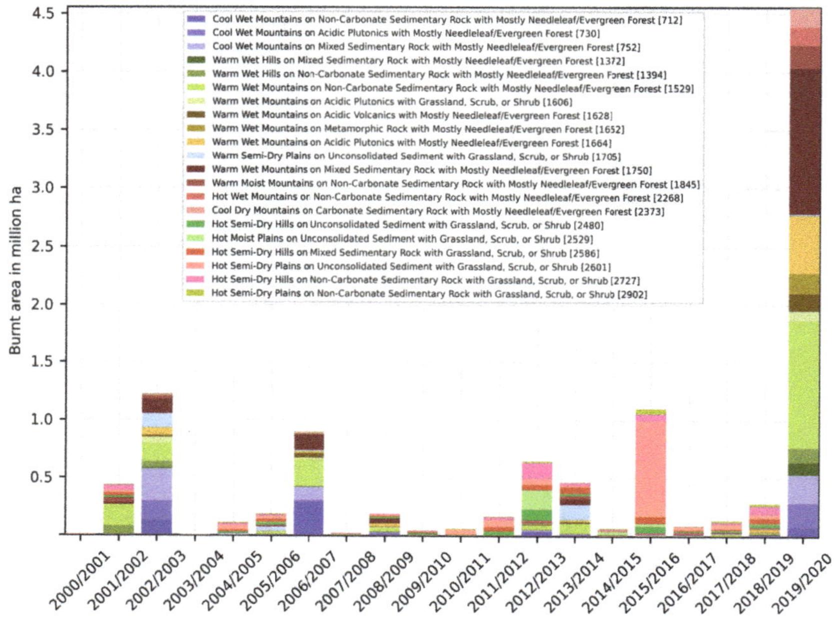

Figure 8. Total yearly burnt area amount in million hectares, subdivided by ecological units.

Figure 9 shows the severity trends regarding ecological units, equivalent to the climate zone analysis above. Similar to the burnt area extent analysis, stable trends can be derived for two classes featuring either needleleaf or evergreen forest. These classes are "Hot Wet Mountains on Non-Carbonate Sedimentary Rock with Mostly Needleleaf/Evergreen Forest (2268)" and "Warm Wet Mountains on Metamorphic Rock with Mostly Needleleaf/Evergreen Forest (1652)". Two further units feature p-values close to statistical significance, and are thus worth being considered: "Warm Semi-Dry Plains on Unconsolidated Sediment with Mostly Cropland (1712)" and "Hot Moist Plains on Unconsolidated Sediment with Grassland, Scrub, or Shrub (2529)". All respective statistical information is found in Table A3 in the Appendix A.

Figure 9. Trends regarding fire severity for the most affected ecological units. The lengths of the bars visualize the strength of the trend. The bars for ecological units featuring a statistically significant trend are drawn with a thick, black border. A gray border is used to identify trends close to statistical significance. The horizontal, black lines represent the error margins.

The increasing burn severity for some classes can be linked to a higher degree of combustion. Other reasons include the higher amount of combustible biomass, the concern of younger, healthier vegetation, or the exposure of different vegetation types.

The cross-check, conducted for areas which have not been affected by fire, shows the expected, generally negative trend. The derived fire severity trends presented above are therefore demonstrably caused by wildfire activity. Figure A3 shows the trend for each ecological unit. All respective statistical information is found in Table A4 in the Appendix A.

Figure 10 shows a section of the two ecological units with a statistically significant increase in fire severity during the analyzed time span. The figure represents a detailed view of the North-Eastern part of New South Wales, where these two ecological units overlap with the area affected by the 2019/2020 wildfires.

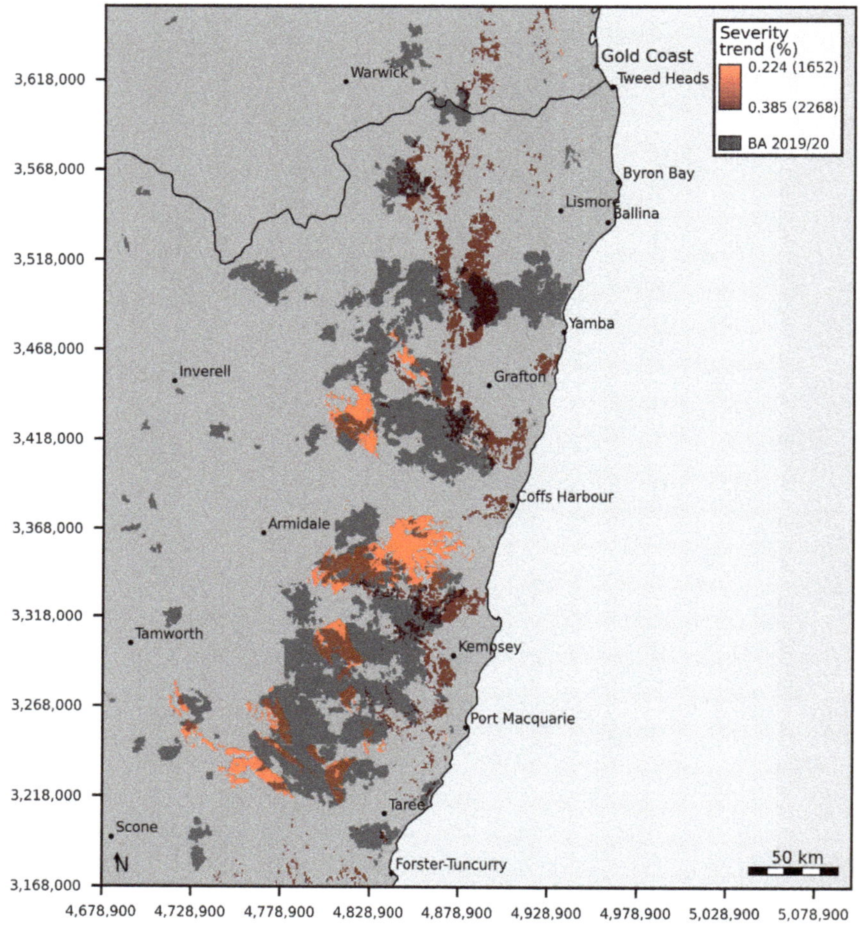

Figure 10. Ecological units featuring a statistically significant severity trend, overlapping 2019/2020 wildfire locations in northern New South Wales.

3.4. Combination of Results from Different Levels

Table 8 lists how much area of New South Wales and Victoria is covered by the relevant climate zones. Features are considered relevant if they show a trend matching or being close to statistical significance regarding burn extent or severity. Two of these zones—BSk and Cfa—cover significant portions of the two states.

Table 8. Percentage of area covered by relevant climate zones, regarding New South Wales and Victoria.

Climate Zone	Area Portion (%)		Label
	NSW	VIC	
BWk	3.8	3.3	Arid, desert, cold
BSk	17.9	36.7	Arid, steppe, cold
Csa	0.16	1.6	Temperate, dry summer, hot summer
Cfa	19.1	1.9	Temperate, no dry season, hot summer

Table 9 lists, equivalent to Table 8, how much of the area of the relevant climate zones is covered by relevant ecological units. It can be seen that unit 1712 covers 36.3% of the area of the BSk climate zone in New South Wales, and even 52.3% of this zone in Victoria. However, this unit contains mostly cropland, and thus the fire activity has to be attributed in large part to agricultural burnings. While it is interesting to note that the burn severity rises on agricultural areas, this study targets the activity of potentially harmful wildfires, and is thus not concerned with controlled, anthropogenic fires. Ecological unit 1712, and with it the BSk climate zone, is therefore considered largely irrelevant for this study. The remaining zone is Cfa, which features a temperate climate with hot summers, and without a dry season. Unit 2268, "Hot Wet Mountains on Non-Carbonate Sedimentary Rock with Mostly Needleleaf/Evergreen Forest", shows the highest trends of all analyzed ecological units, both regarding fire extent and fire severity. The unit features an increasing trend of 0.26% on annual average regarding extent, and 0.39% regarding burn severity. Both trends are shown to be robust, indicated by their statistical significance. A similar trend can also be seen for unit 1652, "Warm Wet Mountains on Metamorphic Rock with Mostly Needleleaf/Evergreen Forest", which shows an increasing burn severity of 0.2% on annual average.

Table 9. Percentage of area covered by relevant ecological units, regarding New South Wales and Victoria.

Ecological Zone	Area Portion (%)				Label
	NSW		VIC		
	BSk	Cfa	BSk	Cfa	
1372	-	0.1	-	-	Warm Wet Hills on Mixed Sedimentary Rock with Mostly Needleleaf/Evergreen Forest
1652	-	0.9	-	-	Warm Wet Mountains on Metamorphic Rock with Mostly Needleleaf/Evergreen Forest
1712	36.3	1.4	52.3	3.7	Warm Semi-Dry Plains on Unconsolidated Sediment with Mostly Cropland
2268	-	2.9	-	-	Hot Wet Mountains on Non-Carbonate Sedimentary Rock with Mostly Needleleaf/Evergreen Forest
2373	-	2.4	-	-	Cool Dry Mountains on Carbonate Sedimentary Rock with Mostly Needleleaf/Evergreen Forest
2529	-	-	-	-	Hot Moist Plains on Unconsolidated Sediment with Grassland, Scrub, or Shrub
2902	-	-	-	-	Hot Semi-Dry Plains on Non-Carbonate Sedimentary Rock with Grassland, Scrub, or Shrub

Apart from the agricultural area, all units located within this climate zone featuring increasing severity trends contain needleleaf or evergreen forest.

This development is also discernible in Figure 11, which shows the affected vegetation types and the respective size in hectares for each of the four exposed states and territories, regarding the period of 2000 to 2020. Yearly land cover information of the ESA CCI-LC (Climate Change Initiative—Land Cover) dataset ([39]) has been used in order to derive the present vegetation types for each respective year. For New South Wales and Victoria, it can be seen that forests represent the predominantly affected vegetation type of the burnings in

2019/2020. The presented figures and statistics indicate that ecologically valuable, woody vegetation is increasingly affected in the study area.

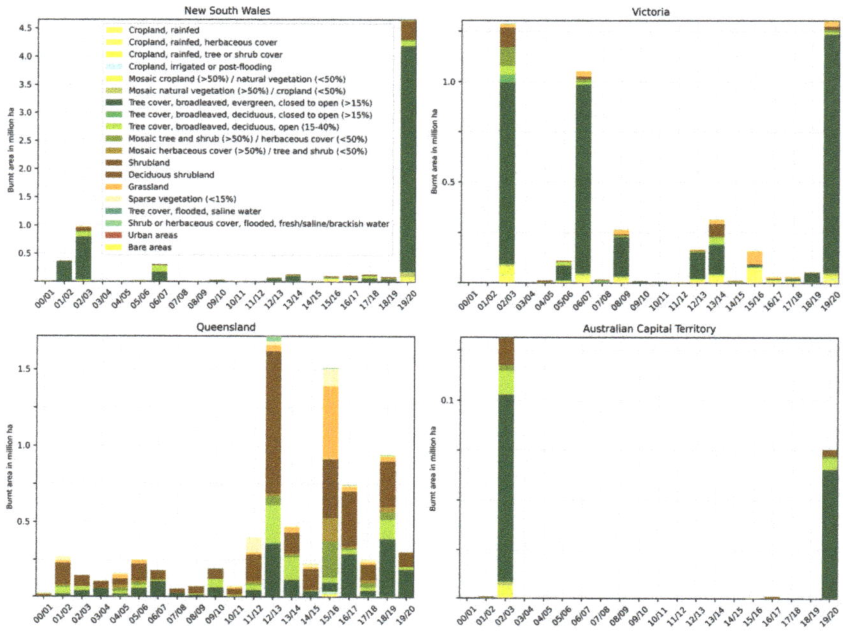

Figure 11. Total yearly burnt area amount in million hectares for each state, subdivided by land cover type.

4. Discussion

While the inter-annual variability in fire activity complicates and in some cases prohibits the derivation of statistically significant trends, several expressive conclusions could be drawn for some of the investigated climate zones and ecological units.

Steady increases regarding burn severity could be found for the climate zones BSk and Cfa, which cover significant parts of New South Wales and Victoria. These development could be traced back to several ecological units, residing inside these climate zones. One of those, number 1712, is mostly characterized by agricultural activities and thus considered less relevant. The other ones, located within the Cfa climate zone, indicate pronounced increases in burn severity regarding needleleaf/evergreen forest. This is supported by a time series study of land use/land cover information. In general, the results show that woody vegetation is increasingly affected in New South Wales and Victoria.

Equivalent conclusions have been drawn by Tran et al., who analyzed fire severity for Victoria [15] regarding the period of 1987 to 2017. The authors furthermore stated that the consequences for ecosystem dynamics might be critical, as temperate forests usually adapted to fire could be damaged irreversibly through higher severity burnings.

Several points need to be taken into account regarding the methodology of this study:

First, note that the inter-comparison of the analyzed classes is only possible in a limited manner. The measure of fire severity has a very different expressiveness between arid, tropical and savannah land cover classes, for example. Hammill et al. also found that determining fire severity from satellite imagery for sedge-swamp or heath surface cover is only possible with lower accuracy compared to forests and woodlands [72]. Results are therefore distorted when study areas cover different ecosystem types, meaning that the robustness of the results increases with rising homogeneity of the study area. While the effect of mixed signals cannot be fully eliminated in a large-scale study, it can be mitigated by analyzing regions of homogeneous climatic conditions or fine-scale ecological regions, as it is done here.

Second, the derivation of burnt area perimeters as well as the assessment of burn severity rely heavily upon the $NDVI_{diff}$. This index has been utilized in numerous investigations, and was validated in a variety of studies, for example [73–75]. The index has been used for decades to assess fire severity, and is also actively used today. For instance, it was recently utilized by Mathews et al. as well as Storey et al. to analyze the burn severity of the wildfires in California in 2020 [76,77]. Tran et al., 2020 investigated indices which are commonly used to assess fire severity, regarding the study area of Victoria/Australia. They identified the $NDVI$ as one of the optimal spectral indices for mapping fire severity, regarding the forest types of this study area [15,75].

Another index frequently utilized is the Normalized Burn Ratio (NBR), which is similar the $NDVI$ but relies on the NIR and Short Wave Infrared (SWIR) band combination instead of red and NIR in case of the $NDVI$. This index could be shown to perform similar to the $NDVI$ regarding high severity fires, but was superior regarding fire events featuring rather low severity [78]. The reason that the DLR burnt area dataset does not utilize the NBR is that this dataset is primarily based on the Sentinel-3 OLCI instrument, which does not feature a band in the SWIR domain. The MODIS instrument does have a SWIR band, however. This one is only available at at a reduced resolution of 500 m, though, opposed to 250 m regarding the red and NIR band. For the conduction of the study, it was decided to utilize the MODIS bands equivalent to the ones available in Sentinel-3 OLCI. This allows a homogeneous methodology at the best available spatial resolution.

Apart from rule-based approaches based on spectral indices, methodologies from the domain of Machine Learning are increasingly used in wildfire science. Collins et al. (2018) [79] used a Random Forest classifier for the determination of burn severity classes, and found a higher detection accuracy compared to index-based approaches. This proceeding, however, requires preceding steps of careful selection and preparation of training data, as well as the actual training of a Neural Network regarding the area of interest and input data to be used. A comprehensive overview of the requirements is given by Collins et al. (2020) [80]. The methodology invoked for the DLR dataset has been designed to be applicable with a variety of optical sensors, and to be operational globally without a preceding training step.

Third, the analyzed time period covers only the months from November to February for each analyzed year, which is the time span the majority of the disastrous burnings happened in the 2019/2020 fire season. The confinement to a subsection of the available input data became necessary because of the massiveness of the complete dataset, which could not have been processed within a reasonable time frame. However, this time range was found to be representative for the fire season regarding the state of Victoria by Tran et al. (2020) [15]. Still, this confinement represents a sub-optimal precondition, as important differences in the seasonality of fire across the study area might be ignored.

Fourth, developments regarding burn severity are dependent on different input factors, and can easily be misinterpreted. These developments can be caused by shifts in the affected vegetation coverage. Woody vegetation features a higher biomass amount compared to shrubland, which will result in a higher severity value when burnt. Furthermore, the spatial extent regarding affected land cover types plays a crucial role, since it proportionally influences the resulting average value. An increase in area of affected woody vegetation can be overcompensated by an even higher increase in area of affected shrubland vegetation.

Finally, it has to be stated that the analyzed time span of 20 years is rather short, with respect to gaining sufficient insight into climate related, long-term developments. This limitation is due to the availability of suitable satellite imagery of the MODIS and OLCI sensors. The available data time range does not allow conclusions regarding the question whether dramatic fire events occur more frequently than in earlier decades. For future studies, it is therefore planned to also incorporate data of the Advanced Very High

Resolution Radiometer (AVHRR) optical satellite sensor [81], which would allow to perform analysis based on a time series covering more than 40 years.

5. Conclusions

The pronounced, inter-annual variability in fire activity together with the spatial dynamics of wildfires often prohibits the derivation of statistically significant trends. The majority of the dramatic burnings occurring mostly in New South Wales and Victoria in the 2019/2020 fire season must be regarded as an exception. However, several meaningful, robust trends regarding fire severity and extent could be derived for some of the affected area, mostly located in the coastal area of northern New South Wales.

Two different climate zones have been found to be responsible for the rising burn severity trends in New South Wales and Victoria. The trends within the BSk zone, which is defined by cold, arid steppe conditions, is mostly due to fire activity in the ecological unit 1712, which contains mostly cropland. The fire activity in this zone is therefore attributed mainly to agricultural burnings, which are not examined in this study. The coastal Cfa climate zone, featuring temperate conditions with hot summers and without a dry season, however, was shown to be increasingly affected by potentially harmful wildfires. The rising trends of fire extent and severity could be traced back to several ecological units. All these units, except for one which is used agriculturally, share the characteristic of being covered by needleleaf/evergreen forest. While the extensive burnings of the 2019/2020 fire season clearly are exceptional, some of the fire activity took place in these forested areas, and is thus regarded to be in parts connected to a steady, long-term upward trend in fire extent and severity.

It is concluded that the forested regions of the Australian East coast residing within the Cfa climate zone (temperate, no dry season, hot summer) will most likely be increasingly affected by wildfire activity in the future. Specifically, this refers to the area covered by, first, ecological unit 2268 (Hot Wet Mountains on Non-Carbonate Sedimentary Rock with Mostly Needleleaf/Evergreen Forest), which features a mean annual increase of 0.26% in fire extent and 0.39% in fire severity. Secondly, this addresses the area covered by ecological unit 1652 (Warm Wet Mountains on Metamorphic Rock with Mostly Needleleaf/Evergreen Forest), which shows a mean annual increase of 0.25% in fire extent and 0.22% regarding fire severity.

The DLR-GZS burnt area dataset, on which this study is based, could be shown to be a valuable asset for wildfire related studies, such as burn severity time series analysis. To the knowledge of the authors, it is the only large-scale, decadal burnt area dataset including detailed burn severity information to this point.

Author Contributions: Conceptualization, M.N., G.S. and T.R.; methodology, M.N.; software, M.N.; validation, M.N. and N.M.; formal analysis, M.N.; investigation, M.N.; resources, M.N.; data curation, M.N.; writing—original draft preparation, M.N.; writing—review and editing, M.N., N.M., G.S. and T.R.; visualization, M.N.; supervision, G.S. and T.R.; project administration, M.N., G.S. and T.R.; funding acquisition, G.S. and T.R. All authors have read and agreed to the published version of the manuscript.

Funding: This research received no external funding.

Institutional Review Board Statement: Not applicable.

Informed Consent Statement: Not applicable.

Data Availability Statement: The MODIS MOD09/MYD09 product is available from the NASA Land Processes Distributed Active Archive Center (LP DAAC), accessed on 8 April 2020 http://e4ftl01.cr.usgs.gov. Sentinel 3A/B OLCI data can be obtained from the Copernicus Open Access Hub, accessed on 10 June 2020 https://scihub.copernicus.eu. Finally, Active Fire data used as auxiliary information in the described methodology can be acquired from the NASA Fire Information for Resource Management System (FIRMS), accessed on 17 May 2020 https://firms.modaps.eosdis.nasa.gov.

Conflicts of Interest: The authors declare no conflict of interest.

Appendix A

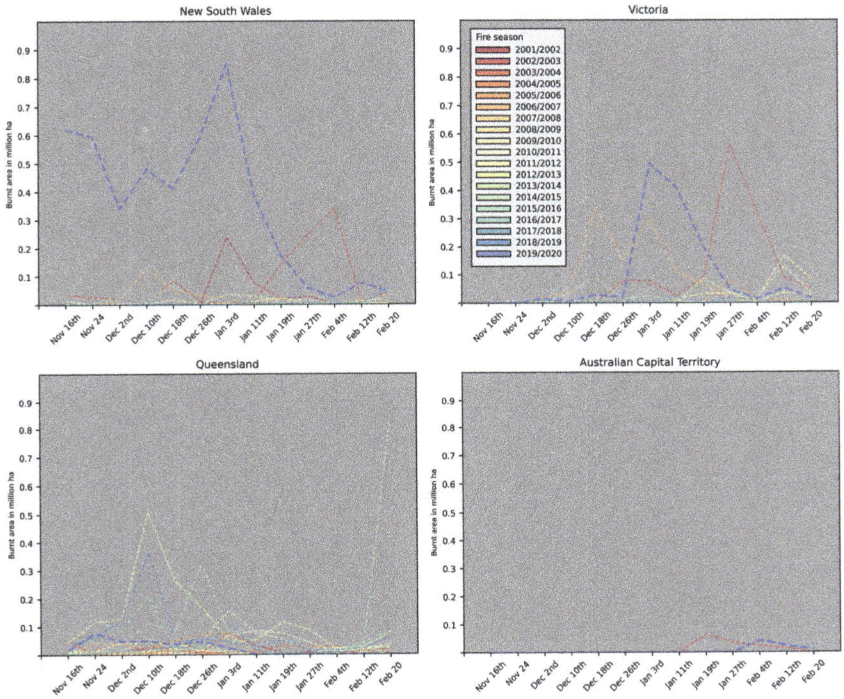

Figure A1. Burnt area extent in million hectares for the investigated states over the past two decades, sub-divided by month of the fire season. The figure illustrates the usual distribution of fire activity throughout the fire season.

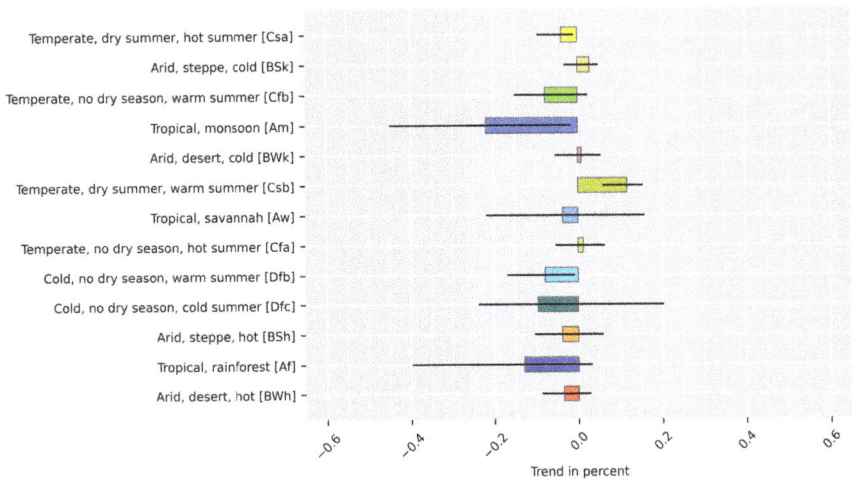

Figure A2. Cross-check regarding fire severity for each climate zone, given in percent of increase/decrease. The length of the bars visualizes the strength of the trend. The horizontal, black lines represent the error margins.

Table A1. Cross-check regarding fire severity for each climate zone.

Class	Slope (%)	Perc 5	Perc 95	Corr. Coef.	*p*-Value	RMSE	Label
Csa	−0.038	−0.093	−0.010	−0.141	0.55	0.015	Temperate, dry summer, hot summer
BSk	0.028	−0.030	0.047	0.158	0.51	0.010	Arid, steppe, cold
Cfb	−0.077	−0.148	0.022	−0.135	0.57	0.032	Temperate, no dry season, warm summer
Am	−0.218	−0.447	−0.018	−0.142	0.55	0.087	Tropical, monsoon
BWk	0.008	−0.052	0.053	0.048	0.84	0.010	Arid, desert, cold
Csb	0.116	0.061	0.152	0.292	0.21	0.022	Temperate, dry summer, warm summer
Aw	−0.037	−0.216	0.156	−0.029	0.9	0.072	Tropical, savannah
Cfa	0.012	−0.051	0.061	0.028	0.91	0.025	Temperate, no dry season, hot summer
Dfb	−0.079	−0.167	−0.009	−0.109	0.65	0.041	Cold, no dry season, warm summer
Dfc	−0.096	−0.235	0.201	−0.111	0.64	0.050	Cold, no dry season, cold summer
BSh	−0.038	−0.102	0.057	−0.072	0.76	0.030	Arid, steppe, hot
Af	−0.129	−0.395	0.030	−0.079	0.74	0.094	Tropical, rainforest
BWh	−0.035	−0.085	0.027	−0.125	0.6	0.016	Arid, desert, hot

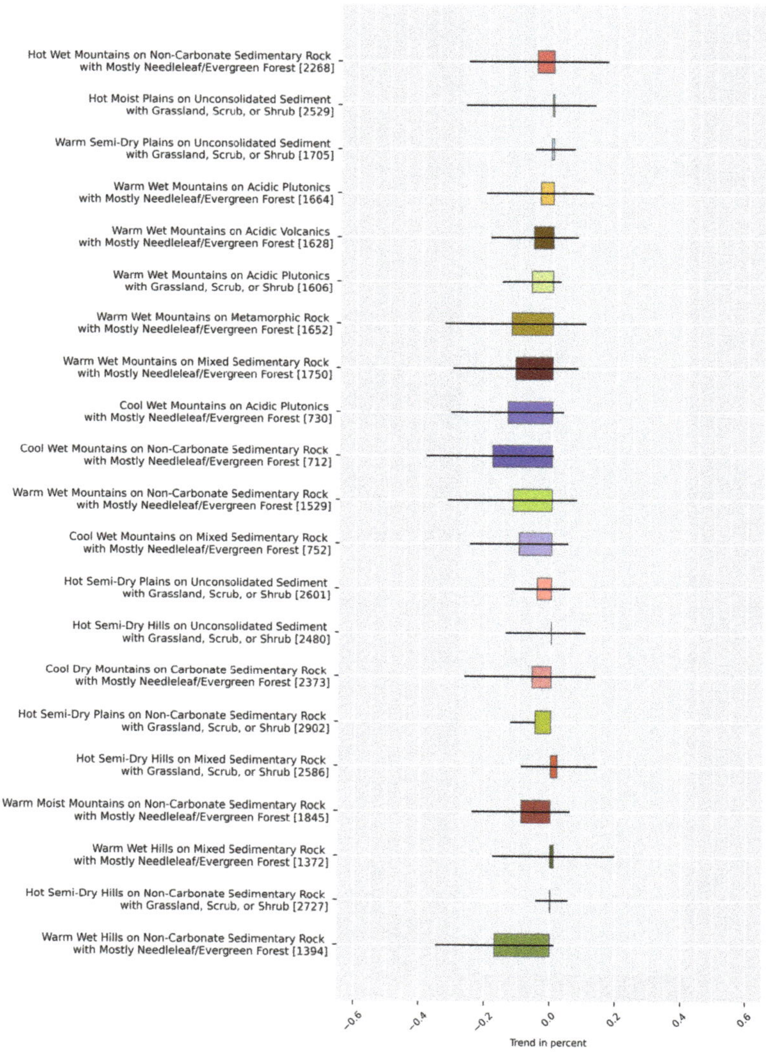

Figure A3. Cross-check regarding fire severity for each ecological unit. The lengths of the bars visualize the strength of the trend. The horizontal, black lines represent the error margins.

Table A2. Trends regarding the burnt area size for each ecological unit in the area of interest. The ecological unit featuring a statistically significant trend is marked in dark gray. Ecological units showing trends close to statistical significance are depicted in light gray.

Class	Slope (%)	Perc 5	Perc 95	Corr. Coef.	*p*-Value	RMSE	Label
2268	0.263	0.150	0.373	0.456	0.04	0.029	Hot Wet Mountains on Non-Carbonate Sedimentary Rock with Mostly Needleleaf/Evergreen Forest
1712	0.058	0.028	0.089	0.418	0.07	0.007	Warm Semi-Dry Plains on Unconsolidated Sediment with Mostly Cropland
2529	0.108	0.029	0.229	0.17	0.47	0.036	Hot Moist Plains on Unconsolidated Sediment with Grassland, Scrub, or Shrub
1705	−0.064	−0.132	0.050	−0.1	0.68	0.037	Warm Semi-Dry Plains on Unconsolidated Sediment with Grassland, Scrub, or Shrub
1664	0.652	0.265	0.998	0.357	0.12	0.098	Warm Wet Mountains on Acidic Plutonics with Mostly Needleleaf/Evergreen Forest
1628	0.184	0.063	0.294	0.329	0.15	0.030	Warm Wet Mountains on Acidic Volcanics with Mostly Needleleaf/Evergreen Forest
1606	0.072	−0.015	0.160	0.189	0.42	0.021	Warm Wet Mountains on Acidic Plutonics with Grassland, Scrub, or Shrub
1734	0.095	0.038	0.150	0.371	0.10	0.013	Warm Wet Mountains on Non-Acidic Volcanics with Mostly Needleleaf/Evergreen Forest
1652	0.253	0.116	0.394	0.38	0.09	0.035	Warm Wet Mountains on Metamorphic Rock with Mostly Needleleaf/Evergreen Forest
1750	1.581	0.567	2.503	0.336	0.14	0.255	Warm Wet Mountains on Mixed Sedimentary Rock with Mostly Needleleaf/Evergreen Forest
730	0.108	−0.103	0.321	0.109	0.64	0.056	Cool Wet Mountains on Acidic Plutonics with Mostly Needleleaf/Evergreen Forest
712	−0.161	−0.417	0.095	−0.136	0.56	0.068	Cool Wet Mountains on Non-Carbonate Sedimentary Rock with Mostly Needleleaf/Evergreen Forest
1529	1.097	0.275	1.968	0.264	0.26	0.231	Warm Wet Mountains on Non-Carbonate Sedimentary Rock with Mostly Needleleaf/Evergreen Forest
752	−0.023	−0.311	0.274	−0.017	0.94	0.079	Cool Wet Mountains on Mixed Sedimentary Rock with Mostly Needleleaf/Evergreen Forest
2601	0.759	0.132	1.083	0.248	0.29	0.171	Hot Semi-Dry Plains on Unconsolidated Sediment with Grassland, Scrub, or Shrub
2480	0.105	0.058	0.153	0.279	0.23	0.020	Hot Semi-Dry Hills on Unconsolidated Sediment with Grassland, Scrub, or Shrub
2373	0.231	0.111	0.355	0.38	0.09	0.032	Cool Dry Mountains on Carbonate Sedimentary Rock with Mostly Needleleaf/Evergreen Forest
2902	0.068	0.015	0.083	0.405	0.08	0.008	Hot Semi-Dry Plains on Non-Carbonate Sedimentary Rock with Grassland, Scrub, or Shrub
2621	0.050	0.030	0.093	0.273	0.24	0.010	Hot Semi-Dry Plains on Unconsolidated Sediment with Sparse Vegetation
2711	4.196	−0.001	0.002	0.259	0.27	0.000	Hot Dry Plains on Unconsolidated Sediment with Bare area
2586	0.106	0.043	0.165	0.313	0.18	0.018	Hot Semi-Dry Hills on Mixed Sedimentary Rock with Grassland, Scrub, or Shrub
2606	0.003	−0.006	0.018	0.072	0.76	0.002	Hot Semi-Dry Plains on Mixed Sedimentary Rock with Sparse Vegetation
1845	0.222	0.074	0.359	0.3	0.19	0.040	Warm Moist Mountains on Non-Carbonate Sedimentary Rock with Mostly Needleleaf/Evergreen Forest
2614	0.019	0.012	0.031	0.337	0.15	0.003	Hot Semi-Dry Plains on Mixed Sedimentary Rock with Grassland, Scrub, or Shrub
1372	0.151	0.066	0.232	0.378	0.10	0.021	Warm Wet Hills on Mixed Sedimentary Rock with Mostly Needleleaf/Evergreen Forest
2822	4.034	−0.003	0.001	0.061	0.8	0.000	Hot Dry Plains on Unconsolidated Sediment with Sparse Vegetation
2784	−0.000	−0.007	0.004	−0.06	0.8	0.000	Hot Semi-Dry Plains on Non-Carbonate Sedimentary Rock with Sparse Vegetation
1849	−0.019	−0.028	−0.007	−0.307	0.19	0.003	Warm Semi-Dry Plains on Unconsolidated Sediment with Sparse Vegetation
2727	0.146	0.092	0.277	0.255	0.28	0.032	Hot Semi-Dry Hills on Non-Carbonate Sedimentary Rock with Grassland, Scrub, or Shrub
2791	0.008	0.007	0.033	0.085	0.72	0.005	Hot Dry Plains on Unconsolidated Sediment with Swampy or Often Flooded Vegetation
1394	0.024	−0.103	0.140	0.045	0.84	0.031	Warm Wet Hills on Non-Carbonate Sedimentary Rock with Mostly Needleleaf/Evergreen Forest

Table A3. Trends regarding the fire severity for each ecological unit in the area of interest. Ecological units featuring statistically significant trends are marked in dark gray. The ones showing trends close to statistical significance are depicted in light gray.

Class	Slope (%)	Perc 5	Perc 95	Corr. Coef.	p-Value	RMSE	Label
2268	0.385	0.269	0.499	0.589	0.006	0.030	Hot Wet Mountains on Non-Carbonate Sedimentary Rock with Mostly Needleleaf/Evergreen Forest
1712	0.326	0.177	0.453	0.385	0.09	0.045	Warm Semi-Dry Plains on Unconsolidated Sediment with Mostly Cropland
2529	0.314	0.251	0.360	0.519	0.019	0.029	Hot Moist Plains on Unconsolidated Sediment with Grassland, Scrub, or Shrub
1705	0.283	0.232	0.413	0.341	0.14	0.044	Warm Semi-Dry Plains on Unconsolidated Sediment with Grassland, Scrub, or Shrub
1664	0.271	0.063	0.480	0.277	0.236	0.054	Warm Wet Mountains on Acidic Plutonics with Mostly Needleleaf/Evergreen Forest
1628	0.262	0.100	0.429	0.318	0.171	0.045	Warm Wet Mountains on Acidic Volcanics with Mostly Needleleaf/Evergreen Forest
1606	0.234	0.056	0.404	0.269	0.251	0.048	Warm Wet Mountains on Acidic Plutonics with Grassland, Scrub, or Shrub
1734	0.224	0.052	0.407	0.259	0.269	0.048	Warm Wet Mountains on Non-Acidic Volcanics with Mostly Needleleaf/Evergreen Forest
1652	0.224	0.120	0.326	0.45	0.046	0.025	Warm Wet Mountains on Metamorphic Rock with Mostly Needleleaf/Evergreen Forest
1750	0.179	−0.038	0.418	0.161	0.498	0.063	Warm Wet Mountains on Mixed Sedimentary Rock with Mostly Needleleaf/Evergreen Forest
730	0.173	−0.012	0.378	0.179	0.450	0.055	Cool Wet Mountains on Acidic Plutonics with Mostly Needleleaf/Evergreen Forest
712	0.163	−0.069	0.393	0.148	0.533	0.062	Cool Wet Mountains on Non-Carbonate Sedimentary Rock with Mostly Needleleaf/Evergreen Forest
1529	0.158	−0.053	0.374	0.158	0.504	0.057	Warm Wet Mountains on Non-Carbonate Sedimentary Rock with Mostly Needleleaf/Evergreen Forest
752	0.142	−0.076	0.344	0.145	0.543	0.056	Cool Wet Mountains on Mixed Sedimentary Rock with Mostly Needleleaf/Evergreen Forest
2601	0.127	0.113	0.265	0.212	0.37	0.034	Hot Semi-Dry Plains on Unconsolidated Sediment with Grassland, Scrub, or Shrub
2480	0.114	−0.029	0.298	0.176	0.46	0.037	Hot Semi-Dry Hills on Unconsolidated Sediment with Grassland, Scrub, or Shrub
2373	0.107	−0.004	0.213	0.217	0.357	0.027	Cool Dry Mountains on Carbonate Sedimentary Rock with Mostly Needleleaf/Evergreen Forest
2902	0.061	−0.004	0.225	0.089	0.71	0.039	Hot Semi-Dry Plains on Non-Carbonate Sedimentary Rock with Grassland, Scrub, or Shrub
2621	0.056	−0.060	0.138	0.113	0.64	0.028	Hot Semi-Dry Plains on Unconsolidated Sediment with Sparse Vegetation
2711	0.035	0.005	0.062	0.259	0.27	0.007	Hot Dry Plains on Unconsolidated Sediment with Bare area
2586	0.031	−0.106	0.057	0.05	0.83	0.036	Hot Semi-Dry Hills on Mixed Sedimentary Rock with Grassland, Scrub, or Shrub
2606	0.022	−0.022	0.128	0.054	0.82	0.024	Hot Semi-Dry Plains on Mixed Sedimentary Rock with Sparse Vegetation
1845	0.017	−0.152	0.199	0.023	0.924	0.045	Warm Moist Mountains on Non-Carbonate Sedimentary Rock with Mostly Needleleaf/Evergreen Forest
2614	0.010	−0.121	0.066	0.019	0.94	0.033	Hot Semi-Dry Plains on Mixed Sedimentary Rock with Grassland, Scrub, or Shrub
1372	0.005	−0.174	0.194	0.007	0.977	0.046	Warm Wet Hills on Mixed Sedimentary Rock with Mostly Needleleaf/Evergreen Forest
2822	−0.018	−0.019	0.006	−0.094	0.69	0.011	Hot Dry Plains on Unconsolidated Sediment with Sparse Vegetation
2784	−0.028	−0.122	0.074	−0.066	0.78	0.025	Hot Semi-Dry Plains on Non-Carbonate Sedimentary Rock with Sparse Vegetation
1849	−0.078	−0.140	0.047	−0.15	0.53	0.029	Warm Semi-Dry Plains on Unconsolidated Sediment with Sparse Vegetation
2727	−0.118	−0.185	0.037	−0.211	0.37	0.031	Hot Semi-Dry Hills on Non-Carbonate Sedimentary Rock with Grassland, Scrub, or Shrub
2791	−0.123	−0.298	−0.087	−0.22	0.35	0.031	Hot Dry Plains on Unconsolidated Sediment with Swampy or Often Flooded Vegetation
1394	−0.231	−0.457	−0.009	−0.218	0.355	0.059	Warm Wet Hills on Non-Carbonate Sedimentary Rock with Mostly Needleleaf/Evergreen Forest

Table A4. Cross-check regarding fire severity for each ecological unit in the area of interest.

Class	Slope (%)	Perc 5	Perc 95	Corr. Coef.	*p*-Value	RMSE	Label
2268	−0.052	−0.259	0.163	−0.053	0.824	0.057	Hot Wet Mountains on Non-Carbonate Sedimentary Rock with Mostly Needleleaf/Evergreen Forest
1712	0.012	−0.006	0.039	0.081	0.73	0.008	Warm Semi-Dry Plains on Unconsolidated Sediment with Mostly Cropland
2529	−0.004	−0.267	0.124	−0.003	0.99	0.070	Hot Moist Plains on Unconsolidated Sediment with Grassland, Scrub, or Shrub
1705	−0.007	−0.055	0.062	−0.034	0.89	0.012	Warm Semi-Dry Plains on Unconsolidated Sediment with Grassland, Scrub, or Shrub
1664	−0.040	−0.204	0.118	−0.055	0.818	0.042	Warm Wet Mountains on Acidic Plutonics with Mostly Needleleaf/Evergreen Forest
1628	−0.060	−0.190	0.073	−0.099	0.679	0.034	Warm Wet Mountains on Acidic Volcanics with Mostly Needleleaf/Evergreen Forest
1606	−0.066	−0.154	0.022	−0.166	0.484	0.022	Warm Wet Mountains on Acidic Plutonics with Grassland, Scrub, or Shrub
1734	−0.110	−0.257	0.030	−0.159	0.501	0.039	Warm Wet Mountains on Non-Acidic Volcanics with Mostly Needleleaf/Evergreen Forest
1652	−0.126	−0.328	0.099	−0.125	0.600	0.057	Warm Wet Mountains on Metamorphic Rock with Mostly Needleleaf/Evergreen Forest
1750	−0.113	−0.303	0.075	−0.13	0.586	0.050	Warm Wet Mountains on Mixed Sedimentary Rock with Mostly Needleleaf/Evergreen Forest
730	−0.136	−0.309	0.032	−0.171	0.469	0.045	Cool Wet Mountains on Acidic Plutonics with Mostly Needleleaf/Evergreen Forest
712	−0.182	−0.383	0.002	−0.2	0.397	0.051	Cool Wet Mountains on Non-Carbonate Sedimentary Rock with Mostly Needleleaf/Evergreen Forest
1529	−0.119	−0.317	0.073	−0.134	0.573	0.050	Warm Wet Mountains on Non-Carbonate Sedimentary Rock with Mostly Needleleaf/Evergreen Forest
752	−0.100	−0.248	0.048	−0.148	0.533	0.038	Cool Wet Mountains on Mixed Sedimentary Rock with Mostly Needleleaf/Evergreen Forest
2601	−0.044	−0.109	0.055	−0.089	0.71	0.028	Hot Semi-Dry Plains on Unconsolidated Sediment with Grassland, Scrub, or Shrub
2480	0.000	−0.137	0.102	0.001	1.0	0.033	Hot Semi-Dry Hills on Unconsolidated Sediment with Grassland, Scrub, or Shrub
2373	−0.058	−0.262	0.134	−0.061	0.799	0.055	Cool Dry Mountains on Carbonate Sedimentary Rock with Mostly Needleleaf/Evergreen Forest
2902	−0.047	−0.121	−0.049	−0.112	0.64	0.024	Hot Semi-Dry Plains on Non-Carbonate Sedimentary Rock with Grassland, Scrub, or Shrub
2621	−0.054	−0.193	0.092	−0.093	0.7	0.033	Hot Semi-Dry Plains on Unconsolidated Sediment with Sparse Vegetation
2711	−0.013	−0.051	0.036	−0.062	0.8	0.012	Hot Dry Plains on Unconsolidated Sediment with Bare area
2586	0.021	−0.087	0.141	0.042	0.86	0.029	Hot Semi-Dry Hills on Mixed Sedimentary Rock with Grassland, Scrub, or Shrub
2606	−0.090	−0.250	0.047	−0.109	0.65	0.047	Hot Semi-Dry Plains on Mixed Sedimentary Rock with Sparse Vegetation
1845	−0.088	−0.237	0.059	−0.123	0.605	0.041	Warm Moist Mountains on Non-Carbonate Sedimentary Rock with Mostly Needleleaf/Evergreen Forest
2614	−0.050	−0.147	0.094	−0.086	0.72	0.033	Hot Semi-Dry Plains on Mixed Sedimentary Rock with Grassland, Scrub, or Shrub
1372	0.011	−0.173	0.195	0.014	0.954	0.048	Warm Wet Hills on Mixed Sedimentary Rock with Mostly Needleleaf/Evergreen Forest
2822	−0.025	−0.071	0.037	−0.096	0.69	0.015	Hot Dry Plains on Unconsolidated Sediment with Sparse Vegetation
2784	−0.096	−0.196	0.024	−0.167	0.48	0.032	Hot Semi-Dry Plains on Non-Carbonate Sedimentary Rock with Sparse Vegetation
1849	−0.012	−0.035	0.008	−0.093	0.7	0.007	Warm Semi-Dry Plains on Unconsolidated Sediment with Sparse Vegetation
2727	0.001	−0.040	0.054	0.002	0.99	0.031	Hot Semi-Dry Hills on Non-Carbonate Sedimentary Rock with Grassland, Scrub, or Shrub
2791	−0.261	−0.350	−0.164	−0.319	0.17	0.044	Hot Dry Plains on Unconsolidated Sediment with Swampy or Often Flooded Vegetation
1394	−0.167	−0.346	0.013	−0.194	0.413	0.049	Warm Wet Hills on Non-Carbonate Sedimentary Rock with Mostly Needleleaf/Evergreen Forest

References

1. Bowman, D.M.; Balch, J.K.; Artaxo, P.; Bond, W.J.; Carlson, J.M.; Cochrane, M.A.; D'Antonio, C.M.; DeFries, R.S.; Doyle, J.C.; Harrison, S.P.; et al. Fire in the Earth system. *Science* **2009**, *324*, 481–484. [CrossRef]
2. Withey, K.; Berenguer, E.; Palmeira, A.F.; Espírito-Santo, F.D.; Lennox, G.D.; Silva, C.V.; Aragão, L.E.; Ferreira, J.; França, F.; Malhi, Y.; et al. Quantifying immediate carbon emissions from El Niño-mediated wildfires in humid tropical forests. *Philos. Trans. R. Soc. B Biol. Sci.* **2018**, *373*, 20170312. [CrossRef]
3. Surawski, N.; Sullivan, A.; Roxburgh, S.; Polglase, P. Estimates of greenhouse gas and black carbon emissions from a major Australian wildfire with high spatiotemporal resolution. *J. Geophys. Res. Atmos.* **2016**, *121*, 9892–9907. [CrossRef]
4. van Wees, D.; van der Werf, G. The contribution of fire to a global increase in forest loss. In Proceedings of the EGU General Assembly Conference Abstracts, Online, 4–8 May 2020; p. 18049.
5. Girardin, M.P.; Mudelsee, M. Past and future changes in Canadian boreal wildfire activity. *Ecol. Appl.* **2008**, *18*, 391–406. [CrossRef] [PubMed]
6. Miller, J.D.; Safford, H.; Crimmins, M.; Thode, A.E. Quantitative evidence for increasing forest fire severity in the Sierra Nevada and southern Cascade Mountains, California and Nevada, USA. *Ecosystems* **2009**, *12*, 16–32. [CrossRef]

7. Werf, G.v.d.; Randerson, J.; Giglio, L.; Wees, D.v.; Andela, N.; Veraverbeke, S.; Morton, D.; Chen, Y. Fire-climate interactions in a warming world. In Proceedings of the EGU General Assembly Conference Abstracts, Online, 4–8 May 2020; p. 10974.
8. Giglio, L.; Justice, C.; Boschetti, L.; Roy, D. MCD64A1 MODIS/Terra+Aqua Burned Area Monthly l3 Global 500 m Sin Grid v006 [Data Set]. 2015. Available online: doi:10.5067/MODIS/MCD64A1.006 (accessed on 5 October 2020).
9. Lizundia-Loiola, J.; Otón, G.; Ramo, R.; Chuvieco, E. A spatio-temporal active-fire clustering approach for global burned area mapping at 250 m from MODIS data. *Remote Sens. Environ.* **2020**, *236*, 111493. [CrossRef]
10. Chuvieco, E.; Yue, C.; Heil, A.; Mouillot, F.; Alonso-Canas, I.; Padilla, M.; Pereira, J.M.; Oom, D.; Tansey, K. A new global burned area product for climate assessment of fire impacts. *Glob. Ecol. Biogeogr.* **2016**, *25*, 619–629. [CrossRef]
11. Giglio, L.; Randerson, J.T.; Van Der Werf, G.R. Analysis of daily, monthly, and annual burned area using the fourth-generation global fire emissions database (GFED4). *J. Geophys. Res. Biogeosci.* **2013**, *118*, 317–328. [CrossRef]
12. Global Fire Emissions Database (GFED). GFED Data. 2020. Available online: ftp://fuoco.geog.umd.edu (accessed on 4 October 2020).
13. San-Miguel-Ayanz, J.; Schulte, E.; Schmuck, G.; Camia, A.; Strobl, P.; Liberta, G.; Giovando, C.; Boca, R.; Sedano, F.; Kempeneers, P.; et al. Comprehensive monitoring of wildfires in Europe: The European Forest Fire Information System (EFFIS). In *Approaches to Managing Disaster-Assessing Hazards, Emergencies and Disaster Impacts*; IntechOpen: London, UK, 2012.
14. Joint Research Center of the European Commission (JRC). Welcome to EFFIS. 2020. Available online: https://effis.jrc.ec.europa.eu (accessed on 5 June 2020).
15. Tran, B.N.; Tanase, M.A.; Bennett, L.T.; Aponte, C. High-severity wildfires in temperate Australian forests have increased in extent and aggregation in recent decades. *PLoS ONE* **2020**, *15*, e0242484. [CrossRef] [PubMed]
16. Nolde, M.; Plank, S.; Riedlinger, T. An Adaptive and Extensible System for Satellite-Based, Large Scale Burnt Area Monitoring in Near-Real Time. *Remote Sens.* **2020**, *12*, 2162. [CrossRef]
17. MOD09A1 MODIS/Terra Surface Reflectance 8-Day l3 Global 500 m Sin Grid v006 [Data Set]. 2015. Available online: doi:10.5067/MODIS/MOD09Q1.006 (accessed on 15 September 2020).
18. Australian Government Bureau of Meteorology (BoM). Annual Climate Statement 2019. 2020. Available online: http://www.bom.gov.au/climate/current/annual/aus/#tabs=Overview (accessed on 1 November 2020).
19. Australian National University (ANU). Australia's Environment Summary Report 2019. 2020. Available online: https://www.wenfo.org/aer/wp-content/uploads/2020/03/AustraliasEnvironment_2019_SummaryReport.pdf (accessed on 2 November 2020).
20. Australian Government Bureau of Meteorology (BoM). Climate Monitoring Graphs. 2020. Available online: http://www.bom.gov.au/climate/enso/indices.shtml?bookmark=iod (accessed on 2 November 2020).
21. Australian Government Bureau of Meteorology (BoM). Map of Climate Zones of Australia. 2020. Available online: http://www.bom.gov.au/climate/how/newproducts/images/zones.shtml (accessed on 2 November 2020).
22. Beck, H.E.; Zimmermann, N.E.; McVicar, T.R.; Vergopolan, N.; Berg, A.; Wood, E.F. Present and future Köppen-Geiger climate classification maps at 1-km resolution. *Sci. Data* **2018**, *5*, 180214. [CrossRef] [PubMed]
23. Peel, M.C.; Finlayson, B.L.; McMahon, T.A. Updated world map of the Köppen-Geiger climate classification. *Hydrol. Earth Syst. Sci.* **2007**, *11*, 1633–1644. [CrossRef]
24. Luke, R.; McArthur, A. *Bushfires in Australia*, 2nd ed.; Australian Government Publishing Service: Canberra, Australia, 1986.
25. Dowdy, A.J.; Mills, G.A.; Finkele, K.; de Groot, W. Australian fire weather as represented by the McArthur forest fire danger index and the Canadian forest fire weather index. *Cent. Aust. Weather Clim. Res. Tech. Rep.* **2009**, *10*, 91.
26. Russell-Smith, J.; Yates, C.P.; Whitehead, P.J.; Smith, R.; Craig, R.; Allan, G.E.; Thackway, R.; Frakes, I.; Cridland, S.; Meyer, M.C.; et al. Bushfires 'down under': Patterns and implications of contemporary Australian landscape burning. *Int. J. Wildland Fire* **2007**, *16*, 361–377. [CrossRef]
27. Allan, G.E.; Southgate, R.I. Fire regimes in the spinifex landscapes of Australia. In *Flammable Australia: The Fire Regimes and Biodiversity of a Continent*; Cambridge University Press: Cambridge, UK, 2002; pp. 145–176.
28. Bradstock, R.A. A biogeographic model of fire regimes in Australia: Current and future implications. *Glob. Ecol. Biogeogr.* **2010**, *19*, 145–158. [CrossRef]
29. Ellis, S.; Kanowski, P.; Whelan, R. *National Inquiry on Bushfire Mitigation and Management*; Council of Australian Governments: Canberra, Australia, 2004.
30. Verdon, D.C.; Kiem, A.S.; Franks, S.W. Multi-decadal variability of forest fire risk—Eastern Australia. *Int. J. Wildland Fire* **2004**, *13*, 165–171. [CrossRef]
31. European Space Agency (ESA). Sentinel-3 OLCI Introduction. 2020. Available online: https://sentinel.esa.int/web/sentinel/user-guides/sentinel-3-olci (accessed on 10 June 2020).
32. European Space Agency (ESA). Copernicus Open Access Hub. 2020. Available online: https://scihub.copernicus.eu (accessed on 10 June 2020).
33. EUMETSAT. Copernicus Online Data Access. 2020. Available online: https://coda.eumetsat.int/#/home (accessed on 15 August 2020).
34. National Aeronautics and Space Administration (NASA). MODIS—Moderate Resolution Imaging Spectroradiometer. 2020. Available online: https://terra.nasa.gov/about/terra-instruments/modis (accessed on 8 April 2020).
35. National Aeronautics and Space Administration (NASA). NASA Land Processes Distributed Active Archive Center (LP DAAC). 2020. Available online: http://e4ftl01.cr.usgs.gov (accessed on 8 April 2020).

36. Giglio, L.; Justice, C. MOD14A2 MODIS/Terra Thermal Anomalies/Fire 8-Day l3 Global 1 km Sin Grid v006 [Data Set]. 2015. Available online: doi:10.5067/MODIS/MOD14A2.006 (accessed on 8 April 2020).
37. Schroeder, W.; Giglio, L. VIIRS/NPP Thermal Anomalies/Fire Daily l3 Global 1 km Sin Grid v001 [Data Set]. 2017. Available online: doi:10.5067/VIIRS/VNP14A1.001 (accessed on 15 May 2020).
38. National Aeronautics and Space Administration (NASA). Fire Information for Resource Management System (FIRMS). 2020. Available online: https://firms.modaps.eosdis.nasa.gov (accessed on 17 May 2020).
39. European Space Agency (ESA). Land Cover Classification Gridded Maps from 1992 to Present Derived from Satellite Observations. 2020. Available online: https://cds.climate.copernicus.eu/cdsapp#!/dataset/satellite-land-cover?tab=overview (accessed on 17 January 2020).
40. Sayre, R.; Dangermond, J.; Frye, C.; Vaughan, R.; Aniello, P.; Breyer, S.; Cribbs, D.; Hopkins, D.; Nauman, R.; Derrenbacher, W.; et al. *A New Map of Global Ecological Land Units—An Ecophysiographic Stratification Approach*; Association of American Geographers: Washington, DC, USA, 2014.
41. University of Maryland/UMD. Archive Download. 2020. Available online: ftp://ba1.geog.umd.edu (accessed on 22 January 2020).
42. Rodrigues, J.A.; Libonati, R.; Pereira, A.A.; Nogueira, J.M.; Santos, F.L.; Peres, L.F.; Santa Rosa, A.; Schroeder, W.; Pereira, J.M.; Giglio, L.; et al. How well do global burned area products represent fire patterns in the Brazilian Savannas biome? An accuracy assessment of the MCD64 collections. *Int. J. Appl. Earth Obs. Geoinf.* **2019**, *78*, 318–331. [CrossRef]
43. Libonati, R.; DaCamara, C.C.; Setzer, A.W.; Morelli, F.; Melchiori, A.E. An algorithm for burned area detection in the Brazilian Cerrado using 4 μm MODIS imagery. *Remote Sens.* **2015**, *7*, 15782–15803. [CrossRef]
44. Department of Planning, Industry and Environment of New South Wales (DPIE). Google Earth Engine Burnt Area Map (GEEBAM). 2020. Available online: https://datasets.seed.nsw.gov.au/dataset/google-earth-engine-burnt-area-map-geebam (accessed on 10 September 2020).
45. Department of Planning, Industry and Environment of New South Wales (DPIE). Google Earth Engine Burnt Area Map (GEEBAM)-Factsheet. 2020. Available online: https://datasets.seed.nsw.gov.au/dataset/f3c6e3da-f356-43f9-b8df-19c2e7fc004a/resource/a3f3f1a4-1758-4551-a005-a243fd26ec4b/download/geebamfactsheetmar2020.pdf (accessed on 11 September 2020).
46. Australian Government. National Indicative Aggregated Fire Extent Datasets. 2021. Available online: https://data.gov.au/dataset/ds-environment-9ACDCB09-0364-4FE8-9459-2A56C792C743/details?q= (accessed on 10 September 2020).
47. Roy, D.P.; Jin, Y.; Lewis, P.; Justice, C. Prototyping a global algorithm for systematic fire-affected area mapping using MODIS time series data. *Remote Sens. Environ.* **2005**, *97*, 137–162. [CrossRef]
48. Zhang, H.K.; Roy, D.P.; Yan, L.; Li, Z.; Huang, H.; Vermote, E.; Skakun, S.; Roger, J.C. Characterization of Sentinel-2A and Landsat-8 top of atmosphere, surface, and nadir BRDF adjusted reflectance and NDVI differences. *Remote Sens. Environ.* **2018**, *215*, 482–494. [CrossRef]
49. Ramo, R.; Chuvieco, E. Developing a Random Forest algorithm for MODIS global burned area classification. *Remote Sens.* **2017**, *9*, 1193. [CrossRef]
50. Petropoulos, G.P.; Kontoes, C.; Keramitsoglou, I. Burnt area delineation from a uni-temporal perspective based on Landsat TM imagery classification using Support Vector Machines. *Int. J. Appl. Earth Obs. Geoinf.* **2011**, *13*, 70–80. [CrossRef]
51. Pinto, M.M.; Libonati, R.; Trigo, R.M.; Trigo, I.F.; DaCamara, C.C. A deep learning approach for mapping and dating burned areas using temporal sequences of satellite images. *ISPRS J. Photogramm. Remote Sens.* **2020**, *160*, 260–274. [CrossRef]
52. Knopp, L.; Wieland, M.; Rättich, M.; Martinis, S. A Deep Learning Approach for Burned Area Segmentation with Sentinel-2 Data. *Remote Sens.* **2020**, *12*, 2422. [CrossRef]
53. Giglio, L.; Loboda, T.; Roy, D.P.; Quayle, B.; Justice, C.O. An active-fire based burned area mapping algorithm for the MODIS sensor. *Remote Sens. Environ.* **2009**, *113*, 408–420. [CrossRef]
54. Humber, M.L.; Boschetti, L.; Giglio, L.; Justice, C.O. Spatial and temporal intercomparison of four global burned area products. *Int. J. Digit. Earth* **2019**, *12*, 460–484. [CrossRef] [PubMed]
55. Padilla, M.; Stehman, S.V.; Ramo, R.; Corti, D.; Hantson, S.; Oliva, P.; Alonso-Canas, I.; Bradley, A.V.; Tansey, K.; Mota, B.; et al. Comparing the accuracies of remote sensing global burned area products using stratified random sampling and estimation. *Remote Sens. Environ.* **2015**, *160*, 114–121. [CrossRef]
56. Oliva, P.; Schroeder, W. Assessment of VIIRS 375 m active fire detection product for direct burned area mapping. *Remote Sens. Environ.* **2015**, *160*, 144–155. [CrossRef]
57. Rouse, J.W.; Haas, R.H.; Schell, J.A.; Deering, D.W.; Harlan, J.C. *Monitoring the Vernal Advancement and Retrogradation (Green Wave Effect) of Natural Vegetation*; NASA/GSFC Type III Final Report; NASA/GSFC: Greenbelt, MD, USA, 1974; Volume 371.
58. Fraser, R.; Li, Z.; Cihlar, J. Hotspot and NDVI differencing synergy (HANDS): A new technique for burned area mapping over boreal forest. *Remote Sens. Environ.* **2000**, *74*, 362–376. [CrossRef]
59. Li, Z.; Nadon, S.; Cihlar, J.; Stocks, B. Satellite-based mapping of Canadian boreal forest fires: Evaluation and comparison of algorithms. *Int. J. Remote Sens.* **2000**, *21*, 3071–3082. [CrossRef]
60. Chan, T.; Vese, L. An active contour model without edges. In Proceedings of the International Conference on Scale-Space Theories in Computer Vision, Corfu, Greece, 26–27 September 1999; Springer: Berlin/Heidelberg, Germany, 1999; pp. 141–151.
61. Chan, T.F.; Vese, L.A. Active contours without edges. *IEEE Trans. Image Process.* **2001**, *10*, 266–277. [CrossRef]
62. Caselles, V.; Kimmel, R.; Sapiro, G. Geodesic active contours. In Proceedings of the IEEE International Conference on Computer Vision, Cambridge, MA, USA, 20–23 June 1995; IEEE: Piscataway, NJ, USA, 1995; pp. 694–699.

63. Liu, Y.; Dai, Q.; Liu, J.; Liu, S.; Yang, J. Study of burn scar extraction automatically based on level set method using remote sensing data. *PLoS ONE* **2014**, *9*, e87480. [CrossRef] [PubMed]
64. Yan, L.; Roy, D.P. Conterminous United States crop field size quantification from multi-temporal Landsat data. *Remote Sens. Environ.* **2016**, *172*, 67–86. [CrossRef]
65. Plank, S.; Martinis, S. A fully automatic burnt area mapping processor based on AVHRR imagery—A timeline thematic processor. *Remote Sens.* **2018**, *10*, 341. [CrossRef]
66. Jaccard, P. The distribution of the flora in the alpine zone. *New Phytol.* **1912**, *11*, 37–50. [CrossRef]
67. Boer, M.M.; de Dios, V.R.; Bradstock, R.A. Unprecedented burn area of Australian mega forest fires. *Nat. Clim. Chang.* **2020**, *10*, 171–172. [CrossRef]
68. Stocks, B.J.; Fosberg, M.; Lynham, T.; Mearns, L.; Wotton, B.; Yang, Q.; Jin, J.; Lawrence, K.; Hartley, G.; Mason, J.; et al. Climate change and forest fire potential in Russian and Canadian boreal forests. *Clim. Chang.* **1998**, *38*, 1–13. [CrossRef]
69. Keeley, J.E. Fire intensity, fire severity and burn severity: A brief review and suggested usage. *Int. J. Wildland Fire* **2009**, *18*, 116–126. [CrossRef]
70. Parks, S.A.; Holsinger, L.M.; Panunto, M.H.; Jolly, W.M.; Dobrowski, S.Z.; Dillon, G.K. High-severity fire: Evaluating its key drivers and mapping its probability across western US forests. *Environ. Res. Lett.* **2018**, *13*, 044037. [CrossRef]
71. United States Geological Survey (USGS). Global Ecosystems Data. 2021. Available online: https://www.usgs.gov/centers/gecsc/science/global-ecosystems-data?qt-science_center_objects=0#qt-science_center_objects (accessed on 4 May 2021).
72. Hammill, K.A.; Bradstock, R.A. Remote sensing of fire severity in the Blue Mountains: Influence of vegetation type and inferring fire intensity. *Int. J. Wildland Fire* **2006**, *15*, 213–226. [CrossRef]
73. Escuin, S.; Navarro, R.; Fernandez, P. Fire severity assessment by using NBR (Normalized Burn Ratio) and NDVI (Normalized Difference Vegetation Index) derived from LANDSAT TM/ETM images. *Int. J. Remote Sens.* **2008**, *29*, 1053–1073. [CrossRef]
74. Chen, X.; Vogelmann, J.E.; Rollins, M.; Ohlen, D.; Key, C.H.; Yang, L.; Huang, C.; Shi, H. Detecting post-fire burn severity and vegetation recovery using multitemporal remote sensing spectral indices and field-collected composite burn index data in a ponderosa pine forest. *Int. J. Remote Sens.* **2011**, *32*, 7905–7927. [CrossRef]
75. Tran, B.N.; Tanase, M.A.; Bennett, L.T.; Aponte, C. Evaluation of spectral indices for assessing fire severity in Australian temperate forests. *Remote Sens.* **2018**, *10*, 1680. [CrossRef]
76. Mathews, L.E.; Kinoshita, A.M. Urban Fire Severity and Vegetation Dynamics in Southern California. *Remote Sens.* **2021**, *13*, 19. [CrossRef]
77. Storey, E.A.; Lee West, K.R.; Stow, D.A. Utility and optimization of LANDSAT-derived burned area maps for southern California. *Int. J. Remote Sens.* **2021**, *42*, 486–505. [CrossRef]
78. Chafer, C.J. A comparison of fire severity measures: An Australian example and implications for predicting major areas of soil erosion. *Catena* **2008**, *74*, 235–245. [CrossRef]
79. Collins, L.; Griffioen, P.; Newell, G.; Mellor, A. The utility of Random Forests for wildfire severity mapping. *Remote Sens. Environ.* **2018**, *216*, 374–384. [CrossRef]
80. Collins, L.; McCarthy, G.; Mellor, A.; Newell, G.; Smith, L. Training data requirements for fire severity mapping using Landsat imagery and Random Forest. *Remote Sens. Environ.* **2020**, *245*, 111839. [CrossRef]
81. United States Geological Survey/USGS. USGS EROS Archive—Advanced Very High Resolution Radiometer (AVHRR)—Sensor Characteristics. 2020. Available online: https://www.usgs.gov/centers/eros/science/usgs-eros-archive-advanced-very-high-resolution-radiometer-avhrr-sensor?qt-science_center_objects=0#qt-science_center_objects (accessed on 11 August 2021).

remote sensing

MDPI

Article

The Surface Velocity Response of a Tropical Glacier to Intra and Inter Annual Forcing, Cordillera Blanca, Peru

Andrew Kos [1,*], Florian Amann [2], Tazio Strozzi [3], Julian Osten [2], Florian Wellmann [4], Mohammadreza Jalali [2] and Anja Dufresne [2]

1 Terrasense Switzerland Ltd., 9470 Buchs, Switzerland
2 Department of Engineering Geology and Hydrogeology, RWTH University, 52062 Aachen, Germany; amann@lih.rwth-aachen.de (F.A.); julian.osten@rwth-aachen.de (J.O.); jalali@lih.rwth-aachen.de (M.J.); dufresne@lih.rwth-aachen.de (A.D.)
3 Gamma Remote Sensing Ltd., 3073 Gümligen, Switzerland; strozzi@gamma-rs.ch
4 Unit of Computational Geoscience & Reservoir Engineering, RWTH University, 52062 Aachen, Germany; florian.wellmann@cgre.rwth-aachen.de
* Correspondence: andrew.kos@terrasense.ch

Abstract: We used synthetic aperture radar offset tracking to reconstruct a unique record of ice surface velocities for a 3.2 year period (15 January 2017–6 April 2020), for the Palcaraju glacier located above Laguna Palcacocha, Cordillera Blanca, Peru. Correlation and spatial cluster analysis of residuals of linear fits through cumulative velocity time series, revealed that velocity variations were controlled by the intra-annual outer tropical seasonality and inter-annual variation in Sea Surface Temperature Anomalies (SSTA), related to the El Niño Southern Oscillation (ENSO). The seasonal signal was dominant, where it was sensitive to altitude, aspect, and slope. The measured velocity variations are related to the spatial and temporal variability of the glacier's surface energy and mass balance, meltwater production, and subglacial water pressures. Evaluation of potential ice avalanche initiation areas, using deviations from linear long-term velocity trends, which were not related to intra- or inter-annual velocities, showed no evidence of imminent avalanching ice instabilities for the observation period.

Keywords: SAR offset tracking; glacier surface velocity; glacier instability; glacier hazards; ice avalanches; ENSO; glacier mass balance; glacier surface energy

Citation: Kos, A.; Amann, F.; Strozzi, T.; Osten, J.; Wellmann, F.; Jalali, M.; Dufresne, A. The Surface Velocity Response of a Tropical Glacier to Intra and Inter Annual Forcing, Cordillera Blanca, Peru. *Remote Sens.* **2021**, *13*, 2694. https://doi.org/10.3390/rs13142694

Academic Editors: Paolo Mazzanti and Saverio Romeo

Received: 10 June 2021
Accepted: 6 July 2021
Published: 8 July 2021

1. Introduction

The Cordillera Blanca (CB) is the most extensively glaciated mountain range in Peru [1] where the present-day climate is characterized by a distinct seasonality, with a dry season from May to September and a wet season from October to April. The seasonality is controlled by the oscillation of the inner-tropic convergence zone [2]. Seasonal temperature variations are small but vary significantly on a daily basis [2]. Wet season snow accumulation at high elevation and meltwater from glaciers are the main source of water [3] for the arid west coast of Peru during the dry seasons. Over the course of the 20th century, glacier mass has significantly contracted (e.g., [1,4–6]). Glacier volume loss may influence the stability state and dynamics of a glacier, as a result of a change of the thermal regime, stress state at the bed-ice interface, and associated changes in geometry. Numerous studies suggest that variations in glacier velocities are related to subglacial water pressures, which are controlled by seasonal meltwaters or infiltration following heavy rainstorms [7–14]. The supply of subglacial water is regulated by the extent of Firn coverage [15], as well as the characteristics of the subglacial drainage system. Pressurized subglacial water reduces the effective normal stress at the bed-ice interface, leading to reduced frictional resistance, and thus sliding rates are enhanced [11,15–17]. Detailed observational changes in a glacier's dynamics as well as its geometry, provide insights into whether a potentially hazardous glacier may develop a critical stability state, or conversely remain harmless [8].

2. Materials and Methods

2.1. SAR Offset Tracking

The calculation of displacement fields using Synthetic Aperture Radar (SAR) data is possible using offset tracking methods. Offset tracking is normally used for analyzing rapidly moving objects, such as glaciers, because it overcomes the loss of signal, due to decorrelation, when using InSAR techniques over long time intervals [18–21]. As the glaciers in the Cordillera Blanca of Peru are relatively small, offset tracking is only feasible using high-resolution SAR images with a resolution of ~2–3 m. Using a unique, continuous time series of archived TerraSAR-X (TSX) radar imagery for the period 15 January 2017–6 April 2020 (3.2 years) we reconstructed ice surface velocities for the Palcaraju glacier. The dataset consists of 51 TSX radar images that were acquired on a descending orbit with a ground resolution of 3 m (Table 1). The radar images were acquired at 10.53 GMT and the distribution of re-visit intervals for the 50 image pairs was 11 days (2 pairs), 22 days (43 pairs), 33 days (1 pair), and 44 days (4 pairs).

Table 1. TSX SAR offset tracking displacement maps time intervals, perpendicular baseline, and resolution.

Scene	Image Pair	Date	Days	Perpendicular Baseline	Resolution
1		6 April 2020			3 m
2	1	15 March 2020	22	74.4	3 m
3	2	31 January 2020	44	1.5	3 m
4	3	9 January 2020	22	−146.8	3 m
5	4	18 December 2019	22	104.7	3 m
6	5	26 November 2019	22	−112.4	3 m
7	6	24 October 2019	33	−44.6	3 m
8	7	02 October 2019	22	136.4	3 m
9	8	10 September 2019	22	−137.0	3 m
10	9	19 August 2019	22	203.5	3 m
11	10	28 July 2019	22	−125.7	3 m
12	11	14 June 2019	44	46.0	3 m
13	12	23 May 2019	22	−14.1	3 m
14	13	1 May 2019	22	29.9	3 m
15	14	9 April 2019	22	−4.2	3 m
16	15	29 March 2019	11	−103.0	3 m
17	16	07 March 2019	22	6.1	3 m
18	17	13 February 2019	22	−78.5	3 m
19	18	22 January 2019	22	174.7	3 m
20	19	31 December 2018	22	−88.2	3 m
21	20	09 December 2018	22	84.8	3 m
22	21	17 November 2018	22	−25.8	3 m
23	22	26 October 2018	22	71.4	3 m
24	23	4 October 2018	22	−222.0	3 m
25	24	12 September 2018	22	280.1	3 m
26	25	21 August 2018	22	−164.2	3 m
27	26	30 July 2018	22	148.7617	3 m
28	27	8 July 2018	22	26.9	3 m
29	28	16 June 2018	22	−17.8	3 m
30	29	25 May 2018	22	4.0	3 m
31	30	3 May 2018	22	−66.2	3 m
32	31	11 April 2018	22	106.8	3 m
33	32	20 March 2018	22	−220.7	3 m
34	33	26 February 2018	22	−65.0	3 m

Table 1. *Cont.*

Scene	Image Pair	Date	Days	Perpendicular Baseline	Resolution
35	34	13 January 2018	44	133.6	3 m
36	35	2 January 2018	11	−149.4	3 m
37	36	11 December 2017	22	30.0	3 m
38	37	19 November 2017	22	−74.9	3 m
39	38	28 October 2017	22	169.1	3 m
40	39	14 September 2017	44	−5.5	3 m
41	40	23 August 2017	22	246.2	3 m
42	41	1 August 2017	22	133.1	3 m
43	42	10 July 2017	22	−206.8	3 m
44	43	18 June 2017	22	182.6	3 m
45	44	27 May 2017	22	−90.0	3 m
46	45	5 May 2017	22	−87.2	3 m
47	46	13 April 2017	22	107.1	3 m
48	47	22 March 2017	22	139.4	3 m
49	48	28 February 2017	22	−111.8	3 m
50	49	6 February 2017	22	−297.0	3 m
51	50	15 January 2017	22	291.0	3 m

Utilizing a normalized cross-correlation of the amplitude component of the SAR images, offsets were measured with rectangular windows at a set of positions uniformly distributed over the scene. To obtain an accurate estimate of subpixel precision of the correlation peak, correlation function values were fitted using a biquadratic polynomial surface. The time interval of the image pairs can be adjusted according to the expected maximum displacement over the glaciers from 11 days to several months. Mismatches or errors were filtered by applying a threshold to the correlation coefficient [22], by (1) iteratively discarding spurious matches based on the angle and size of displacement vectors in the surrounding areas, (2) applying a low-pass filter on the resultant displacement fields, and (3) applying a 2–98% cumulative cutoff to remove potentially uncertain velocity values and outliers. Such values cannot be realistically validated and are likely to be artifacts of the radar data processing method. Hence they have no physical meaning.

Slant range and azimuth offset displacement fields were geocoded and transformed to 3D displacements along the terrain surface using the Advanced Land Observing Satellite (ALOS) World 3D (AW3D30), Digital Elevation Model (DEM) [23]. TSX images were processed in series with offset-tracking procedures [18–21] to 3D ice surface displacement maps. This involved combining slant-range and azimuth offsets by assuming that flow occurs parallel to the ice surface, as estimated from the DEM. Matching window sizes of 128 × 96 pixels (e.g., 202 m × 192 m) were applied with steps of 16 × 12 pixels (e.g., 25 m × 24 m). The displacement maps in m/a were geocoded to a posting of ca. 60 m. For an estimation of the uncertainties in the ice displacement maps, a precision of 1/10th of a pixel in the offset estimation can be assumed [24,25]. The displacement error of TSX data with pixel sizes in ground-range and azimuth direction of 0.9 m × 2.0 m respectively, and a time interval of 11 days, are thus on the order of 10 m/year. For longer time intervals, the noise level increases, hence, a similar displacement error of about 10 m/year can be assumed for other image pairs. The displacement values were extracted from individual pixels from the spatial displacement maps.

2.2. Satellite Image Analysis

Surface features and glacier extent were mapped in a geographic information software system [26] using high-resolution satellite Pléiades imagery (Table 2). In addition, we visually inspected Google Earth and ESRI World Atlas imagery (Table 2) captured between 1999 and 2020. Aspect and slope were calculated using the AW3D30 DEM, and, together with velocity and elevation data, exported with one data point per pixel for further statistical analyses.

Table 2. Satellite imagery analyzed in this study.

Image Capture Date	Image Source	Resolution (m)	Accuracy (m)
3 July 2020	Google Earth		
24 February 2020	Pleiades		
22 June 2019	Google earth		
08 June 2019	Pleiades		
24 June 2018	Maxar/DigitalGlobe (WV02) *	0.5	8.47
18 June 2017	Google Earth		
30 May 2016	Google Earth		
30 May 2016	Pleiades		
01 June 2016	DigitalGlobe (WV02) *	0.5	10.2
28 April 2013	Google Earth		
28 July 2013	Google Earth		
5 May 2012	Google Earth		
15 January 1999	Earthstar Geographic (Terracolor) *	15	50
31 December 1985	Google Earth		
1 January 1970	Google Earth		

* as published on https:\livingatlas.argis.com/wayback, accesed on 8 July 2021.

2.3. *Statistical Analysis of Time Series*

The residuals from a linear fit of the glacier surface velocities over the 3.2 year observation period were used to conduct various correlation analyses. The analyses were based on about 1300 measurement points. Data time series with less than 50% completeness were removed from the statistical analysis. For the remaining data, where gaps were evident, a linear interpolation was applied to ensure consistent time intervals. The filtered and interpolated data allowed us to perform spatial and temporal correlation analysis between the residuals, precipitation, and SST. The Pearson correlation coefficient was determined for all correlation analyses.

As an independent test, the processed time series were also subjected to a time series analysis, without a prior assumption of correlations with climate or weather data. The entire data processing workflow was performed in Python using standard packages for transformation and optimization. The dominant wavelength was extracted for each time series using a Fourier transformation. From this dominant signal, a fit to the time series was performed with a Levenberg-Marquardt or dogleg algorithm, resulting in a single fitted trigonometric function with an estimate of best-fitting values for amplitude, phase, offset, and period, as well as the covariance matrix. Physically unreasonable outliers were removed based on the following criteria:

- Period > 2000 days (beyond the observation time; no periods below Nyquist frequency)
- Amplitude <–20 and >20 (beyond the data range)
- Phase: no obvious outliers
- Offset <–4 and >4 (beyond the data range)

In total, 28 data points were excluded based on the above criteria.

2.4. *Cluster Analysis of the Time Series*

We employed a Bayesian unsupervised machine learning approach [27], to extract clusters of similar behavior from our surface displacement variations (i.e., residuals). The primary aim of the employed method was to extract clusters from the data based on feature similarity, with consideration of the spatial configuration in the physical space. This combination is enabled through a combination of a Gaussian Mixture Model (GMM) in feature space with a Hidden Markov Random Field (HMRF) model in physical space. Both models are integrated into a Bayesian model, and MCMC sampling is performed to obtain the posterior distribution of the GMM parameters.

The GMM was applied to samples of N data points (image pixels) in M-dimensional feature space to determine a set of multivariate Gaussian distributions (L classes). The distributions were parametrized by their mean, μ, and covariance, Σ for each cluster. As features originate from spatial data sets, the HMRF was then used to consider spatial dependencies of sampling points, with a smoothing coefficient, which parameterized the strength of spatial correlation, independently for each class.

The model parameters (μ, Σ, β), as well as the latent field values, were obtained through a Bayesian optimization, iterative sampling process using a Markov Chain Monte Carlo (MCMC) approach, following an initial Expectation-Maximization (EM) step. The final association of sampling points to clusters were obtained from the Maximum a posteriori (MAP) distribution. In addition, cluster assignment probabilities were obtained from the MCMC chain and combined using information entropy, to obtain a spatial estimate of uncertainty [28], for the assigned cluster values.

The segmentation procedure is described in Wang et al. [27] and has previously been applied to geophysical [29] and arctic sea ice [30] data sets. It is implemented in the open-source Python package bayseg, available on https://github.com/cgre-aachen/bayseg, accessed on 10 June 2021. The clustering workflow is presented in an exemplary form in the supplementary information section (Figure S1a–d).

3. Results

3.1. Spatio-Temporal Distribution of Glacier Velocities

We calculated the spatial distribution of average daily surface velocities of the Palcaraju glacier over 50 measurement intervals (i.e., velocity maps with 1300 pixels per interval), for the 3.2 year observation period. As the spatial distribution of individual velocity maps were similar for all 50 intervals, the velocity distribution shown in Figure 1 is an appropriate representation for the observation period. Daily surface velocities averaged over 3.2 years ranged between 0.01 and 0.47 m/day, where both their spatial distribution and magnitude are consistent with results obtained from Sentinel-2 data [31]. The displacement time series, representing the residuals of the linear fit and cumulative surface displacements, for selected points shown in Figure 1, are available in the supplementary materials (Figures S2–S10).

The resulting surface velocities are a combination of basal sliding as well as creep and ablation of the glacier, which cannot be individually quantified without independent measurements. Such measurements are outside the scope of this study. Values from the literature [13] and field measurements [7,9,32] suggest that the contribution of basal sliding to the total surface velocities varies between 50–90%, and is spatially variable across the glacier bed [32].

Higher mean annual surface velocities are observed in areas with a high frequency of crevassing and or steep glacier topography (>30–40°) (see zones S2, S4a, S4b, E1, E2, and E3 Figures 1a and 2). Based on the occurrence of ridgelines and terrain steps in the 30 m DEM, we infer that steep glacier sections directly reflect the underlying topography of the glacier bed. Although the slope gradient in Zone S4a (Figures 1a and 2) is comparatively flat (20–35°), such higher velocity rates can be related to larger ice thickness at this location [9,33].

West exposed slopes (Figure 3) typically showed the highest mean annual surface velocities, followed by south and east exposed slopes over the 3.2 year observation period. Surface velocities of south exposed slopes tend to decrease with altitude, while surface velocities of East exposed slopes increase with altitude. West exposed slopes also showed a general increase with altitude, however, the trend is reversed at an elevation range of approximately 5400–5700 m.a.s.l.

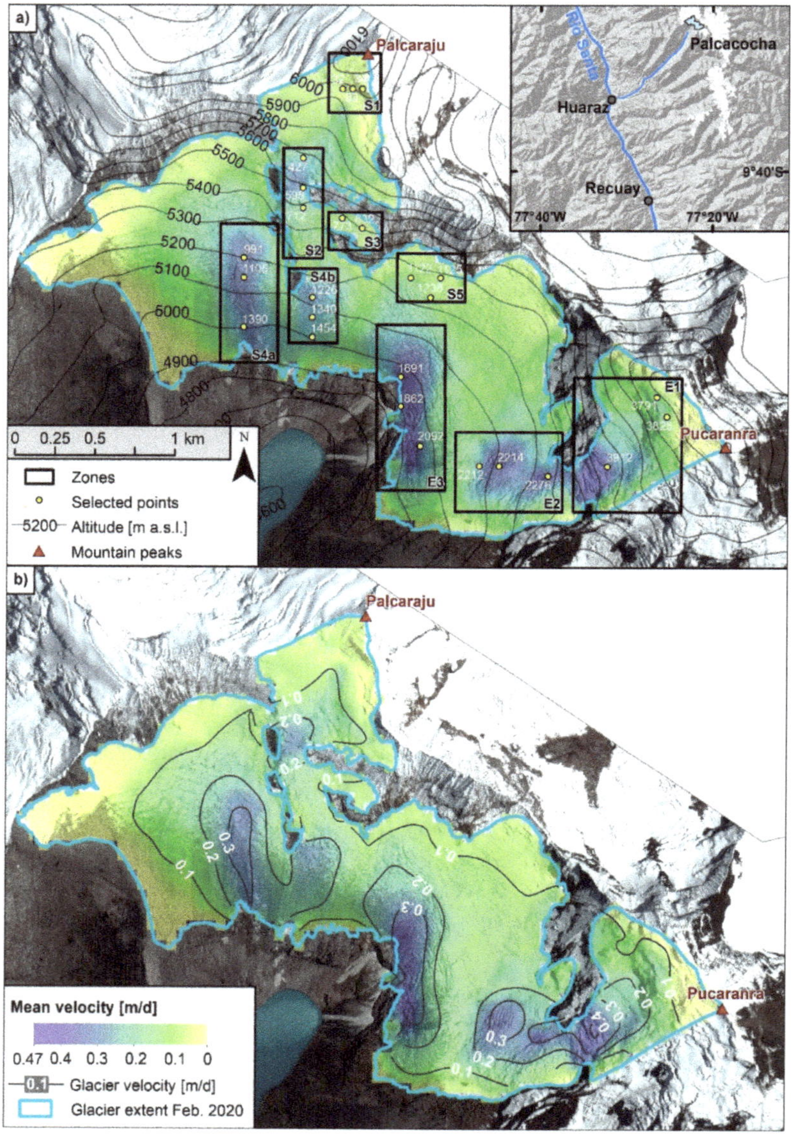

Figure 1. (**a**) Spatial distribution of mean daily surface velocities with respect to elevation, showing the zones and point selections used in this study (for location see insert map), (**b**) Spatial distribution of contoured mean daily surface velocities on the Palcaraju glacier. Elevation range of zone E1 is 5′400–6050 m a.s.l, and 5′900–6150 m a.s.l for zone S1.

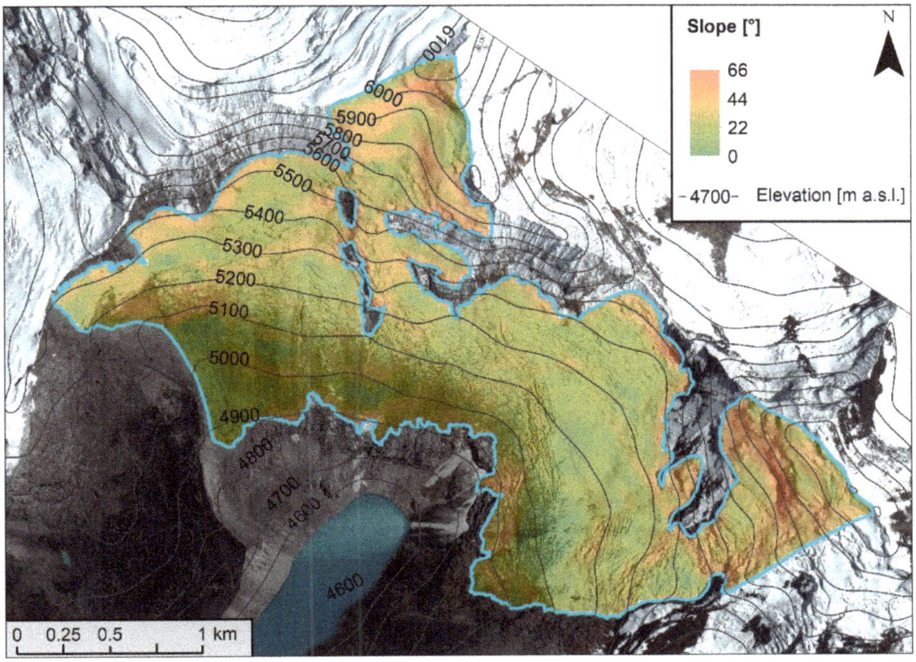

Figure 2. Glacier slope map.

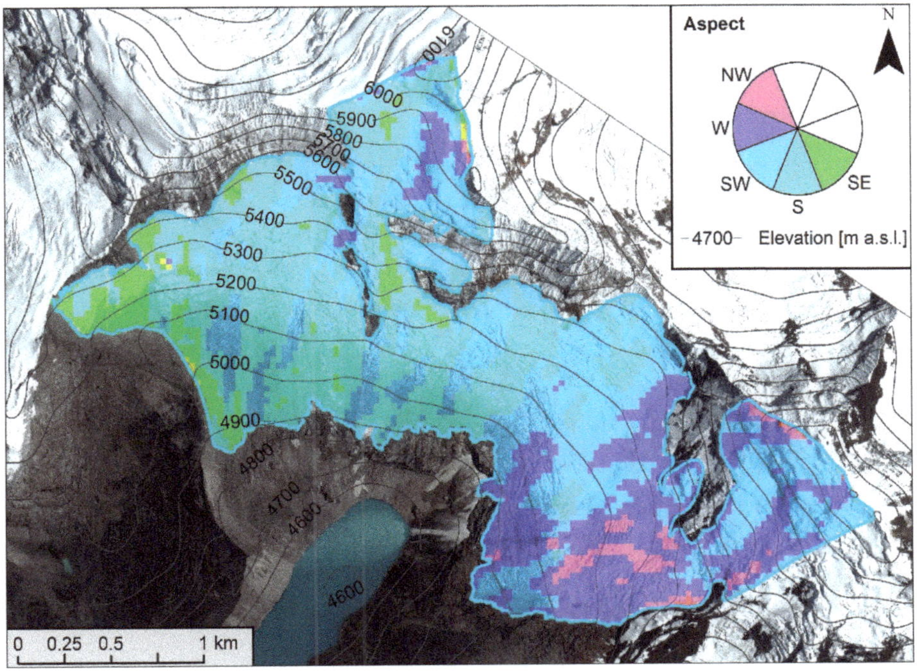

Figure 3. Glacier aspect map.

Zones with higher surface velocities for all slope aspects were only found below 5700 m a.s.l., suggesting that the steep hanging glacier in E1 and S1 (Figures 1 and 3) are cold glaciers [34], likely to be frozen to the glacier bed. This is supported by estimates of the cold-temperate transition line at approximately 5500 m a.s.l., elevation in the region [35]. Hence, in these areas, the surface velocity field mainly represents creep deformation within the ice body. Below 5700 m a.s.l., the glacier is likely to be polythermal to temperate (i.e., subglacial liquid water is temporarily present), and therefore the measured surface velocities are likely controlled by creep processes, basal sliding along the ice-bed interface, and an indeterminable error due to ablation (e.g., a shift in the slant range geometry between image acquisitions and consequent projection into DEM predating the radar image acquisition, resulting in an over-estimation of surface velocities).

3.2. Velocity Variations

Linear regression analysis (see methods section) reveals a coefficient of determination $R^2 > 0.90$–0.99 for all cumulative time series (Figure 4), indicating steady surface movements with no obvious acceleration phases within the observation period. Approximately 2.6% of the analyzed time series showed an R^2 ranging between 0.90–0.96, however, this is primarily due to under-sampling of the time series where data points were missing (Figure 5).

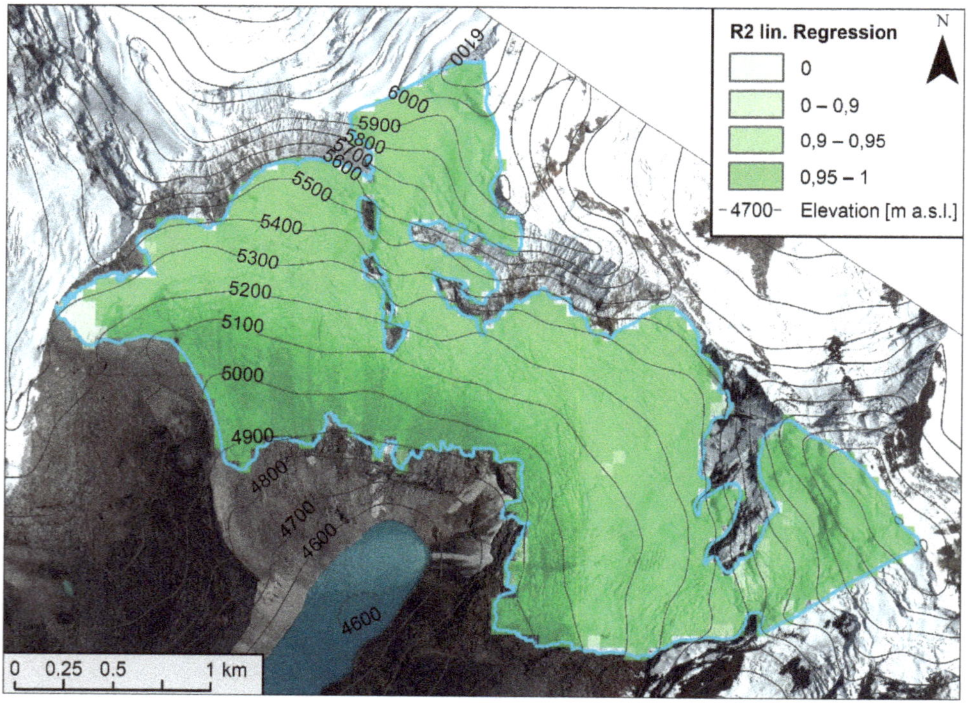

Figure 4. Spatial distribution of coefficient of determination (R^2) from the linear correlation analysis.

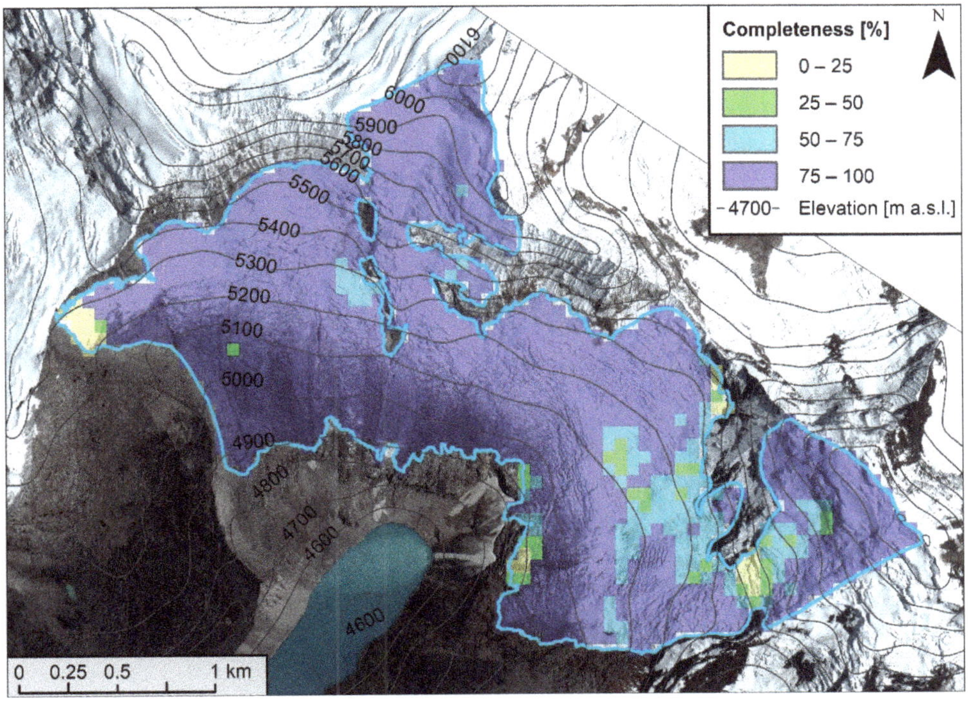

Figure 5. Spatial distribution of sampling completeness.

Extracted residuals of the time series from linear fitting revealed that the residuals of the glacier surface velocities fluctuate around the long-term (3.2 years) linear trend. Spatial correlation and cluster analysis of the residuals shows that these variations in surface velocity have a spatial and temporal variability with respect to zones in the central, eastern, and western parts of the glacier, as well as zones above 5900 m a.s.l. (Figure 6).

Surface energy and glacier mass balance models (SEMB) across the glaciated Andes [1,36,37] collectively show that the glacier response has a strong spatial variability due to its sensitivity to regional and local meteorological and topographic factors [2,37,38]. In this connection, the observed spatio-temporal variation in our glacier velocity residuals may be interpreted as being closely linked to meteorological and topographical variables, which in turn influence the variability of the glacier's surface energy, mass balance, meltwater production, and effective stress states at the glacier bed.

3.3. Spatial Correlation and Cluster Analysis

A spatial correlation and cluster analysis following the procedure described in Sections 2.3 and 2.4 were applied to elucidate details of the factors influencing the spatio-temporal velocity patterns in different parts of the glacier (Figure 7). It is important to note that this analysis is performed independently of the previous correlation evaluations and is performed directly on the fitted time series as an additional investigation using an unsupervised machine learning algorithm. From these analyses, three clusters, each with a distinct behavior, were identified from the residuals (Figure 6b). These clusters may be considered as two end-members; an intra-annual and inter-annual cluster, with a transition cluster between the two (Figure 7c), revealing a behavior similar to the results of the correlation analysis (Figure 7a,b).

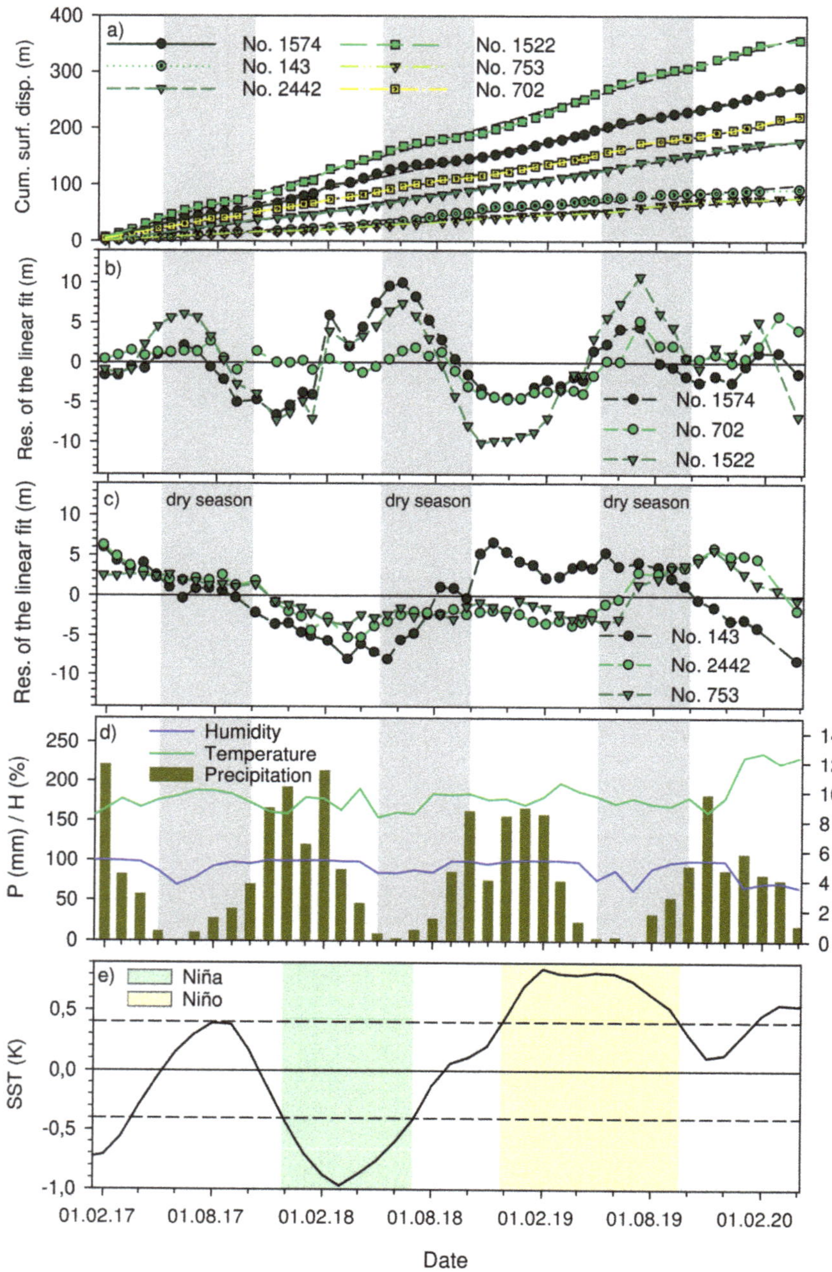

Figure 6. (**a**) Cumulative surface velocities from six representative points, (**b**) Residuals of the linear fit displaying an intra-annual (seasonal) pattern, from three representative points above 5500 m a.s.l. (No. 143), in the East (No. 2442), and West (No. 752) of the glacier, (**c**) Residuals of the linear fit displaying an inter-annual pattern, from three representative points for the glacier response in the central, south-exposed part of the glacier (Nos. 702, 1522, 1574), (**d**) meteorological data; Temperature T, Humidity H, Precipitation P (Palcacocha weather station, altitude 4607 m.a.s.l., latitude 09°24′09.3211″S, 77°23′07.0258″, Instituto Nacional de Investigación en Glaciares y Ecosistemas de Montaña INAIGEM), and (**e**) Sea Surface Temperature Anomalies (SSTA) shifted by 3 months (see main text) for the NOAA Niño 3.4 region.

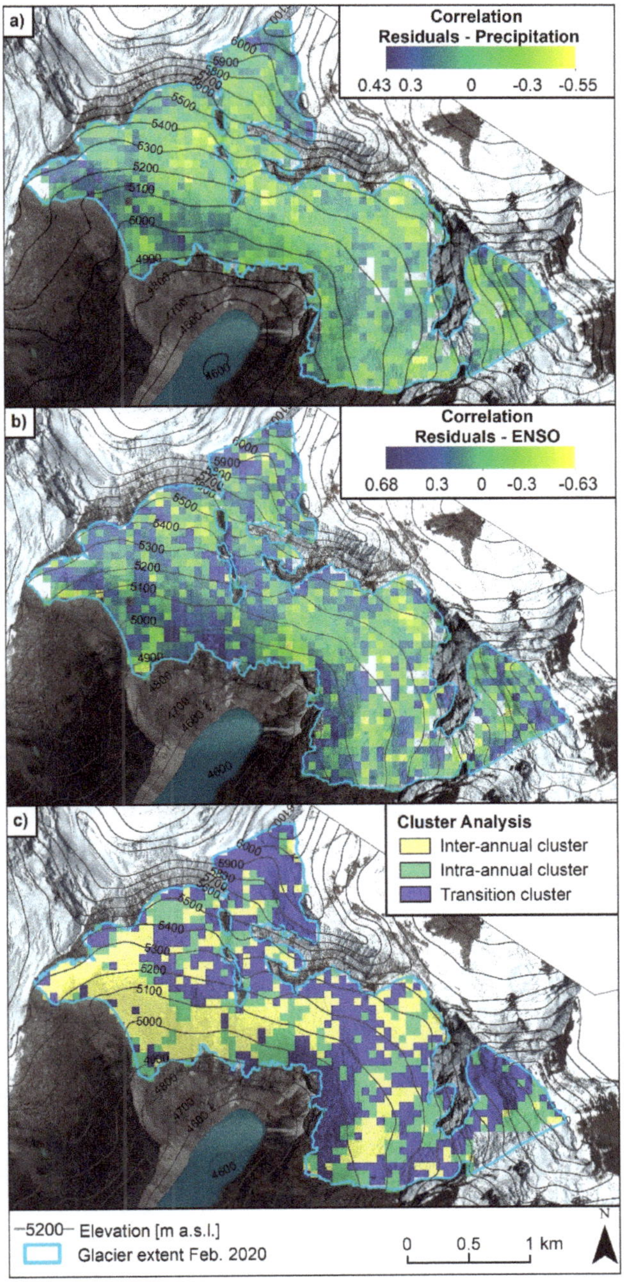

Figure 7. (**a**) Map of correlation coefficients between residuals of the long-term linear fit through the velocity time series and precipitation, measured from the Palcacocha weather station, altitude 4607 m.a.s.l., latitude 09°24′09.3211″S, 77°23′07.0258″, INAIGEM, (**b**) Map of correlation coefficients between residuals of the long-term linear fit through the velocity time series and SST, (**c**) Results of the cluster analysis show the distribution of intra, inter and transitional clusters.

4. Discussion

4.1. Intra-Annual Dominant Cluster

The velocities belonging to the intra-annual cluster are found below 5900 m.a.s.l., on slopes with flat to medium steepness (<35°). The slopes are typically south-southeast and steeply southwest exposed. On relatively flat southwest and steep southeast exposed glacier slopes, at altitudes above 5900 m.a.s.l., the intra-annual signal weakens and becomes dominated by a pattern characteristic of the inter-annual cluster.

This cluster is characterized by glacier surface velocities that are in phase with the outer tropical seasons. The velocities generally peak in the dry season and subsequently reach a minimum in the following wet season (Figure 6b). Meteorological data from the region provided by INAIGEM (Palcacocha weather station 4607 m.a.s.l.) show relatively constant mean monthly temperatures and a distinct seasonal fluctuation in monthly precipitation (Figure 6d). Statistical analysis of the residuals from this cluster shows an anti-correlation with monthly precipitation. Higher precipitation during the wet season corresponds to lower surface velocities and vice versa. Palcaraju glacier is predominantly located above the Equilibrium Line Altitude (ELA) (i.e., 4850–4950 m.a.s.l.) where precipitation during the wet season falls as snow [38], and a temperature-sensitive increase of the snowline occurs towards the dry season. For a slightly different aspect (e.g., ~20°), but similar altitude at the Shallap glacier, (i.e., a glacier about 10 km south of the Palcaraju glacier), Gurgiser et al. [38] found surface Albedo up to 80% during the wet season and a significantly reduced Albedo during the dry season. Annual differences in the gross glacier mass balance were solely related to ablation below the ELA, which were driven by net shortwave radiation in both seasons (with stronger surface energy fluxes during the dry season), the temperature-affected altitude of the snow line, and snow quantity in the wet season (i.e., reduced albedo).

Above the ELA, the mass balance was primarily sensitive to annual cumulative precipitation. Gurgiser et al. [38] suggested that above 5000 m a.s.l. seasonal fluctuations of surface and subsurface melting may change gradually with time and altitude between the seasons. They associate this with gradual changes in the snow line, precipitation sum and type, snow cover, snow age, albedo conditions, and the net shortwave radiation budget. High ablation rates during the dry season coincide with the shortwave budget and surface/subsurface melting peaks. Thus, our surface velocity variations in the intra-annual cluster correspond with seasonal variations in ice-melt water production, which in turn modifies subglacial pressure conditions and basal motion.

4.2. Inter-Annual Dominant Cluster

The inter-annual cluster is characterized by glacier surface velocity variations that are most likely influenced by factors controlling subglacial water pressure conditions over a longer temporal scale. We compared the Sea Surface Temperature (SST) curve for the Niño 3.4 region (National Oceanic and Atmospheric Administration, NOAA) to the residuals of our surface velocity time series. Similar to Maussion et al. [39] and Francou et al. [40], we shifted the SST time series by three months to account for the time lag in the glacier mass balance response, and only consider $0.4\ K \leq SST \leq -0.4\ K$, with duration > 6 months [36]. Figure 6e shows the SST curve for the observation period. From January to December 2017 the SST intensity and duration do not qualify as either a La Niña or El Niño event. Between December 2017 and July 2018, the SST intensity and duration qualify this period as a La Niña event, which is followed by an El Niño event commencing in December 2019, extending through to September 2020. The duration of SST > 0.4 K from 20 January, to 15 June 2020, does not qualify as an El Niño event. We note that prior to our observation period, SST was 1.5 K for more than 6 months indicating a strong El Niño event in 2016 (not shown in Figure 6e).

We find that the observed Sea Surface Temperature Anomalies (SSTA) show a strong correlation ($r > 0.5$, $p < 0.05$) with the residuals obtained from surface velocity time series in the inter-annual cluster (Figure 6c,e). The general decrease in surface velocities at the beginning of the time series corresponds to insignificant SST variations verging into a La Niña event starting in December 2017. From March 2017, SST increases gradually. This is associated with a gradual increase in surface velocities and corresponds to the subsequent El Niño (December 2019–September 2020). Maussion et al. [39] used an SEB/SMB model calibrated against a 4-year temperature time series obtained on the Shallap glacier. Reanalysis of downscaled atmospheric variables were used to estimate the monthly SEB/SMB for the period 1980–2013. Correlation analysis revealed a strong anti-correlation ($r < -0.5$, $p < 0.05$) between SSTA and individual fluxes of the SEB/SMB. During an El Niño event, the correlation was attributed to increased air temperatures, leading to an elevated snowfall line, an increase in short-wave radiation, and reduced precipitation [39]. At higher altitudes, the ENSO correlation was found to be weaker and mostly related to changes in total precipitation. Therefore, during an El Niño event, meltwater production is strongly enhanced, where subglacial water pressure conditions influence glacier surface velocities, concomitant to the approximate duration of the El Niño event.

Our results also show that steep slopes with both East and West orientations, as well as relatively flat areas (regardless of altitude), are prone to inter-annual glacier surface velocity changes as a result of ENSO. This is supported by the findings of Kaser and Georges [2], where regional-scale differences in the ELA between E- and W-facing glaciers were observed in the CB. They found a zonal asymmetry caused by convective cloud development in the afternoon, which influences the radiation balance and ablation. In the morning, east exposed slopes received a high amount of shortwave solar radiation, whilst during the day, clouds develop, and west exposed glaciers receive a significantly lower amount of shortwave solar radiation. Further, Gurgiser et al. [38] found a topographic signature in surface energy and mass balance above 5000 m a.s.l at the Shallap glacier, where the combination of aspect and increased slope steepness results in a smaller incidence for solar radiation and a slower decrease in Albedo.

4.3. Implications for Avalanching Glacier Instabilities

Based on our continuous time series of surface velocities, and high-resolution satellite imagery we evaluated whether avalanching glacier instabilities were evident on the Palcaraju glacier for our period of observation. Various authors have suggested a high glacier lake outburst flood (GLOF) susceptibility at Laguna Palcacocha due to the potential of avalanching glaciers, that could induce an impulse wave and result in moraine overtopping [41–43]. A GLOF susceptibility approach, such as those of Wang et al. [44] and Bolch et al. [45] were applied by Emmer and Vilimek [46] for Laguna Palcacocha. Those first-order assessments may only be used as a basis for prioritizing more comprehensive hazard assessments at the local scale [44,45], and require region-specific adjustments to the criteria used [46]. Somos-Valenzuela et al. [42], and Frey et al. [41] defined hazard scenarios for modelling the expected hazard cascade at Laguna Palcacocha. The characteristics of the ice avalanche scenarios, for example, volume and area of initiation, were arbitrarily selected, and lack the detailed analysis concerning the glacier state and dynamics, which are required to understand ice avalanche potential. The scenario definition, in terms of volume for the nearby Laguna 513 can be traced back to those reported by Schneider et al. [47]. The ice avalanche scenarios, selected for the Palcacocha GLOF modeling are therefore poorly constrained, and may ultimately lead to a misrepresentation of the GLOF risk (i.e., particularly when considering pessimistic or worst case hazard scenarios).

We focus our evaluation of potential areas of ice avalanche initiation by identifying deviations from the linear long-term velocity trends that are not related to intra or inter-annual velocities (e.g., indications for an imminent glacier instability), as well as qualitative features of the glacier from high-resolution satellite imagery. The evaluation of ice avalanching potential is primarily concerned with the characteristics of the (thermal) contact between a glacier and the bedrock [48–50] as well as the type of starting zones recognized as being important for their initiation [49]. Type I starting zones have a relatively large area of bedrock with an almost constant slope, where glacier stability depends upon altitude [48,49], and the proportion of the glacier frozen to the bedrock [50]. Failure of cold glaciers (Type IA), which are those that are frozen to the glacier bed, are primarily controlled by intra-glacial rupture processes, and typically occur on bedrock inclinations > 45° [8,50,51]. Detachment in steep, cold glaciers is associated with crevasse formation and accelerating surface velocities. For polythermal or temperate glaciers (Type IB), the bedrock inclination for starting zones are >25°, where several factors, such as adhesion to the glacier bed, proportion of glacier frozen to the bed, lateral support, and subglacial water pressure influence glacier stability [8,49,50]. Type II starting zones are associated with abrupt changes in bedrock inclination [49], where glaciers may develop a steep cliff, and unstable ice slabs form parallel to the cliff.

At the Palcaraju glacier, Type IA starting zones are expected to occur at altitudes above ~5500 m a.s.l. with a slope angle > 45°. On the Palcaraju glacier, those areas include: (1) the South face of the Palcaraju summit, which is a hanging glacier suggested as a potential ice avalanche scenario by Vilimek et al. [43] (Figure 1, zone S1), (2) the upper part of the steep glaciated ridge in zone S2 (Figure 1), and (3) the hanging glaciers on the SW-slope of the Pucaranra West face (Figure 1, zone E1). At these locations, satellite imagery indicates evidence for relatively small snow avalanches and/or ice falls, some of which reach the relatively flat glacier area above ~5200 m a.s.l. The time series of surface velocities from these locations all show a linear trend ($R^2 > 0.97$), and the scale of acceleration/deceleration phases of the glacier correspond to seasonal dominated oscillations in S2 and inter-annual dominated oscillations in zone S1 and E2.

The steep (up to 50°) glaciated ridge between 5200 and 5750 m a.s.l. in zones S2, show mean daily surface velocities of >0.3 m/d. Mean surface velocities of 0.2–0.3 m/d are solely found above 5500 m (Figure 1). Below 5′500 m a.s.l., where the glacier is likely to be polythermal to temperate, daily surface velocities decrease to >0.2 m/d. All velocity time series in this region exhibit a linear trend ($R^2 > 0.98$), and acceleration/deceleration phases are associated with seasonal-scale dominated oscillations.

The potential ice avalanche release area suggested by Frey et al. [41], is located 200–300 m east of the steep ridge, on a 25–30° inclined terrace, at approximately 5430 to 5525 m a.s.l. (Figure 1, zone S3). Satellite imagery shows that the ice thickness and area located on this terrace reduced significantly between 2013 and 2020 (Figure 8). The remaining glacier ice at this location shows mean surface velocities < 0.1 m/d with seasonal dominated oscillation in the West and inter-annual dominated oscillation in the East (Figure S1).

Below ~5400 m a.s.l., the slope of the glacier surface ranges between 1°–35° with some steeper sections in the Northwest and Southeast. For zone S4a (Figure 1), where higher surface velocities are observed (i.e., >0.3 m/d), the slope parallel to the glacier flow lines range between 20–35° (<30° on average). As mentioned earlier, this zone is characterized by an elevated ice thickness and hence elevated velocities [9,33]. The surface velocity trends over the observation period are linear ($R^2 > 0.99$), and acceleration/deceleration phases correspond to inter-annual scale-dominated oscillations (Figure S2). The two zones showing higher velocities in the eastern part of the glacier (E2 and lower part of zone E1 in Figure 1) are likely associated with terrain steps in the glacier bed and multiple crevasses perpendicular to the glacier flow line. The maximum average velocity is 0.2–0.4 m/d. Below the higher velocity zones, the average daily velocity decreases to < 0.2 m/d. Linear regression analysis shows a linear velocity trend (E1 $R^2 > 0.97$, E2 $R^2 > 0.98$), again, characterized by a strong seasonal-scale oscillation (Figures S3 and S4).

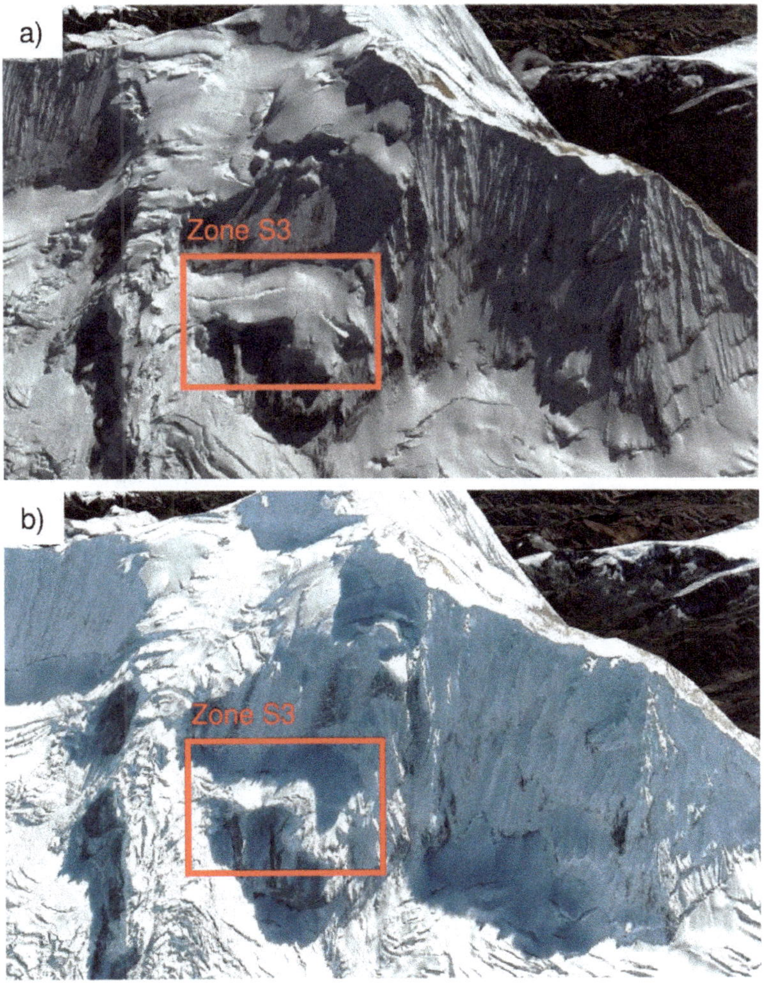

Figure 8. Satellite image of Zone S3 on (**a**) 20 August 2013 and (**b**) 3 July 2020 (Google Earth).

Somos-Valenzuela et al. [42] suggested a potential ice avalanche starting zone at 5′202 m.a.s.l. in zone S5 (Figure 1), located at the toe of a steep glaciated flank. Within the area delineated by Somos-Valenzuela et al. [42], the slope of the glacier varies between 10–30°. Satellite imagery shows that the lower end of this zone is in close contact with the surrounding glacier without any visible terrain step, which renders any locally enhanced process kinematically unlikely. Further, the surface velocities are between 0.1–0.2 m/d, exhibiting a linear trend ($R^2 > 0.97$) with acceleration/deceleration phases corresponding to distinct seasonal dominated oscillations (Figure S5).

Despite all accumulated surface velocity time series exhibiting linear trends (e.g., $R^2 > 0.95$), potential instabilities may form at the glacier terminus, which are characterized by glacier flow directions perpendicular to steep terrain steps (Type IB or Type II ice avalanche starting zones). The velocity-time series for the glacier terminus did not suggest the development of significant ice instabilities within the observation period. However, as our surface velocity field has a 60 m spatial resolution, instabilities smaller than the resolution may not be clearly detectable. Satellite images obtained from 2012 and 2020 show that two distinct zones at the terminus exhibit frequent ice falls or snow avalanches. One zone is located at the terminus of zone S4a and on the terminus of E3 (Figure 1). Due to the dynamics of the glacier, small icefall events at these locations are expected. Those types of events may occasionally reach the Laguna, causing small-scale disturbances, an example of which was captured by cameras operated by INAIGEM (e.g., small icefall on 5 February 2019 and a snow avalanche on 17 January 2021). However, based on available data it is not possible to ascertain whether they are single or multiple icefall events. Evaluation of ice avalanche potential for the Palcaraju glacier, based on our velocity time series indicates that an imminent failure of a glacier instability was not evident during the observation period. However, continuous observations, for which our study provides an important baseline, would be necessary to assess whether the behavior of the glacier changes over time, and to detect indications for an imminent ice avalanche.

Supplementary Materials: The following are available online at https://www.mdpi.com/article/10.3390/rs13142694/s1, Figures S1–S10 Conceptual representation of the clustering approach and displacement time series residuals for selected points.

Author Contributions: Conceptualization, A.K. and F.A.; methodology, A.K., F.A., F.W. and M.J.; software, T.S. and J.O.; investigation, A.K., F.A., A.D., F.W., M.J. and T.S.; data curation, J.O.; writing—original draft preparation, A.K. and F.A.; writing—review and editing, A.K., F.A. T.S., J.O., F.W. M.J. and A.D.; visualization, F.W., M.J., and J.O. All authors have read and agreed to the published version of the manuscript.

Funding: This research received no external funding.

Data Availability Statement: The data presented in this study are available on request from the corresponding author.

Acknowledgments: The TerraSAR-X data are copyrighted by the German Aerospace Center (DLR) and were provided through a commercial data purchase. Permission to use the results for this publication was granted by the firm RWE AG. We acknowledge the detailed review and suggestions provided by Martin Funk, VAW ETH Zurich. We thank two anonymous reviewers for their improvements to the manuscript.

Conflicts of Interest: The authors declare no conflict of interest.

References

1. Vuille, M.; Francou, B.; Wagnon, P.; Juen, I.; Kaser, G.; Mark, B.G.; Bradley, R.S. Climate change and tropical Andean glaciers: Past, present and future. *Earth-Sci. Rev.* **2008**, *89*, 79–96. [CrossRef]
2. Kaser, G.; Georges, C. Changes of the equilibrium-line altitude in the tropical Cordillera Blanca, Peru, 1930–1950, and their spatial variations. *Ann. Glaciol.* **1997**, *24*, 344–349. [CrossRef]
3. Gascoin, S.; Kinnard, C.; Ponce, R.; Lhermitte, S.; MacDonell, S.; Rabatel, A. Glacier contribution to streamflow in two headwaters of the Huasco River, Dry Andes of Chile. *Cryosphere* **2011**, *5*, 1099–1113. [CrossRef]
4. Bolch, T.; Kulkarni, A.; Kääb, A.; Huggel, C.; Paul, F.; Cogley, J.G.; Frey, H.; Kargel, J.S.; Fujita, K.; Scheel, M.; et al. The state and fate of Himalayan glaciers. *Science* **2012**, *336*, 310–314. [CrossRef]
5. Georges, C. 20th-Century Glacier Fluctuations in the Tropical Cordillera Blanca, Perú. *Arctic, Antarct. Alp. Res.* **2004**, *36*, 100–107. [CrossRef]
6. Kaser, G.; Ames, A.; Zamora, M. Glacier Fluctuations and Climate in the Cordillera Blanca, Peru. *Ann. Glaciol.* **1990**, *14*, 136–140. [CrossRef]
7. Hooke, R.L.; Pohjola, V.A.; Jansson, P.; Kohler, J. Intra-seasonal changes in deformation profiles revealed by borehole studies, Storglaciären, Sweden. *J. Glaciol.* **1992**, *38*, 348–358. [CrossRef]
8. Faillettaz, J.; Funk, M.; Vincent, C. Avalanching glacier instabilities: Review on processes and early warning perspectives. *Rev. Geophys.* **2015**, *53*, 203–224. [CrossRef]
9. Vincent, C.; Morceau, L.U. Sliding velocity fluctuations and subglacial hydrology over the last two decades on Argentière glacier, Mont Blanc area. *J. Glaciol.* **2016**, *62*, 805–815. [CrossRef]
10. Iken, A.; Bindschadler, R.A. Combined measurements of Subglacial Water Pressure and Surface Velocity of Findelengletscher, Switzerland: Conclusions about Drainage System and Sliding Mechanism. *J. Glaciol.* **1986**, *32*, 101–119. [CrossRef]

11. Bindschadler, R. The Importance of Pressurized Subglacial Water in Separation and Sliding at the Glacier Bed. *J. Glaciol.* **1983**, *29*, 3–19. [CrossRef]
12. Engelhardt, H.F. Water in glaciers: Observations and theory of the behaviour of water levels in boreholes. *Z. Gletsch. Glaziologie* **1978**, *14*, 35–60.
13. Weertman, J. The Theory of Glacier Sliding. *J. Glaciol.* **1964**, *5*, 287–303. [CrossRef]
14. Agassiz, L. *Système Glaciaire, ou Recherches sur les Glaciers: Pt. 1. Novelles Études et Expériences sur les Glaciers Actuels*. Mason, V., Ed.; 1847. Available online: http://www.google.cat/books?id=MdQWAAAAQAAJ&printsec=frontcover&vq=froid&hl=ca&output=html_text&source=gbs_ge_summary_r&cad=0 (accessed on 10 June 2021).
15. Fountain, A.G.; Walder, J.S. Water flow through temperate glaciers. *Rev. Geophys.* **1998**, *36*, 299–328. [CrossRef]
16. Kääb, A.; Jacquemart, M.; Gilbert, A.; Leinss, S.; Girod, L.; Huggel, C.; Falaschi, D.; Ugalde, F.; Petrakov, D.; Chernomorets, S.; et al. Sudden large-volume detachments of low-angle mountain glaciers–more frequent than thought? *Cryosphere* **2021**, *15*, 1751–1785. [CrossRef]
17. Harper, J.T.; Humphrey, N.; Pfeffer, W.T.; Lazar, B. Two modes of accelerated glacier sliding related to water. *Geophys. Res. Lett.* **2007**, *34*. [CrossRef]
18. Paul, F.; Bolch, T.; Kääb, A.; Nagler, T.; Nuth, C.; Scharrer, K.; Shepherd, A.; Strozzi, T.; Ticconi, F.; Bhambri, R.; et al. The glaciers climate change initiative: Methods for creating glacier area, elevation change and velocity products. *Remote Sens. Environ.* **2015**, *162*, 408–426. [CrossRef]
19. Werner, C.; Wegmuller, U.; Strozzi, T.; Wiesmann, A. Precision estimation of local offsets between pairs of SAR SLCs and detected SAR images. In Proceedings of the 2005 IEEE International Geoscience and Remote Sensing Symposium, Seoul, Korea, 25–29 July 2005; pp. 4803–4805, ISBN 0-7803-9050-4.
20. Strozzi, T.; Luckman, A.; Murray, T.; Wegmuller, U.; Werner, C.L. Glacier motion estimation using SAR offset-tracking procedures. *IEEE Trans. Geosci. Remote Sens.* **2002**, *40*, 2384–2391. [CrossRef]
21. Gray, A.L.; Mattar, K.E.; Vachon, P.W.; Bindschadler, R.; Jezek, K.C.; Forster, R.; Crawford, J.P. InSAR results from the RADARSAT Antarctic Mapping Mission data: Estimation of glacier motion using a simple registration procedure. In Proceedings of the IGARSS '98, Sensing and Managing the Environment, IEEE International Geoscience and Remote Sensing, Symposium, Seattle, WA, USA, 6–10 July 1998; Volume 3, pp. 1638–1640, ISBN 0-7803-4403-0.
22. Friedl, P.; Seehaus, T.; Braun, M. Global time series and temporal mosaics of glacier surface velocities, derived from Sentinel-1 data. *Earth Syst. Sci. Data Discuss.* **2021**. in review.
23. Mohr, J.J.; Reeh, N.; Madsen, S.N. Three-dimensional glacial flow and surface elevation measured with radar interferometry. *Nature* **1998**, *391*, 273–276. [CrossRef]
24. Strozzi, T.; Paul, F.; Wiesmann, A.; Schellenberger, T.; Kääb, A. Circum-Arctic Changes in the Flow of Glaciers and Ice Caps from Satellite SAR Data between the 1990s and 2017. *Remote Sens.* **2017**, *9*, 947. [CrossRef]
25. Paul, F.; Bolch, T.; Briggs, K.; Kääb, A.; McMillan, M.; McNabb, R.; Nagler, T.; Nuth, C.; Rastner, P.; Strozzi, T.; et al. Error sources and guidelines for quality assessment of glacier area, elevation change, and velocity products derived from satellite data in the Glaciers_cci project. *Remote Sens. Environ.* **2017**, *203*, 256–275. [CrossRef]
26. Pellikka, P.K.E.; Rees, G. *Remote Sensing of Glaciers: Techniques for Topographic, Spatial and Thematic Mapping of Glaciers*; CRC Press: Boca Raton, FL, USA, 2010; ISBN 9780367384647.
27. Wang, H.; Wellmann, J.F.; Li, Z.; Wang, X.; Liang, R.Y. A Segmentation Approach for Stochastic Geological Modeling Using Hidden Markov Random Fields. *Math. Geol.* **2017**, *49*, 145–177. [CrossRef]
28. Wellmann, J.F.; Regenauer-Lieb, K. Uncertainties have a meaning: Information entropy as a quality measure for 3-D geological models. *Tectonophysics* **2012**, *526–529*, 207–216. [CrossRef]
29. Wang, H.; Wellmann, F.; Zhang, T.; Schaaf, A.; Kanig, R.M.; Verweij, E.; Hebel, C.; Kruk, J. Pattern Extraction of Topsoil and Subsoil Heterogeneity and Soil-Crop Interaction Using Unsupervised Bayesian Machine Learning: An Application to Satellite-Derived NDVI Time Series and Electromagnetic Induction Measurements. *J. Geophys. Res. Biogeosci.* **2019**, *124*, 1524–1544. [CrossRef]
30. Herbert, C.; Camps, A.; Wellmann, F.; Vall-Llossera, M. Bayesian Unsupervised Machine Learning Approach to Segment Arctic Sea Ice Using SMOS Data. *Geophys. Res. Lett.* **2021**, *48*, e2020GL091285. [CrossRef]
31. Millan, R.; Mouginot, J.; Rabatel, A.; Jeong, S.; Cusicanqui, D.; Derkacheva, A.; Chekki, M. Mapping Surface Flow Velocity of Glaciers at Regional Scale Using a Multiple Sensors Approach. *Remote Sens.* **2019**, *11*, 2498. [CrossRef]
32. Blake, W.; Fischer, U.H.; Bentley, C.R.; Clarke, G.K.G. Instruments and Methods: Direct measurement of sliding at the glacier bed. *J. Glaciol.* **1994**, *40*, 595–599. [CrossRef]
33. Cuffey, K.M.; Paterson, W.S.B. *The physics of Glaciers*, 4th ed.; Elsevier: Amsterdam, The Netherlands, 2010; ISBN 978-0-123-69461-4.
34. Pralong, A.; Funk, M. On the instability of avalanching glaciers. *J. Glaciol.* **2006**, *52*, 31–48. [CrossRef]
35. Vimeux, F.; Ginot, P.; Schwikowski, M.; Vuille, M.; Hoffmann, G.; Thompson, L.G.; Schotterer, U. Climate variability during the last 1000 years inferred from Andean ice cores: A review of methodology and recent results. *Palaeogeogr. Palaeoclim. Palaeoecol.* **2009**, *281*, 229–241. [CrossRef]
36. Favier, V. One-year measurements of surface heat budget on the ablation zone of Antizana Glacier 15, Ecuadorian Andes. *J. Geophys. Res.* **2004**, *109*, 1–15. [CrossRef]
37. Sagredo, E.A.; Rupper, S.; Lowell, T.V. Sensitivities of the equilibrium line altitude to temperature and precipitation changes along the Andes. *Quat. Res.* **2014**, *81*, 355–366. [CrossRef]

38. Gurgiser, W.; Marzeion, B.; Nicholson, L.; Ortner, M.; Kaser, G. Modeling energy and mass balance of Shallap Glacier, Peru. *Cryosphere* **2013**, *7*, 1787–1802. [CrossRef]
39. Maussion, F.; Gurgiser, W.; Großhauser, M.; Kaser, G.; Marzeion, B. ENSO influence on surface energy and mass balance at Shallap Glacier, Cordillera Blanca, Peru. *Cryosphere* **2015**, *9*, 1663–1683. [CrossRef]
40. Francou, B. New evidence for an ENSO impact on low-latitude glaciers: Antizana 15, Andes of Ecuador, 0°28′S. *J. Geophys. Res.* **2004**, *109*, 1–17. [CrossRef]
41. Frey, H.; Huggel, C.; Chisolm, R.E.; Baer, P.; McArdell, B.; Cochachin, A.; Portocarrero, C. Multi-Source Glacial Lake Outburst Flood Hazard Assessment and Mapping for Huaraz, Cordillera Blanca, Peru. *Front. Earth Sci.* **2018**, *6*, 1–16. [CrossRef]
42. Somos-Valenzuela, M.A.; Chisolm, R.E.; Rivas, D.S.; Portocarrero, C.; McKinney, D.C. Modeling a glacial lake outburst flood process chain: The case of Lake Palcacocha and Huaraz, Peru. *Hydrol. Earth Syst. Sci.* **2016**, *20*, 2519–2543. [CrossRef]
43. Vilímek, V.; Zapata, M.L.; Klimeš, J.; Patzelt, Z.; Santillán, N. Influence of glacial retreat on natural hazards of the Palcacocha Lake area, Peru. *Landslides* **2005**, *2*, 107–115. [CrossRef]
44. Wang, W.; Yao, T.; Gao, Y.; Yang, X.; Kattel, D.B. A First-order Method to Identify Potentially Dangerous Glacial Lakes in a Region of the Southeastern Tibetan Plateau. *Mt. Res. Dev.* **2011**, *31*, 122. [CrossRef]
45. Bolch, T.; Peters, J.; Yegorov, A.; Pradhan, B.; Buchroithner, M.; Blagoveshchensky, V. Identification of potentially dangerous glacial lakes in the northern Tien Shan. *Nat. Hazards* **2011**, *59*, 1691–1714. [CrossRef]
46. Emmer, A.; Vilímek, V. Review Article: Lake and breach hazard assessment for moraine-dammed lakes: An example from the Cordillera Blanca (Peru). *Nat. Hazards Earth Syst. Sci.* **2013**, *13*, 1551–1565. [CrossRef]
47. Schneider, D.; Huggel, C.; Cochachin, A.; Guillén, S.; García, J. Mapping hazards from glacier lake outburst floods based on modelling of process cascades at Lake 513, Carhuaz, Peru. *Adv. Geosci.* **2014**, *35*, 145–155. [CrossRef]
48. Huggel, C.; Haeberli, W.; Kääb, A.; Bieri, D.; Richardson, S. An assessment procedure for glacial hazards in the Swiss Alps. *Can. Geotech. J.* **2004**, *41*, 1068–1083. [CrossRef]
49. Alean, J. Ice Avalanches: Some Empirical Information about their Formation and Reach. *J. Glaciol.* **1985**, *31*, 324–333. [CrossRef]
50. Röthlisberger, H. *(Translated from German) Ice Avalanches and Outburst Floods of Glacial Lakes: Offprint, Yearbook of the Swiss Natural Research Society, Scientific Section*; 1978.
51. Alean, J. *(Translated from German) Studies on the Conditions of Formation and Range of Ice Avalanches. Communications of the Laboratory of Hydraulics. Hydrology and Glaciology*; ETH Zurich, 1984; Volume 74.

remote sensing

Communication

A Survey of Remote Sensing and Geographic Information System Applications for Flash Floods

Lisha Ding [1,2], Lei Ma [3,4], Longguo Li [1,2], Chao Liu [1,2], Naiwen Li [1,2], Zhengli Yang [1,2], Yuanzhi Yao [5] and Heng Lu [1,2,*]

1 State Key Laboratory of Hydraulics and Mountain River Engineering, Sichuan University, Chengdu 610065, China; dinglisha@stu.scu.edu.cn (L.D.); lilongguo@scu.edu.cn (L.L.); liuchao@scu.edu.cn (C.L.); linaiwen@scu.edu.cn (N.L.); yangzhengli@scu.edu.cn (Z.Y.)
2 College of Hydraulic and Hydroelectric Engineering, Sichuan University, Chengdu 610065, China
3 School of Geography and Ocean Science, Nanjing University, Nanjing 210093, China; maleinju@nju.edu.cn
4 Signal Processing in Earth Observation, Technical University of Munich (TUM), 80333 Munich, Germany
5 School of forestry and wildlife sciences, Auburn University, Auburn, AL 36830, USA; yzy0040@auburn.edu
* Correspondence: luheng@scu.edu.cn

Abstract: Flash floods are among the most dangerous natural disasters. As climate change and urbanization advance, an increasing number of people are at risk of flash floods. The application of remote sensing and geographic information system (GIS) technologies in the study of flash floods has increased significantly over the last 20 years. In this paper, more than 200 articles published in the last 20 years are summarized and analyzed. First, a visualization analysis of the literature is performed, including a keyword co-occurrence analysis, time zone chart analysis, keyword burst analysis, and literature co-citation analysis. Then, the application of remote sensing and GIS technologies to flash flood disasters is analyzed in terms of aspects such as flash flood forecasting, flash flood disaster impact assessments, flash flood susceptibility analyses, flash flood risk assessments, and the identification of flash flood disaster risk areas. Finally, the current research status is summarized, and the orientation of future research is also discussed.

Keywords: remote sensing; geographic information system; flash floods; visual analysis

Citation: Ding, L.; Ma, L.; Li, L.; Liu, C.; Li, N.; Yang, Z.; Yao, Y.; Lu, H. A Survey of Remote Sensing and Geographic Information System Applications for Flash Floods. *Remote Sens.* **2021**, *13*, 1818. https://doi.org/10.3390/rs13091818

Academic Editor: Paolo Mazzanti

Received: 24 March 2021
Accepted: 4 May 2021
Published: 7 May 2021

1. Introduction

A flash flood is a rapid response to a severe thunderstorm that occurs in a short period of time (usually only a few minutes) [1]. Rapid snowmelt and sudden releases of impounded water may also cause flash floods [2]. In recent years, increasingly severe flash floods have occurred due to increased rainfall caused by climate change [3–5]. Moreover, the risk of flash floods increases with an increase in the impervious area in a given catchment area. Urbanization and reductions in rural land areas have led to declines in drainage capacity and increased numbers of flash floods [6]. As population structures change and the degree of urbanization increases, more people will be exposed to flash floods [7–9]. Many countries have carried out research on flash floods, such as China [10], the United States [11], Saudi Arabia [1], Egypt [12], and Italy [13].

The current research on flash floods mainly involves flash flood forecasting [14–21], flash flood disaster impact assessments [22–25], identifications of flash flood hazard areas [26–35], flash flood susceptibility assessments [36–39], and flash flood risk assessments [40–45]. Therefore, the motivation for our study is to comprehensively review the major areas related to the application of remote sensing and GIS technologies to flash flood disasters. Based on the analysis, the literature is classified, the study trends and hot spots of the current study on the application of remote sensing and GIS to flash floods in the last 20 years are summarized, and the main scientific progress in the literature is also summarized. Finally, the whole paper is summarized, commented on, and prospected.

2. Data and Methods

2.1. Retrieval Strategy

Articles with flash floods and remote sensing as research topics that were published from 2000 to October 2020 were retrieved, these articles were included in the Web of Science (WOS) Core Collection. A total of 248 articles were retrieved.

2.2. Literature Visual Analysis

CiteSpace is software developed by Chaomei Chen with an information visualization function based on the Java environment. Through keywords, authors, institutions, etc., one can perform visual analysis and generate various knowledge graphs, which can be used to show current research hotspots and trends to help people better understand research in a certain field. To date, many people have used CiteSpace for data mining and visual analysis [46–48]. In this paper, a total of 248 articles were included to generate citation analysis reports (such as node size, keyword co-occurrence, time zone view, etc.) by CiteSpace 5.6. R4(64-bit).

(1) First, we conducted a keyword co-occurrence analysis as follows: using the remove duplicates (WOS) function of CiteSpace to remove duplicates, we merged words with similar meanings and deleted meaningless words to generate a word cloud image (keyword co-occurrence network map). The size of the word indicates the frequency of the keyword, the larger the size of the keyword is, the more frequently the keyword appears.
(2) Second, we generated a time zone map of keywords appearing in 248 articles, revealing the dynamic evolution of research hotspots.
(3) Third, we conducted a co-citation analysis of references so that we could obtain landmark articles in the 248 articles, and could analyze the changes in research trends. If two articles (A and B) appear in the reference list of the third cited article (C) at the same time, the two documents constitute a cited relationship. If two articles (A and B) refer to the same article (C), there is a coupling relationship between the two articles (A and B).
(4) Fourth, we generated a keyword burst map to find hot words that, in the field of flash flood research, use remote sensing. We define a keyword with sudden changes in frequency within a certain period of time as a burst word, which represents the hotspot of research in that stage.

2.3. Explanation of Visual Map Icons in Maps

(1) Tree ring history: this represents the citation history of an article, and the overall size of the annual ring reflects the number of times the paper has been cited. The color of the citation ring indicates the corresponding citation time. The thickness of an annual ring is proportional to the number of citations in the corresponding time zone.
(2) Node circles: in the author's coauthored network and the institutional coauthored network, the size of the node circle represents the number of publications.
(3) In the keyword co-occurrence network, the size of the node circle represents the frequency of keywords.
(4) Connections between nodes: the connection between nodes indicates that they have a common copyright or have appeared at the same time, and the color of the connection indicates the time of the first cooperation or the first common appearance.
(5) Node colors: in the keyword co-occurrence network, the colors of nodes indicate different years, the color in the center of the node represents the time when the keyword first appeared, and the thickness of the circle represents the frequency of the keyword in the corresponding year. The higher the frequency, the more often it appears.
(6) Cluster#: in this paper, the clusters are based on the generated map, the keywords in the toolbar are clicked to cluster, and the clusters are marked by the keywords. The names of the clusters are #0, #1, #2...

3. Results

3.1. Citation Frequency of Remote Sensing and GIS Applied to Flash Flood

From 2000 to 2020, the citation frequency of articles on remote sensing used in flash floods increased year by year. Therefore, interest in research hotspots related to flash floods using remote sensing is increasing year by year (Figure 1).

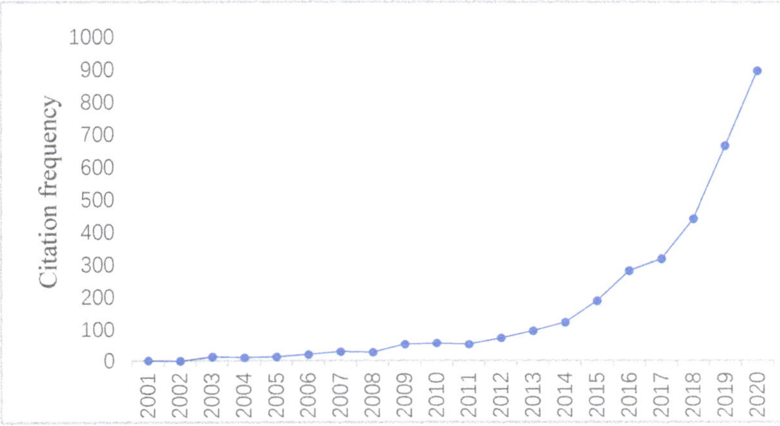

Figure 1. Trends in the citation frequency of the 248 included articles from 2000 to 2020.

Representative examples of highly cited articles: the 5 most cited articles from 248 articles are selected, listed in Table 1. These articles mainly involve rainfall estimation, methods of determining flood occurrence, and estimating the risk of flash floods.

The first three of these five highly cited articles are review articles. They are about radar rainfall estimation, flash flood warning systems, and flash flood forecasting modeling technology, which are all around the key issues of flash floods. The last two articles both researched a specific watershed in Egypt and used models to predict locations that are vulnerable to flash floods. The difference lies in the different models used by the two. The research of Youssef et al. [49] was conducted in a GIS environment. The amount of data is greater, the research scope is wider, and the parameters used are greater. The research of Foody et al. [50] has less data and fewer parameters, so the results obtained are less and simpler compared to Youssef's research.

The most cited articles are usually landmarks, they are groundbreaking or forward-looking. The study performed by Krajewski et al. [51] proposed some suggestions, including the establishment of long-term monitoring and verification stations to provide detailed information about precipitation, and believed that radar rain products have great development potential in flash flood forecasting. In recent years, radar has been widely used in precipitation estimation [7,51,52], confirming the prediction of Krajewski et al. [51] One of the scientific advances proposed by Borga et al. [53] is integrating multiple early warning methods, which has not been achieved until now. Whether the flash flood forecasting methods proposed in each research area can be realized in other areas still needs further discussion and verification [54]. For areas with similar topography, climate, soil, geology, land use, land cover, etc., it seems possible to use the same method for risk assessment [41]. Since cities have large impervious areas and a large population, once flash floods occur, they will cause many economic losses and casualties. Therefore, special attention should be given to flash flood forecasts in urban areas by Hapuarachchi et al. [7].

Table 1. Hot spot analysis of highly cited articles, classified from 248 articles using remote sensing to study flash floods.

Author	Cite Frequency	Title	Research Contents
Youssef, AM et al. [49]	177	Flash flood risk estimation along the St. Katherine road, southern Sinai, Egypt using GIS based morphometry and satellite imagery	The biggest influencing factors of flash flood disasters and key sensitive zones was discussed, and a detailed map of the most dangerous sub-basin was drawn.
Giles M. Foody et al. [50]	89	Predicting locations sensitive to flash flooding in an arid environment	The hydrological model was used to predict the location of sites that are particularly vulnerable to the threat of flooding, and peak flow was proven.
W.F. Krajewski et al. [51]	344	Radar hydrology: rainfall estimation	The problems of radar rainfall product development and the framework of rainfall estimation based on reflectivity were discussed, and the theoretical and practical requirements of radar rainfall maps and new radar technology were verified.
Marco Borga et al. [53]	192	Hydrogeomorphic response to extreme rainfall in headwater systems: flash floods and debris flows	The latest research on flash floods and debris flows was comprehensively summarized, and the progress in three areas that will produce important results were proposed.
H. A. P. Hapuarachchi et al. [7]	178	A review of advances in flash flood forecasting	The new modeling techniques and data used in flash flood forecasting from 2000 to 2010 were introduced.

In Youssef's research [49], GIS software was used to process remote sensing data, as well as to address terrain and field data to assess the risk of flash floods. Morphometrics were used to estimate the risk level of flash floods in the research basin. In subsequent research, there were a large number of articles that referred to the method in this article for the evaluation of flash flood hazards [1,41,55,56].

The research of Foody was published in 2004 [50]. The study used hydrological models to predict locations that are particularly vulnerable to flash floods under limited data conditions. This result has promoted the development of related fields and has guiding significance for subsequent research on the use of hydrological models to forecast flash floods [25,57–60].

3.2. Keyword Co-Occurrence

CiteSpace software was used to analyze the 248 selected articles, the software's own remove duplicates (WOS) function was used to remove duplicates; one was selected for time slicing, the keyword node type was used, and pathfinder and pruning sliced were selected for the pruning option to improve the network readability networks. After merging words with the same meaning and deleting meaningless words, keywords with a frequency of more than five times are retained, and a keyword co-occurrence map was generated, as shown in Figure 2. The list of the frequency of keywords that appear more than 10 times is shown in Figure 3.

In Figure 2, the larger the word, the more frequently it appears. In Figure 2, it can be seen that, in addition to remote sensing and flash floods, the size of GIS, model, rainfall, risk, and other words is larger. The GIS is a method discovered by Correia et al. [61] that can be used to integrate and investigate information about flood disasters and is widely used to reproduce the research results of different cases. This is not surprising. As the tool most frequently used in the research of flash floods using remote sensing data is GIS, GIS can provide powerful tools for risk assessment and can integrate a variety of remote sensing data in the GIS environment. To forecast flash floods and evaluate the risk value of flash flood disasters, a variety of models are generated and frequently used. It is worth noting that, because the digital elevation model (DEM) can be used for hydrological analysis such as rainfall analysis, inundation analysis, and water system network analysis, the DEM is the most important factor in the hydrological model used to draw the flash flood disaster index of the study area [62]. The susceptibility of hydraulic modeling results was influenced by DEM accuracy [63]. DEM is often used in the study of flash floods [64].

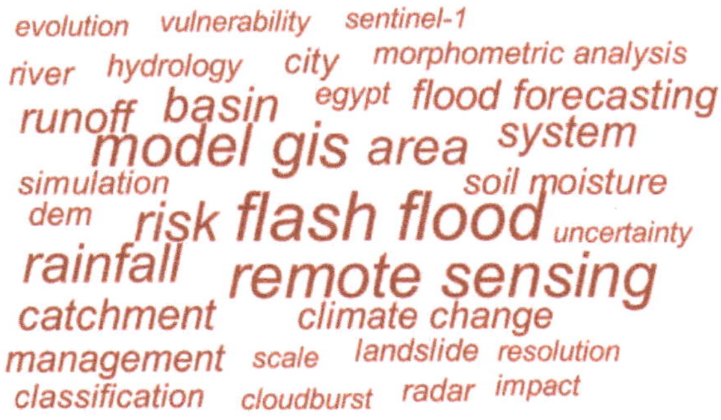

Figure 2. The keyword co-occurrence map.

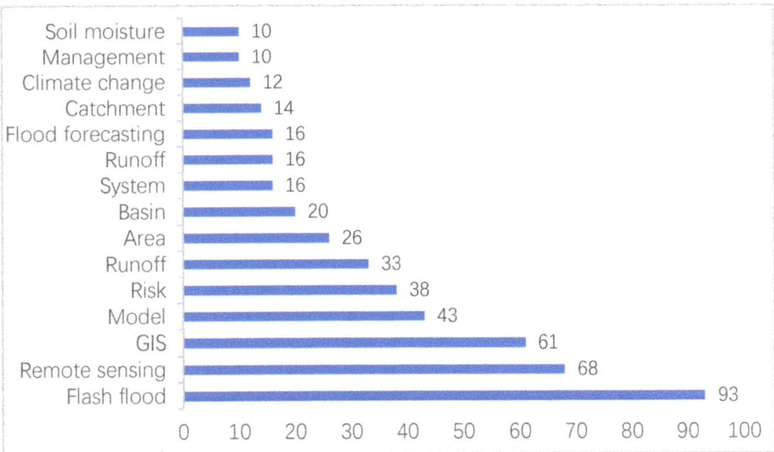

Figure 3. Keywords with more than 10 occurrences in 248 articles.

The most commonly used remote sensing data are Sentinel-1 and radar. People are paying more attention to flash floods in Egypt. Risk appears more frequently, indicating that there are more articles on the evaluation of the risk value of flash floods, indicating that more people are concerned about where flash floods may occur in order to take preventive measures in advance.

3.3. A Time Zone Map of Keywords

To further explore the dynamic evolution of flash flood research hotspots using remote sensing from 2000 to 2020, and to understand the key points of international research in different periods based on the generated keyword co-occurrence map, we use CiteSpace software and select "Time Zone View" in "Layout" to generate a keyword time zone map, as shown in Figure 4. (Selected keywords that appear more than five times are displayed in the figure).

The abscissa corresponding to the keyword in the figure indicates the year when it first appeared. The node colors of red, orange, yellow, green, blue, and purple are 2000 to 2020. The color of the line between the nodes indicates the year when the two keywords first appeared at the same time. Similar to the color of the node, the line colors of red, orange, yellow, green, blue, and purple correspond to the years 2000 to 2020.

From Figure 4, we can see that, in the study of flash floods using remote sensing, radar data have been used in the study of flash floods in 2000 or even before 2000, and since 2018, Sentinel-1 has been frequently used in research, the cumulative number of Sentinel-1 appearing in the article has reached six times in just three years. The images collected by Sentinel-1 can be used to obtain high-resolution images, regardless of weather conditions, so that they can be used to monitor floods. The synthetic aperture radar (SAR) data were collected from the Sentinel-1 sensor. Using SAR images, it can be used to distinguish water from other objects. Therefore, Sentinel-1 has often been used in the research of flash floods in recent years [65–68].

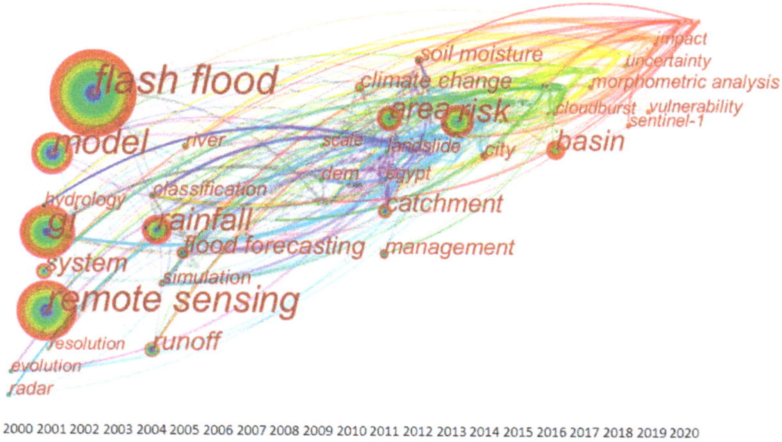

Figure 4. A time zone map of keywords.

The appearance of keywords in the picture can be roughly divided into five stages. The first stage was from 2000 to 2005. Keywords such as GIS, model, rainfall, runoff, etc. that appeared in this stage are still hotspots of current research. GIS as a research tool is effective and reliable. Research using models is also a common method. Among the factors that cause flash floods, rainfall is the most common cause of flash floods and has been studied the most. Soil erosion caused by flash floods has always been a concern. The second stage is from 2006 to 2009. During this stage, few new research hotspots appeared, and the research hotspots were mainly focused on the previous stage. The third stage is from 2010 to 2013. In this stage, climate change, risk, catchment, Egypt, etc. have become new research hotspots. People are beginning to pay attention to the increasing frequency of flash floods due to the frequent occurrence of extreme weather such as climate change and heavy rains [69,70]. Egypt is a typical area suffering from severe flash floods [71–74], this result is the same as Figure 2, but since 2010, the use of remote sensing to study flash floods has become popular. The fourth stage is from 2014 to 2015. There are few hot words in this stage, and the research hotspots are still based on previous research hotspots. In the fifth stage, from 2016 to 2019, research on basins increased, and the use of morphometric analysis and Sentinel-1 greatly promoted research on flash floods [65–68,75–77].

3.4. A Map of Burst Keywords from 248 Articles

Figure 5 lists the five keywords with the highest emergence intensity. From the figure, we can see that there was no keyword emergence before 2011, which means that before 2011, there were no issues that received more attention in the research on flash floods using remote sensing data. Egypt has the longest burst time. The increase in research on Egypt from 2011 to 2015 shows that there were more flash flood disasters in Egypt, and many places were affected by flash floods. The keyword with the strongest burst intensity is basin. The emergence time is 2017, and the stop time is 2020. This shows that since 2017, people's attention to the basin has increased. The burst time closest to the current burst keyword is uncertainty. The existing hydrological forecast chain is affected by many uncertain factors [78]. In recent years, with the continuous development of remote sensing technology, the pursuit of nearly real-time accurate simulation is about to become a global standard to ensure improved flash flood forecast and warning systems and ensure that models can be used in more areas and reduce the uncertainty of the model's output value. N. S. Bartsotas, Rouya Hdeib, and Hossein Mojaddadi Rizeei reduced the error of satellite precipitation estimation by optimizing algorithms and calibrating models [79–81].

Figure 5. Keywords with the strongest citation bursts in 248 articles.

3.5. Co-cited Results of Cited References

Clustering analysis of the cited references of 248 articles published from 2000 to 2020, the results can be divided into 8 clusters, using A (Abstract) to extract nominal terms to name the clusters. The results are shown in Figure 6: #0 eastern desert, #1 debris flow, #2 Najran area, #5 flood susceptibility map, #6 flash-flood predictor, #8 ground radar, and #9 flood susceptibility map.

The color of each cluster block represents the year when the co-citation relationship first appeared in each cluster. The colors of cluster blocks range from gray to purple, blue, green, yellow, and red, representing the years from 2000 to 2020. The color of each cluster block indicates the year when the co-citation relationship first appeared in each cluster. The connecting line between the nodes indicates the path of the reference. The connecting lines between the nodes indicate the path being cited, and the color of each line indicates the time when it was first cited. A few references are highly co-cited, so here, we set a threshold to show them.

The timeline map reveals changes in reference co-citations over time. According to the generated cluster diagram (Figure 6), a timeline map of cited references can be generated by the layout function. The *Y*-axis is defined as the cluster name defined by A (Abstract), and the *X*-axis is defined as the year of publication. The timeline chart shows the time span and research progress of the development and evolution of the eight clusters, as shown in Figure 7.

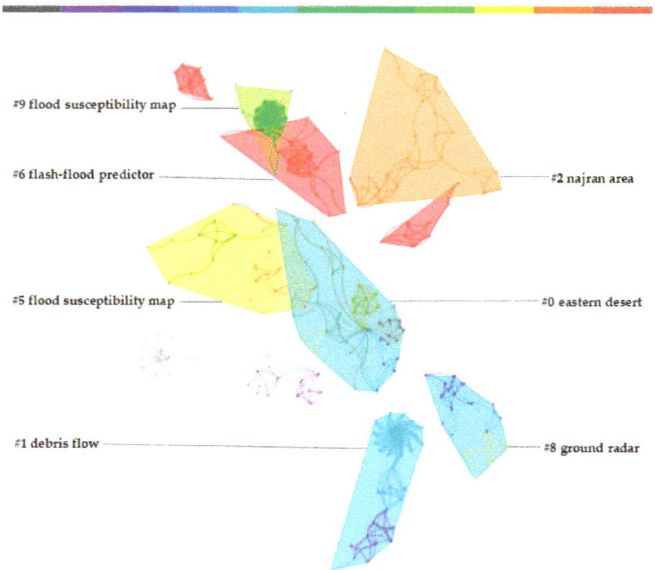

Figure 6. Reference co-citation network for the 248 included articles (clustered according to index terms).

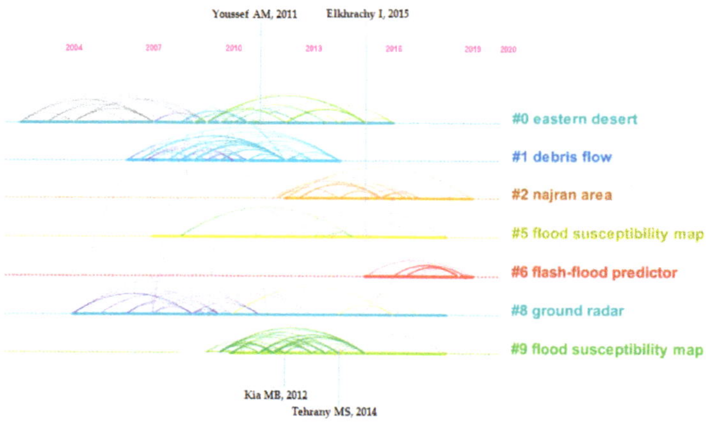

Figure 7. Timeline view of the citation trends identified in the 248 included articles.

In the timeline view, the references of the same cluster are placed on the same horizontal line. In the timeline view, the number of references in each cluster can be clearly seen. More references in the cluster representing the cluster are more important. The cluster labelling on the right shows the research hotspot category associated with each reference.

The circle in the figure represents the circle of the citation directory tree, the color at the center of the circle represents the year of the reference publication (the color corresponds to the year at the top of the view in the figure), and the size of the circle represents the frequency of citations. The cluster label name on the right indicates the research hotspot category related to the references in the cluster. Gray represents the earlier publication year, and red represents the most recent publication year. The longer color line segment indicates that the citation has a large time span, and its research hotspot is the subject that people have paid attention to for a long time. The red color at the outermost layer of the node's annual ring indicates that the citation frequency has increased rapidly or continues to increase rapidly.

Figure 7 shows four highly cited landmark articles, which are authoritative studies in the corresponding clusters. Research on the eastern desert took a long time, and a landmark article appeared in this cluster: Youssef et al. [49], which has been introduced before.

Elkhrachy et al. [1] were co-cited 14 times (Figure 8a,b), which is the most co-cited article among 248 articles. Tehrany et al. [80] were co-cited 10 times (Figure 8c,d), which is the second most co-cited article among 248 articles. Masoud Bakhtyari Kia et al. were co-cited six times (Figure 8e,f), and they are very representative articles in the field of flood susceptibility maps. Elkhrachy et al. [1] provided an accurate assessment by using SPOT and SRTM DEMs data. The analytical hierarchical process (AHP) was used to determine the relative impact weight of flash flood causative factors to obtain the composite flood hazard index (FHI). Finally, all the used data were integrated into ArcMap to generate the final flood disaster map of the study area. In previous studies, researchers have proposed many methods to perform flood susceptibility mapping, but these methods have certain shortcomings. To find a more accurate method, Tehrany et al. [80] proposed a new integration method that combines weights-of-evidence (WoE) and the support vector machine (SVM) model, not only solving the shortcomings of WoE but also enhancing the performance of SVM. The results are compared with the results obtained by using WoE and SVM alone, and the results obtained through integration are more ideal. Kia et al. [64] used artificial neural network (ANN) technology, which is one of the machine learning methods, to develop a flood model using various flood causative factors (including slope, flow accumulation, rainfall, soil, elevation, geology, and land use) to model and simulate flood-prone areas in the southern part of peninsular Malaysia. The ANN is more robust than other statistical and deterministic methods and has high computational efficiency. However, when using ANN modeling, there may be disadvantages such as errors caused by the length of the dataset.

Figure 8a,c,e are the pennant diagrams of Elkhrachy et al. [1], Tehrany et al. [80], and Kia et al. [64], respectively, which can be used to view the information for the references directly connected to a node. These figures show the distribution of articles that have a citation relationship with these articles. The closer the position of the reference article is to the bottom, the more times it has been cited. Figure 8b,d,f show the time trends, respectively, and show the number of times that Elkhrachy et al. [1], Tehrany et al. [80], and Kia et al. [64] were co-cited.

Combining Figures 7 and 8, we can see that Elkhrachy et al. [1], Tehrany et al. [80], and Kia et al. [64] were cited twice in 2016, and then, in 2019–2020, the number of citations increased suddenly, indicating that in 2016 and from 2019 to 2020, the research on flood susceptibility maps was relatively concentrated.

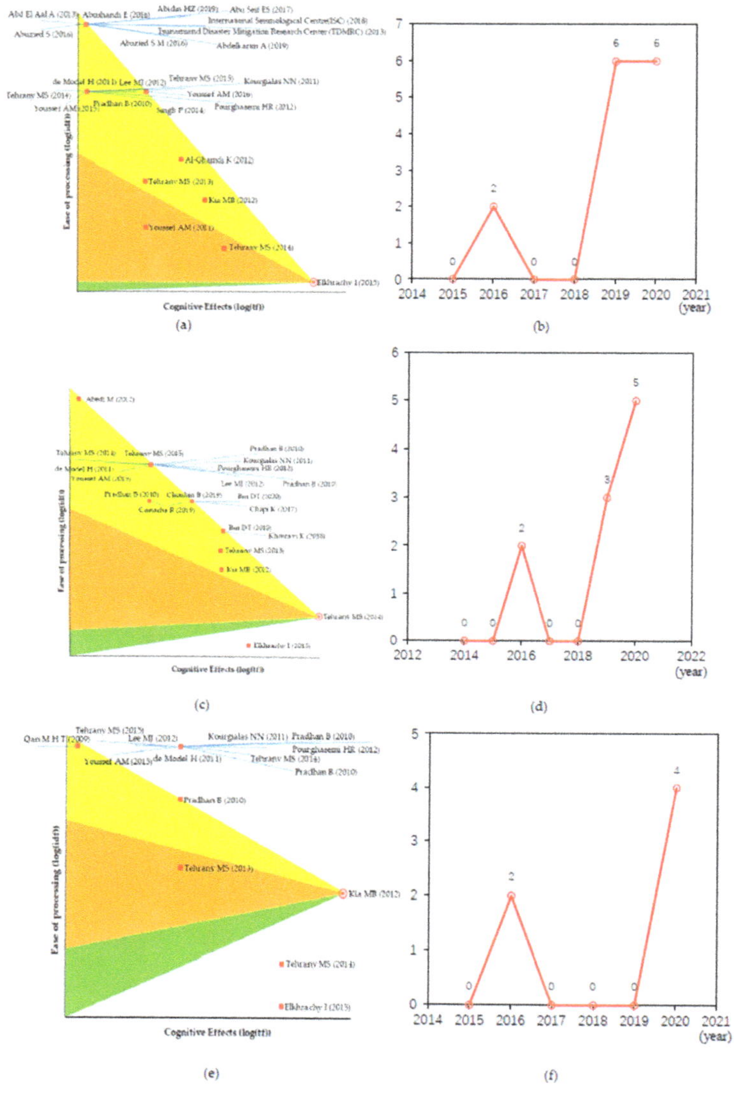

Figure 8. Co-citation status of four representative articles in the #6 flood susceptibility map in the timeline view. (**a**,**c**,**e**) are the pennant diagrams of Elkhrachy et al. [1], Tehrany et al. [80], and Kia et al. [64], respectively, which can be used to view the information for the refer-ences directly connected to a node. (**b**,**d**,**f**) show the time trends, respectively, and show the num-ber of times that Elkhrachy et al. [1], Tehrany et al. [80], and Kia et al. [64] were co-cited.

4. Main Subfields of Remote Sensing and Geographic Information Systems for Flash Floods

In the past two decades, due to the continuous development of science and technology, there have been many subfields in the application of remote sensing and GIS to flash floods. This article introduces five main subfields of the application of remote sensing and GIS to flash floods.

4.1. Flash Flood Forecasting

Since a flash flood may occur suddenly and the time to reach the peak is short, the accuracy of any early warning of flash floods depends largely on the accuracy of precipitation monitoring and prediction [14,64].

Accurate and timely measurement of the temporal and spatial distribution of rainfall is the starting point for flash flood forecasting [64]. Due to the wide coverage of satellites, satellite data are regarded as an important data source for areas with sparse and uneven distributions of measurement stations. Satellite data have been widely used in meteorological research, and the ability to estimate rainfall directly affects the ability to observe flash flood time. Table 2 lists six representative studies on the evaluation of satellite precipitation products in recent years. These studies combined multiple precipitation products and evaluated them with multiple statistical indicators, abundant precipitation products, and relatively rich types of research areas covered.

Table 2. For evaluating satellite precipitation products.

Study	Product Name	Study Area
Haonan Chen et al. [17]	Quantitative precipitation estimation (QPE), National Weather Service (NWS) single-polarization rainfall product, NWS dual-polarization rainfall products	America
N. S. Bartsotas et al. [79]	GSMaP (v.7), Climate Prediction Center morphing method (CMORPH)	Ethiopia and Italy
Mohamed Salem Nashwan et al. [81]	Global Satellite Mapping of Precipitation (GSMaP (v. 6)), Tropical applications of meteorology using satellite data and ground-based observations (TAMSAT (v. 3)), Precipitation estimation from remotely sensed information using artificial neural networks-cloud classification system (PERSIANN-CCS)	Egypt
Mengye Chen et al. [82]	Multi-radar multi-sensor system (MRMS), Global Precipitation Measurement Mission (GPM), National Centers for Environmental Prediction (NCEP)	America
Vincenzo Levizzani et al. [83]	Advanced microwave humidity sounder-unit B (AMSU-B) onboard the National Oceanic Microwave Humidity Sounder (MHS) on board the EUMETSAT Metop-A satellite and Atmospheric Administration (NOAA) polar satellites	The Island of Madeira
Ali Behrangi et al. [84]	Rain estimation using forward adjusted-advection of microwave estimates (REFAME), REFAMEgeo, PERSIANN, PERSIANN-CCS	America

The performance of precipitation products was evaluated for arid areas, mountain areas, and urban areas [17,79,81]. Satellite precipitation products can accurately detect and estimate extreme precipitation events. There are some uncertainties in the results obtained by using satellite precipitation products [82,83]. On the one hand, the algorithm can be optimized, and the inherent deviations in the precipitation calculation can be corrected to reduce the uncertainty. On the other hand, the integration of high-resolution and multisource precipitation analysis can be considered to compensate for the deficiency of a single precipitation product [85].

4.2. Impact of Flash Flood Assessment

The analysis of the disaster area after the occurrence of flash floods can better provide suggestions for regional development, flood prevention, and disaster reduction [33,86]. Table 3 lists four representative studies about the impact of flash flood assessment, which describe the impact of flash floods in terms of vegetation, agricultural products, topographic changes, and land cover.

Table 3. For impact of flash floods.

Study	Analytical Method	Factors
Mohammed Sadek et al. [22]	Sentinel-1 and Sentinel-2 satellite data, geolocated terrestrial photos and GIS technology, and hydrologic and hydraulic modeling were integrated to evaluate the impact of flash floods.	Catchment slope, relief ratio, drainage density, basin ruggedness number, land cover types
Bilal Ahmad Munir et al. [23]	The hydrological engineering center river analysis system (HEC-RAS) 2D hydraulic modeling was used to analyses the impact of flash floods in downstream Piedmont plains. Personal computer storm water management model PCSWMM (hydrologic) and HEC-RAS 5.x (hydraulic) models were integrated to monitor the flash flood.	Rainfall, peak events discharge, land use, land cover, soil, curve number, runoff, water surface elevation, sub-catchment width, slope, water depth, dry time, lag time, storm duration
Takahiro Sayama et al. [24]	The backpack-mounted mobile mapping system (MMS) was used to investigate and estimate landform changes.	Ground elevation, inundation depths, ground height, inundation level, latitude, sediment, rainfall
Joan Estrany et al. [33]	The meteorological, hydrological, geomorphological, damage, and risk data analyses were integrated to damage assessment based on field-based remote sensing and modeling.	Rainfall, runoff, slope, land use/cover, soil type

It can be concluded from Table 3 that hydrologic and hydraulic modeling are commonly used methods to study the effects of flash floods. Modern remote sensing technology can already use spaceborne imageries, airborne imageries, and unmanned aerial vehicle (UAV) systems to quickly and accurately map during or after a flood event. Free satellite data (Sentinel-2 images) were used to determine the impact of flash floods on Ras Ghareb city and the Wadi EI-Natryn region in Egypt [22,40]. Landsat-8 and MODIS data were used to describe the impact of flash floods on rice [87,88], Landsat TM data were used to map the extent of coastal floodplain flooding [89], and multispectral lkonos data were applied to a land use/land cover classification [90], all of which are useful for assessing the impact of flash floods. The combination of UAV data and field surveys can be used as observational data in conjunction with hydraulic models, which greatly promotes the understanding of the mechanism of flash floods [91]. Different from other studies, the backpack type MMS has been proven to be used for post-flood surveys and can ideally reproduce the flooding situation in mountainous areas [24].

4.3. Identification of Flash Flood Hazard Areas

Typically, due to the remote location of the flash flood area and the harsh weather, it is difficult to arrive at the scene to analyze the behavior of mountain torrents. In GIS environments, the most commonly used method involves drawing hazard maps of flash floods using hydrological and hydrodynamic models [30–33]. Table 4 lists five articles that use hydrological models or hydraulic models to map flash flood hazards.

Table 4. For identification of flash flood hazard areas.

Study	Analytical Method	Factors
Aneesha Satya Bandi et al. [30]	The multiple-criteria decision-making tools were used to generate the composite flood hazard index (FHI).	Runoff, type of soil, slope percentage, surface roughness, flow accumulation, distance to main channel in the stream network, land use
Jehan Mashaly et al. [34]	The hydrological model and the fused ASTER multispectral and ALOS-PALSAR synthetic aperture radar (SAR) data were combined to predict flash flood hazard.	Surface topology variables, land use, land cover data, soil texture properties, curve number, lithology, ground surface type
Hossein Mojaddadi Rizeei et al. [35]	A 2D high-resolution sub-grid model was performed to simulate FF probability and hazard. GIS and physics-based random forest (RF) models optimized by particle swarm optimization algorithm (PSO-RF) were used to model pluvial flash flood (PFF) hazard.	Curvature, SPI, TRI, TWI, DSM, surface slope, surface runoff, maximum precipitation intensity, LULC
Mohamed Saber et al. [58]	A physics-based distributed hydrological model for flash floods simulation was proposed.	Rainfall, land use, soil types, topography, storage amount, inflow, outflow, curve number, depth of rainfall, depth of runoff, excess rainfall
Eman Ghoneim et al. [92]	The hydrological response of the study basin to a rainfall event was explored, and the hydrological model approach was used to predict flash flood hazard in the research area.	Soil texture, curve number, channel slope, longest flow path, lag time for each sub-watershed, rainfall

Hydrological models can be used to predict the spatial ranges, depths, and speeds of flash flood disasters to determine the areas with high flash flood risks [34]. Two-dimensional hydrodynamic models are considered to be the most promising model for accurate flash flood mapping [35], but such models usually require large amounts of input data. The AHP and soil conservation service curve number (CN) methods are commonly used methods for drawing flash flood hazard maps. The AHP is used to assign grades and weights and is usually used to assign weights to the causes of mountain torrents in the study of flash flood hazards [1,53,93]. The SCS model is commonly used in distributed hydrological models and research in arid and semiarid regions, which is a method developed by the U.S. Department of Agriculture (USDA) to estimate runoff and peak discharge [94]. According to specific circumstances, the hazard factors of flash floods selected by researchers are not exactly the same, but many hazard factors are recognized as necessary.

4.4. Flash Flood Susceptibility Assessment

Identifying areas susceptible to flash floods is one of the most effective measures to reduce losses caused by floods and achieve flood management [95,96]. For large-scale flash flood susceptibility analysis, machine learning methods, bivariate statistics, and multicriteria decision-making methods are mainly used [97]. The machine learning method is considered to be the most advanced and first considered method [36]. Table 5 lists four representative studies that use machine learning methods, bivariate statistics, and multicriteria decision-making methods to map susceptibility to flash floods.

Table 5. For susceptibility mapping of flash flood.

Study	Analytical Method	Factors
Romulus Costache et al. [36]	The K-nearest neighbor (kNN) and K-star (KS) stand-alone models and kNN–AHP and KS–AHP ensemble models were used to define and calculate FFPI (flash flood potential index) in flash flood susceptibility mapping.	Slope, angle, TPI, TWI, curve number, lithology, profile curvature, plan curvature, convergence index, modified Fourier index
Viet-Nghia Nguyen et al. [37]	The chi-square automatic interaction detector (CHAID) random subspace, optimized by biogeography-based optimization (the CHAID-RS-BBO model) was proposed for the spatial prediction of flash floods.	Land use, land cover, soil type, lithology, river density, rainfall, topographic wetness index (TWI), elevation, slope, curvature, aspect
Khosravi, Khabat et al. [39].	Three multi-standard decision analysis techniques (vlse kriterijuska optamizacija I komoromisno resenje (VIKOR), technique for order preference by similarity to ideal solution (TOPSIS), and simple additive weighting (SAW)), and two machine learning methods (naïve Bayes trees (NBT) and naïve Bayes (NB) were tested for their ability to model flash flood susceptibility.	NDVI, lithology, land use, distance from river, curvature, altitude, stream transport index (STI), (TWI), SPI, soil type, slope, rainfall
Quang-Thanh Bui et al. [98]	A hybrid model for susceptibility mapping that combines swarm intelligence algorithms and deep learning neural networks was proposed.	Aspect, slope, curvature, TWI, stream power index (SPI), distance to river, river density, NDVI, NDBI, rainfall

From Table 5, the conclusion that the flash flood susceptibility mapping technologies rely on various adjustment factors representing the physical characteristics of the study area can be obtained. The choice of conditional factors depends on the scale of the studied area because it is more difficult to obtain data of the same scale or the same resolution. Therefore, if the study area is larger, the number of factors selected may be smaller, which seems reasonable. Researchers should select factors for research based on actual conditions. Of course, using more extensive data and impact factors can more accurately define the flash flood susceptibility of the study area [99,100]. Land use, slope, rainfall, TWI, and distance to the river are the most commonly considered factors. Logistic regression, bivariate statistical analysis, and AHP are the most commonly used methods to calculate factor weights. The combination of AHP and GIS can also define the flash flood susceptibility zones. The machine learning method is considered to be the most advanced and first considered method [101]. The effect of the mixed model is better than that of the single model, as proven by a large number of examples. The K-nearest neighbor (kNN) and K-star (KS) stand-alone models and kNN–AHP and KS–AHP ensemble models were used to define and calculate the FFPI (flash flood potential index) in flash flood susceptibility mapping. The Bayesian belief network (BBN) model was combined with an extreme learning machine (ELM) and back propagation (BP) structure to develop a new ensemble learning model for predicting flash flood susceptibility [102]. This fact has also been emphasized by Wang et al. [3].

4.5. Flash Flood Risk Assessment

The risk proposed by the United Nations refers to the expected loss of people's lives, property, and economic activities caused by a specific natural disaster in a certain area and a given time period [103]. Therefore, the flash flood risk analysis is obtained by combining hazard analysis and vulnerability analysis. Different from flash flood susceptibility analysis and flash flood disaster analysis, some flash flood risk analyses considered the factors of city and climate change [41–43]. The different geomorphic processes and hydraulic behaviors of the watershed are controlled by its morphometric characteristics [49]. Therefore, morphometric analyses are frequently used in flash flood risk analysis [31,44,45]. Table 6 summarizes the representative literature on flash flood risk assessment in terms of analytical methods and factors.

Table 6. For flash flood risk assessment.

Study	Analytical Method	Factors
Ahmed M. Youssef et al. [43]	The flash flood risk map was generated using GIS based morphometry and satellite data.	Area, total stream number, total stream length, elongation ratio, circulation ratio, shape factor, slope degree, length of over land flow, ruggedness degree, relief ratio, drainage density, drainage frequency, total drainage number
Shuvasish Karmokar et al. [41]	The flash flood risk map was achieved by the susceptibility map obtained by analyzing three satellite images in a GIS environment and using morphometric parameters to assign the relative susceptibility of flash floods.	Topography, climatological, soil, geological, hydrology, land use and land cover, digitized drainage network, rainfall, geomorphological map
Ram Nagesh Prasad et al. [104]	The flash flood risk map was generated by using the weighted sum analysis (WSA) model results and Snyder synthetic hydrological parameters.	Basin perimeter, basin length, stream order, stream length, area, drainage density, stream frequency, elongation ratio, circularity ratio, form factor, shape, basin relief, relief ratio
Sara Abuzied et al. [45]	The Soil Conservation Service (SCS) rainfall-runoff model was used to estimate the hydrological response of the catchments, and all risk factors were spatially integrated; the morphometric and SCS analyses were integrated to create the risk map.	Basin dimensions, basin shape, basin surface, drainage network

5. Discussion

In the past 20 years, the application of remote sensing and GIS technology in flash flood research has made great progress, mainly reflected in the increasingly abundant multisource remote sensing data sources, GIS strong spatial analysis ability, and coupling ability with hydrological and hydrodynamic models. However, the uncertainty of the data and model is still a huge challenge for future research. How to obtain real-time or quasi-real-time accurate simulations and reduce the uncertainty of data input (such as precipitation, land use, evaluation unit division, etc.) and model output is the goal of future research. To date, in many areas, through remote sensing data sources, GIS, and hydrological coupling models, a large number of studies and analyses have been carried out on flash flood susceptibility analysis, flash flood disaster impact assessment, and flash flood hazard identification. Most of the experimental results show that the established or improved model is effective for the experimental area, but as to whether the model can be applied in other areas, the universality of the model needs further verification.

For flash flood forecasting, with the development of meteorological satellite technology and radar-based rainfall forecast technology, more accurate and real-time precipitation data can be used in flash flood forecasting, and after precipitation data from multiple sources are acquired, the precipitation data can be corrected via the correction model.

For the impact of flash flood disaster assessment, with the development of data association analysis and multimodel coupling technology, the impact of flash floods on the regional ecology and environment can be rapidly and quantitatively assessed.

For flash flood susceptibility assessment, at present, most of the susceptibility zoning maps belong to static mapping and cannot show the inundation depth and advance speed. Future studies should combine machine learning with the hydrodynamic model to complete the dynamic susceptibility mapping of flash flood disasters. Then, a 2D model will be researched and developed to obtain the inundation depth and advance speed.

For flash flood risk assessment and hazard area identification, mapping flash flood disaster maps and flash flood risk maps relies on various adjustment factors that represent the physical characteristics of the study area. Due to the influence of data precision, data volume, size of the study area, and the authors' subjective choices, there are some differences among the adjustment factors selected in these papers, and the weights of the adjustment factors are not always the same. Even in regions with similar geological conditions, whether the adjustment factors selected in other areas can be used and their weights need to be further discussed and verified. It is hoped that there will be a set of systematic rules in the future so that adjustment factors and corresponding weights can be selected for regions with different sizes and different physical characteristics, according to their conditions, to obtain better results.

6. Conclusions

In this study, the related literature on remote sensing and GIS applied in the field of flash flood disasters was systematically analyzed. Then, a visualization analysis of the literature was adopted to perform keyword co-occurrence analysis, time zone chart analysis, keyword burst analysis, and literature co-citation analysis. Finally, several main subfields of the application of remote sensing and GIS in flash floods were summarized, including flash flood forecasting, the impact of flash flood assessment, flash flood susceptibility assessment, flash flood risk assessment, and the identification of flash flood hazard areas, which makes our study different from the previous review of remote sensing and geographical information application to natural disasters. The main conclusions are as follows: (1) through the analysis of the time zone map, the appearance of keywords can be roughly divided into five stages. (2) Analyzing the burst of keywords in 248 articles, we found that current research focuses on reducing uncertainty, and reducing the uncertainty of flash flood forecasting is the basis for real-time accurate simulation. (3) Through the co-cited analysis of 248 articles, 7 clusters were obtained. Among them, there were three highly co-cited articles from 2012 to 2015, which are landmark studies. Therefore, from this review, various applications of remote sensing and GIS in the field of flash floods and specific opportunities and challenges in different fields can be found.

Author Contributions: L.D. and H.L. drafted the manuscript and were responsible for the research design, experiment, and analysis. L.M. and C.L. reviewed and edited the manuscript. L.L., N.L., Z.Y., and Y.Y. supported the data preparation and the interpretation of the results. All of the authors contributed to editing and reviewing the manuscript. All authors have read and agreed to the published version of the manuscript.

Funding: This research was supported by the National Key R&D Program of China(2019YFC1510700), the National Natural Science Foundation of China (41701499), the funding provided by the Alexander von Humboldt-Stiftung, the Sichuan Science and Technology Program (2018GZ0265), the Geomatics Technology and Application Key Laboratory of Qinghai Province, China (QHDX-2018-07), the Major Scientific and Technological Special Program of Sichuan Province, China (2018SZDZX0027), and the Key Research and Development Program of Sichuan Province, China (2018SZ027, 2019-YF09-00081-SN).

Institutional Review Board Statement: Not applicable for studies not involving humans or animals.

Informed Consent Statement: Not applicable for studies not involving humans.

Data Availability Statement: The data that support the findings of this study are available from the corresponding author upon reasonable request.

Conflicts of Interest: The authors declare no conflict of interest.

References

1. Elkhrachy, I. Flash flood hazard mapping using satellite images and GIS tools: A case study of Najran City, Kingdom of Saudi Arabia (KSA). *Egypt. J. Remote Sens. Space Sci.* **2015**, *18*, 261–278. [CrossRef]
2. Bonacci, O.; Ljubenkov, I.; Roje-Bonacci, T. Karst flash floods: An example from the Dinaric karst (Croatia). *Nat. Hazards Earth Sys.* **2006**, *6*, 195–203. [CrossRef]
3. Wang, G.; Liu, Y.; Hu, Z.; Lyu, Y.; Zhang, G.; Liu, J.; Liu, Y.; Gu, Y.; Huang, X.; Zheng, H.; et al. Flood risk assessment based on fuzzy synthetic evaluation method in the Beijing-Tianjin-Hebei metropolitan area, China. *Sustainability* **2020**, *12*, 1451. [CrossRef]
4. Suarez, P.; Anderson, W.; Mahal, V.; Lakshmanan, T.R. Impacts of flooding and climate change on urban transportation: A systemwide performance assessment of the Boston Metro Area. *Transp. Res. Part D Transp. Environ.* **2005**, *10*, 231–244. [CrossRef]

5. Mohanty, M.P.; Simonovic, S.P. Understanding dynamics of population flood exposure in Canada with multiple high-resolution population datasets. *Sci. Total Environ.* **2021**, *759*, 143559. [CrossRef]
6. Mustafa, A.; Szydłowski, M. The impact of spatiotemporal changes in land development (1984–2019) on the increase in the runoff coefficient in Erbil, Kurdistan region of Iraq. *Remote Sens.* **2020**, *12*, 1302. [CrossRef]
7. Hapuarachchi, H.A.P.; Wang, Q.J.; Pagano, T.C. A review of advances in flash flood forecasting. *Hydrol. Process.* **2011**, *25*, 2771–2784. [CrossRef]
8. Beniston, M.; Stoffel, M.; Hill, M. Impacts of climatic change on water and natural hazards in the Alps: Can current water governance cope with future challenges? Examples from the European "ACQWA" project. *Environ. Sci. Policy* **2011**, *14*, 734–743. [CrossRef]
9. Kleinen, T.; Petschel-Held, G. Integrated assessment of changes in flooding probabilities due to climate change. *Clim. Chang.* **2007**, *81*, 283–312. [CrossRef]
10. Liang, W.; Yongli, C.; Hongquan, C.; Daler, D.; Jingmin, Z.; Juan, Y. Flood disaster in Taihu Basin, China: Causal chain and policy option analyses. *Environ. Earth Sci.* **2011**, *63*, 1119–1124. [CrossRef]
11. Mukherjee, F.; Singh, D. Detecting flood prone areas in Harris County: A GIS based analysis. *GeoJournal* **2020**, *85*, 647–663. [CrossRef]
12. El Bastawesy, M.; Attwa, M.; Abdel Hafeez, T.H.; Gad, A. Flash floods and groundwater evaluation for the non-gauged dryland catchment using remote sensing, GIS and DC resistivity data: A case study from the Eastern Desert of Egypt. *J. Afr. Earth Sci.* **2019**, *152*, 245–255. [CrossRef]
13. Miglietta, M.M.; Regano, A. An observational and numerical study of a flash-flood event over south-eastern Italy. *Nat. Hazards Earth Syst.* **2008**, *8*, 1417–1430. [CrossRef]
14. Collier, C.G. Flash flood forecasting: What are the limits of predictability? *Q. J. R. Meteorol. Soc.* **2007**, *133*, 3–23. [CrossRef]
15. Hong, Y.; Hsu, K.; Sorooshian, S.; Gao, X. Precipitation estimation from remotely sensed imagery using an artificial neural network cloud classification system. *J. Appl. Meteorol.* **2004**, *43*, 1834–1852. [CrossRef]
16. Nocholas, S.N.; Wassila, M.T. African rainfall climatology version 2 for famine early warning systems. *Am. Meteorol. Soc.* **2013**, *3*, 588–606.
17. Chen, H.; Chandrasekar, V. The quantitative precipitation estimation system for Dallas–Fort Worth (DFW) urban remote sensing network. *J. Hydrol.* **2015**, *531*, 259–271. [CrossRef]
18. Chung, H.; Liu, C.; Cheng, I.; Lee, Y.; Shieh, M. Rapid response to a typhoon-induced flood with an SAR-derived map of inundated area case study and validation. *Remote Sens.* **2015**, *7*, 11954–11973. [CrossRef]
19. Kannaujiya, S.; Chattoraj, S.L.; Jayalath, D.; Ray, P.K.C.; Bajaj, K.; Podali, S.; Bisht, M.P.S. Integration of satellite remote sensing and geophysical techniques (electrical resistivity tomography and ground penetrating radar) for landslide characterization at Kunjethi (Kalimath), Garhwal Himalaya, India. *Nat. Hazards* **2019**, *97*, 1191–1208. [CrossRef]
20. Baltaci, H. The role of atmospheric processes associated with a flash-flood event over Northwestern Turkey. *Pure Appl. Geophys.* **2020**, *177*, 3513–3526. [CrossRef]
21. Boluwade, A. Remote sensed-based rainfall estimations over the East and West Africa regions for disaster risk management. *ISPRS J. Photogramm. Remote Sens* **2020**, *167*, 305–320. [CrossRef]
22. Sadek, M.; Li, X.; Mostafa, E.; Freeshah, M.; Kamal, A.; Sidi Almouctar, M.A.; Zhao, F.; Mustafa, E.K. Low-cost solutions for assessment of flash flood impacts using sentinel-1/2 data fusion and hydrologic/hydraulic modeling: Wadi El-Natrun Region, Egypt. *Adv. Civ. Eng.* **2020**, *2020*, 1–21. [CrossRef]
23. Munir, B.A.; Ahmad, S.R.; Hafeez, S. Integrated hazard modeling for simulating torrential stream response to flash flood events. *ISPRS Int. J. Geo Inf.* **2020**, *9*, 1. [CrossRef]
24. Sayama, T.; Matsumoto, K.; Kuwano, Y.; Takara, K. Application of backpack-mounted mobile mapping system and rainfall–runoff–inundation model for flash flood analysis. *Water* **2019**, *11*, 963. [CrossRef]
25. Abdelkarim, A.; Gaber, A.; Youssef, A.; Pradhan, B. Flood hazard assessment of the urban area of Tabuk City, Kingdom of Saudi Arabia by integrating spatial-based hydrologic and hydrodynamic modeling. *Sensing* **2019**, *19*, 1024. [CrossRef] [PubMed]
26. Abuzied, S.M.; Mansour, B.M.H. Geospatial hazard modeling for the delineation of flash flood-prone zones in Wadi Dahab basin, Egypt. *J. Hydroinform.* **2019**, *21*, 180–206. [CrossRef]
27. Kamel, M.; Arfa, M. Integration of remotely sensed and seismicity data for geo-natural hazard assessment along the Red Sea Coast, Egypt. *Arab. J. Geosci.* **2020**, *13*, 1–23. [CrossRef]
28. Hadihardaja, I.K.; Vadiya, R. Identification of flash flood hazard zones in mountainous small watershed of Aceh Besar Regency, Aceh Province, Indonesia. *Egypt. J. Remote Sens. Space Sci.* **2019**, *19*, 143–160.
29. Lamovec, P.; Veljanovski, T.; Mikoš, M.; Oštir, K. Detecting flooded areas with machine learning techniques: Case study of the Selška Sora river flash flood in September 2007. *J. Appl. Remote Sens.* **2013**, *7*, 073564. [CrossRef]
30. Bandi, A.S.; Meshapam, S.; Deva, P. A geospatial approach to flash flood hazard mapping in the city of Warangal, Telangana, India. *Environ. Socio-Econ. Stud.* **2019**, *7*, 1–13. [CrossRef]
31. El Alfy, M. Assessing the impact of arid area urbanization on flash floods using GIS, remote sensing, and HEC-HMS rainfall–runoff modeling. *Hydrol. Res.* **2016**, *47*, 1142–1160. [CrossRef]
32. Psomiadis, E.; Tomanis, L.; Kavvadias, A.; Soulis, K.X.; Charizopoulos, N.; Michas, S. Potential dam breach analysis and flood wave risk assessment using HEC-RAS and remote sensing data: A multicriteria approach. *Water* **2021**, *13*, 364. [CrossRef]

33. Estrany, J.; Ruiz-Pérez, M.; Mutzner, R.; Fortesa, J.; Nácher-Rodríguez, B.; Tomàs-Burguera, M.; García-Comendador, J.; Peña, X.; Calvo-Cases, A.; Vallés-Morán, F.J. Hydrogeomorphological analysis and modelling for a comprehensive understanding of flash-flood damage processes: The 9 October 2018 event in northeastern Mallorca. *Nat. Hazards Earth Syst.* **2020**, *20*, 2195–2220. [CrossRef]
34. Mashaly, J.; Ghoneim, E. Flash flood hazard using optical, radar, and stereo-pair derived DEM: Eastern Desert, Egypt. *Remote Sensing* **2018**, *10*, 1204. [CrossRef]
35. Rizeei, H.M.; Pradhan, B.; Saharkhiz, M.A. An integrated fluvial and flash pluvial model using 2D high-resolution sub-grid and particle swarm optimization-based random forest approaches in GIS. *Complex Intell. Syst.* **2019**, *5*, 283–302. [CrossRef]
36. Costache, R.; Pham, Q.B.; Sharifi, E.; Linh, N.T.T.; Abba, S.I.; Vojtek, M.; Vojteková, J.; Nhi, P.T.T.; Khoi, D.N. Flash-flood susceptibility assessment using multi-criteria decision making and machine learning supported by remote sensing and GIS techniques. *Remote Sens.* **2020**, *12*, 106. [CrossRef]
37. Nguyen, V.-N.; Yariyan, P.; Amiri, M.; Tran, A.D.; Pham, T.D.; Do, M.P.; Ngo, P.T.T.; Nhu, V.-H.; Long, N.Q.; Bui, D.T. A New modeling approach for spatial prediction of flash flood with biogeography optimized CHAID tree ensemble and remote sensing data. *Remote Sens.* **2020**, *12*, 1373. [CrossRef]
38. Khosravi, K.; Pourghasemi, H.R.; Chapi, K.; Bahri, M. Flash flood susceptibility analysis and its mapping using different bivariate models in Iran: A comparison between Shannon's entropy, statistical index, and weighting factor models. *Environ. Monit. Assess.* **2016**, *188*, 1–21. [CrossRef] [PubMed]
39. Khosravi, K.; Shahabi, H.; Pham, B.T.; Adamowski, J.; Shirzadi, A.; Pradhan, B.; Dou, J.; Ly, H.; Gróf, G.; Ho, H.L.; et al. A comparative assessment of flood susceptibility modeling using multi-criteria decision-making analysis and machine learning methods. *J. Hydrol.* **2019**, *573*, 311–323. [CrossRef]
40. Sadek, M.; Li, X. Low-cost solution for assessment of urban flash flood impacts using Sentinel-2 Satellite images and Fuzzy Analytic Hierarchy process: A case study of Ras Ghareb City, Egypt. *Adv. Civ. Eng.* **2019**, *2019*, 2561215. [CrossRef]
41. Karmokar, S.; De, M. Flash flood risk assessment for drainage basins in the Himalayan foreland of Jalpaiguri and Darjeeling Districts, West Bengal. *Modeling Earth Syst. Environ.* **2020**, *6*, 2263–2289. [CrossRef]
42. Barasa, B.N.; Perera, E.D.P. Analysis of land use change impacts on flash flood occurrences in the Sosiani River basin Kenya. *Int. J. River Basin Manag.* **2018**, *16*, 179–188. [CrossRef]
43. Youssef, A.M.; Sefry, S.A.; Pradhan, B.; Alfadail, E.A. Analysis on causes of flash flood in Jeddah city (Kingdom of Saudi Arabia) of 2009 and 2011 using multi-sensor remote sensing data and GIS. *Geomat. Nat. Hazards Risk* **2016**, *7*, 1018–1042. [CrossRef]
44. Abuzied, S.; Yuan, M.; Ibrahim, S.; Kaiser, M.; Saleem, T. Geospatial risk assessment of flash floods in Nuweiba area, Egypt. *J. Arid Environ.* **2016**, *133*, 54–72. [CrossRef]
45. Abdel-Lattif, A.; Sherief, Y. Morphometric analysis and flash floods of Wadi Sudr and Wadi Wardan, Gulf of Suez, Egypt: Using digital elevation model. *Arab. J. Geosci.* **2012**, *5*, 181–195. [CrossRef]
46. Li, X.; Li, C.; Bai, D.; Leng, Y. Insights into stem cell therapy for diabetic retinopathy: A bibliometric and visual analysis. *Neural Regen. Res.* **2021**, *16*, 172–178.
47. Jia, G.; Ma, R.; Hu, Z. Review of urban transportation network design problems based on citespace. *Math. Probl. Eng.* **2019**, *2019*, 1–22. [CrossRef]
48. Fang, Y.; Yin, J.; Wu, B. Climate change and tourism: A scientometric analysis using citespace. *J. Sustain. Tour.* **2017**, *26*, 108–126. [CrossRef]
49. Youssef, A.M.; Pradhan, B.; Hassan, A.M. Flash flood risk estimation along the St. Katherine road, southern Sinai, Egypt using GIS based morphometry and satellite imagery. *Environ. Earth Sci.* **2011**, *62*, 611–623. [CrossRef]
50. Foody, G.M.; Ghoneim, E.M.; Arnell, N.W. Predicting locations sensitive to flash flooding in an arid environment. *J. Hydrol.* **2004**, *292*, 48–58. [CrossRef]
51. Krajewski, W.F.; Smith, J.A. Radar hydrology: Rainfall estimation. *Adv. Water Resour.* **2002**, *25*, 1387–1394. [CrossRef]
52. Nhu, V.-H.; Ngo, P.-T.T.; Pham, T.; Dou, J.; Song, X.; Hoang, N.-D.; Tran, D.; Cao, D.; Aydilek, I.; Amiri, M.; et al. A new hybrid Firefly–PSO optimized random subspace tree intelligence for torrential rainfall-induced flash flood susceptible mapping. *Remote Sens.* **2020**, *12*, 2688. [CrossRef]
53. Borga, M.; Stoffel, M.; Marchi, L.; Marra, F.; Jakob, M. Hydrogeomorphic response to extreme rainfall in headwater systems: Flash floods and debris flows. *J. Hydrol.* **2014**, *518*, 194–205. [CrossRef]
54. Psomiadis, E.; Soulis, K.; Zoka, M.; Dercas, N. Synergistic approach of remote sensing and GIS techniques for flash-flood monitoring and damage assessment in Thessaly plain area, Greece. *Water* **2019**, *11*, 448. [CrossRef]
55. Ali, S.A.; Khatun, R.; Ahmad, A.; Ahmad, S.N. Application of GIS-based analytic hierarchy process and frequency ratio model to flood vulnerable mapping and risk area estimation at Sundarban region, India. *Modeling Earth Syst. Environ.* **2019**, *5*, 1083–1102. [CrossRef]
56. Rahmati, O.; Zeinivand, H.; Besharat, M. Flood hazard zoning in Yasooj region, Iran, using GIS and multi-criteria decision analysis. *Geomat. Nat. Hazards Risk* **2016**, *7*, 1000–1017. [CrossRef]
57. Abdelfattah, M.; Saber, M.; Kantoush, S.A.; Khalil, M.F.; Sumi, T.; Sefelnasr, A.M. A Hydrological and Geomorphometric Approach to Understanding the Generation of Wadi Flash Floods. *Water* **2017**, *9*, 553.
58. Saber, M.; Hamaguchi, T.; Kojiri, T.; Tanaka, K.; Sumi, T. A physically based distributed hydrological model of wadi system to simulate flash floods in arid regions. *Arab. J. Geosci.* **2015**, *8*, 143–160. [CrossRef]

59. Eslami, Z.; Shojaei, S.; Hakimzadeh, M.A. Exploring prioritized sub-basins in terms of flooding risk using HEC_HMS model in Eskandari catchment, Iran. *Spat. Inf. Res.* **2017**, *25*, 677–684. [CrossRef]
60. Ezz, H. Integrating GIS and HEC-RAS to model Assiut plateau runoff. *Egypt. J. Remote Sens. Space Sci.* **2018**, *21*, 219–227. [CrossRef]
61. Correia, F.N.; Da Graça Saraiva, M.; Da Silva, F.N.; Ramos, I. Floodplain management in urban developing areas. Part, I. urban growth scenarios and land-use controls. *Water Resour. Manag.* **1999**, *13*, 1–21. [CrossRef]
62. Hamid, H.T.A.; Wenlong, W.; Qiaomin, L. Environmental sensitivity of flash flood hazard using geospatial techniques. *Glob. J. Environ. Sci. Manag.* **2020**, *6*, 31–46.
63. Akter, A.; Tanim, A.H.; Islam, M.K. Possibilities of urban flood reduction through distributed-scale rainwater harvesting. *Water Sci. Eng.* **2020**, *13*, 95–105. [CrossRef]
64. Kia, M.B.; Pirasteh, S.; Pradhan, B.; Mahmud, A.R.; Sulaiman, W.N.A.; Moradi, A. An artificial neural network model for flood simulation using GIS: Johor River Basin, Malaysia. *Environ. Earth Sci.* **2012**, *67*, 251–264. [CrossRef]
65. Elhag, M.; Abdurahman, S.G. Advanced remote sensing techniques in flash flood delineation in Tabuk City, Saudi Arabia. *Nat. Hazards* **2020**, *103*, 3401–3413. [CrossRef]
66. Elkhrachy, I.; Pham, Q.B.; Costache, R.; Mohajane, M.; Rahman, K.U.; Shahabi, H.; Linh, N.T.T.; Anh, D.T. Sentinel-1 remote sensing data and Hydrologic Engineering Centres River Analysis System two-dimensional integration for flash flood detection and modelling in New Cairo City, Egypt. *J. Flood Risk Manag.* **2021**, e12692. [CrossRef]
67. Kocaman, S.; Tavus, B.; Nefeslioglu, H.A.; Karakas, G.; Gokceoglu, C. Evaluation of floods and landslides triggered by a meteorological catastrophe (Ordu, Turkey, August 2018) using optical and radar data. *Geofluids* **2020**, *2020*, 1–18. [CrossRef]
68. Hakdaoui, S.; Emran, A.; Pradhan, B.; Lee, C.; Nguemhe Fils, S.C. A collaborative change detection approach on multi-sensor spatial imagery for desert Wetland monitoring after a flash flood in Southern Morocco. *Remote Sens.* **2019**, *11*, 1042. [CrossRef]
69. Chen, S.; Hong, Y.; Cao, Q.; Kirstetter, P.; Gourley, J.J.; Qi, Y.; Zhang, J.; Howard, K.; Hu, J.; Wang, J. Performance evaluation of radar and satellite rainfalls for Typhoon Morakot over Taiwan: Are remote-sensing products ready for gauge denial scenario of extreme events? *J. Hydrol.* **2013**, *506*, 4–13. [CrossRef]
70. Moeyersons, J.; Trefois, P.; Nahimana, L.; Ilunga, L.; Vandecasteele, I.; Byizigiro, V.; Sadiki, S. River and landslide dynamics on the western Tanganyika rift border, Uvira, D.R. Congo: Diachronic observations and a GIS inventory of traces of extreme geomorphologic activity. *Nat. Hazards* **2010**, *53*, 291–311. [CrossRef]
71. Arnous, M.O.; Aboulela, H.A.; Green, D.R. Geo-environmental hazards assessment of the north western Gulf of Suez, Egypt. *J. Coast. Conserv.* **2011**, *15*, 37–50. [CrossRef]
72. Arnous, M.O.; Green, D.R. GIS and remote sensing as tools for conducting geo-hazards risk assessment along Gulf of Aqaba coastal zone, Egypt. *J. Coast. Conserv.* **2011**, *15*, 457–475. [CrossRef]
73. Soussa, H.; El Feel, A.A.; Alfy, S.Z.; Yousif, M.S.M. Flood hazard in Wadi Rahbaa area, Egypt. *Arab. J. Geosci.* **2012**, *5*, 45–52. [CrossRef]
74. Masoud, A.A. Runoff modeling of the wadi systems for estimating flash flood and groundwater recharge potential in Southern Sinai, Egypt. *Arab. J. Geosci.* **2011**, *4*, 785–801. [CrossRef]
75. Asode, A.N.; Sreenivasa, A.; Lakkundi, T.K. Quantitative morphometric analysis in the hard rock Hirehalla sub-basin, Bellary and Davanagere Districts, Karnataka, India using RS and GIS. *Arab. J. Geosci.* **2016**, *9*, 381. [CrossRef]
76. Al-Saady, Y.I.; Al-Suhail, Q.A.; Al-Tawash, B.S.; Othman, A.A. Drainage network extraction and morphometric analysis using remote sensing and GIS mapping techniques (Lesser Zab River Basin, Iraq and Iran). *Environ. Earth Sci.* **2016**, *75*, 1–23. [CrossRef]
77. Jahan, C.S.; Rahaman, M.F.; Arefin, R.; Ali, S.; Mazumder, Q.H. Morphometric analysis and hydrological inference for water resource management in Atrai-Sib River Basin, NW Bangladesh using remote sensing and GIS technique. *J. Geol. Soc. India* **2018**, *91*, 613–620. [CrossRef]
78. Senatore, A.; Furnari, L.; Mendicino, G. Impact of high-resolution sea surface temperature representation on the forecast of small Mediterranean catchments' hydrological responses to heavy precipitation. *Hydrol. Earth Syst. Sci.* **2020**, *24*, 269–291. [CrossRef]
79. Bartsotas, N.S.; Anagnostou, E.N.; Nikolopoulos, E.I.; Kallos, G. Investigating satellite precipitation uncertainty over complex terrain. *J. Geophys. Res. Atmos.* **2018**, *123*, 5346–5359. [CrossRef]
80. Tehrany, M.S.; Pradhan, B.; Jebur, M.N. Flood susceptibility mapping using a novel ensemble weights-of-evidence and support vector machine models in GIS. *J. Hydrol.* **2014**, *512*, 332–343. [CrossRef]
81. Nashwan, M.S.; Shahid, S.; Dewan, A.; Ismail, T.; Alias, N. Performance of five high resolution satellite-based precipitation products in arid region of Egypt: An evaluation. *Atmos. Res.* **2020**, *236*, 104809. [CrossRef]
82. Chen, M.; Nabih, S.; Brauer, N.S.; Gao, S.; Gourley, J.J.; Hong, Z.; Kolar, R.L.; Hong, Y. Can remote sensing technologies capture the extreme precipitation event and its cascading hydrological response? A case study of hurricane harvey using EF5 modeling framework. *Remote Sens.* **2020**, *12*, 445. [CrossRef]
83. Levizzani, V.; Laviola, S.; Cattani, E.; Costa, M.J. Extreme precipitation on the Island of Madeira on 20 February 2010 as seen by satellite passive microwave sounders. *Eur. J. Remote Sens.* **2013**, *46*, 475–489. [CrossRef]
84. Behrangi, A.; Sorooshian, S.; Hsu, K. Summertime evaluation of REFAME over the Unites States for near real-time high resolution precipitation estimation. *J. Hydrol.* **2012**, *456–457*, 130–138. [CrossRef]
85. Levizzani, V.; Cattani, E. Satellite Remote sensing of precipitation and the terrestrial water cycle in a changing climate. *Remote Sens.* **2019**, *11*, 2301. [CrossRef]

86. Li, X.; Lin, J.; Zhao, W.; Wen, F. Approximate calculation of flash flood maximum inundation extent in small catchment with large elevation difference. *J. Hydrol.* **2020**, *590*, 125195. [CrossRef]
87. Ahmed, M. Remote sensing-based quantification of the impact of flash flooding on the rice production: A case study over Northeastern Bangladesh. *Sensing* **2017**, *17*, 2347.
88. Dao, P.; Liou, Y. Object-based flood mapping and affected rice field estimation with landsat 8 OLI and MODIS data. *Remote Sensing* **2015**, *7*, 5077–5097. [CrossRef]
89. Wang, Y.; Colby, J.D.; Mulcahy, K.A. An efficient method for mapping flood extent in a coastal floodplain using Landsat TM and DEM data. *Int. J. Remote Sens.* **2002**, *23*, 3681–3696. [CrossRef]
90. Gerl, T.; Bochow, M.; Kreibich, H. Flood damage modeling on the basis of urban structure mapping using high-resolution remote sensing data. *Water* **2014**, *6*, 2367–2393. [CrossRef]
91. Kastridis, A.; Kirkenidis, C.; Sapountzis, M. An integrated approach of flash flood analysis in ungauged Mediterranean watersheds using post-flood surveys and unmanned aerial vehicles. *Hydrol. Process.* **2020**, *34*, 4920–4939. [CrossRef]
92. Ghoneim, E.; Foody, G.M. Assessing flash flood hazard in an arid mountainous region. *Arab. J. Geosci.* **2013**, *6*, 1191–1202. [CrossRef]
93. Singh, S.; Dhote, P.R.; Thakur, P.K.; Chouksey, A.; Aggarwal, S.P. Identification of flash-floods-prone river reaches in Beas river basin using GIS-based multi-criteria technique: Validation using field and satellite observations. *Nat. Hazards* **2021**, *105*, 2431–2453. [CrossRef]
94. Wahid, A.; Madden, M.; Khalaf, F.; Fathy, I. Geospatial analysis for the determination of hydro-morphological characteristics and assessment of flash flood potentiality in arid coastal plains: A case in Southwestern Sinai, Egypt. *Earth Sci. Res. J.* **2016**, *20*, 1–9. [CrossRef]
95. Elkhrachy, I. Assessment and management flash flood in Najran Wady using GIS and remote sensing. *J. Indian Soc. Remote Sens* **2018**, *46*, 297–308. [CrossRef]
96. Radwan, F.; Alazba, A.A.; Mossad, A. Flood risk assessment and mapping using AHP in arid and semiarid regions. *Acta Geophys.* **2019**, *67*, 215–229. [CrossRef]
97. Costache, R.; Bao Pham, Q.; Corodescu-Roșca, E.; Cîmpianu, C.; Hong, H.; Thi Thuy Linh, N.; Ming Fai, C.; Najah Ahmed, A.; Vojtek, M.; Muhammed Pandhiani, S.; et al. Using GIS, remote sensing, and machine learning to highlight the correlation between the land-use/land-cover changes and flash-flood potential. *Remote Sens.* **2020**, *12*, 1422. [CrossRef]
98. Bui, Q.; Nguyen, Q.; Nguyen, X.L.; Pham, V.D.; Nguyen, H.D.; Pham, V. Verification of novel integrations of swarm intelligence algorithms into deep learning neural network for flood susceptibility mapping. *J. Hydrol.* **2020**, *581*, 124379. [CrossRef]
99. Santangelo, N.; Santo, A.; Di Crescenzo, G.; Foscari, G.; Liuzza, V.; Sciarrotta, S.; Scorpio, V. Flood susceptibility assessment in a highly urbanized alluvial fan: The case study of Sala Consilina (southern Italy). *Nat. Hazard. Earth Syst.* **2011**, *11*, 2765–2780. [CrossRef]
100. Cao, C.; Xu, P.; Wang, Y.; Chen, J.; Zheng, L.; Niu, C. Flash flood hazard susceptibility mapping using frequency ratio and statistical index methods in coalmine subsidence areas. *Sustainability* **2016**, *8*, 948. [CrossRef]
101. Vojtek, M.; Vojteková, J. Flood susceptibility mapping on a national scale in Slovakia using the analytical hierarchy process. *Water* **2019**, *11*, 364. [CrossRef]
102. Shirzadi, A.; Asadi, S.; Shahabi, H.; Ronoud, S.; Clague, J.J.; Khosravi, K.; Pham, B.T.; Ahmad, B.B.; Bui, D.T. A novel ensemble learning based on Bayesian Belief Network coupled with an extreme learning machine for flash flood susceptibility mapping. *Eng. Appl. Artif. Intell.* **2020**, *96*, 103971. [CrossRef]
103. Meng, J.; Fenglin, L.; Huxing, L. Research on the division of risk areas of mountain flood disasters in Henan Province based on GIS. *Flood Control Drought Relief China* **2017**, *27*, 54–59.
104. Prasad, R.N.; Pani, P. Geo-hydrological analysis and sub watershed prioritization for flash flood risk using weighted sum model and Snyder's synthetic unit hydrograph. *Modeling Earth Syst. Environ.* **2017**, *3*, 1491–1502. [CrossRef]

remote sensing

MDPI

Article

Watch Out for the Tailings Pond, a Sharp Edge Hanging over Our Heads: Lessons Learned and Perceptions from the Brumadinho Tailings Dam Failure Disaster

Deqiang Cheng [1], Yifei Cui [2,*], Zhenhong Li [3,4,5] and Javed Iqbal [6,7,8]

1 Key Research Institute of Yellow River Civilization and Sustainable Development & Collaborative Innovation Center on Yellow River Civilization Jointly Built by Henan Province and Ministry of Education, Henan University, Kaifeng 475001, China; 10340052@henu.edu.cn
2 State Key Laboratory of Hydroscience and Engineering, Tsinghua University, Beijing 100084, China
3 College of Geological Engineering and Geomatics, Chang'an University, Xi'an 710054, China; zhenhong.li@chd.edu.cn
4 Key Laboratory of Western China's Mineral Resource and Geological Engineering, Ministry of Education, Xi'an 710054, China
5 Big Data Center for Geosciences and Satellites (BDCGS), Xi'an 710054, China
6 Key Laboratory of Mountain Hazards and Earth Surface Process, Institute of Mountain Hazards and Environment, Chinese Academy of Sciences, Chengdu 610041, China; javediqbalgeo@uoh.edu.pk
7 Department of Earth Sciences, The University of Haripur, Haripur 22620, Pakistan
8 China-Pakistan Joint Research Centre on Earth Sciences, Chinese Academy of Sciences, Islamabad 45320, Pakistan
* Correspondence: yifeicui@mail.tsinghua.edu.cn; Tel.: +86-010-62788585

Abstract: A catastrophic tailings dam failure disaster occurred in Brumadinho, Brazil on 25 January 2019, which resulted in over 270 casualties, 24,000 residents evacuated, and a huge economic loss. Environmental concerns were raised for the potential pollution of water due to tailings waste entering the Paraopeba River. In this paper, a detailed analysis has been carried out to investigate the disaster conditions of the Brumadinho dam failure using satellite images with different spatial resolutions. Our in-depth analysis reveals that the hazard chain caused by this failure contained three stages, namely dam failure, mudflow, and the hyperconcentrated flow in the Paraopeba River. The variation characteristics of turbidity of the Rio Paraopeba River after the disaster have also been investigated using high-resolution remote sensing images, followed by a qualitative analysis of the impacts on the downstream reservoir of the Retiro Baixo Plant that was over 300 km away from the dam failure origin. It is believed that, on the one hand, the lack of dam stability management at the maintenance stage was the main cause of this disaster. On the other hand, the abundant antecedent precipitation caused by extreme weather events should be a critical triggering factor. Furthermore, the spatiotemporal pattern mining of global tailings dam failures revealed that the Brumadinho dam disaster belonged to a Consecutive Hot Spot area, suggesting that the regular drainage inspection, risk assessment, monitoring, and early warning of tailings dam in Consecutive Hot Spot areas still need to be strengthened for disaster mitigation.

Keywords: hazard chain; turbidity; suspended sediment detection; extreme climate events; tailing dam risk management; spatiotemporal pattern mining; El Niño

Citation: Cheng, D.; Cui, Y.; Li, Z.; Iqbal, J. Watch Out for the Tailings Pond, a Sharp Edge Hanging over Our Heads: Lessons Learned and Perceptions from the Brumadinho Tailings Dam Failure Disaster. *Remote Sens.* **2021**, *13*, 1775. https://doi.org/10.3390/rs13091775

Academic Editors: Paolo Mazzanti and Saverio Romeo

Received: 18 January 2021
Accepted: 26 April 2021
Published: 2 May 2021

1. Introduction

Tailings are the material left after the valuable parts have been separated from the uneconomic or low-economic ore. A tailings dam is typically an earth-fill embankment dam which is usually designed for permanent containment by intercepting valleys or enclosing lands in order to form a tailings pond used to store metal or non-metal ore-separation and discharging tailings or other industrial waste residues. Consequently, the tailings dam is a dangerous source of man-made mudflow with high potential energy. The failure of the tailings dam is one of the most dangerous disasters causing serious accidents. Tailings dams are considered more vulnerable than hydraulic dams due to the lack of regulations on specific design criteria, stability requirements regarding monitoring during the construction and maintenance process, and high potential of pollution due to its filled material (solid waste) [1].

The mining byproducts collected in tailings dams may damage the environment by releasing toxic metals and poisoning the aquatic wildlife that relies on clear water [2]. In past decades, many researchers have investigated tailings dam failures using different research methods, e.g., site investigation, numerical simulation, and remote sensing technology. A field investigation has been proven to be a good method to obtain first-hand data when accessibility is possible to the disaster site. For example, Macklin et al. [3] collected sediment samples affected by tailings dam failures to assess the long-term fate and environmental significance of contaminant metals. Porsani et al. [4] used Ground-Penetrating Radar (GPR) on an iron mining area after the collapse of the tailings dam I at the Córrego do Feijão Mine in Brumadinho-MG, Brazil to map bodies, structural buildings, and equipment buried in the mud. A numerical simulation is another effective means, which can help us understand the tailing flow characteristics and assess the possible extents of the affected areas [5,6]. With the advance in remote sensing technology, remote sensing has been becoming a more and more important means of information acquisition in disaster investigation [7–10]. Many scholars have applied remote sensing technology to the study of tailings dam failure disasters. For example, Silveira et al. [11] used semivariogram indices derived from NDVI images to obtain an object-based change detection caused by the Mariana dam disaster. Grenerczy and Wegmüller [12] performed a Persistent Scatterer InSAR (PSI) analysis to examine the embankment failure of a red mud reservoir. The tailings dam failure is a kind of disaster which could affect wide areas, especially those along rivers, and different disaster characteristics often appear in different regions. Therefore, different remote sensing techniques are usually requested to be employed to analyze a series of remote sensing images to reflect the whole disaster process. In this study, we attempt to make full use of the available remote sensing images to examine the Brumadinho tailings failure disaster in order to make people realize the great harm of tailings dam failures. In addition, this paper demonstrates how RS techniques can be used to characterize and monitor the evolution of such complex processes, which provides a reference for disaster prevention and mitigation.

2. Background of the Brumadinho Tailings Dam Failure Disaster

Brumadinho is a Brazilian municipality, located near the Paraopeba River at an altitude of 880 m. It belongs to the microregion of Belo Horizonte, Metropolitana de Belo Horizonte, Minas Gerais, Brazil (Figure 1). Just after noon on 25 January 2019, the Brumadinho dam disaster occurred when Dam I (Figures 2 and 3), a tailings dam at the Córrego do Feijão iron ore mine owned by Vale, 9 km (5.6 mi) east of Brumadinho (Figure 1b), suffered a catastrophic failure [13].

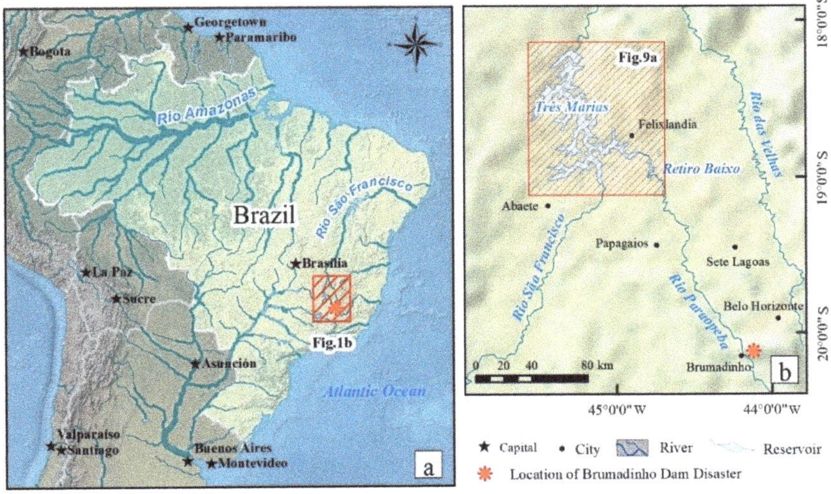

Figure 1. Location of Brumadinho dam disaster. (**a**) Location of the disaster in Brazil; (**b**) relative relationship between the disaster location and two reservoirs. Note that in Figure 1b, the position of two reservoirs were marked using a red box whose extent corresponds to Figure 9a.

Figure 2. Video screenshots of the Brumadinho tailings dam failure [14]; (**a**) 25 January 2019, 12:28:21; (**b**) 25 January 2019, 12:28:36; (**c**) 25 January 2019, 12:28:43; (**d**) 25 January 2019, 12:28:52 (local time).

In this disaster, at least 248 people were confirmed dead, and 22 missing. Most of the victims were Vale's employees. At a railroad branch, in the Córrego do Feijão region, three locomotives and 132 wagons were buried, and four railway men were missing. Two sections of the railway bridge (Figure 3) and about 100 m of railway track were also struck and destroyed by the mud [15]. Due to the potential hazards, about 24,000 residents from several districts of Brumadinho were evacuated [16]. Many agricultural areas were affected or totally destroyed, and the local livestock industry suffered damages due to the loss of animals such as cattle and poultry [17]. In addition, the tailings dam failure spilt about 12 million cubic meters of mud and sludge [18] and some came into and ran along the

Paraopeba River. The metals in the tailings may be adsorbed by the river sediments or may pollute the soil in the floodplain, and would end up affecting the region's ecosystem.

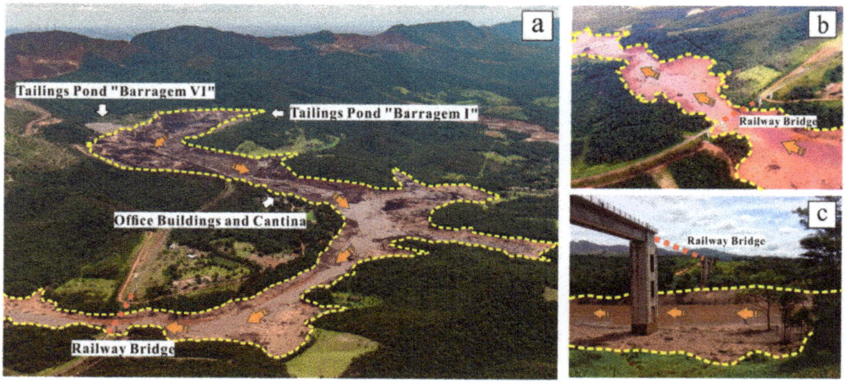

Figure 3. Disaster pictures. (**a**) Aerial view of Brumadinho dam disaster taken at 11:46 on 27 January 2019 [19]; (**b**) video screenshot of the destroyed railway bridge, 3 km downstream from the collapsed dam, on 26 January 2019 from YouTube [20]; (**c**) picture of iron ore railway bridge destroyed by mudflow taken on 26 January 2019 [21]. Note that the yellow line represents the extents of the areas affected by the mudflow, the red line represents the railway bridge which was damaged, and the brown arrow indicates the movement direction of the mudflow.

3. Materials and Methods

3.1. Materials

3.1.1. High-Resolution Remote Sensing Images

Freely available remote sensing (RS) imagery can be used to investigate natural hazards such as landslides [9,22], debris flows [8,9] and mountain fires [23]. Google Earth is an important open source of high-resolution remote sensing imagery, and more importantly, it can provide multi-temporal remote sensing datasets [24–26]. In this study, seven high-resolution remote sensing images from Google Earth were used to track the movement of sediment in the channel after the tailings entered the Paraopeba River: Three images at the confluence of the debris flow gully and the Paraopeba River (collected at different times), one image at the stage of sediment transport in the river, and three others covering the area where the river enters the Retiro Baixo reservoir (different times).

3.1.2. Medium-Resolution Remote Sensing Images

Landsat satellite images of National Aeronautics and Space Administration (NASA) are important medium-resolution image datasets that can be used to investigate natural hazards. Landsat 8, as an American Earth observation satellite, is the eighth satellite launched on 11 February 2013 in the Landsat program. It has two sensors including the Operational Land Imager (OLI) and the Thermal Infrared Sensor (TIRS). The OLI consists of 8 bands with a spatial resolution of 30 m, and a 15-m panchromatic band. The TIRS can provide 100-m thermal infrared images.

In this study, Landsat 8 images were used to investigate the hazard chain. They were true color synthesized images of the pre- and post-disaster and released by the researchers of NASA without copyright protection [27,28]. Landsat 8 images were also used to examine the sediment concentration of pre- and post-disaster in the reservoirs. Three Landsat 8 images of the pre- and post-disaster (Table 1) were downloaded from the USGS Global Visualization Viewer (GloVis) website [29], and used to examine the diffusion of waste in the reservoirs of two hydroelectric plants.

Table 1. Landsat 8 remote sensing images used for the examination of sediment concentration [29].

Type	File Name	Resolution	Date
Pre-disaster	LC08_L1TP_219073_20181220_20181227_01_T1	15/30/100 m (panchromatic/multispectral/thermal)	20 December 2018
Post-disaster	LC08_L1TP_219073_20190222_20190222_01_RT	15/30/100 m (panchromatic/multispectral/thermal)	22 February 2019
	LC08_L1TP_219073_20190427_20190508_01_T1	15/30/100 m (panchromatic/multispectral/thermal)	27 April 2019

3.1.3. Global Tailings Dam Failures Database

There are about 3500 active tailings ponds in the world, among which 2000 experience about two to five known "major" failures, and 35 "minor" failures annually [30]. During the period from 2007 to 2017, there were at least 10 very serious mine tailings dam failures involving multiple loss of life, with approximately 20 lives per incident, a release of at least 1 million cubic meters of waste each time, and a travel of 20 km or longer every waste movement [31]. Based on the world mine tailings failure data with more than 300 records during the period from 1915 to 2019 [14], a global spatial geographic database of tailings dam failure was made using ArcGIS. Using the "Natural Break" classification method, the tailings dam failure records could be divided into five categories (Figure 4).

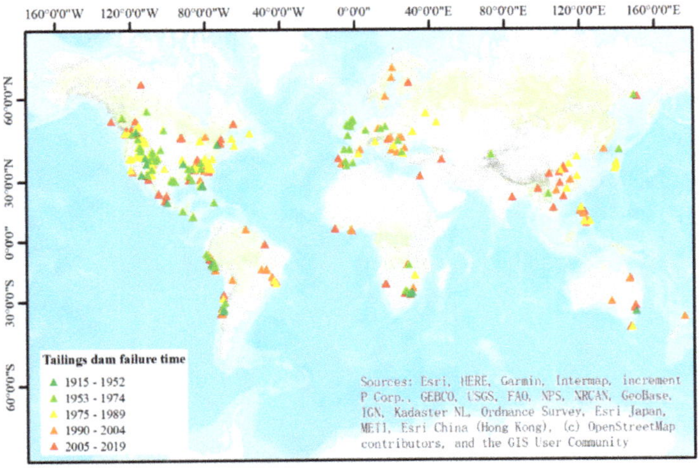

Figure 4. Tailings dam failures (1915–2019) with their occurrence times.

3.2. Methods

3.2.1. FLAASH Atmospheric Correction and Remote Sensing Image Fusion

Solar radiation needs to pass through the atmosphere before it is collected by satellites [32]. Due to this, remote sensing images include complex information derived from the atmosphere and the Earth's surface. As this research is focused on the quantitative analysis of surface reflectance, we need to mitigate the influence from the atmosphere. Using the Atmospheric Correction Module, we can compensate for atmospheric effects.

Atmospheric correction can be realized using many available software tools. For example, the Atmospheric Correction Module in the ENVI software [33] provides two atmospheric correction modeling tools for retrieving spectral reflectance from multispectral and hyperspectral radiance images: Quick Atmospheric Correction (QUAC) and Fast Line-of-sight Atmospheric Analysis of Spectral Hypercubes (FLAASH). The accuracy of FLAASH model is higher than that of QUAC model. The application of QUAC model is simpler than that of FLAASH, and it has less dependence on input parameters and calibration accuracy of instruments [34,35]. FLAASH is a first-principle atmospheric correction tool that corrects wavelengths in the visible through near-infrared and shortwave infrared regions. In this study, the atmospheric correction of Landsat images was carried out using the FLAASH tool within the ENVI software [36].

Image fusion in remote sensing has several application domains. An important domain is multi-resolution image fusion [37]. Many different multi-resolution image fusion methods are available with different characteristics [38], including Gram–Schmidt Pan Sharpening [39], HSV Transformation [40], and Brovey Transformation [41]. Using these image fusion methods, important information from multiple images can be gathered together to form a new image with both high spatial resolution and multispectral characteristics. The OLI has two types of images including panchromatic images and multispectral images. On the basis of comparing different methods, we selected the Gram–Schmidt Pan Sharpening method to fuse images due to its superiority to maintain spatial texture information, especially to keep spectral features with high fidelity [42].

3.2.2. Waterbody Extraction

In order to carry out the research of suspended sediment information in the reservoirs, it is necessary to obtain an accurate extent of the reservoirs. Due to the water spectral characteristics of the near-infrared band absorbing strongly, but reflecting highly in the green band (Figure 5), the Normalized Difference Water Index (NDWI) was proposed by Mcfeeters [43] as follows:

$$\text{NDWI} = (\text{Green} - \text{NIR})/(\text{Green} + \text{NIR}) \quad (1)$$

where Green and NIR are reflectance factors in green and near-infrared bands, corresponding to Bands 3 and 5 of Landsat 8 imagery. After calculating the NDWI, we used 0 as the segmentation threshold to extract the water body, and the water body boundaries were manually extracted in the ArcGIS software.

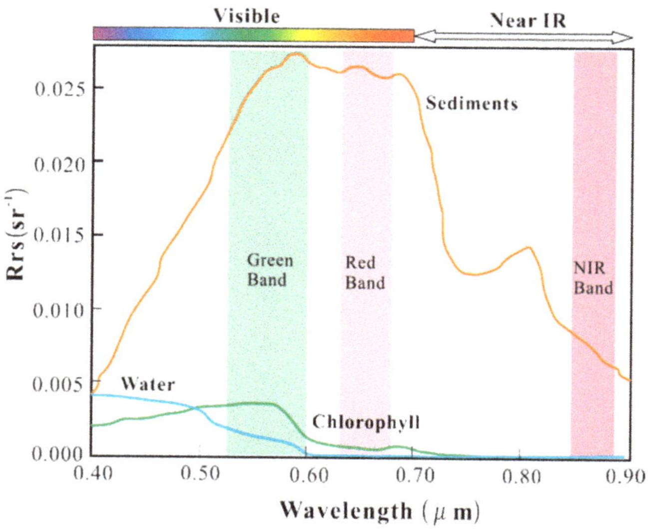

Figure 5. Remote sensing reflectance (Rrs) of clear water (blue), water with chlorophyll content (green), and water with sediments (orange). The green, red, and NIR bands of Landsat 8 images are drawn in the above figure. Note that the figure is modified from Sherry's research [44].

3.2.3. Suspended Sediment Detection

Remote sensing techniques have been widely used to measure qualitative parameters of water bodies [45], including turbidity [46], chlorophyll-a [47], Colored Dissolved Organic matters (CDOM) [48], Secchi disk depth [49], and water temperature [50]. Here, the concentration of suspended sediment is chosen to examine the degree of waste pollution of Paraopeba River after the Brumadinho dam disaster. The suspended sediment is one of the most important water quality parameters, which directly affects the optical properties of water, such as transparency, turbidity, watercolor, and aquatic ecological conditions [51]. In particular, the level of water turbidity is dependent on the concentration of suspended sediment in the water body. With the increase of suspended particles, it is more difficult for light to travel through the water, and as a result the turbidity of the water increases accordingly. To date, many remote sensing quantitative models have been developed to monitor suspended sediments in water bodies, and several researchers used both the single and double band algorithms to calculate the concentration of suspended sediment of the water body [52–54]. The reflectivity of suspended sediment water is higher in the green and red bands (Figure 5). According to the above band reflectance characteristics, Wang et al. [55] proposed the concept of sediment index as follows:

$$SI = (Green + Red)/(Green/Red) \tag{2}$$

where SI is the sediment parameter, Green and Red are the reflectance in green and red bands, corresponding to 30 m resolution bands 3 and 4 of the Landsat 8 OLI image. Compared with the field-measured data, the correlation coefficient between the measured data and SI value is 0.89 [55], which shows that this method can directly and quantitatively reflect the relative concentration distribution of suspended sediments. The following indicators (Table 2) were used as the criteria to divide different suspended sediment water bodies (M represents the average, D represents the standard deviation, and MIN represents the minimum value):

Table 2. Level of sediment concentration in water bodies.

Indicator Criteria	Level of Sediment Concentration
$SI > M + D$	High suspended sediments
$M < SI \leq M + D$	Medium suspended sediments
$M - D < SI \leq M$	Low suspended sediments
$MIN < SI \leq M - D$	Clean water

3.2.4. Spatiotemporal Pattern Mining

Spatiotemporal pattern mining is often used to analyze data distribution and patterns in space and time. The emerging spatiotemporal hot spot analysis regards data cubes as input and identifies statistically significant hot and cold point trends over time. Using this method, the spatiotemporal hot spots of tailings dam failure database were analyzed. In this study, five main hot spots including New Hot Spot, Consecutive Hot Spot, Sporadic Hot Spot, Oscillating Hot Spot, and No Pattern were detected. Their definitions are listed in Table 3.

3.3. Technical Route

In order to make the structure of the article clearer, the technical route is shown in Figure 6. In this study, we made full use of the available remote sensing images to examine the Brumadinho tailings dam failure disaster. Firstly, we collected different remote sensing data from different data sources. Secondly, we used the true color remote sensing images from NASA to investigate the hazard chain along the gully where the dam failure occurred. Thirdly, considering that the river width is narrow and the medium resolution remote sensing image cannot meet the needs, we used the multi-temporal high-resolution remote sensing images from Google Earth to interpret the transport process of waste along the Paraopeba River. Fourthly, we used the original Landsat 8 images to carry out the analysis of waste diffusion in the reservoirs. Through the above procedure, the whole disaster process was clearly recovered using RS techniques. Last but not least, we used the global tailings dam failures database to examine tailings-dam-failure trends based on spatiotemporal pattern mining, and found that this area where the Brumadinho tailings dam failure occurred belonged to the Consecutive Hot Spot area with a relatively high risk.

Table 3. Definitions of different hot spots [56].

Name	Meaning
New Hot Spot	A location that is a statistically significant hot spot for the final time step, and has never been a statistically significant hot spot before.
Consecutive Hot Spot	A location that is a single uninterrupted run of statistically significant hot spot in the final time-step intervals.
Sporadic Hot Spot	A location that is an on-again then off-again hot spot.
Oscillating Hot Spot	A location that is a statistically significant hot spot for the final time-step interval with a history of also being a statistically significant cold spot during a prior time step.
No Pattern Detected	A location that does not fall into any of the hot or cold spot patterns defined above.

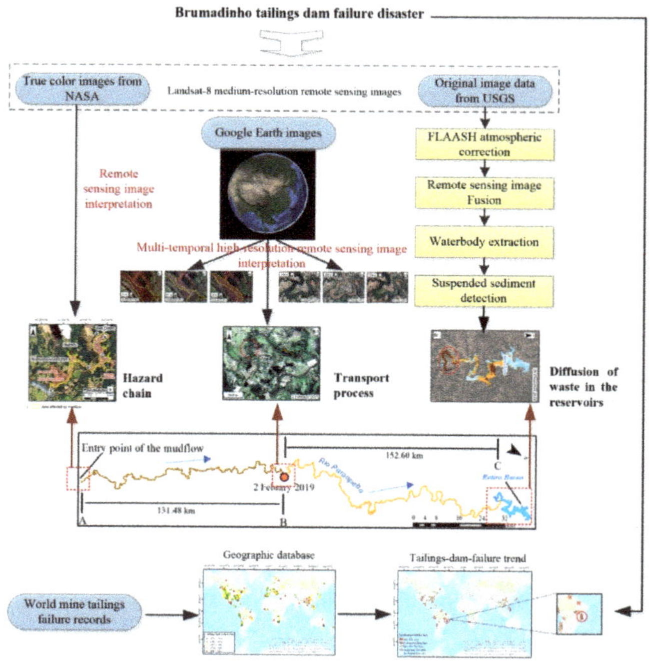

Figure 6. Technical route.

4. Results

4.1. Hazard Chain Caused by This Event

Hazard chains are the hot topic in the broader realm of natural hazards [57] (e.g., earthquake-induced chains [58], glacial-outburst-induced chains [59], and volcano-eruption-induced chains [60]). In this disaster, the dam released a mudflow of tailings after the dam failure (Figure 7b-A). The high-speed mudflow struck the mine's administrative area (Figure 7b-C) [61], destroyed the railway bridge (Figure 3b,c and Figure 7b-D), and continued to move downstream. At about 3:50 pm on 25 January 2019, the mud reached and came into the Paraopeba River (Figure 7b-E). On 27 January 2019, around 5:30 am, sirens were sounded for the stability of the mine's adjacent Dam VI [16] (Figure 7b-B), where increased water levels were observed.

This hazard chain contained three stages including dam failure, mudflow, and hyperconcentrated flow with tailings waste (Figure 7b). The occurrence of this hazard chain is the result of a combination of many factors. Since the tailings contained a certain amount of water, it created conditions for the mudflow after the tailings dam failure. The tailings waste entered the river following the original branch channel, which in turn enlarged the impact of this disaster with extremely high turbidity and metal concentrations, lower dissolved oxygen, and change of microbial communities which would impact the growth and reproduction of aquatic creatures [62,63].

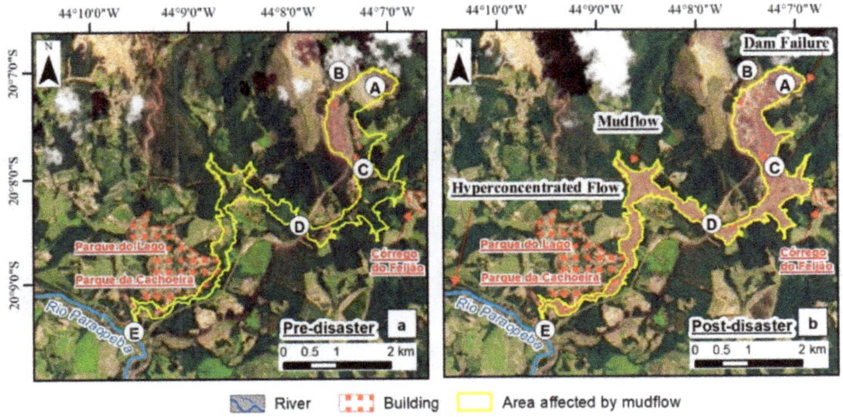

Figure 7. Comparison of pre- and post-disaster Landsat 8 images [64]. (**a**) Pre-disaster remote sensing image (14 January 2019) [27]; (**b**) post-disaster remote sensing image (30 January 2019) [28]. A = location of the destroyed tailings dam and the tailings pond "Barragem I" on 25 January 2019. B = location of tailings pond "Barragem VI" which appeared as an early warning of stability on 27 January 2019. C = location of the destroyed cantina and office buildings. D = location of the destroyed railway bridge (Figure 3b,c). E = location of the entry point of the mudflow into the Paraopeba River.

4.2. Transport Process of Waste in the Rio Paraopeba River

Through the examination of multi-temporal Google Earth images, the transport process of waste in the Paraopeba River can be observed (Figure 8). Comparing Figure 8A-1 with 8A-2, a large amount of waste entered the river several days after the failure, which might block the river for a certain period. Figure 8A-3 shows that due to the increase of precipitation in the later period (Figure 13), the water level of the river increased and eroded a new channel.

Figure 8B-1 shows the location of the waste as of 2 February 2019. It can be observed that the color of the right river section is vermeil, compared with the left section (red circle). The length of AB reach is 131.48 km with a height difference of about 60 m. It took less than a week for the waste to transport from A to B. After the waste entered the Paraopeba River, the transport speed of waste in the water was affected by many factors, such as concentration and stream gradients [65,66]. In addition, according to remote sensing images of Google Earth, it appears that the barrier of some small river dams in the Paraopeba River might also slow down the movement of the waste. Figure 8B-1 is a true-color image of tailings in rivers, and the change of water color in the circle position can be observed. In order to make the watercolor contrast more obvious, considering that the green and red bands are sensitive to the sediment [67], the ratio between these bands were calculated. As can be seen in Figure 8B-2, there is an obvious change in water color at the red circle.

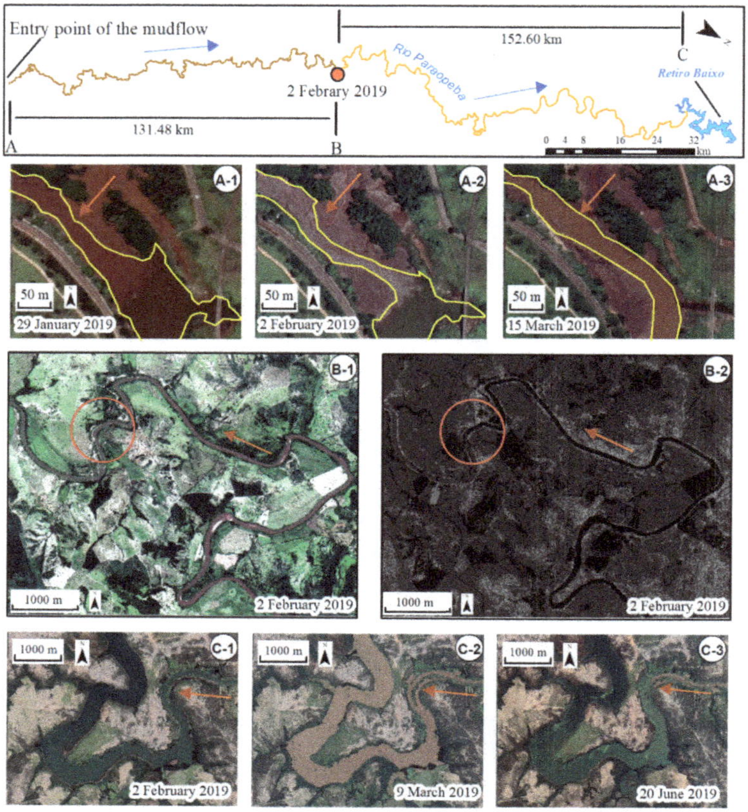

Figure 8. Transport process of waste in the Paraopeba River. A—1, 2, 3 = images of the entry point of the mudflow into the Paraopeba River, the yellow lines are used to mark the river boundary; B—1, 2 = images on 2 February 2019 around point B in the Paraopeba River, the watercolor change can be seen; C—1, 2, 3 = images of the entry point into the Retiro Baixo.

Comparing Figure 8C-1 with 8C-2, the influence of waste on the water body was obvious after it entered the reservoir of hydropower station—Retiro Baixo. A few months later, the watercolor recovered due to the deposition of sediments. Based on the analysis of Figure 8 and multi-temporal Google Earth images, a table of sediment transport time node was generated (Table 4) in order to make the interpretation of sediment transport clear.

Table 4. Sediment transport time nodes.

Locations	Dates					
	25 January	**29 January**	**2 February**	**9 March**	**15 March**	**20 June**
A		A-1	A-2		A-3	Unknown
B			B-1 and B-2			
C			C-1	C-2		C-3

Note: The brown color indicates that the location exhibits sediments on that specific date, while the blue color implies that there was no sediment transported to this location or sediments had settled down.

4.3. *Diffusion of Waste in the Reservoirs of Two Hydroelectric Plants*

(1) Determination of reservoir boundaries

Using the Atmospheric Correction Module, it can be accurately compensated for atmospheric effects. In this study, the atmospheric correction using the FLAASH model [34,35,68] was performed and Landsat 8 images were fused using the Gram—Schmidt Pan Sharpening method [69–72]. After calculating the NDWI using the images after atmospheric correction and image fusion, zero was used as the segmentation threshold to extract the water body, and the manual editing was used to complete the extracted water body boundaries in the ArcGIS software. Figure 9 shows the boundaries of two reservoirs, Retiro Baixo and Três Marias.

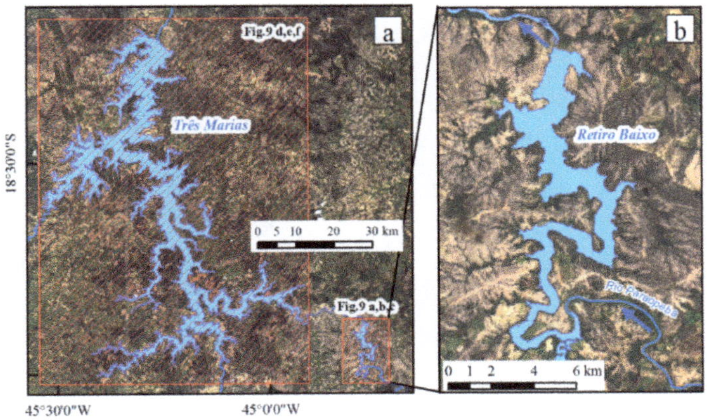

Figure 9. (**a**) Retiro Baixo and Três Marias reservoirs. The larger reservoir on the left is Três Marias, and the smaller one in the lower right corner is Retiro Baixo (**b**).

(2) Sediment index used to estimate the level of sediment concentration

The sediment index was obtained using the SI calculating method. Figure 10 shows the pre- and post-disaster sediment concentrations. Contrasting the area of high sediment concentration where the river enters into the Retiro Baixo reservoir (Figure 10a–c), it had a larger area of 3.66 km^2 compared with 1.89 km^2 on 20 December 2018 and 2.49 km^2 on 27 April 2019 with a 2-month interval spanning this disaster. It can be observed that the sediment plume had a great impact on the reservoir of the Retiro Baixo Plant, over 300 km from the failure location, while less impact on the reservoir of the Retiro Baixo Plant. This result is consistent with Vale's evaluation [73]. It appeared that small river gradients and obstruction of the reservoir barriers played an important role in slowing down the tailings waste moving into the São Francisco River. It is not hard to find out that this disaster event had little impact on the Três Marias reservoir (Figure 10d–f).

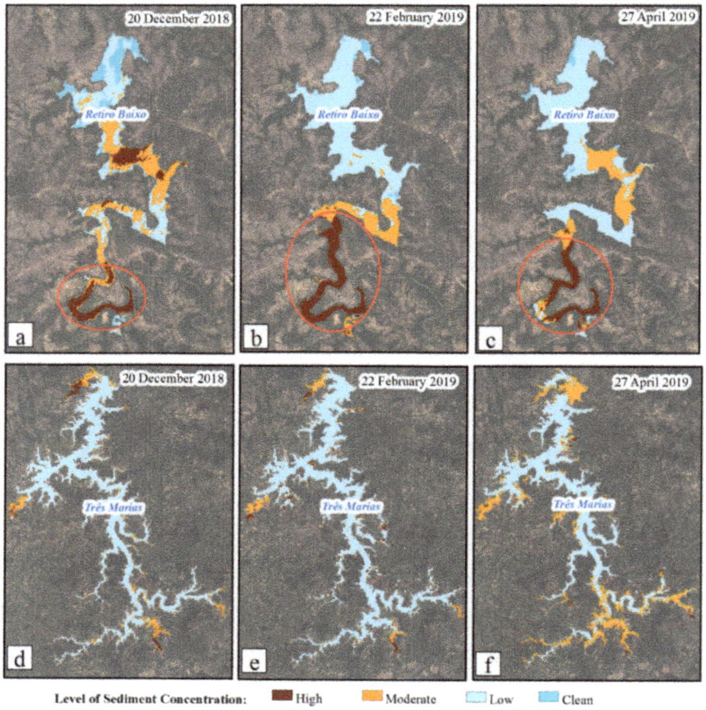

Figure 10. Level of sediment concentration of Retiro Baixo and Três Marias. (**a**) Level of sediment concentration of Retiro Baixo on 20 December 2018; (**b**) level of sediment concentration of Retiro Baixo on 22 February 2019; (**c**) level of sediment concentration of Retiro Baixo on 27 April 2019; (**d**) level of sediment concentration of Três Marias on 20 December 2018; (**e**) level of sediment concentration of Três Marias on 22 February 2019; (**f**) level of sediment concentration of Três Marias on 27 April 2019.

4.4. Tailings-Dam-Failure Trend Analysis Based on Spatiotemporal Pattern Mining

Using the emerging spatiotemporal hot spot analysis method, it can be found that the Brumadinho dam disaster in Brazil belongs to the Consecutive Hot Spot area (Figure 11). This disaster happened 3 years and 2 months after the Mariana dam disaster (5 November 2015), which was considered the worst environmental disaster in Brazil [74,75]. The Brumadinho and Mariana dam disasters both occurred in Minas Gerais, Brazil, and the two dams were both owned by Vale, a Brazilian multinational corporation engaged in metals and mining. Furthermore, based on the world mine tailings failure records [14], tailings dam failures have been recorded several times in this area. As a result, this area belongs to the Consecutive Hot Spot area, and the risk of tailings dam failure in this area would be relatively high if the necessary pond's management and the engineering safety measures were not carried out.

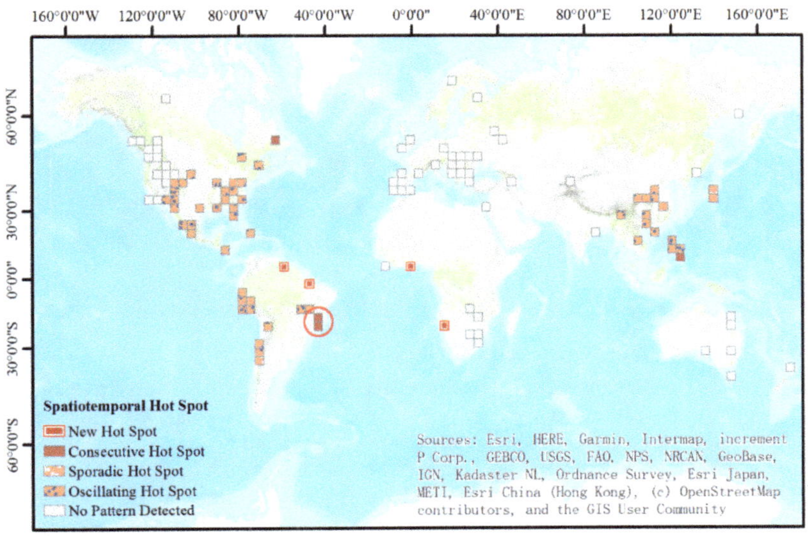

Figure 11. Spatiotemporal hot spot analysis of tailings dam failures. The location of Brumadinho dam disaster is marked with red circles.

According to Figure 11, New Hot Spots appear in Africa and South America which belong to developing regions, where the mining industry has been an important economic pillar in recent years [76,77]. Consecutive Hot Spots mainly lie in Eastern South America and Western Pacific islands where there are a lot of tailings left by mining, but due to poor management, tailings failure is easy to occur. The Sporadic Hot Spot is in Southwest South America and Oscillating Hot Spots mainly lie in Asia and America. China has many Oscillating Hot Spots and there have been some particularly serious tailings failure disasters, such as the 8 September 2008 dam break accident in Shanxi [78]. Different hot spots have different characteristics of disaster occurrences. These characteristics can be influenced by the mining history, mining features (e.g., man-made or natural), etc. [79–81]. The recommendation is that different hot spots should be treated differently. The areas of New Hot Spots and Consecutive Hot Spots are the ones that deserve the most attention. Local governments should adopt appropriate risk management strategies to monitor and change the trend. The risk assessment and monitoring of tailings reservoirs should be adopted and implemented. In this regard, some risk assessment and monitoring methods of mountain disasters can be used as references [82–85].

5. Discussion

5.1. Cause Analysis of This Disaster

5.1.1. Lack of Stability Management during the Maintenance Stage

Some experts believed that Brazil's weak regulatory structures and regulatory gaps allowed the dam's failure [86]. This dam was built in 1976 using the "upstream" method, in which coarse rubble, compacted soil, and dried tailings were used to build the dam (Figure 12). This is similar to the Fundão dam which failed in November 2015, killing 19 people and causing an environmental catastrophe, compared with a more expensive and strong method using solid rocks to contain the waste. The water leak was first observed near its base in July 2018, and then repairs were carried out [87]. José de Gouveia, the worker of Vale, said that the dam exhibited a small leak soon after the rainy season, and leaking water was observed in several places at the bottom [87]. The possible pore pressure build-up would have resulted in a decrease in effective stress and initiated a failure.

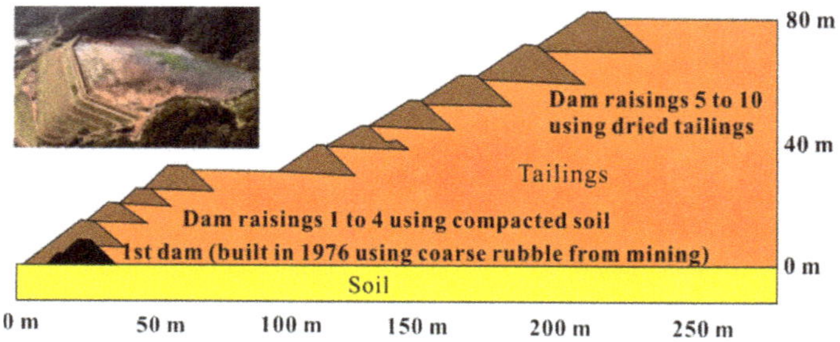

Figure 12. Cross section of Brumadinho dam "Barragem I" from the west to the east [88,89].

5.1.2. Extreme Weather Effects

In this section, weather effects are examined. Daily precipitation data in Brumadinho were obtained from the World Weather Online website [90]. It appeared that precipitation increased the water content of the tailings pond before the dam failure event, increased the pore pressure, and thus induced failure initiation, which could be an important triggering factor for the tailings dam collapse (Figure 13). Although there was less rainfall in January than in February and March 2019, the rainfall in January 2019 sometimes reached the peak of monthly rainfall in some years from the perspective of multi-year rainfalls (Figure 14).

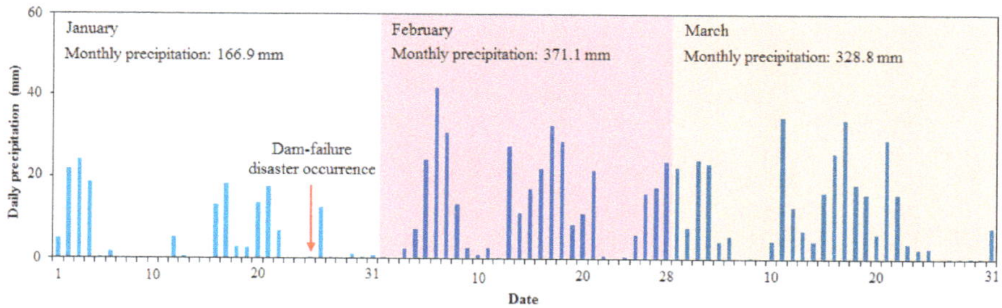

Figure 13. Daily precipitation in Brumadinho during the period from January to March 2019 [90].

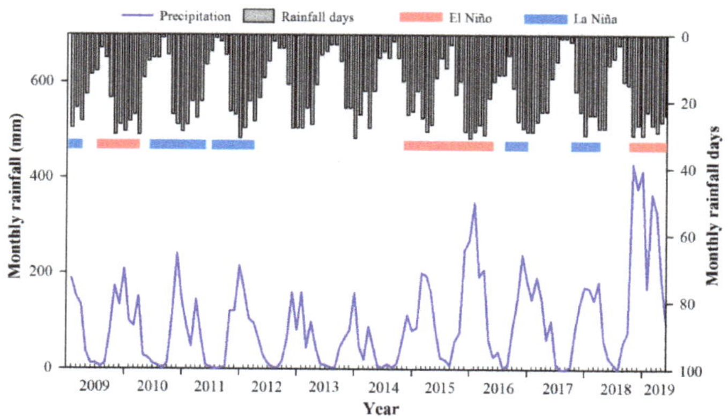

Figure 14. Monthly precipitation in Brumadinho during the period from 2009 to 2019 [90].

El Niño and La Niña events are complex weather patterns resulting from variations in ocean temperatures in the Equatorial Pacific [91]. Their circulation is a global scale climate oscillation. Generally, the impacts of most El Niño events include above-average rainfalls in southeastern South America, eastern equatorial Africa, and the southern USA [92].

Extreme weathers in Brumadinho are linked to El Niño conditions from September to January typically [93]. A major El Niño event has been recorded since October 2018, and there were unusual rainfalls before the disaster (Figure 14) [94]. The increase of soil water content created conditions for the dam instability.

5.2. Lessons Learned and Perceptions about Safety Management of Tailings Ponds

The catastrophic tailings dam failure disaster that occurred in Brumadinho, Brazil is worthy of reviewing the safety management of tailings ponds in order to reduce the occurrence of such incidents. Tailings pond management has the characteristics of a heavy task and wide involvement, and requires the cooperation of different departments and joint law enforcement. The participation of multiple departments also illustrates that tailings pond management is a comprehensive and complex task involving disaster, development, industry, finance and taxation, resources, ecology, water conservancy, meteorology, and other fields [95–97]. At the same time, the successful completion of this task requires interdisciplinary integration and close cooperation of the related fields. Based on the above considerations, the following recommendations are made as a reference:

(1) **Set up a joint working group for tailings pond management, and build a unified management platform, in order to guide tailings pond management**. The comprehensive management work of tailings ponds involves many departments and disciplines. Relevant personnel should be selected from each department for the docking of the management work, and a joint working group should be established to actively and steadily promote the management work of tailings ponds. In addition, according to the needs of governance work, researchers from relevant disciplines of scientific research institutes and universities should be invited to join the joint working group as consulting experts to carry out academic exchanges. Through multidisciplinary exchanges and interdisciplinary integration, theoretical support and technical support could be provided for tailings pond management [98].

(2) **Build the basic geographic information database of tailings ponds, and obtain the basic data of tailings ponds.** The geographic database is an effective way to scientifically organize and manage geographic data. To find out the stock and spatial distribution of tailings ponds is the premise to realize the comprehensive and efficient management of tailings ponds in the later period. During the treatment period of tailings ponds, the number of tailings ponds fluctuates greatly, and some of the abandoned tailings ponds with a long history and a small size may cause statistical omissions. Therefore, it is urgent to find out the basic data of tailings ponds through a detailed investigation and real-time dynamic update and adjustment. Based on the above problems, remote sensing image interpretation [99], literatures and field investigation [100], telephone polls, etc. could be used to obtain the location of the tailings pond, year of construction, condition of use, storage of the pond, height of the tailings dam, type of tailings, and the geographic database could be used for the unified organization and management of the above date [101]. Later, a new investigation information could be used to dynamically update the database.

(3) **The sites selection of new tailings ponds should take into consideration many factors such as safety, ecology, sustainability, and land planning, so as to realize "whole-chain" planning and guide the whole life cycle of tailings ponds with a system engineering theory** [102,103]. A complete life cycle of the tailings pond should start from site selection, go through the process of construction, operation and management, and finally take "reduction" as the end of the mission of a tailings pond. Therefore, the problems needed to be considered in the planning stage of the new tailings ponds include: First, whether geological, geomorphic, environmental, and other factors are suitable for the construction of tailings ponds [104]; second, whether the tailing dams could meet the relevant standards and requirements [105]; third, how to monitor and simulate the stability in the running stage of the tailings pond [106,107]; fourth, how to carry out the comprehensive treatment after the tailings

pond is stopped, the follow-up treatment of tailings, the ecological restoration, and land use planning of the mining area after the tailings treatment [108,109]. The above plans finally form a "whole chain" plan, and the system engineering theory would be introduced into it. On the basis of interdisciplinary integration, the overall optimal operation of the tailings pond system could guide the entire life cycle of the tailings pond. During the management of tailings ponds, the problem of tailing pond failures is the most important part. Such failures are mainly related to geotechnical engineering. Engineering measures should be taken to evaluate and control the water content in the reservoir and pore water pressure. To evaluate the slope stability a range of geotechnical datasets are necessary, including tailing granulometry, hydraulic conductivity, effective porosity, water level in the pond, and pore pressure under meteorological stresses. Based on these datasets, hydraulic simulations, followed by slope stability simulations will lead to the establishment of design requirements.

(4) **Integrate multi-discipline to carry out the comprehensive safety assessment of the built tailings ponds, make clear the management sequence of tailings ponds**. The safety of the tailings pond and its impact on the ecological environment are the two most concerned issues of the built tailings pond. Therefore, it is necessary to organize engineering researchers to evaluate the stability of the built tailings pond and complete the safety risk and ecological risk assessment of the tailings pond together with ecological researchers [110,111]. Comprehensively considering safety risks and ecological risks, priority treatment objects for the tailings pond treatment could be selected in order to improve the utilization efficiency of governance funds [112]. Some engineering measures to improve the stability of the tailing ponds are necessary, such as gentle slopes, norms regarding the maximum height of dams, drainage systems and simulation of their influence on the slope stability, mathematical simulations to evaluate the slope stability under different meteorological conditions, projects for closing the ponds, etc.

(5) **The study on efficient utilization of tailings ponds should be carried out in order to clarify the concept that "tailings are the resource in the wrong place".** The definition of tailing indicates that tailing is the part with a too low content of the target component to be used in production. With the progress of science and technology and the improvement of the efficient utilization ability of mineral resources, on the one hand, the target components in the tailings pond have the possibility of being re-extracted and utilized. On the other hand, other components in the tailings may become effective components in other industries and could be utilized. Nowadays, tailings reuse has made great progress in many aspects, such as heavy separation of useful materials, production of building materials, production of fertilizers, and filling of mine goaf [113–115]. The efficient utilization of tailings ponds in the later stage needs further research and new technology support, but the cognition of tailings from "waste" to "resource" also needs to be changed.

(6) **Mine tailings reservoir potential tourism value, broaden the tourism resources of industrial heritage.** Tailings ponds are the product of the industrial age, but also the unique brand of the industrial age, with obvious characteristics. Tailings ponds and their surrounding mining industry remains constitute an organism of history, technology, society, architecture, and industrial heritage with a scientific value. The organism becomes a witness of history, and has become one of the important tourism contents today [116,117]. After transforming, industrial sites can be transformed into beneficial scenic spots with the function of education. There is a long way for the transformation of industrial heritages into successful tourism products, but there have been many successful cases to learn from, such as Ruhr area in Germany [118] and Beijing 798 Art Zone [119]. The development of industrial heritage tourism resources of tailings ponds needs the full cooperation of scholars such as planning, tourism, and heritage protection, and also needs the strong support of government departments.

6. Conclusions

Tailings reservoir materials are easy to cause harm to the environment, and dam-failure disasters often occur in production mining areas where there are population and production equipment. Therefore, the damages caused by such disasters are often more serious than others, and more attention should be paid. In this study, we carried out a disaster investigation of the Brumadinho tailings failure event. A detailed analysis of the Brumadinho tailing dam failure disaster was carried out using medium to high-resolution satellite images covering the entire affected areas of the event, including the place where the disaster occurred to the transport of tailings in the river and its impact on downstream reservoirs. Especially, the research of the diffusion of sediment in the reservoir was done in order to assess the impact of tailing waste, and discuss whether the waste reached the dams of two hydroelectric plants: Retiro Baixo and Três Marias or even the São Francisco River. Different from those caused by common landslides and debris flows, the disasters caused by the tailings dam failure are more serious and could affect larger areas due to the tailing waste pollution, and they should be paid more attention. On the other hand, the impacts of climatic factors on this event were also discussed in order to make people pay attention to the relationship between extreme weather events and nature disasters. Most importantly, the temporal and spatial characteristics of tailings dam failure were analyzed and summarized by building a global tailings dam failure database. The following conclusions are drawn:

1. The analysis of disaster characteristics revealed that the Brumadinho disaster could be identified as a hazard chain caused by dam failure, mudflow, and hyperconcentrated flow. Especially, the tailings made a great impact on the reservoir of the Retiro Baixo Plant.
2. The Brumadinho disaster is the result of weak regulatory structures and regulatory gaps. However, the influence of weather factors cannot be ignored.
3. The in-depth analysis and interpretation of rainfall data over 11 years revealed that the El Niño event which started in 2018 increased the rainfall, and in turn played an important role and affected the stability of tailings soil.
4. Based on the spatiotemporal analysis of the global tailings dam failure disaster events, different types of hot spots were found. Different hot spots should be dealt with different coping strategies.
5. This disaster also shows that the risk assessment, monitoring, and early warning of tailings ponds in mining areas are necessary for disaster prevention and mitigation.

Author Contributions: Conceptualization, Y.C.; methodology, D.C. and Y.C.; formal analysis, D.C.; supervision, Y.C.; validation, Y.C. and Z.L.; writing—original draft, D.C. and Y.C.; writing—review and editing, Z.L. and J.I.; funding acquisition, Z.L. and Y.C. All authors have read and agreed to the published version of the manuscript.

Funding: This research was funded by the National Natural Science Foundation of China (reference numbers 41941019 and 42077238), and the Chinese Academy of Sciences through the International partnership program (reference number 131551KYSB20160002). Part of this work was also supported by the Shaanxi Province Science and Technology Innovation team (reference number 2021TD-51), the Fundamental Research Funds for the Central Universities, CHD (reference numbers 300102260301 and 300102261108), and by the European Space Agency through the ESA-MOST DRAGON-5 project (reference number 59339). The authors are grateful for the Key Research Program of Frontier Sciences, CAS (grant number QYZDY-SSW-DQC006), and the Chinese Academy of Sciences President's International Fellowship Initiative (grant number 2021VCB0003).

Data Availability Statement: Our research data are from relevant open data websites, which can be obtained according to the links listed in our references.

Acknowledgments: The authors would like to thank all colleagues who participated in this study.

Conflicts of Interest: The authors declare no conflict of interest.

References

1. Rico, M.; Benito, G.; Salgueiro, A.R.; Diez-Herrero, A.; Pereira, H.G. Reported tailings dam failures. A review of the European incidents in the worldwide context. *J. Hazard. Mater.* **2008**, *152*, 846–852. [CrossRef] [PubMed]
2. Franks, D.M.; Boger, D.V.; Côte, C.M.; Mulligan, D.R. Sustainable development principles for the disposal of mining and mineral processing wastes. *Resour. Policy* **2011**, *36*, 114–122. [CrossRef]
3. Macklin, M.G.; Brewer, P.A.; Balteanu, D.; Coulthard, T.J.; Driga, B.; Howard, A.J.; Zaharia, S. The long term fate and environmental significance of contaminant metals released by the January and March 2000 mining tailings dam failures in Maramureş County, upper Tisa Basin, Romania. *Appl. Geochem.* **2003**, *18*, 241–257. [CrossRef]
4. Porsani, J.L.; Jesus, F.A.N.d.; Stangari, M.C. GPR survey on an iron mining area after the collapse of the tailings dam I at the Córrego do Feijão Mine in Brumadinho-MG, Brazil. *Remote Sens.* **2019**, *11*, 860. [CrossRef]
5. Yu, D.; Tang, L.; Chen, C. Three-dimensional numerical simulation of mud flow from a tailings dam failure across complex terrain. *Nat. Hazards Earth Syst. Sci.* **2020**, *20*, 727–741. [CrossRef]
6. Mizani, S.; He, X.; Simms, P. Application of lubrication theory to modeling stack geometry of high density mine tailings. *J. Non Newton. Fluid Mech.* **2013**, *198*, 59–70. [CrossRef]
7. Al-Husseinawi, Y.; Li, Z.; Clarke, P.; Edwards, S. Evaluation of the stability of the Darbandikhan Dam after the 12 November 2017 Mw 7.3 Sarpol-e Zahab (Iran–Iraq border) earthquake. *Remote Sens.* **2018**, *10*, 1426. [CrossRef]
8. Cheng, D.; Cui, Y.; Su, F.; Jia, Y.; Choi, C.E. The characteristics of the Mocoa compound disaster event, Colombia. *Landslides* **2018**, *15*, 1223–1232. [CrossRef]
9. Cui, Y.; Cheng, D.; Choi, C.E.; Jin, W.; Lei, Y.; Kargel, J.S. The cost of rapid and haphazard urbanization: Lessons learned from the Freetown landslide disaster. *Landslides* **2019**, *16*, 1167–1176. [CrossRef]
10. Metternicht, G.; Hurni, L.; Gogu, R. Remote sensing of landslides: An analysis of the potential contribution to geo-spatial systems for hazard assessment in mountainous environments. *Remote Sens. Environ.* **2005**, *98*, 284–303. [CrossRef]
11. Silveira, E.M.d.O.; Júnior, F.W.A.; Mello, J.M.d.; Bueno, I.T. Object-based change detection using semivariogram indices derived from NDVI images: The environmental disaster in Mariana, Brazil. *Ciência e Agrotecnologia* **2017**, *41*, 554–564. [CrossRef]
12. Grenerczy, G.; Wegmüller, U. Persistent scatterer interferometry analysis of the embankment failure of a red mud reservoir using ENVISAT ASAR data. *Nat. Hazards* **2011**, *59*, 1047–1053. [CrossRef]
13. Schvartsman-F. Fabio Schvartsman—Annoucement about Brumadinho Breach Dam. Available online: http://www.vale.com/brasil/EN/aboutvale/news/Pages/fabio-schvartsman-annoucement-about-brumadinho-breach-dam.aspx (accessed on 17 August 2019).
14. WMTF. World Mine Tailings Failures—From 1915. Available online: https://worldminetailingsfailures.org/ (accessed on 17 August 2019).
15. G1. Tragédia em Brumadinho: Lista da Vale de Pessoas não Encontradas. Available online: https://g1.globo.com/mg/minas-gerais/noticia/2019/01/26/vale-divulga-lista-de-pessoas-sem-contato-em-brumadinho.ghtml (accessed on 17 August 2019).
16. Sky-news. Brazil Dam Rescue Resumes after Second Barrier Ruled Safe. Available online: https://news.sky.com/story/brazil-dam-collapse-imminent-risk-of-further-rupture-sparks-evacuation-11619213 (accessed on 17 August 2019).
17. World-Animal-Protection. Helping Animals after Brazil Dam Collapse. Available online: https://www.worldanimalprotection.ca/news/helping-animals-after-brazil-dam-collapse (accessed on 17 August 2019).
18. Lempriere, M. Time to Talk about Tailings Dams. Available online: https://www.mining-technology.com/features/time-to-talk-about-tailings-dams/ (accessed on 17 August 2019).
19. Ibama. Brumadinho, Minas Gerais. Available online: https://commons.wikimedia.org/w/index.php?curid=76936072 (accessed on 17 August 2019).
20. TV-NBR. Resumo da Agenda do Presidente da República 26.01.2019 (Video Screenshot) 01. Available online: https://pt.wikipedia.org/wiki/Ficheiro:Resumo_da_agenda_do_Presidente_da_Rep%C3%BAblica_26.01.2019_(video_screenshot)_01_(cropped).png (accessed on 17 August 2019).
21. Venaglia, G. Iron Ore Railway Bridge Destroyed by Mudflow, 3 Kilometres (1.9 mi) Downstream from Collapsed Dam. Available online: https://en.wikipedia.org/wiki/Brumadinho_dam_disaster#/media/File:Brumadinho4.jpg (accessed on 17 August 2019).
22. Gao, K.; Cui, P.; Zhao, C.; Wei, F. Landslide hazard evaluation of Wanzhou based on GIS information value method in the Three Gorges Reservoir. *Yanshilixue Yu Gongcheng Xuebao Chin. J. Rock Mech. Eng.* **2006**, *25*, 991–996.
23. Cui, Y.; Cheng, D.; Chan, D. Investigation of Post-Fire Debris Flows in Montecito. *ISPRS Int. J. Geo Inf.* **2019**, *8*, 5. [CrossRef]
24. Malarvizhi, K.; Kumar, S.V.; Porchelvan, P. Use of high resolution Google Earth satellite imagery in landuse map preparation for urban related applications. *Procedia Technol.* **2016**, *24*, 1835–1842. [CrossRef]
25. Rabby, Y.W.; Li, Y. An integrated approach to map landslides in Chittagong Hilly Areas, Bangladesh, using Google Earth and field mapping. *Landslides* **2018**, *16*, 633–645. [CrossRef]
26. Cheng, D.; Gao, C.; Shao, T.; Iqbal, J. A landscape study of Sichuan University (Wangjiang Campus) from the perspective of campus tourism. *Land* **2020**, *9*, 499. [CrossRef]
27. NASA-Earth-Observatory. Another Deadly Dam Collapse in Brazil 01.Jpg. Available online: https://commons.wikimedia.org/wiki/File:Another_Deadly_Dam_Collapse_in_Brazil_01.jpg (accessed on 10 August 2019).
28. NASA-Earth-Observatory. Another Deadly Dam Collapse in Brazil 02.Jpg. Available online: https://commons.wikimedia.org/wiki/File:Another_Deadly_Dam_Collapse_in_Brazil_02.jpg (accessed on 10 August 2019).

29. USGS. Glovis. Available online: https://glovis.usgs.gov/ (accessed on 17 August 2019).
30. Edumine. Advance Your Career. Available online: http://www.infomine.com/publications/docs/Martin2000.pdf (accessed on 17 August 2019).
31. UN-Environment. Dam or be Damned: Mining Safety under Scrutiny. Available online: https://www.unenvironment.org/news-and-stories/story/dam-or-be-damned-mining-safety-under-scrutiny (accessed on 17 August 2019).
32. Martins, V.S.; Barbosa, C.C.F.; De Carvalho, L.A.S.; Jorge, D.S.F.; Lobo, F.D.L.; Novo, E.M.L.d.M. Assessment of atmospheric correction methods for Sentinel-2 MSI images applied to Amazon floodplain lakes. *Remote Sens.* **2017**, *9*, 322. [CrossRef]
33. ENVI. Atmospheric Correction Module: QUAC and FLAASH User's Guide. Available online: http://www.exelisvis.com/portals/0/pdfs/envi/Flaash_Module.pdf (accessed on 11 July 2019).
34. Guo, Y.; Zeng, F. Atmospheric correction comparison of SPOT-5 image based on model FLAASH and model QUAC. *Int. Arch. Photogramm. Remote Sens. Spat. Inf. Sci.* **2012**, *39*, 21–23.
35. Vibhute, A.D.; Kale, K.; Dhumal, R.K.; Mehrotra, S. Hyperspectral imaging data atmospheric correction challenges and solutions using QUAC and FLAASH algorithms. In Proceedings of the 2015 International Conference on Man and Machine Interfacing (MAMI), Bhubaneswar, India, 17–19 December 2015; pp. 1–6.
36. Blog ENVI-IDL. Atmospheric Correction of Landsat 8 Data in ENVI5.1. Available online: http://blog.sina.com.cn/s/blog_764b1e9d0101cb2l.html (accessed on 11 July 2019).
37. Nencini, F.; Garzelli, A.; Baronti, S.; Alparone, L. Remote sensing image fusion using the curvelet transform. *Inf. Fusion* **2007**, *8*, 143–156. [CrossRef]
38. Deng, S.; Chen, Q.; Du, H.; Xu, E. *ENVI Remote Sensing Image Processing Method*; Higher Education Press: Beijing, China, 2014; p. 91.
39. Maurer, T. How to pan-sharpen images using the Gram-Schmidt pan-sharpen method—A recipe. *ISPRS Int. Arch. Photogramm. Remote Sens. Spat. Inf. Sci.* **2013**, *239–244*. [CrossRef]
40. Pekel, J.; Vancutsem, C.; Bastin, L.; Clerici, M.; Vanbogaert, E.; Bartholome, E.; Defourny, P. A near real-time water surface detection method based on HSV transformation of MODIS multi-spectral time series data. *Remote Sens. Environ.* **2014**, *140*, 704–716. [CrossRef]
41. Tu, T.; Lee, Y.; Chang, C.; Huang, P.S. Adjustable intensity-hue-saturation and Brovey transform fusion technique for IKONOS/QuickBird imagery. *Opt. Eng.* **2005**, *44*, 116201. [CrossRef]
42. Arenas-Castro, S.; Julien, Y.; Jiménez-Munoz, J.C.; Sobrino, J.A.; Fernández-Haeger, J.; Jordano-Barbudo, D. Mapping wild pear trees (pyrus bourgaeana) in Mediterranean forest using high-resolution QuickBird satellite imagery. *International Journal of Remote Sens.* **2013**, *34*, 3376–3396. [CrossRef]
43. McFeeters, S.K. The use of the Normalized Difference Water Index (NDWI) in the delineation of open water features. *Int. J. Remote Sens.* **2007**, *17*, 1425–1432. [CrossRef]
44. Sherry, L. Fundamentals of Aquatic Remote Sensing. Available online: https://arset.gsfc.nasa.gov/sites/default/files/users/fundamentals/fundamentals-aquatic-web.pdf (accessed on 17 August 2019).
45. Gholizadeh, M.H.; Melesse, A.M.; Reddi, L. A comprehensive review on water quality parameters estimation using remote sensing techniques. *Sensors* **2016**, *16*, 1298. [CrossRef] [PubMed]
46. Wass, P.; Marks, S.; Finch, J.; Leeks, G.J.L.; Ingram, J. Monitoring and preliminary interpretation of in-river turbidity and remote sensed imagery for suspended sediment transport studies in the Humber catchment. *Sci. Total Environ.* **1997**, *194*, 263–283. [CrossRef]
47. Lim, J.; Choi, M. Assessment of water quality based on Landsat 8 operational land imager associated with human activities in Korea. *Environ. Monit. Assess.* **2015**, *187*, 384. [CrossRef]
48. Antoine, D.; d'Ortenzio, F.; Hooker, S.B.; Bécu, G.; Gentili, B.; Tailliez, D.; Scott, A.J. Assessment of uncertainty in the ocean reflectance determined by three satellite ocean color sensors (MERIS, SeaWiFS and MODIS-A) at an offshore site in the Mediterranean Sea (BOUSSOLE project). *J. Geophys. Res. Ocean.* **2008**, *113*. [CrossRef]
49. Lindell, L.; Steinvall, O.; Jonsson, M.; Claesson, T. Mapping of coastal-water turbidity using Landsat imagery. *Int. J. Remote Sens.* **1985**, *6*, 629–642. [CrossRef]
50. Haakstad, M.; Kogeler, J.; Dahle, S. Studies of sea surface temperatures in selected northern Norwegian fjords using Landsat TM data. *Polar Res.* **1994**, *13*, 95–110. [CrossRef]
51. Ryan, P.A. Environmental effects of sediment on New Zealand streams: A review. *N. Z. J. Mar. Freshw. Res.* **1991**, *25*, 207–221. [CrossRef]
52. Doxaran, D.; Froidefond, J.-M.; Lavender, S.; Castaing, P. Spectral signature of highly turbid waters: Application with SPOT data to quantify suspended particulate matter concentrations. *Remote Sens. Environ.* **2002**, *81*, 149–161. [CrossRef]
53. Feng, L.; Hu, C.; Chen, X.; Song, Q. Influence of the Three Gorges Dam on total suspended matters in the Yangtze Estuary and its adjacent coastal waters: Observations from MODIS. *Remote Sens. Environ.* **2014**, *140*, 779–788. [CrossRef]
54. Nechad, B.; Ruddick, K.G.; Park, Y. Calibration and validation of a generic multisensor algorithm for mapping of total suspended matter in turbid waters. *Remote Sens. Environ.* **2010**, *114*, 854–866.
55. Wang, X.; Li, W.; Yan, X.; Lu, Y.; Zhou, X.; Li, X.; Xia, L.; Tang, X.; Mei, C. Information extraction for suspended sediment in Lake Chaohu and its distribution based on Landsat TM/ETM+ data. *J. Lake Sci.* **2007**, *19*, 255–260.

56. Esri. Emerging Hot Spot Analysis. Available online: https://pro.arcgis.com/en/pro-app/tool-reference/space-time-pattern-mining/emerginghotspots.htm (accessed on 17 August 2019).
57. Xu, L.; Meng, X.; Xu, X. Natural hazard chain research in China: A review. *Nat. Hazards* **2013**, *70*, 1631–1659. [CrossRef]
58. Fan, X.; Scaringi, G.; Korup, O.; West, A.J.; Westen, C.J.; Tanyas, H.; Hovius, N.; Hales, T.C.; Jibson, R.W.; Allstadt, K.E.; et al. Earthquake-Induced chains of geologic hazards: Patterns, mechanisms, and impacts. *Rev. Geophys.* **2019**, *57*, 421–503. [CrossRef]
59. Wei, R.; Zeng, Q.; Davies, T.; Yuan, G.; Wang, K.; Xue, X.; Yin, Q. Geohazard cascade and mechanism of large debris flows in Tianmo gully, SE Tibetan Plateau and implications to hazard monitoring. *Eng. Geol.* **2018**, *233*, 172–182. [CrossRef]
60. Walter, T.R.; Haghighi, M.H.; Schneider, F.M.; Coppola, D.; Motagh, M.; Saul, J.; Babeyko, A.; Dahm, T.; Troll, V.R.; Tilmann, F.; et al. Complex hazard cascade culminating in the Anak Krakatau sector collapse. *Nat. Commun.* **2019**, *10*, 4339. [CrossRef]
61. Rotta, L.H.S.; Alcantara, E.; Park, E.; Negri, R.G.; Lin, Y.N.; Bernardo, N.; Mendes, T.S.G.; Filho, C.R.S. The 2019 Brumadinho tailings dam collapse: Possible cause and impacts of the worst human and environmental disaster in Brazil. *Int. J. Appl. Earth Obs. Geoinf.* **2020**, *90*, 102119. [CrossRef]
62. Cordeiro, M.C.; Garcia, G.D.; Rocha, A.M.; Tschoeke, D.A.; Campeao, M.E.; Appolinario, L.R.; Soares, A.C.; Leomil, L.; Froes, A.; Bahiense, L.; et al. Insights on the freshwater microbiomes metabolic changes associated with the world's largest mining disaster. *Sci. Total Environ.* **2019**, *654*, 1209–1217. [CrossRef] [PubMed]
63. Queiroz, H.M.; Nobrega, G.N.; Ferreira, T.O.; Almeida, L.S.; Romero, T.B.; Santaella, S.T.; Bernardino, A.F.; Otero, X.L. The Samarco mine tailing disaster: A possible time-bomb for heavy metals contamination? *Sci. Total Environ.* **2018**, *637–638*, 498–506. [CrossRef] [PubMed]
64. ESA. Path of Mudflow after Dam Failure. Available online: https://en.wikipedia.org/wiki/Brumadinho_dam_disaster#/media/File:Brumadinho_dam_catastrophy_2.jpg (accessed on 17 August 2019).
65. Comiti, F.; Mao, L.; Wilcox, A.; Wohl, E.E.; Lenzi, M.A. Field-derived relationships for flow velocity and resistance in high-gradient streams. *J. Hydrol.* **2007**, *340*, 48–62. [CrossRef]
66. Liljegren, L. The effect of a mean fluid velocity gradient on the streamwise velocity variance of a particle suspended in a turbulent flow. *Int. J. Multiph. Flow* **1993**, *19*, 471–484. [CrossRef]
67. Huang, H.B.; Ma, H.Z.; Sha, Z.J.; Cao, G.C.; Ou, L.Y.; Yang, H.Z. Research of distribution of suspended sediments in Longyangxia reservoir based on RS. *J. Salt Lake Res.* **2004**, *12*, 34–37.
68. Kumar, V.; Garg, R.D. Comparison of different mapping techniques for classifying hyperspectral data. *J. Indian Soc. Remote Sens.* **2012**, *40*, 411–420. [CrossRef]
69. Zhang, H.K.; Roy, D.P. Computationally inexpensive Landsat 8 Operational Land Imager (OLI) pansharpening. *Remote Sens.* **2016**, *8*, 180. [CrossRef]
70. Fatemi, S.B.; Gholinejad, S. Assessing the effectiveness of Google Earth images for spatial enhancement of rapidEye multi-spectral imagery. *Int. J. Remote Sens.* **2019**, *40*, 4526–4543. [CrossRef]
71. Delleji, T.; Kallel, A.; Hamida, A.B. Iterative scheme for MS image pansharpening based on the combination of multi-resolution decompositions. *Int. J. Remote Sens.* **2016**, *37*, 6041–6075. [CrossRef]
72. Guo, Q.; Chen, S.; Leung, H.; Liu, S. Covariance intersection based image fusion technique with application to pansharpening in remote sensing. *Inf. Sci.* **2010**, *180*, 3434–3443. [CrossRef]
73. Vale. Clarifications Regarding Dam I of the Córrego do Feijão Mine. Available online: http://www.vale.com/brasil/EN/aboutvale/news/Pages/Clarifications-regarding-Dam-I-of-the-Corrego-do-Feijao-Mine.aspx (accessed on 18 April 2019).
74. Fernandes, G.W.; Goulart, F.F.; Ranieri, B.D.; Coelho, M.S.; Dales, K.; Boesche, N.; Bustamante, M.; Carvalho, F.A.; Carvalho, D.C.; Dirzo, R.; et al. Deep into the mud: Ecological and socio-economic impacts of the dam breach in Mariana, Brazil. *Nat. Conserv.* **2016**, *14*, 35–45. [CrossRef]
75. Francini-Filho, R.B.; Cordeiro, M.C.; Omachi, C.Y.; Rocha, A.M.; Bahiense, L.; Garcia, G.D.; Tschoeke, D.; de Almeida, M.G.; Rangel, T.P.; De Oliveira, B.C.V.; et al. Remote sensing, isotopic composition and metagenomics analyses revealed Doce River ore plume reached the southern Abrolhos Bank Reefs. *Sci. Total Environ.* **2019**, *697*, 134038. [CrossRef] [PubMed]
76. Festin, E.S.; Tigabu, M.; Chileshe, M.N.; Syampungani, S.; Odén, P.C. Progresses in restoration of post-mining landscape in Africa. *J. For. Res.* **2018**, *30*, 381–396. [CrossRef]
77. Robertson, P.K.; da Fonseca, A.V.; Ulrich, B.; Coffin, J. Characterization of unsaturated mine waste: A case history. *Can. Geotech. J.* **2017**, *54*, 1752–1761. [CrossRef]
78. Yu, G.; Song, C.; Pan, Y.; Li, L.; Li, R.; Lu, S. Review of new progress in tailing dam safety in foreign research and current state with development trent in China. *Chin. J. Rock Mech. Eng.* **2014**, *33*, 3238–3248.
79. Agurto-Detzel, H.; Bianchi, M.; Assumpção, M.; Schimmel, M.; Collaço, B.; Ciardelli, C.; Barbosa, J.R.; Calhau, J. The tailings dam failure of 5 November 2015 in SE Brazil and its preceding seismic sequence. *Geophys. Res. Lett.* **2016**, *43*, 4929–4936. [CrossRef]
80. Fourie, A.; Blight, G.; Papageorgiou, G. Static liquefaction as a possible explanation for the Merriespruit tailings dam failure: Reply. *Can. Geotech. J.* **2002**, *39*, 1441. [CrossRef]
81. Kossoff, D.; Dubbin, W.; Alfredsson, M.; Edwards, S.; Macklin, M.; Hudson-Edwards, K.A. Mine tailings dams: Characteristics, failure, environmental impacts, and remediation. *Appl. Geochem.* **2014**, *51*, 229–245. [CrossRef]
82. Cardinali, M.; Galli, M.; Guzzetti, F.; Ardizzone, F.; Reichenbach, P.; Bartoccini, P. Rainfall induced landslides in December 2004 in south-western Umbria, central Italy: Types, extent, damage and risk assessment. *Nat. Hazards Earth Syst. Sci.* **2006**, *6*, 237–260. [CrossRef]

83. Cui, P.; Dang, C.; Zhuang, J.-q.; You, Y.; Chen, X.-q.; Scott, K.M. Landslide-dammed lake at Tangjiashan, Sichuan province, China (triggered by the Wenchuan Earthquake, May 12, 2008): Risk assessment, mitigation strategy, and lessons learned. *Environ. Earth Sci.* **2010**, *65*, 1055–1065. [CrossRef]
84. Cui, P.; Zou, Q.; Xiang, L.-z.; Zeng, C. Risk assessment of simultaneous debris flows in mountain townships. *Prog. Phys. Geogr. Earth Environ.* **2013**, *37*, 516–542. [CrossRef]
85. Lu, P.; Catani, F.; Tofani, V.; Casagli, N. Quantitative hazard and risk assessment for slow-moving landslides from Persistent Scatterer Interferometry. *Landslides* **2013**, *11*, 685–696. [CrossRef]
86. Globo, O. Empresas Envolvidas em Desastres Ambientais Quitaram só 3.4% de R$785 Milhões em Multas. Available online: https://oglobo.globo.com/economia/empresas-envolvidas-em-desastres-ambientais-quitaram-so-34-de-785-milhoes-em-multas-22657874 (accessed on 18 April 2019).
87. Guardian, T. 'That's Going to Burst': Brazilian Dam Workers Say They Warned of Disaster. Available online: https://www.theguardian.com/world/2019/feb/06/brazil-dam-collapse-workers-say-they-warned-owners (accessed on 17 August 2019).
88. Agricolax. Schematic Cross Section Showing Design of Failed Dam. Available online: https://en.wikipedia.org/wiki/Brumadinho_dam_disaster#/media/File:Brumadinho_dam_-_cross_section_-_en.png (accessed on 14 October 2019).
89. AGU-Blogosphere. Brumadinho Disaster: The Extent of the Environmental Impact. Available online: https://blogs.agu.org/landslideblog/page/13/ (accessed on 23 October 2019).
90. World-Weather-Online. Brumadinho Historical Weather. World Weather Online. Available online: https://www.worldweatheronline.com/brumadinho-weather-history/minas-gerais/br.aspx (accessed on 17 August 2019).
91. NOAA. What are El Niño and La Niña? Available online: https://oceanservice.noaa.gov/facts/ninonina.html (accessed on 17 October 2020).
92. Barnston, A. How ENSO Leads to a Cascade of Global Impacts. Available online: https://www.climate.gov/news-features/blogs/enso/how-enso-leads-cascade-global-impacts (accessed on 17 August 2019).
93. IRI. What Changes in Rainfall are Typical during ei Niño? Available online: http://iridl.ldeo.columbia.edu/maproom/IFRC/FIC/elninorain.html (accessed on 17 August 2019).
94. IRI. Are the Next 3 Months Likely to be Unusually Wet or Dry? Available online: http://iridl.ldeo.columbia.edu/maproom/IFRC/FIC/prcp_fcst.html?F=Oct%202018&L=2 (accessed on 17 August 2019).
95. Hatje, V.; Pedreira, R.M.A.; de Rezende, C.E.; Schettini, C.A.F.; de Souza, G.C.; Marin, D.C.; Hackspacher, P.C. The environmental impacts of one of the largest tailing dam failures worldwide. *Sci. Rep.* **2017**, *7*, 10706. [CrossRef] [PubMed]
96. de Carvalho, D.W. The ore tailings dam rupture disaster in Mariana, Brazil 2015: What we have to learn from anthropogenic disasters. *Nat. Resour. J.* **2019**, *59*, 281.
97. Adiansyah, J.S.; Rosano, M.; Vink, S.; Keir, G. A framework for a sustainable approach to mine tailings management: Disposal strategies. *J. Clean. Prod.* **2015**, *108*, 1050–1062. [CrossRef]
98. Laasonen, M.J. European Working Group "Management of dam incidents" Case study: Finland. In Proceedings of the International Symposium on Dams in a Global Environmental Challenges, Bali, Indonesia, 1–6 June 2014.
99. Balaniuk, R.; Isupova, O.; Reece, S. Mining and tailings dam detection in satellite imagery using deep learning. *Sensors* **2020**, *20*, 6936. [CrossRef]
100. Wu, J.; Wu, Y.; Lu, J.; Lee, L. Field investigations and laboratory simulation of clogging in Lixi tailings dam of Jinduicheng, China. *Environ. Geol.* **2007**, *53*, 387–397.
101. Mulligan, M.; van Soesbergen, A.; Sáenz, L. GOODD, a global dataset of more than 38,000 georeferenced dams. *Sci. Data* **2020**, *7*, 1–8. [CrossRef]
102. Blanchard, B.S. *System Engineering Management*; John Wiley & Sons: Hoboken, NJ, USA, 2004.
103. Blanchard, B.S.; Fabrycky, W.J.; Fabrycky, W.J. *Systems Engineering and Analysis*; Prentice Hall: Englewood Cliffs, NJ, USA, 1990; Volume 4.
104. Yang, Y.; Wei, Z.; Cao, G.; Yang, Y.; Wang, H.; Zhuang, S.; Lu, T. A case study on utilizing geotextile tubes for tailings dams construction in China. *Geotext. Geomembr.* **2019**, *47*, 187–192. [CrossRef]
105. Wei, Z.; Yin, G.; Wang, J.-G.; Wan, L.; Li, G. Design, construction and management of tailings storage facilities for surface disposal in China: Case studies of failures. *Waste Manag. Res.* **2013**, *31*, 106–112. [CrossRef]
106. Wang, K.; Yang, P.; Hudson-Edwards, K.A.; Lyu, W.; Yang, C.; Jing, X. Integration of DSM and SPH to model tailings dam failure run-Out slurry routing across 3D real terrain. *Water* **2018**, *10*, 1087. [CrossRef]
107. Yin, G.; Li, G.; Wei, Z.; Wan, L.; Shui, G.; Jing, X. Stability analysis of a copper tailings dam via laboratory model tests: A Chinese case study. *Miner. Eng.* **2011**, *24*, 122–130. [CrossRef]
108. Li, M. Ecological restoration of mineland with particular reference to the metalliferous mine wasteland in China: A review of research and practice. *Sci. Total Environ.* **2006**, *357*, 38–53. [CrossRef]
109. Aires, U.R.V.; Santos, B.S.M.; Coelho, C.D.; da Silva, D.D.; Calijuri, M.L. Changes in land use and land cover as a result of the failure of a mining tailings dam in Mariana, MG, Brazil. *Land Use Policy* **2018**, *70*, 63–70. [CrossRef]
110. Xin, Z.; Xiaohu, X.; Kaili, X. Study on the risk assessment of the tailings dam break. *Procedia Eng.* **2011**, *26*, 2261–2269. [CrossRef]
111. Buch, A.C.; Niemeyer, J.C.; Marques, E.D.; Silva-Filho, E.V. Ecological risk assessment of trace metals in soils affected by mine tailings. *J. Hazard. Mater.* **2020**, *403*, 123852. [CrossRef]

112. Li, W.; Ye, Y.; Hu, N.; Wang, X.; Wang, Q. Real-time warning and risk assessment of tailings dam disaster status based on dynamic hierarchy-grey relation analysis. *Complexity* **2019**, *2019*, 5873420. [CrossRef]
113. Kiventerä, J.; Golek, L.; Yliniemi, J.; Ferreira, V.; Deja, J.; Illikainen, M. Utilization of sulphidic tailings from gold mine as a raw material in geopolymerization. *Int. J. Miner. Process.* **2016**, *149*, 104–110. [CrossRef]
114. Choi, Y.W.; Kim, Y.J.; Choi, O.; Lee, K.M.; Lachemi, M. Utilization of tailings from tungsten mine waste as a substitution material for cement. *Constr. Build. Mater.* **2009**, *23*, 2481–2486.
115. Bakken, A.; Gautneb, H.; Sveistrup, T.; Myhr, K. Crushed rocks and mine tailings applied as K fertilizers on grassland. *Nutr. Cycl. Agroecosystems* **2000**, *56*, 53–57. [CrossRef]
116. Ballesteros, E.R.; Ramírez, M.H. Identity and community—Reflections on the development of mining heritage tourism in Southern Spain. *Tour. Manag.* **2007**, *28*, 677–687. [CrossRef]
117. Chon, K.-S.; Evans, M.R. Tourism in a rural area—A coal mining-county experience. *Tour. Manag.* **1989**, *10*, 315–321. [CrossRef]
118. Ćopić, S.; ĐorđevićA, J.; Lukić, T.; Stojanović, V.; Đukičin, S.; Besermenji, S.; Stamenković, I.; Tumarić, A. Transformation of industrial heritage: An example of tourism industry development in the Ruhr area (Germany). *Geogr. Pannonica* **2014**, *18*, 43–50. [CrossRef]
119. Wang, K.; Tang, C.; Liu, J. A measuring model on tourist satisfaction index in cultural creative-type tourism destination—A case of 798 Art Zone in Beijing. *Tour. Trib. Lvyou Xuekan* **2011**, *26*, 36–44.

remote sensing

Article

Assessment of k-Nearest Neighbor and Random Forest Classifiers for Mapping Forest Fire Areas in Central Portugal Using Landsat-8, Sentinel-2, and Terra Imagery

Admilson da Penha Pacheco [1], Juarez Antonio da Silva Junior [1], Antonio Miguel Ruiz-Armenteros [2,3,4,*] and Renato Filipe Faria Henriques [5]

1 Center for Technology and Geosciences, Department of Cartographic and Surveying Engineering, Federal University of Pernambuco, Av. Prof. Moraes Rego, 1235, Cidade Universitária, Recife 50670-901, Brazil; admilson.pacheco@ufpe.br (A.d.P.P.); juarez.silvajunior@ufpe.br (J.A.d.S.J.)
2 Department of Cartographic, Geodetic and Photogrammetry Engineering, University of Jaén, Campus Las Lagunillas s/n, 23071 Jaén, Spain
3 Microgeodesia Jaén Research Group (PAIDI RNM-282), University of Jaén, Campus Las Lagunillas s/n, 23071 Jaén, Spain
4 Center for Advanced Studies on Earth Sciences, Energy and Environment CEACTEMA, University of Jaén, Campus Las Lagunillas, s/n, 23071 Jaén, Spain
5 Department of Earth Sciences, Institute of Earth Sciences (ICT), University of Minho (UMinho), Campus de Gualtar, 4710-057 Braga, Portugal; rhenriques@dct.uminho.pt
* Correspondence: amruiz@ujaen.es

Abstract: Forest fires threaten the population's health, biomass, and biodiversity, intensifying the desertification processes and causing temporary damage to conservation areas. Remote sensing has been used to detect, map, and monitor areas that are affected by forest fires due to the fact that the different areas burned by a fire have similar spectral characteristics. This study analyzes the performance of the k-Nearest Neighbor (kNN) and Random Forest (RF) classifiers for the classification of an area that is affected by fires in central Portugal. For that, image data from Landsat-8, Sentinel-2, and Terra satellites and the peculiarities of each of these platforms with the support of Jeffries–Matusita (JM) separability statistics were analyzed. The event under study was a 93.40 km^2 fire that occurred on 20 July 2019 and was located in the districts of Santarém and Castelo Branco. The results showed that the problems of spectral mixing, registration date, and those associated with the spatial resolution of the sensors were the main factors that led to commission errors with variation between 1% and 15.7% and omission errors between 8.8% and 20%. The classifiers, which performed well, were assessed using the receiver operating characteristic (ROC) curve method, generating maps that were compared based on the areas under the curves (AUC). All of the AUC were greater than 0.88 and the Overall Accuracy (OA) ranged from 89 to 93%. The classification methods that were based on the kNN and RF algorithms showed satisfactory results.

Keywords: k-Nearest Neighbor; Random Forest; fires; Landsat 8; Sentinel 2; Terra; ASTER; MODIS; burned; mapping

Citation: Pacheco, A.d.P.; Junior, J.A.d.S.; Ruiz-Armenteros, A.M.; Henriques, R.F.F. Assessment of k-Nearest Neighbor and Random Forest Classifiers for Mapping Forest Fire Areas in Central Portugal Using Landsat-8, Sentinel-2, and Terra Imagery. *Remote Sens.* **2021**, *13*, 1345. https://doi.org/10.3390/rs13071345

Academic Editor: Teodosio Lacava

Received: 14 February 2021
Accepted: 27 March 2021
Published: 1 April 2021

1. Introduction

Forests are subject to a variety of disturbances, which are strongly influenced by climate change and human activities [1]. Forest disturbance due to fires is a major challenge for forest management in various ecosystems due to the loss of life and infrastructure, emissions of greenhouse gases, degradation, soil erosion, and the destruction of species, biomass, and biodiversity [1–30]. According to the Intergovernmental Panel on Climate Change (IPCC), climate change tends to increase the risks of major fires on Earth.

Accurate information that is related to the impact of fire on the environment is a key factor in quantifying the consequences of fires on the landscape, planning and monitoring restoration and recovery activities, and providing relevant data for understanding the dynamics of fire, serving as a basis for future monitoring [31]. After a fire, detailed and rapid knowledge of the level of damage and its spatial distribution are the first desirable information. Accurate and complete data on fire sites and burned areas are important for a variety of applications, including quantifying trends and patterns of occurrences in a variety of natural and social systems [32–41].

The understanding of fire regimes and forest recovery patterns in different environmental and climatic conditions improves the management of sustainable forests, facilitating the process of forest resilience, according to Chu and Guo [42].

In the last decades, the use of remote sensing has allowed unprecedented advances in mapping fire dynamics, mainly to locate the occurrence of fire in time and space, and to quantify the total extent of the burned area. Several remote sensing studies have been carried out to map burned areas on a global and regional scale [10,12,38,39,43–56]. In particular, some authors have studied burned areas in Portugal using remote sensing techniques by [12,47,49,51–53,57–59].

The availability of well-calibrated global remote sensing data since the late 1990s has enabled the production of a variety of global and multi-annual products for burned areas, which are now freely available [60]. Several of these products are based on data from orbital sensor systems with different spatial resolutions (coarse, medium, and high), such as: Operational Land Imager (OLI)/Landsat-8, MultiSpectral Instrument (MSI)/Sentinel-2, Advanced Spaceborne Thermal Emission and Reflection Radiometer (ASTER)/Terra, or Moderate Resolution Image Spectroradiometer (MODIS)/Terra. According to Libonati et al. [61], the development of a precise algorithm to detect changes in surfaces that are caused by fires on a global scale is hampered by the complexity, diversity, and high number of biomes involved. The limitations of estimating burned areas, on a global scale, can be reduced with the development of algorithms that consider characteristics, such as vegetation type, soil, and climate, and where validation and calibration exercises are less complex to implement [61].

Mapping burned areas using remote sensing techniques is based on post-fire changes due to the burns [57]. The approaches include supervised and unsupervised classification techniques at the pixel level. The quality of the classification of the natural environment is associated with the precision and reliability derived from satellite data, which are determined by the classification algorithm. This involves the image resolution (pixel, window, or segment size) that is used in the classification process. To evaluate the classifiers and obtain thematic precision, it is necessary to take the different classes of forest identified into account [62]. In the last decades, non-parametric methods, algorithms that are based on machine learning (MLAs), have gained great attention from applications based on remote sensing [63,64], although some of them, such as the k-Nearest Neighbor (kNN), have been used since the 1950's [65–71]. MLAs have become widely accepted as evidenced by their use in mapping burned areas [44,46,72]. They perform well in situations that involve category prediction of spatially dispersed training data and are especially useful when the process under investigation is complex and/or represented by a high-dimensional input space [73].

In recent years, Landsat, Sentinel-2, and Terra data have been used in conjunction with MLAs to distinguish and map fires in different types of biomes, anthropogenic types of land use (including plantations), and degraded forests ([61,74,75]). Many of the classification algorithms have been compared with standard products from burned areas and active fires derived from satellite data, such as MCD64A1 [76], MCD14DL [75], Landsat Burn Area [77], or Fire_cci [78].

MLAs have also been implemented in satellite data to map fires, examine spectral properties, accurately delineate the area affected by the fire [79], analyze fire severity [72], and carry out precision analysis of the product [43,61]. Some of the most common MLAs for classifying and mapping burned areas include support vector machines (SVM), kNN, and Random Forest (RF) [80,81]. RF, for example, allows for integrating data from different scales and sources, which explains its wide use in many mapping applications based on satellite images [72]. In particular, several studies show the RF potential that is applied to satellite images for the detection of forest fires [82–88].

The ability of MLAs to distinguish and map different forest types, which have suffered varying levels of fire severity and their consequences across the planet, needs to be further assessed by different orbital sensors. This will support conservation management, being able to serve in places of different territorial extension. However, it should be noted, that there are few published studies on the performance of kNN and RF using different orbital platforms in areas burned by fire at the local scale, especially in Portugal [81,89–92].

In this work, the feasibility of kNN and RF classification algorithms to map areas that are burned by forest fires in a region of native pine vegetation in the municipalities of Santarém and Castelo Branco (central Portugal) is evaluated using Landsat-8, Sentinel-2, and Terra satellite data. The main aims are: (i) to examine the effectiveness of different remote sensing data sources for delineating the area affected by the fire; (ii) to compare, while considering the advantages and limitations of the sensors used, the performance of two MLAs (kNN and RF) that are commonly used to delineate and map forests that suffered fires; and, (iii) to evaluate the structural and spectral properties of the burned area and its influence on the classification.

We found that no significant differences in the burned area are obtained with each algorithm for each image sensor. The classifications carried out using both kNN and RF algorithms mapped the burned areas with high accuracy for the different sensors, regardless of the spatial resolutions and the spectral characteristics of each source data.

2. Materials and Methods

2.1. Study Area

Portugal is characterized by a mild Mediterranean climate with climatic variability, involving droughts and desertification in the southern sector, according to Miranda et al. [93]. The majority of burned areas in Portugal (80%) are due to fires, which occur in a small number of summer days (10%) when the atmospheric circulation forms a prominent ridge over the Iberian Peninsula with a strong flow to the south [94].

The study area (Figure 1) covers a 93.4 km^2 fire that occurred on 20 July 2019 in the districts of Santarém and Castelo Branco (central Portugal). In this area, the vegetation of maritime pine and microclimate predominate with prolonged summers, having very limited rainfall. High temperatures reduce the moisture content of forest fuels, often resulting in large fires when combined with strong winds [95].

According to Nunes et al. [96], who analyzed a set of 506 fires that occurred in Portugal in 1991, large fires (greater than 1500 ha) mainly occur in posts of Pinus pinaster, Eucalyptus globulus Labill., and Eucalyptus/Pine trees mixture, and later by bush. On the other hand, as these types of vegetation are sowers, which respond to fire through the rapid dispersion of seeds, post-fire regeneration in the central region of Portugal will crucially depend on the destruction of seeds that are present on the soil surface during the fire episode [97]. Therefore, it can be predicted that the magnitude of fire damage will play an important role in the dynamics of vegetation in this region.

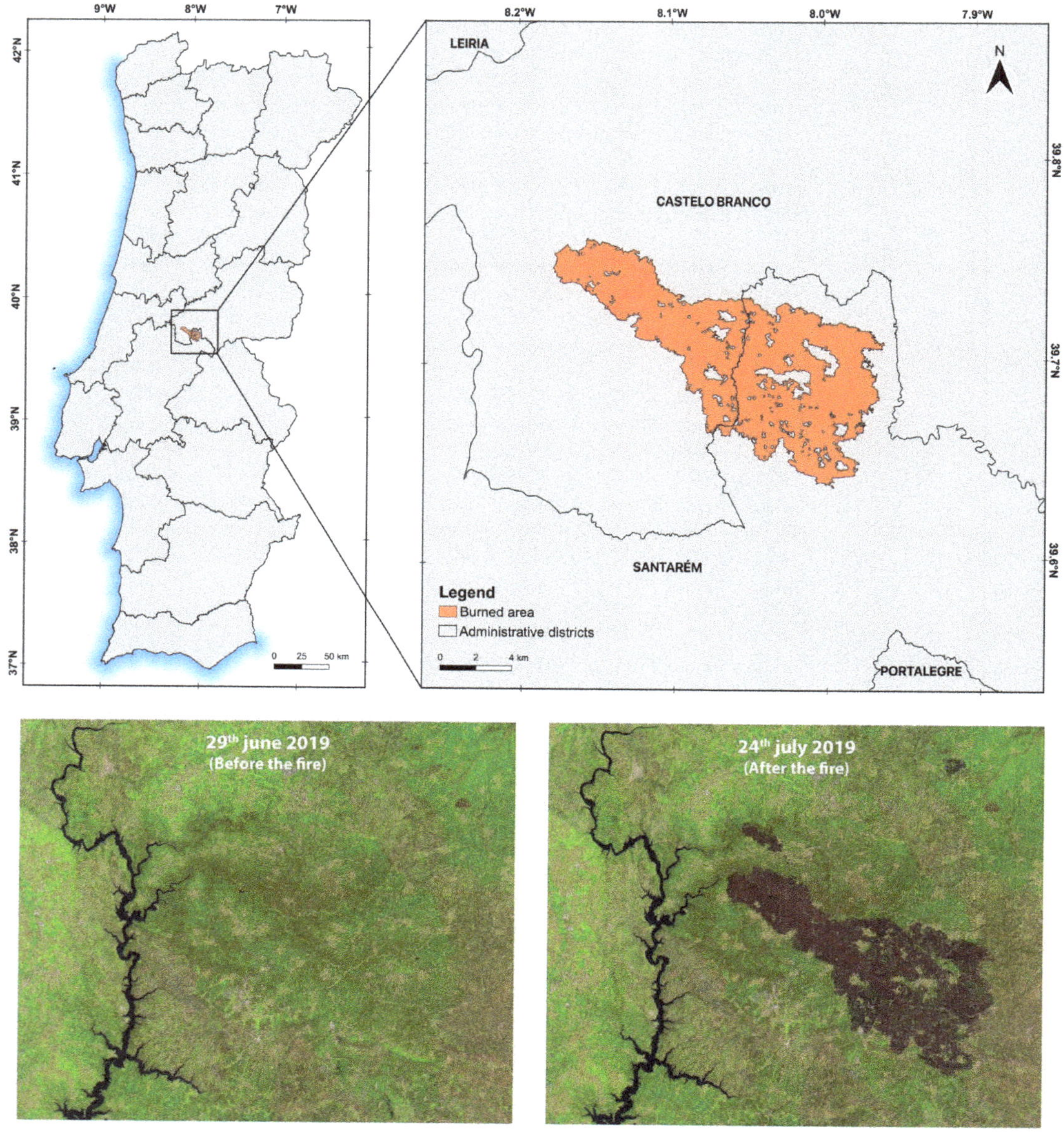

Figure 1. Location of the study area at the Santarém and Castelo Branco discricts of Portugal. The analyzed burned area is represented on the right map with a brown pattern. At the bottom are two satellite images from Sentinel 2, corresponding to the RGB mosaics of 29 June 2019 (L1C_T29SND_A012075_20190629T112256), before the fire, and 24 July 2019 (L1C_T29SND_A021341_20190724T112448), after the fire.

2.2. Data and Image Processing

In this work, the following multisensor satellite images fully covering the study area, including Landsat-8, Sentinel-2, and Terra, as well as their spectral bands, were selected and used to discriminate the area that is affected by the fire in the pixel distribution format on the digital number (DN) scale:

(i) A Landsat-8 scene acquired on 1 August 2019 by the OLI sensor (LC08_L1TP_203033_20190801_20190801, orbit/point: 203/033) with a spatial resolution of 30 m obtained from the Earth Resources Observation and Science Center of the US Geological Survey (USGS) [98]. This is a product of level 1T (corrected terrain) and adjusted with the solar angle with the processing steps described in [99].

(ii) A Sentinel-2 scene acquired on 3 August 2019 through the cloudless MSI sensor (S2A_MSIL1C_20190803T112121_N0208_R037_T29SND_20190803T132806) with 20 m spatial resolution obtained from the European Space Agency (Copernicus Open Access Hub). It is a Level 1C Top of Atmosphere (TOA) Reflectance product, which includes radiometric and geometric corrections (UTM projection with Geodetic Reference System WGS84), together with orthorectification [100].

(iii) For Terra satellite, one scene acquired on 25 July 2019 by the ASTER sensor. It is a cloud-free 1T level product with 15 m spatial resolution obtained from the USGS EROS Center [98]. For ASTER, unfortunately, shortwave infrared (SWIR) bands were not available for the study region, as they are no longer usable since 2008.

(iv) Additionally, for Terra satellite, one scene was acquired on 28 July 2019 by the MODIS sensor using the surface reflectance product (product MOD09A1). We also used the MODIS Terra MOD09A1 (Version 6) product from the Oak Ridge National Laboratory's Distributed Active Archive Center (ORNL DAAC) (Global Subset Tool: MODIS/VIIRS Land Products: https://modis.ornl.gov/cgi-bin/MODIS/global/subset.pl (accessed on 14 February 2021)). This product, with 500 m spatial resolution, provides spectral surface reflectance of the MODIS 1–7 Terra bands corrected for atmospheric conditions (for example, gases, aerosols, and Rayleigh scattering) at eight-days interval. For each pixel, a value is selected from all acquisitions within the eight-day compounding period. The criteria for choosing the pixel include cloud and solar zenith. When several acquisitions meet the criteria, the pixel with the minimum value of channel 3 (blue) is used [101].

Table 1 summarizes the bands that were used in this study for the different sensors. In the case of MSI, an image composition with all bands (10 and 20 m) was performed, resulting in a product of 20 m of Ground Sampling Distance (GSD).

Table 1. Landsat-8/Operational Land Imager (OLI), Sentinel-2/MultiSpectral Instrument (MSI), Terra/Advanced Spaceborne Thermal Emission and Reflection Radiometer (ASTER), and Terra/Moderate Resolution Image Spectroradiometer (MODIS) spectral band numbers, wavelength ranges (λ), and spatial resolutions used in this study.

Band	OLI			MSI			ASTER			MODIS		
	λ (μm)	Res.	(m)	λ (μm)	Res.	(m)	λ (μm)	Res.	(m)	λ (μm)	Res.	(m)
B1	–				–		0.52–0.60	Green	15	–		
B2	0.45–0.51	Blue	30	0.45–0.52	Blue	10	0.63–0.69	Red	15	–		
B3	0.53–0.59	Green	30	0.54–0.57	Green	10	0.78–0.86	NIR	15	0.45–0.47	Blue	500
B4	0.64–0.67	Red	30	0.65–0.68	Red	10		–		0.54–0.56	Green	500
B5	0.85–0.88	NIR	30	0.69–0.71	Red edge	20		–		1.23–1.25	NIR2	500
B6	1.57–1.65	SWIR1	30	0.73–0.74	Red edge	20		–		1.62–1.65	SWIR1	500
B7	2.11–2.29	SWIR2	30	0.77–0.79	Red edge	20		–		2.10–2.15	SWIR2	500
B8		–		0.78–0.89	NIR	10		–			–	
B11		–		1.56–1.65	SWIR	20		–			–	
B12		–		2.10–2.28	SWIR	20		–			–	

2.2.1. Flowchart

Figure 2 summarizes the classification scheme and analysis followed in this work. Fire area classification methods using kNN and RF algorithms were used to explain the effects of different satellite images on both classifiers.

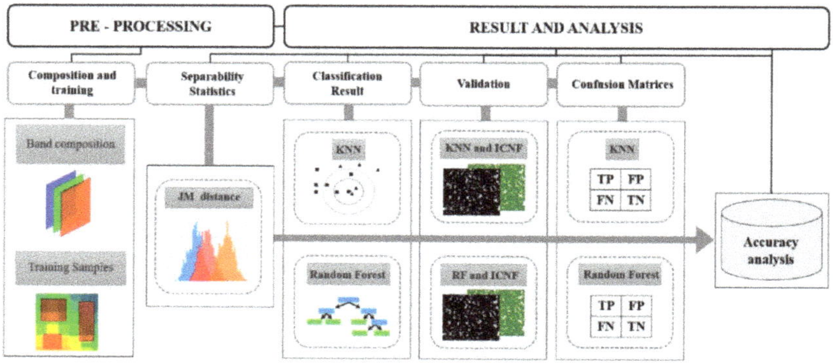

Figure 2. The flowchart of the methodology used in this study.

The workflow for the supervised classification of burned vegetation using kNN and RF algorithms was implemented with multispectral images from Landsat 8/OLI, Sentinel-2/MSI, and Terra (ASTER/MODIS) through training samples using photointerpretation features. The classification accuracy was determined making use of validation data and the results obtained from the analysis of the classification parameters using the generated confusion matrices. After image composition, the procedure includes the following steps: training samples, spectral separability analysis, classification with kNN and RF algorithms, validation, and accuracy analysis.

2.2.2. Training Samples

The initialization of a supervised classification process requires composite images and training samples (polygons). The sample polygons selected in the composite images are used to obtain the burned and unburned areas class descriptors. The training areas were randomly collected, with 30 polygons with an area of 65 km^2 for each class, respecting the separation limits that are based on the ICNF burned area product.

2.2.3. Separability Analysis

The purpose of the separability analysis was to evaluate the spectral separability in all of the bands used in the classification of burned and unburned areas and contribute, for instance, to the decision of which bands have greater classification properties in supervised classification algorithms. The separability of each pair between classes can be quantitatively measured by the average distance between the class density distributions of the pairs or histograms of the values of each band [75]. The Jeffries–Matusita (JM) distance is one of the most widely criterion used in remote sensing in the field of pattern recognition and feature selection. In comparison with other separability indices, JM distance has been suggested as more reliable in separability measures, and also more suitable for less homogeneous main classes [102]. Therefore, we chose the JM distance to indicate the separability between the burned and unburned vegetation. It is calculated according to Equation (1), as [103]:

$$JM_{ij} = \sqrt{2(1 - e^{-B})} \tag{1}$$

where B is the Bhattacharyya distance given by Equation (2), as:

$$B_{ij} = \frac{1}{8}(\mu_j - \mu_i)^T \left[\frac{\Sigma_i + \Sigma_j}{2}\right]^{-1} (\mu_j - \mu_i) + \frac{1}{2} ln \frac{\left|\frac{\Sigma_i + \Sigma_j}{2}\right|}{\sqrt{|\Sigma_i||\Sigma_j|}} \tag{2}$$

For classes i and j, μ is the mean vector of the reflectance values and Σ is the variance-covariance matrix. Previous research has shown that the JM distance can provide a more accurate classification than other distance measures, such as the Euclidean distance or divergence [104]. It ranges between 0 (completely inseparable) and 2 (completely separable) [102].

2.2.4. kNN Classifier

The kNN algorithm proposed by Aha et al. [105] is an instance-based learning method that classifies elements based on the closest k training samples in the resource space. These data play important roles in spatial forecasting, in addition to being the main adjustment parameter of the kNN algorithm. kNN is a common classification tool used in remote sensing data mining applications [63,105], and it is widely used for mapping burned areas [106,107]. kNN is a non-parametric MLA, which makes no assumptions regarding the main data set. This is important when classifying processes of change in territory, such as floods and fires, for which there is little or no prior knowledge of data distribution. In kNN, the pixel whose class is unknown is a member of a class, as described by its spectrally closest neighbors whose class identities are recognized. Figure 3 shows the scheme of the kNN algorithm.

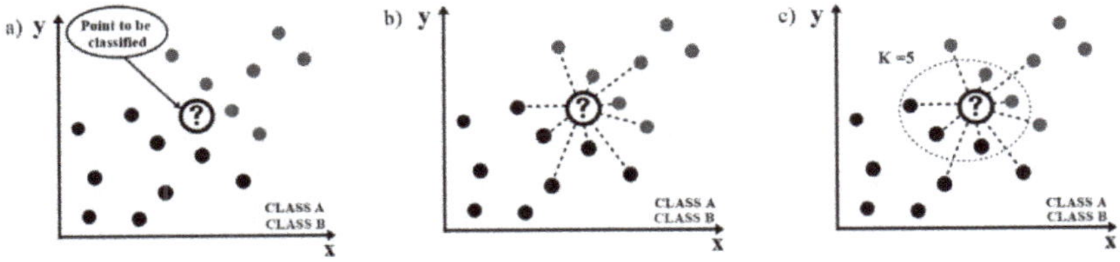

Figure 3. k-Nearest Neighbor (kNN) classification scheme (**a**–**c**).

Initially, the parameter k, which represents the number of closest neighbors, must be selected. This parameter will direct the number of neighbors. In the case of k = 5 in a binary group, the five closest points are identified by the Euclidean distance. In this way, through the shortest distance between the existing k neighbors, which is, the point to be classified and all points in the data set, it is possible to know which class is most similar to. Thus, the classification is completed, and the unknown point is classified. The parameter k plays an important role in the performance of the kNN, being the main kNN adjustment parameter. In this study, we tested different k values (5 to 20) to select the ideal parameter for the kNN classifier based on the lowest estimate of the Root Mean Square Error (RMSE), using different subsets of data. However, in previous studies, as in Cariou et al. [108] and Noi and Kappas [63], it was revealed that this is not the only criterion for selecting an appropriate k value because a small or large k value has characteristics that are suitable for each case. We used SNAP (Sentinel Application Software, ESA) software for this classification.

2.2.5. RF Classifier

The RF algorithm is based on the creation of several decision trees, combining them to obtain a more accurate and stable forecast. According to Rodríguez-Galiano et al. [109], the RF algorithm has advantages in remote sensing area, as it generates an internal unbiased estimate of the generalization that is represented by the Out of Bag (OOB) error, which is a way of validating the RF model. Therefore, it is relatively robust for outliers and noise, in addition to being computationally lighter than other tree set methods. The RF is trained using bootstrap aggregation, where each new tree is adjusted based on a bootstrap sample from the training observations. OOB is the average error for each calculated tree using predictions from trees that do not contain it in their respective bootstrap sample. This allows for the RF classifier to be adjusted and validated while being trained [110].

The Information Gain Rate criterion [111] and the Gini Index [112] are the attribute selection measures most frequently used to induce the decision tree. We chose the Gini

Index, which measures the impurity of an attribute in relation to the classes. For a given T training set, it randomly selects a case (pixel) and determines the class that it belongs to.

In this work, the RF classification was tested for 10 to 400 trees for the set of images composed for each sensor. One-third of the training number of trees was used to test the error that is associated with the predictions, the above-mentioned OOB error. In RF, the parameter MTRY, the optimal trees at each node, controls the number of variables available to split at each node of a tree [113]. In this study, a default value was used as provided by the SNAP software.

2.2.6. Validation and Accuracy Analysis

The validation of remote sensing data is generally based on measurements that were obtained in field campaigns, which are seen as a reference on site. In many cases, the validation process is carried out by remote sensing products provided by official institutions or by sensors with high spectral or spatial resolutions. In this work, the validation product that was used as a reference was the 2019 annual burned area of the atlas provided by the National Institute for Conservation of Nature and Forests (ICNF) of Portugal.

The data were made available on the website http://www.icnf.pt/ (accessed on 14 February 2021) in an ESRI shapefile format covering the entire national territory through the representation of polygons from areas that are affected by fires, coupled with information such as area, date, duration, and the cause that started the fire. The elaboration of the national mapping of the burned areas through the compilation of all geospatial files comes from semiautomatic classification processes using Landsat, Sentinel, or other satellite images [114].

The quality of a given thematic map that is derived from remote sensing data is generally assessed by systematic comparison with other maps also derived from remote sensing [6]. Quality assessment is generally carried out based on verification measures derived from confusion matrices [115]. The choice of validation methods and objectives must be guided by the end use of the products. The cross-tabulation approach is the most common way to assess thematic accuracy. In this context, the comparison and analysis of the quality of the burned area maps that were obtained by the kNN and RF classifications in the different tested sensors were carried out.

The burned area polygon that was obtained by the ICNF map was used as a spatial reference in this study. The pixel-based analysis was based on a confusion matrix (Table 2). Following the terminology that was presented by Fawcett [116], the reference data (true class) will be referred to as positive or negative (burned or unburned). If the instance is positive (burned) and classified as positive (burned), it will be counted as a true positive (TP); if it is classified as negative (unburned), it will be counted as false positive (FP). On the other hand, if the instance is negative (unburned) and it is classified as negative (unburned), it will be counted as true negative (TN); if it is classified as positive (burned), it will be counted as false negative (FN) (Table 2).

Table 2. Confusion matrix between the reference product and the burned/unburned classified areas.

		Reference Map (True Class)		
		Burned	**Unburned**	**Total**
Classified Product	Burned	TP	FN	TP + FN
	Unburned	FP	TN	FP + TN
	Total	TP + FP	FN + TN	TP + FN + FP + TN

The confusion matrices aim to determine the probability of detection of burned areas in the different sizes of fractions of this area at the study site. This explains the error inherent in the burned areas due to the difference between the reference product and the resolutions between the sensor images. According to Cohen [117], the classification methods are evaluated while using statistical parameters, such as the Omission Error (OE), Commission Error (CE), Overall Accuracy (OA), and Dice Coefficient (DC).

OE is related to the producer's accuracy, which is, when a pixel is classified as unburned area being really burned area. CE is related to the user's accuracy, which is, when a pixel is attributed to a class of burned area to which it does not really belong. OA is defined as the fraction of pixels correctly classified as burned or unburned [61]. Finally, DC is a measure of similarity between the classifier and reference map in terms of the number of common burned pixels.

OE and CE vary on a reverse scale of (0–100%), where the lowest values indicate the best estimates. For OA and DC, on the contrary, the largest values indicate the best estimates.

2.2.7. ROC Curve and AUC

The ROC curve has been used in studies of burned areas analysis to verify the general performance of classifiers and models. The ROC curve and a useful statistic calculated from it, the area under the curve (AUC), are mainly used to compare diagnostic tests and act as a performance measure for classifying binary data. The AUC value, as in Equation (3), shows the success rate of the model through the analysis of the training data set and its forecast rate for the tested data set.

$$\mathrm{AUC} = \frac{\sum \mathrm{TP} + \sum \mathrm{TN}}{\mathrm{M} + \mathrm{N}} \tag{3}$$

where M and N are the total number of pixels in the burned and unburned areas. An AUC value that is close to 1 indicates a better performance. An AUC value of 1 indicates a perfect model, while an AUC value of 0 indicates a poor performance model. Between these values, the model performance is classified as excellent (0.9–1), very good (0.8–0.9), good (0.7–0.8), medium (0.6–0.7), and poor (0–0.6).

3. Results

3.1. *Spectral Separability Analysis*

Table 3 summarizes the JM separability values at the study site, where the burned and unburned pixels were analyzed for each spectral band used between the OLI, MSI, ASTER, and MODIS sensors.

Table 3. Jeffries–Matusita (JM) separability values and band for the OLI, MSI, ASTER, and MODIS sensor bands used in the classification.

Band	JM Separability							
	OLI		MSI		ASTER		MODIS	
B1	–		–		0.02	Green	–	
B2	0.31	Blue	0.39	Blue	0.07	Red	–	
B3	0.18	Green	0.19	Green	1.84	NIR	0.15	Blue
B4	0.26	Red	0.36	Red	–		0.53	Green
B5	1.91	NIR	0.45	Red edge	–		1.65	NIR2
B6	0.24	SWIR1	1.82	Red edge	–		0.50	SWIR1
B7	0.70	SWIR2	1.83	Red edge	–		0.75	SWIR2
B8	–		1.75	NIR	–		–	
B11	–		0.14	SWIR	–		–	
B12	–		0.81	SWIR	–		–	

In general, less separability is observed for the visible bands in all sensors, mainly for the bands B1 and B2 for ASTER, and especially in the green range for OLI, MSI, and ASTER, where the bands presented low separability values, with the exception of MODIS, which presented slightly greater separability in this range.

The near infrared (NIR) is the spectral region where the sign of recent fire scars is the strongest, being generally considered to be the best spectral region for detection and mapping burned areas [118] and, therefore, of crucial contribution to image digital classification processes. This is seen in the results of Table 3 with the high values of separability in all sensors, even with some existing spectral and spatial resolution disparities. In addition, the results corroborate the spectral resolution of the sensors, where the thinner infrared range of MODIS and OLI (Table 1) ensured greater separability, very different from the sparse range of MSI and ASTER, even with a slight difference in the spatial resolution and methods of pre-processing.

In the visible–NIR transition bands, there was high separability, as shown in bands B6 = 1.82 and B7 = 1.83 for MSI sensor, except in band B5 = 0.45. However, the band B5 presented low separability, because it is closer to the red band in relation to bands B6 and B7.

The short-wavelength infrared SWIR bands showed low JM separability values.

3.2. *kNN Training*

In this study, we tested different k values (5 to 20) to select the ideal kNN classifier parameter for each set of images. The lowest RMSE value was used as a criterion to select the best k parameter. Thus, despite the low RMSE, from Figure 4 we can see that, after tests, the k parameter was set to 5. It shows that, the lower the value of k, the higher the accuracy of the classification.

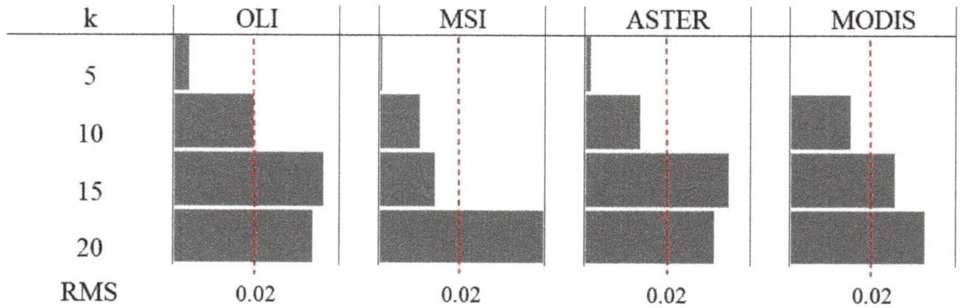

Figure 4. Evaluation of the performance of the kNN classifier with RMSE in relation to k value.

3.3. *RF Training*

Figure 5 shows the distribution of OOB errors for a different number of trees from 10 to 400. It is observed that the classification error between the sensors in the same tree does not change significantly. However, with the increase in the number of trees, the error decreases considerably. In this study, we used the number of trees that had the lowest OOB error. It can be seen that 400 is the best value for trees. One of the advantages of using the RF classifier is its versatility with the processing time, and this can be verified in this work. The classification performed with 10 trees took 10 s, while for 400 trees it took two minutes, a moderately acceptable time interval.

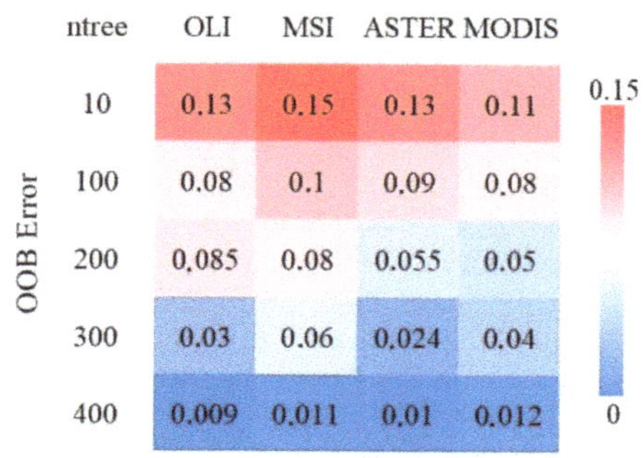

Figure 5. Evaluation of the performance of the RF classifier with the Out of Bag (OOB) error in relation to the number of trees (ntree).

3.4. Burned Area Analysis

Figure 6 and Table 4 show the pixel distribution and size of the burned area for the classification provided by the different sensors with both kNN and RF algorithms. The finer spatial resolution of OLI, MSI, and ASTER showed a burned area with greater spatial detail, but with less density of features. In turn, the map that was generated by MODIS presented, as expected, a burned area with less detail at the edges and a high distribution of overestimated features within the burned area. When comparing the classifiers, the maps visually showed no significant differences with variations in the burned areas ranging between 0.36 and 1.43 km^2, with the lowest differences being for MODIS (0.36 km^2) and the largest for MSI (1.43 km^2). However, they presented important errors in the total burned area when compared to the ICNF reference map. The errors in the total burned area are not constant, ranging between 4.3% and 51.1% (Table 4), and being the difference sensitive to the technical specifications of the images.

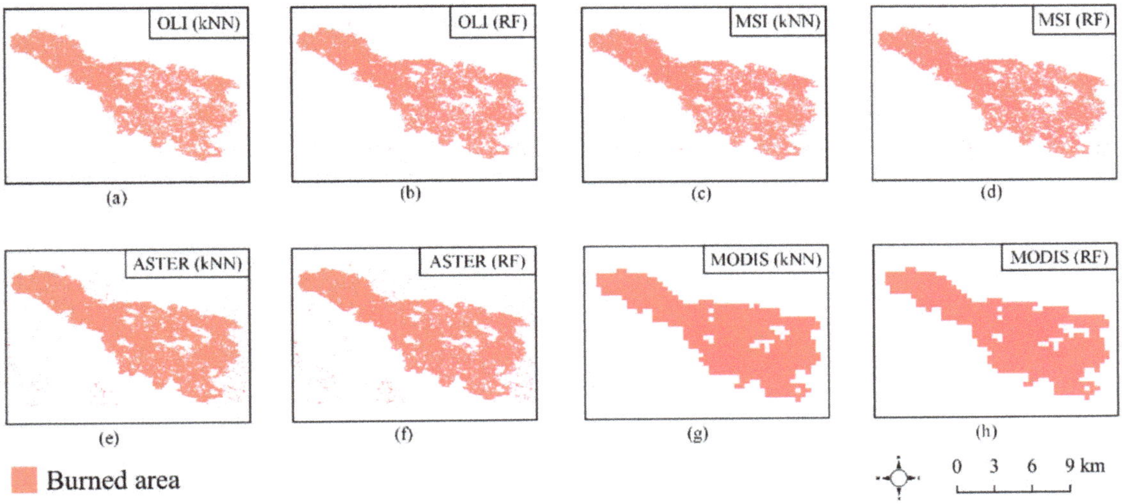

Figure 6. Spatial distribution of the burned area for the Random Forest (RF) and kNN classifiers in the OLI, MSI, ASTER, and MODIS classifications.

Table 4. Size of the burned area obtained with each classifier for each sensor as well as the size of the burned area in the reference map (INCF), differences between the areas obtained with each classifier, and errors when compared with the reference map together to the percentage that they represent with respect to the reference area.

Classifiers	Area (km^2)				
	OLI	**MSI**	**ASTER**	**MODIS**	**ICNF (Reference Map)**
kNN	75.39	76.95	88.02	45.99	93.40
RF	76.36	78.38	89.37	45.63	
|kNN − RF|	0.97	1.43	1.35	0.36	
kNN − ICNF	−18.01 (19.3%)	−16.45 (17.6%)	−5.38 (5.8%)	−47.41 (50.8%)	
RF − ICNF	−17.04 (18.2%)	−15.02 (16.1%)	−4.03 (4.3%)	−47.77 (51.1%)	

3.5. Classification Errors

Figure 7 shows the spatial distribution of the OE and CE for the classification of burned and unburned areas from OLI, MSI, ASTER, and MODIS sensors using the kNN and RF algorithms.

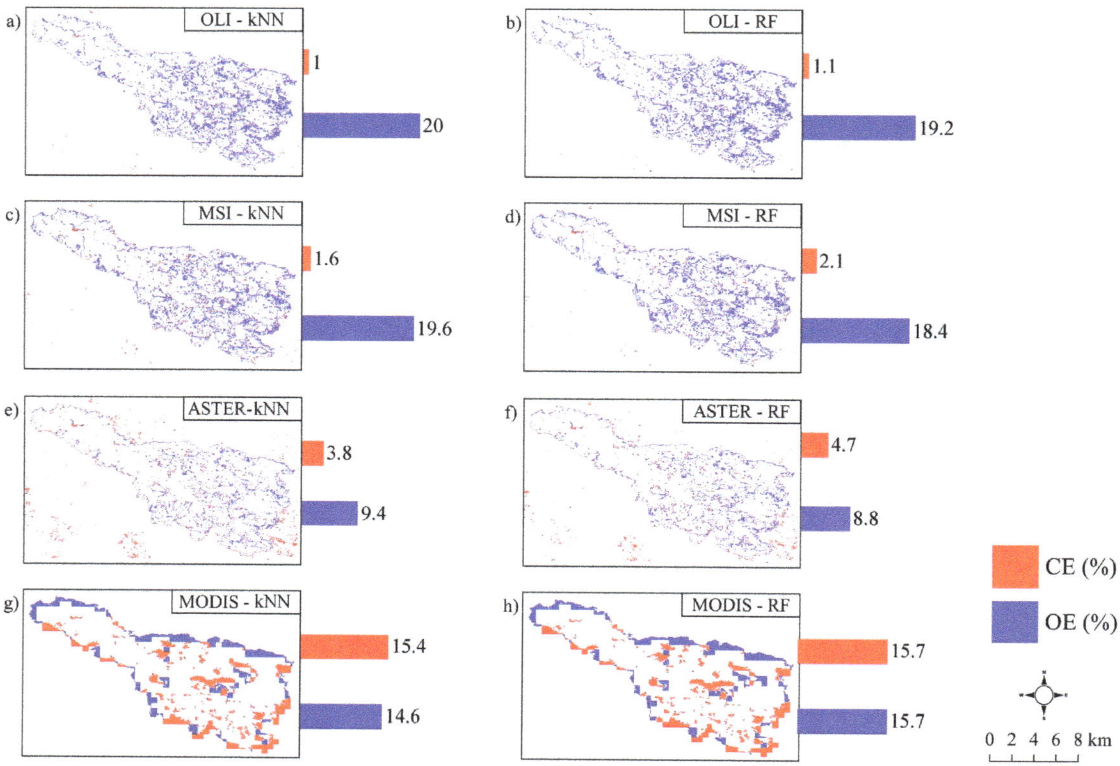

Figure 7. Spatial distributions of Omission Error (OE) and Commission Error (CE) for each classification. (**a**) OLI with kNN, (**b**) OLI with RF, (**c**) MSI with kNN, (**d**) MSI with RF, (**e**) ASTER with kNN, (**f**) ASTER with RF, (**g**) MODIS with kNN, and (**h**) MODIS with RF.

It is observed that, in general, all of the classifications have low CE more frequently within the perimeter that is affected by the fire, although, for ASTER, there is a significant presence of missing mixing pixels and CE outside the burned area (Figure 7e,f).

For ASTER images, the classifications present the smallest OE, with a spatial distribution of 8.73 km^2 of areas with missing pixels for kNN and 8.19 km^2 for RF. In contrast, despite the lower spatial resolution of MODIS, there was a moderate frequency of missing pixels within the burned area when compared to the other sensors, which decreased the sensors OE reaching ~13–14 km^2. It is more evident in the upper border, as shown in Figure 7g,h, the place of transition between burned and unburned areas, which, in turn, is more susceptible to errors that are caused by low spatial resolution.

3.6. Overall Accuracy (OA)

The differences in areas that were classified as burned in our classifications and the reference map were the lowest for ASTER (4 km^2) and the highest for MODIS (47 km^2).

This result is consistent for the images with better spatial resolution and greater proximity to the date of the reference product, such as OLI, MSI, and ASTER, resulting in a stable thematic quality. When the time interval between the data is too long, it is difficult to know exactly what period the pixel finally extracted from the image refers to. This statement is disconnected from the results that were obtained by the MODIS sensor, which, despite the proximity of the day of the burning occurrence, its spatial resolution, and its eight-days compaction form, was an important factor as mentioned above.

In terms of algorithms, RF was the classification method that presented the smallest error in the total burned area in relation to the ICFN reference area with values of the order of 4 to 17 km^2 for the finer spatial resolution sensors (Table 4) and good estimates of OA and DC, as can be seen in Table 5.

Table 5. Values of OA and Dice Coefficient (DC) for the products generated by kNN and RF classifiers in the different sensors used.

Classifiers	Parameters	OLI	MSI	ASTER	MODIS
kNN	OA (%)	92.95	93.09	93.62	89.83
	DC	0.88	0.88	0.93	0.85
RF	OA (%)	93.24	93.35	93.52	89.45
	DC	0.89	0.89	0.93	0.84

The results show that the classification based on kNN and RF for the different sensors mapped the burned area with a very high accuracy (OA > 89% and DC > 0.8) and without significant variations in the computed OA and DC values for all of the sensors.

3.7. Algorithms Errors

A ROC curve analysis was performed to graphically assess the sensitivity and specificity of the classifications carried out. From the analysis of Figure 8, it can be seen that, as the score point increases, the discriminating power also increases, which is, the curve is closer to the upper left corner and, consequently, a greater area is obtained below the ROC curve. In both classifiers, the largest value was recorded for ASTER and the lowest for MODIS, corroborating the results obtained by the OE and CE.

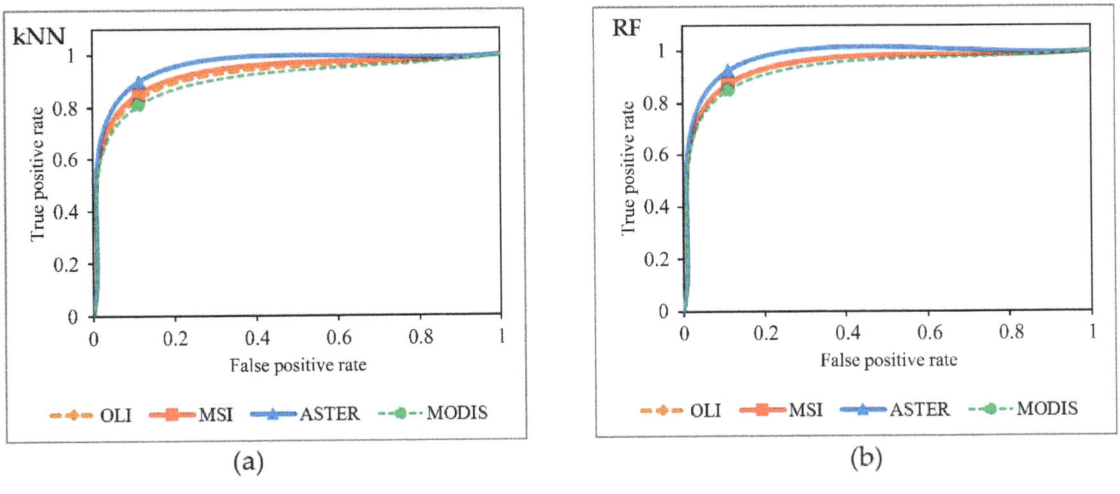

Figure 8. Receiver operating characteristic (ROC) curve graphs for (**a**) kNN and (**b**) RF.

4. Discussion

This study assesses the application of an automatic methodology for mapping burned areas in Portugal through the supervised classification algorithms kNN and RF using multispectral satellite images of different technical specifications. The integration of the use of these images increases the temporal accuracy of imaging a target that is susceptible to extreme events, which often require intense monitoring.

In this study we show, in detail, the quality, errors, and incompatibilities in the classification of a burned area at a local scale, which, in turn, can be used to explain phenomena of non-resistance (edge effects, unexpected artifacts, or underestimation related to low intensity fires) that are often propagated or masked when applied on a continental or global scale. This is widely discussed in Randerson et al. [99] who observed an underestimation of 4–15% of the burned area missing in global products, slightly below the 30% that is normally assumed. This underestimation occurs due to the absence or small overlap in the detection of small fires (<270 ha) derived from different global burned area products. Such problems are also found in Nogueira et al. [119], Chuvieco et al. [120], and Roteta et al. [56]). Therefore, our analysis demonstrates the importance of accurate mapping of a burned area at a local scale, which still remains the most accurate base of reference data in protocols for validations of global burned area, after evaluation by photointerpretation [121,122] or in the field [61].

4.1. Separability Analysis

The errors that were found in the classification of burned areas were caused by several factors, one of which was the spectral similarity of burned areas with other surface elements, mainly darker bodies, in addition to the technical disparities of the kNN and RF classifiers. However, the spatial accuracy of the images was the most important agent in reducing the performance of the products. This behavior can be seen in the maps that are generated by MODIS sensor, due to its coarse spatial resolution.

The assessment of the ability to detect burned areas was performed using the JM separability index in the different bands (Equations (1) and (2) and Table 3) and the results of the confusion matrices represented by OE, CE, OA, DC, and AUC (Figures 7 and 8 and Table 5).

In agreement with previous studies [123–125], less separability is observed for the visible bands in all sensors in our results, mainly for the bands B1 and B2 for ASTER [126] and especially in the green range. This occurred because forest fires affect the leaf structure and photosynthetic capacity. They also decrease the green pigment of the leaf (chlorophyll) and increase the brown-yellow pigment (carotenoids, pheophytin, and xanthophyll) [124]. In the visible-NIR transition bands, there was high separability corroborating the studies conducted by Fernández-Manso et al. [127]. The authors proved that recent fires in healthy vegetation show a characteristic increase in the reflectance from red to NIR, associated with variations in chlorophyll content.

The analysis was able to show good discrimination of the burned areas. This approach improved the spatial homogeneity of the affected areas (even if random) of the classification thresholds, as shown by the high values of AUC (>0.88), reducing the dependence on having information on land cover, usually used in automatic burned area algorithms. Although it is important to emphasize that the lack of information on land use for adapting the algorithms behavior can imply the recurrence of systematic errors, increasing the uncertainty of the final burned area classification, as shown in Figure 9. As already mentioned, we note the presence of features that presented spectral behavior that was similar to the burned area (for example, low reflectance values in the NIR), which can be caused by topography shadows and changes in land cover not associated with fires, such as very humid soils. Therefore, it is recommended to take special care in regions where these characteristics and events occur close to the area that is affected by the fire, in addition to controlling the photointerpretation with the size of the samples of interest, especially in applications with sensors of different spatial and spectral resolutions [128]. Thus, as a future study in the study area, assessing the separability for different classes of land use and the influence of sample size may be a good alternative.

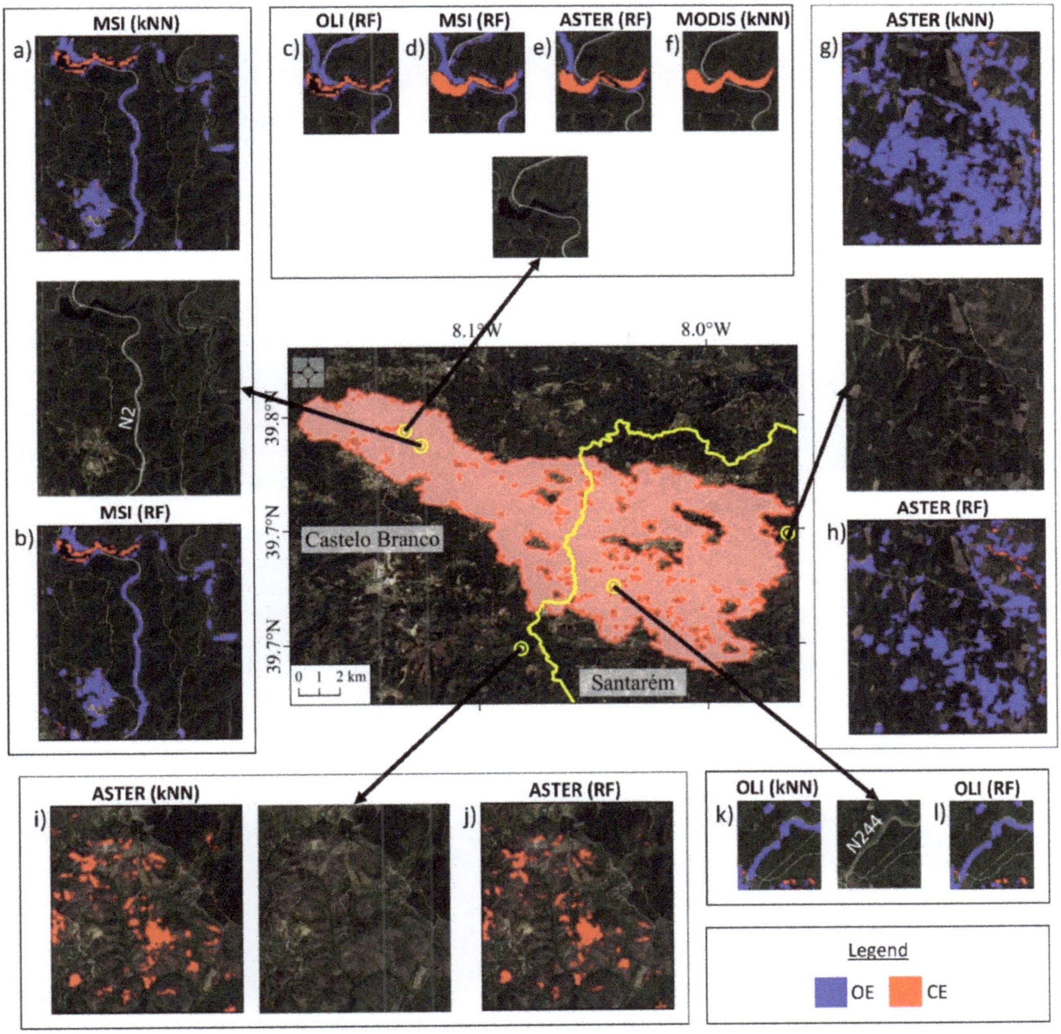

Figure 9. Visual analysis of the errors presented in the different land cover in the study area: highways (**a**,**b**,**k**,**l**), pasture and agriculture (**g**,**h**), soil degradation (**i**,**j**), and water bodies (**c**–**f**).

4.2. Validation Product

The reference mapping that is generated by ICNF proves to be quite efficient in the generation of geospatial data, providing a database that is rich in accurate information of the burned areas throughout the national territory and of open access. However, here we list some advantages and limitations of the product, based on a visual comparison after the classification process. Initially, we emphasize the thoroughness of the delimitation at the edges of the burned area that is generated by the ICNF product, in front of a complex landscape, where the study area is inserted. This could be proven in both product classifiers that are generated by the MODIS sensor, as expected, with a high frequency of omission and commission pixels at the edges of the burned area and in urban areas, as shown in Figure 7. However, according to Mouillot et al. [45], OE and CE found at the edges of burned areas cannot be strictly seen as false or omitted alarms. For a given level of CE and OE, it is acceptable as long as both are similar. We can also see that the influence of the spatial validation product provided by ICNF was crucial for the errors that are shown in Figure 7, since this product is based on the 10 m resolution bands of MSI [129], thus causing the lowest error estimates for the ASTER sensor (15 m) due to its greater proximity to spatial detail.

Even with the absence of the blue and SWIR bands in the ASTER images, this sensor showed the highest accuracy parameters in both of the classifiers, although the good results found in OLI and MSI can be attributed to the use of these bands. Therefore, the 15 m GSD of ASTER was responsible for this good performance, although its proximity to the date of the reference mapping must also be taken into account.

4.3. kNN and RF Classifiers

The RF algorithm presented the highest quality values of the classification among all of the sensors and a greater stability in relation to the data change in the attribution of burned and unburned classes. This result corroborates the low complexity of its application, low cost of time, and memory. Despite the variations found in the OOB error with the number of trees, this parameter may not be very relevant in binary classifications, since the use of two classes reduces the voting options of each set of trees in the data set. In general, empirically, the error in the classification with this algorithm depends on the strength of the individual tree and the correlation between two trees in the forest. Strength can be interpreted as a performance measure for each tree. Increasing the correlation increases the error rate of the forest, and increasing the strength of an individual tree decreases the error rate of the forest, since a tree with a low error rate is a strong classifier. On the other side, reducing the number of selected random attributes reduces correlation and strength [85,130]. In our study, we selected 400 trees. In several studies of buried areas by the RF classifier, the largest number of trees commonly used ranges from 100 to 1500 [83,84,131].

However, kNN, even with accuracy values very close to the RF, mainly in the AUC parameter, has a direct relationship with the k parameter, time, and memory. Once a k value is given, more training samples are needed to improve the performance, but more time and storage memory were needed. In this study, the k value that was based on the RMSE was used. Therefore, the disparities found in the quality of the mapping of this classifier can be attributed to other parameters not tested here. The values are consistent with the studies conducted by Meng et al. [132]. The value of k may not present significant differences in relation to the final result of the mapping; however, this value directly influences the processing time. It is worth mentioning that, for $k = 5$, the processing time was 0.56 min, while, for $k = 20$, the time was approximately 70 min. This time interval depends significantly on the resources of the computer used and are common for kNN classifications, depending on the size and composition of the data set [133]. Blanzieri and Melgani [134] show that the best values of k were found empirically below $k = 5$ using SAR data, which could be explained by the image filtering applied to the true soil homogeneity. This indicates that the decrease in k is associated with the registration of optically active elements in the images. This statement is also related to the location of the pixels to be classified in relation to the training samples. When the k neighboring pixels are close enough, the precision will naturally tend to the value very close to the sample pixel set, consisting of a decrease in time and error.

4.4. Accuracy Analysis

It is observed that, in general, all of the classifications have low CE more frequently within the perimeter that is affected by the fire, although, for ASTER, there is a significant presence of missing mixing pixels and CE outside the burned area. This behavior may be related to the confusion of the classifiers in distinguishing between burned areas and dark soils with little vegetation. As said, this was quite evident in the classification with ASTER images, since, for this sensor, only three bands (green, red, and NIR) were used in the classification, which is, less resources for feature detection, which also favored the increase of false alarms pixels in relation to the other sensors for both classification algorithms, as also found in [59]. This is spectrally true while taking the results found with OLI and MSI sensors into account because of NIR and SWIR bands used in the classification probably influenced the presence of low CE. These channels strongly reflect the spectral signal of change detection in the vegetation state in addition to having high separability between burned and unburned areas, as shown in Table 3 and in several works [46,135–137], who also used this region of the spectrum for the separation of burned areas obtaining satisfactory results. According to Lambin et al. [138], reflectance generally decreases in the NIR range after the fire event due to the removal of vegetation retained by water due to the fire. The decrease in brightness is more substantive than in the visible, which makes the NIR range more suitable for discriminating burned areas. The low CE for OLI, MSI, and ASTER can be attributed to the higher spatial resolution, since this condition improves the performance of classifying algorithms mainly in places with homogeneous and more compact distribution of the burned area [31,139]. For the classifications that were performed with MODIS, the largest CE of the data series were observed with pixels well distributed throughout the affected perimeter. In this case, the low spatial resolution of this sensor was the main cause of the errors, causing a high frequency of underestimated pixels inside and outside the burned area.

OE, being represented by pixels mistakenly classified as unburned areas, presented significant and well distributed values on the maps, with emphasis on the east sector of the burned area in both classifiers. These errors are related to the high frequency of pixels referring to small urban centers that are inserted in the burned area, which, in turn, were correctly classified as unburned areas, but, due to problems of pixels spectrally mixed at the edges of these features, there was a high presence of pixels of burned areas omitted from their assignment in the classification. This problem was also found in [46,140], who showed moderate performance in mapping burned areas in optically complex locations, caused by ambiguity problems in the classification and spectral mixing.

In the western sector, the same problem occurred, but, more frequently, because, in addition to the housing polygons, agricultural areas also caused confusion in the classifiers. This directly influenced the results of the spatial distribution of the missing pixels in burned areas, where both of the sensors presented area variations between 17 and 18 km^2 for kNN classifier and between 16 and 17 km^2 for RF, which is, a high frequency of pixels incorrectly classified as unburned areas.

It is observed that for ASTER images, the classifications presented the smallest OE, with a spatial distribution of 8.73 km^2 of areas with missing pixels for kNN and 8.19 km^2 for RF. These values were already expected, since this sensor has the best spatial resolution of the set of images and, consequently, reduced spectral mixing problems, even using only visible bands. In addition, the use of ASTER images limited the overestimation of the burned areas due to the pixel size, most suitable for classifying unburned areas that are inserted in the investigated fire polygon [141].

In contrast, despite the lower spatial resolution of MODIS, there was a moderate frequency of missing pixels within the burned area when compared to the other sensors, which decreased the sensors OE reaching ~13–14 km^2. It is more evident in the upper border, as shown in Figure 9g,h, the place of transition between burned and unburned areas, which, in turn, is more susceptible to errors that are caused by low spatial resolution. Another influencing factor can be explained by the process of creating the image composition of the MODIS sensor with the acquisition of the best pixel within the eight-day period. The result generates an image with moderate quality once some information is lost.

For the classifications that are generated by the best spatial resolution sensors (OLI, MSI, and ASTER), errors were found in the different elements of land use in the study area, for example, in the products generated by the OLI and MSI scenes. We detected a high frequency of OE on the main highways that cut the area that is affected by the fire, especially the highways N2 and N244 (Figure 9a,b,k,l), thus showing the limitation of the ICNF product in the detection of burned areas in these characteristics. In addition, these errors were also found in the ASTER images, but more frequently in the areas of pasture and agriculture with approximately ~0.06–0.1 km^2 (Figure 9g,h). However, the reference product proved to be advantageous in the classification of areas of soil degradation in kNN and RF classifiers, erroneously classifying these areas as burned areas, as shown in Figure 9i,j.

Finally, kNN and RF classifiers were not efficient in differentiating water bodies and burned areas in all sensors, causing several CE pixels, as shown in Figure 9c–f. This result is in accordance with Roy et al. [74], Palomino-Ángel et al. [142], and Shimabukuro et al. [143], who reported classification errors in burned areas caused by the spectral similarity with water bodies.

4.5. OA and Algorithms Errors

Overall, the classifications present good estimates of OA and DC. These OA values are also related to the correct classification of unburned areas and, for this reason, particular attention needs to be paid to this parameter, not using it as the only thematic quality parameter [61]. The high DC values, as summarized in Table 5, show a good performance in continuous adherence with the reference data for the class of presence of burned area, even when considering the sensitivity of this parameter to false alarms and missing pixels shown in the maps of Figure 7.

Although Tanase et al. [144], in studies of burned areas in Tropical Africa, suggested that temporarily short sample units may underestimate the accuracy of the detection of burned areas, Schroeder et al. [145] showed, in their studies in the Brazilian Amazon, that the date of the imaging must be as close as possible with respect to the spatial reference data, which may have intensified the OE or increased areas with different time on the hour scale. The methods of detecting changes based on the application of temporal metrics to assess sudden variations in the pixel signature of moderate and coarse resolution sensors are gaining importance as better-quality satellite data sets become available [146,147].

In general, ASTER presented the highest values of OA and DC in relation to the values of the other sensors, because its spatial resolution may have a greater influence in detecting the details of fire scars. MODIS sensor showed the lowest values of OA and DC of all the sensors, being, however, large values. These data are important, as they show that even the low spatial accuracy of MODIS in relation to the reference map as well as OE and CE greater than 10% did not drastically decrease the estimates of OA and DC, because, with both classifiers and sensors, the maps were considered to be excellent according to Cohen's classification [117]. The same behavior was seen in Lanorte et al. [141], who showed, in applications of ASTER and MODIS sensors in burned areas in southern Italy, that these data were efficient in allowing the detection of burned areas and discriminating the severity of the fire.

The OLI and MSI sensors did not show significant variations in OA and DC, displaying MSI the best results, which is attributed to the low OE that was made in the classification. An identical result was found in [71,148], who reiterated that the reason why the classification provided by Sentinel-2 is more accurate than by Landsat 8 is due to the higher spatial resolution of Sentinel-2 images. Because of this, the burned areas obtained with the classification process on Landsat-8 may have been overestimated. Other studies following this approach also found similar OA values, for example, 90% in Axel [149], 79.2% in Liu et al. [75], 95% in Libonati et al. [61], 94.7% in Zhang et al. [150], 99% in Alonso-Cañas and Chuvieco [46], and 96% in Roy et al. [74].

It is worth noting that both of the classifiers require that choices be made by the modeler concerning numerous parameters under different performances. In general, the classifiers based on kNN and RF brought high quality in the classification of burned areas with AUC values above 0.88, DC above 76%, and OA above 89%, in addition to the ability to process data efficiently and enable parallel training of the same samples in different orbital data sets.

Therefore, the results show a statistically significant ROC curve with an AUC varying between 0.88 and 0.94 for both algorithms, showing that, even in the case of supervised classifications, approximately 90% of the burned areas were well classified by the algorithms in the different sensors. This result agrees with the initial study by Chou et al. [151], where the classification improvement was significant when accounting for spatial autocorrelation in logistic fire probability models in Southern California. Likewise, Siljander [152] found values of AUC in the order of 0.86–0.94, indicating that the fire classification models that were responsible for the spatial distribution of the affected areas showed themselves to be superior in the estimate of burned area on a regional scale when compared with products of global scale burning. In addition, Dlamini [153] found high precision with AUCs of 0.94 and 0.97 in models of Bayesian networks for data of active fire and burned area in ASTER images, respectively. The author also stressed the validity of the Bayesian networks and that the probability estimation based on the data from the burned area can estimate the fire risk a little better than from the active fire data.

5. Conclusions

Based on kNN and RF classifiers and using Landsat-8, Sentinel-2, and Terra imagery, a methodology for assessing their performance in the classification of burned areas in a forest fire occurred in central Portugal is proposed. The main conclusions are as follows:

(i) Less separability is observed for the visible and SWIR bands in all sensors, particularly in the green range, and high separability for NIR region.

(ii) For kNN classification algorithm, k = 5 was found as the best parameter. In the same line, for RF, 400 trees were selected as an optimal value.

(iii) No significant differences in the burned areas that were obtained with each classifier for each sensor were found.

(iv) When compared with ICNF validation data, the lower errors in the total burned area were found in the classifications that were performed with ASTER and the largest errors with MODIS.

(v) Contrary to expectation, the classification that was performed by OLI had greater precision but lower accuracy when compared to MSI. In general, high precision and accuracy were found in the classifications.

(vi) The lowest CE (<5%) were found in the classifications carried out with kNN and RF in OLI, MSI, and ASTER, and large CE, of the order of 15%, with MODIS, with a significant presence in ASTER outside the burned areas. Related to OE, significant and well distributed values were found in all sensors (8–20%), with emphasis on the eastern sector of the burned area, being the low values for ASTER.

(vii) The classification that was based on kNN and RF for the different sensors mapped the burned area with a very high accuracy (OA > 89% and DC > 0.8). The results show a statistically significant ROC curve with an AUC varying between 0.88 and 0.94 for both classifiers, showing that, even in the case of supervised classifications, approximately 90% of the burned areas were well classified by the algorithms in the different sensors.

It is possible to observe that the visible, intermediate, and SWIR bands showed low values of separability, which corresponds to the results that were found by Pereira et al. [118], who stated that the spectral changes induced by fire in the SWIR are similar to those in the visible range, since the burned areas are generally more reflective than green vegetation, but darker than vegetation predominantly in savannas during the dry season. It is important to note that the SWIR band has the advantage of having low interference with atmospheric scattering during the scene recording process. Following this premise, there may be no significant reduction in the spectral contrast of the surface in the images, consequently resulting in increased separability indices. However, this behavior was not observed in our experiments.

This methodology can be useful for mapping the burned areas in regions of native vegetation and the improvement of methods for monitoring the burned areas in Portugal, in addition to assisting in the management of fire in the region and estimating the impacts that are generated by it. The availability of detailed information on the spatial and temporal distributions of the burned areas is currently crucial. Therefore, the applied method makes it possible to survey the scars of fires while using geospatial data with the greatest possible accuracy, assisting in the maintenance of an information bank, serving not only the management of the territory, but also the comparison with related future events.

In general, the errors that were found in both kNN and RF classifiers can also be related to the creation of very heterogeneous objects, even in a region with a predominance of sparse vegetation. Despite the similar results of OE and CE and the differences in the processing of each algorithm, it was shown that the spectral resolution and, especially the spatial resolution, is a more important factor in the process of classification of burned areas. OE and CE are directly linked to the burned areas used as reference mapping, as product incompatibility can generate low generalization capacity and, consequently, OE and CE close to 100% as found in Lizundia-Laiola et al. [154].

Finally, this study opens up the possibility of using multiple Earth Observation data to assess environmental disturbances, increasing the range of possibilities for implementing these data when, for example, there is no scene or a specific band for a given period or problems with cloud cover.

Author Contributions: Conceptualization, A.d.P.P., J.A.d.S.J., A.M.R.-A. and R.F.F.H.; Data curation, A.d.P.P., J.A.d.S.J., A.M.R.-A. and R.F.F.H.; Formal analysis, A.d.P.P., J.A.d.S.J., A.M.R.-A. and R.F.F.H.; Investigation, A.d.P.P., J.A.d.S.J., A.M.R.-A. and R.F.F.H.; Methodology, A.d.P.P., J.A.d.S.J., A.M.R.-A. and R.F.F.H.; Supervision, A.d.P.P., J.A.d.S.J., A.M.R.-A. and R.F.F.H.; Validation, A.d.P.P., J.A.d.S.J., A.M.R.-A. and R.F.F.H.; Writing—original draft, A.d.P.P., J.A.d.S.J., A.M.R.-A. and R.F.F.H.; Writing—review & editing, A.d.P.P., J.A.d.S.J., A.M.R.-A. and R.F.F.H. All authors have read and agreed to the published version of the manuscript.

Funding: This research received no external funding.

Acknowledgments: Research was supported by PAIUJA-2019/2020 and CEACTEMA from University of Jaén (Spain), and RNM-282 research group from the Junta de Andalucía (Spain). Special thanks to the four anonymous reviewers for their insightful comments.

Conflicts of Interest: The authors declare no conflict of interest.

References

1. Food and Agriculture Organization (FAO). *Global Forest Resources Assessment 2010—Main Report*; FAO Forestry Paper 163; FAO: Rome, Italy, 2010. Available online: http://www.fao.org/3/i1757e/i1757e.pdf (accessed on 24 January 2021).
2. Carmenta, R.; Parry, L.; Blackburn, A.; Vermeylen, S.; Barlow, J. Understanding human-fire interactions in tropical forest regions: A case for interdisciplinary research across the natural and social sciences. *Ecol. Soc.* **2011**, *16*, 53–75. [CrossRef]
3. Mabuhay, J.; Nakagoshi, N.; Horikoshi, T. Microbial biomass and abundance after forest fire in pine forests in Japan. *Ecol. Res.* **2003**, *18*, 431–441. [CrossRef]
4. Lauk, C.; Erb, K.-H. Biomass consumed in anthropongenic vegetation fires: Global patterns and processes. *Ecol. Econ.* **2009**, *69*, 301–309. [CrossRef]
5. Chandra, K.; Bhardwaj, A.K. Incidence of Forest Fire in India and Its Effect on Terrestrial Ecosystem Dynamics, Nutrient and Microbial Status of Soil. *Int. J. Agric. For.* **2015**, *5*, 69–78. [CrossRef]
6. Vicente, F.; Cesari, M.; Serrano, A.; Bertolani, R. The impact of fire on terrestrial tardigrade biodiversity: A first case-study from Portugal. *J. Limnol.* **2013**, *72*, 152–159. [CrossRef]
7. Vandermeer, J.; Perfecto, I. *Breakfast of Biodiversity. The Political Ecology of Rain Forest Destruction*, 2nd ed.; Food First Books: New York, NY, USA, 2013.
8. Fearnside, P.M. Biodiveristy as an Environmental Service in Brazil's Amazonian Forests: Risks, Value and Conservation. *Environ. Conserv.* **1999**, *26*, 305–321. [CrossRef]
9. Griffith, D.M. Agroforestry: A refuge for tropical biodiversity after fire. *Conserv. Biol.* **2000**, *14*, 325–326. [CrossRef]
10. Alkhatib, A.A.A. A review of forest fire detection techniques. *Int. J. Distrib. Sens. Netw.* **2014**, 597368. [CrossRef]
11. Bonazountas, M.; Kallidromitou, D.; Kassomenos, P.A.; Passas, N. Forest fire risk analysis. *Hum. Ecol. Risk Assess.* **2004**, *11*, 617–626. [CrossRef]
12. Miranda, A.I.; Coutinho, M.; Borrego, C. Forest fire emissions in Portugal: A contribution to global warming? *Environ. Pollut.* **1994**, *83*, 121–123. [CrossRef]
13. Randerson, J.T.; Liu, H.; Flanner, M.G.; Chambers, S.D.; Jin, Y.; Hess, P.G.; Pfister, G.; Mack, M.C.; Treseder, K.K.; Welp, L.R.; et al. The impact of Boreal forest fire on climate warming. *Science* **2006**, *314*, 1130–1132. [CrossRef]
14. Dixon, R.K.; Krankina, O.N. Forest fires in Russia: Carbon dioxide emissions to the atmosphere. *Can. J. For. Res.* **1993**, *23*. [CrossRef]

15. Gillett, N.A.; Weaver, A.J.; Zwiers, F.W.; Fiannigan, M.D. Detecting the effect of climate change on Canadian forest fires. *Geophys. Res. Lett.* **2004**, *31*, L18211. [CrossRef]
16. Pribadi, A.; Kurata, G. Greenhouse gas and air pollutant emissions from land and forest fire in Indonesia during 2015 based on satellite data. *IOP Conf. Ser. Earth Environ. Sci.* **2017**, *54*, 012060. [CrossRef]
17. Matricardi, E.A.T.; Skole, D.L.; Pediowski, M.A.; Chomentowski, W.; Fernandes, L.C. Assessment of tropical forest degradation by selective logging and fire using Landsat imagery. *Remote Sens. Environ.* **2020**, *114*, 1117–1129. [CrossRef]
18. Dahn, C.N.; Candelaria-Ley, R.I.; Reale, C.S.; Reale, J.K.; Van Horn, D.J. Extreme water quality degradation following a catastrophic forest fire. *Freshwater Biol.* **2015**, *60*, 2584–2599. [CrossRef]
19. Pearson, T.R.H.; Brown, S.; Murray, L.; Sidman, G. Greenhouse gas emissions from tropical forest degradation: An underestimated source. *Carbon Balance Manag.* **2017**, *3*. [CrossRef] [PubMed]
20. Morris, S.E.; Moses, T.A. Forest fire and the natural soil erosion regime in the Colorado front range. *Ann. Am. Assoc. Geogr.* **1987**, *77*, 245–254. [CrossRef]
21. Kutiel, P.; Inbar, M. Fires impacts on soil nutrients and soil erosion in a Mediterranean pine forest plantation. *Catena* **1993**, *20*, 129–139. [CrossRef]
22. Salesa, D.; Minervino Amodio, A.; Rosskopf, C.M.; Garfi, V.; Terol, E.; Cerdà, A. Three topographical approaches to survey soil erosion on a mountain trail affected by a forest fire. Barranc de la Manesa, Llutxent, Eastern Iberian Peninsula. *J. Environ. Manag.* **2020**, *264*, 110491. [CrossRef]
23. Fernández, C.; Vega, J.A.; Jiménez, E.; Fonturbel, T. Effectiveness of three post-fire treatments at reducing soil erosion in Galicia (NW Spain). *Int. J. Wildland Fire* **2011**, *20*, 104–114. [CrossRef]
24. Myronidis, D.I.; Emmanouloudis, E.A.; Mitsopoulos, I.A.; Riggos, E.E. Soil erosion potential after fire and rehabilitation treatments in Greece. *Environ. Model. Assess.* **2010**, *15*, 239–250. [CrossRef]
25. Kozlowski, T.T. *Fires and Ecosystems*; Academic Press, Inc.: Cambridge, MA, USA, 1974.
26. Cochrane, M.A.; Schulze, M.D. Fire as a recurrent event in tropical forests of the eastern Amazon: Effects on forest structure, biomass, and species composition. *Biotropica* **1999**, *31*, 2–16. [CrossRef]
27. Wallenius, T.; Niskanen, L.; Virtanen, T.; Hottola, J.; Brumelis, G.; Angervouri, A.; Julkumen, J.; Pihiström, M. Loss of hábitats, naturalness and species diversity in Eurasian forest landscapes. *Ecol. Indic.* **2010**, *10*, 1093–1101. [CrossRef]
28. Bowman, D.M.J.S.; Murphy, B.P.; Boer, M.M.; Bradstock, R.A.; Cary, G.J.; Cochrane, M.A.; Fensham, R.J.; Krawchck, M.A.; Price, O.F.; Williams, R.J. Forest fire management, climate change, and the risk of catastrophic carbon losses. *Front. Ecol. Environ.* **2013**. [CrossRef]
29. Bonazountas, M.; Kalidromitou, D.; Kassomenos, P.; Passas, N. A decision support system for managing forest fires casualties. *J. Environ. Manag.* **2007**, *84*, 412–418. [CrossRef]
30. Diakakis, A.; Xanthopoulos, G.; Gregos, L. Analysis of forest fire fatalities in Greece: 1977–2013. *Int. J. Wildland Fire* **2016**, *25*, 797–809. [CrossRef]
31. Polychronaki, A.; Gitas, I.Z. Burned Area Mapping in Greece Using SPOT-4 HRVIR Images and Object-Based Image Analysis. *Remote Sens.* **2012**, *4*, 424–438. [CrossRef]
32. Chuvieco, E.; Yue, C.; Heil, A.; Mouillot, F.; Alonso-Canas, I.; Padilla, M.; Pereira, J.M.; Oom, D.; Tansey, K. A new global burned area product for climate assessment of fire impacts. *Glob. Ecol. Biogeogr.* **2016**, *25*, 619–629. [CrossRef]
33. Domenikiotis, C.; Loukas, A.; Dalezios, N.R. The use of NOAA/AVHRR satellite data for monitoring and assessment of forest fires and floods. *Nat. Hazards Earth Syst. Sci.* **2003**, *3*, 115–128. [CrossRef]
34. San Miguel Ayanz, J.; Barbosa, P.; Schmuck, G.; Liberta, G.; Schulte, E. Towards a coherent forest fire information system in Europe: The European Forest Fire Information System (EFFIS). In *Environmental Monitoring in the South-Eastern Mediterranean Region Using RS/GIS Techniques*; Gitas, I.Z., San Miguel Ayanz, J., Eds.; CIHEAM Options Méditerranéennes: Série B; Etudes et Recherches: Chania, Greece, 2003; Volume 46, pp. 5–16.
35. Gutiérrez, E.; Lozano, S. Avoidable damage assessment of forest fires in European countries: An efficient frontier approach. *Eur. J. For. Res.* **2013**, *132*, 9–21. [CrossRef]
36. Souza, C.M.; Roberts, D.A.; Cochrane, M.A. Combining spectral and spatial information to map canopy damage from selective logging and forest fires. *Remote Sens. Environ.* **2005**, *98*, 329–343. [CrossRef]
37. Plank, S. Rapid damage assessment by means of multi-temporal SAR—A comprehensive review and outlook to Sentinel-1. *Remote Sens.* **2014**, *6*, 4870–4906. [CrossRef]
38. Sunar, F.; Özkan, C. Forest Fire Analysis with Remote Sensing Data. *Int. J. Remote Sens.* **2001**, *22*, 2265–2277. [CrossRef]
39. Novo, A.; Fariñas-Álvarez, N.; Martínez-Sánchez, J.; González-Jorge, H.; Fernández-Alonso, J.M.; Lorenzo, H. Mapping forest fire risk—A case study in Galicia (Spain). *Remote Sens.* **2020**, *12*, 3705. [CrossRef]
40. Adaktylou, N.; Stratoulias, D.; Landenberger, R. Wildfire risk assessment based on geospatial open data: Application on Chios, Greece. *ISPRS Int. J. Geo-Inf.* **2020**, *9*, 516. [CrossRef]
41. Chuvieco, E.; Congalton, R.G. Mapping and inventory of forest fires from digital processing of tm data. *Geocarto Int.* **2008**, *3*, 41–53. [CrossRef]
42. Chu, T.; Guo, X. Remote Sensing Techniques in Monitoring Post-Fire Effects and Patterns of Forest Recovery in Boreal Forest Regions: A Review. *Remote Sens.* **2014**, *6*, 470–520. [CrossRef]

43. Pereira, A.A.; Pereira, J.M.C.; Libonati, R.; Oom, D.; Setzer, A.W.; Morelli, F.; Machado-Silva, F.; De Carvalho, L.M.T. Burned Area Mapping in the Brazilian Savanna Using a One-Class Support Vector Machine Trained by Active Fires. *Remote Sens.* **2017**, *9*, 1161. [CrossRef]
44. Loboda, T.V.; Hoy, E.E.; Giglio, L.; Kasischke, E.S. Mapping burned area in Alaska using MODIS data: A data limitations-driven modification to the regional burned area algorithm. *Int. J. Wildland Fire* **2011**, *20*, 487–496. [CrossRef]
45. Mouillot, F.; Schultz, M.G.; Yue, C.; Cadule, P.; Tansey, K.; Ciais, P.; Chuvieco, E. Ten years of global burned area products from spaceborne remote sensing—A review: Analysis of user needs and recommendations for future developments. *Int. J. Appl. Earth Obs. Geoinf.* **2014**, *26*, 64–79. [CrossRef]
46. Alonso-Canas, I.; Chuvieco, E. Global Burned Area Mapping from ENVISAT-MERIS data. *Remote Sens. Environ.* **2015**, *163*, 140–152. [CrossRef]
47. Llorens, R.; Sobrino, J.A.; Cristina Fernández, C.; Fernández-Alonso, J.M.; José, J.V. A methodology to estimate forest fires burned areas and burn severity degrees using Sentinel-2 data. Application to the October 2017 fires in the Iberian Peninsula. *Int. J. Appl. Earth Obs. Geoinf.* **2021**, *95*, 102243. [CrossRef]
48. Arellano, S.; Vega, J.A.; Rodríguez y Silva, F.; Fernández, C.; Vega-Nieva, D.; Álvarez-González, J.G.; Ruiz-González, A.D. Validación de los índices de teledetección dNBR y RdNBR para determinar la severidad del fuego en el incendio forestal de Oia-O Rosal (Pontevedra) en 2013. *Rev. Teledetección* **2017**, *49*, 49–61. [CrossRef]
49. Teodoro, A.; Amaral, A. A Statistical and Spatial Analysis of Portuguese Forest Fires in Summer 2016 Considering Landsat 8 and Sentinel 2A Data. *Environments* **2019**, *6*, 36. [CrossRef]
50. Huesca, M.; de Miguel, S.M.; Alonso, F.G.; García, C.V. An intercomparison of satellite burned area maps derived from MODIS, MERIS, SPOT-VEGETATION and ARST images. An application to the August 2006 Galicia (Spain) forest fires. *For. Syst.* **2013**, *22*, 222–231. [CrossRef]
51. Brown, A.R.; Petropoulos, G.P.; Ferentinos, K.P. Appraisal of the Sentinel-1 & 2 use in a large-scale wildfire assessment: A case study from Portugal's fires of 2017. *Appl. Geogr.* **2018**, *100*, 78–89. [CrossRef]
52. Navarro, G.; Caballero, I.; Silva, G.; Parra, P.C.; Vázquez, A.; Caldeira, R. Evaluation of forest fire on Madeira Island using Sentinel-2A MSI imagery. *J. Appl. Earth Obs. Geoinf.* **2017**, *58*, 97–106. [CrossRef]
53. Oliveira, S.L.; Pereira, J.M.; Carreiras, J.M. Fire frequency analysis in Portugal (1975–2005), using Landsat-based burnt area maps. *Int. J. Wildland Fire* **2012**, *21*, 48–60. [CrossRef]
54. Vallejo-Villalta, I.; Rodríguez-Navas, E.; Márquez-Pérez, J. Mapping forest fire risk at a local scale—A case study in Andalusia (Spain). *Environments* **2019**, *6*, 30. [CrossRef]
55. Fernández, A.; Illera, P.; Casanova, J.L. Automatic mapping of surfaces affected by forest fires in Spain using AVHRR NDVI composite image data. *Remote Sens. Environ.* **1997**, *60*, 153–162. [CrossRef]
56. Roteta, E.; Bastarrika, A.; Padilla, M.; Storm, T.; Chuvieco, E. Development of a Sentinel-2 burned area algorithm: Generation of a small fire database for sub-Saharan Africa. *Remote Sens. Environ.* **2019**, *222*, 1–17. [CrossRef]
57. Chuvieco, E.; Mouillot, F.; van der Werf, G.R.; Miguel, J.; Tanasse, M.; Koutsias, N.; García, M.; Yebra, M.; Padilla, M.; Gitas, I.; et al. Historical background and current developments for mapping burned area from satellite Earth observation. *Remote Sens. Environ.* **2019**, *225*, 45–64. [CrossRef]
58. Teodoro, A.C.; Amaral, A. Evaluation of forest fires in Portugal mainland during 2016 summer considering different satellite datasets. In Proceedings of the SPIE 10421, Remote Sensing for Agriculture, Ecosystems, and Hydrology XIX, Warsaw, Poland, 2 November 2017; 104211R. [CrossRef]
59. Panisset, J.; Dacamara, C.C.; Libonati, R.; Peres, L.F.; Calado, T.J.; Barrios, A. Assigning dates and identifying areas affected by fires in Portugal based on MODIS data. *An. Acad. Bras. Ciênc.* **2017**, *89*. [CrossRef]
60. Humber, M.L.; Boschetti, L.G.; Giglio, L.; Justice, C.O. Spatial and temporal intercomparison of four global burned area products. *Int. J. Digit. Earth* **2019**, *12*, 460–484. [CrossRef]
61. Libonati, R.; Dacamara, C.C.; Setzer, A.W.; Morelli, F.; Melchiori, A.E. An Algorithm for Burned Area Detection in the Brazilian Cerrado Using 4 μm MODIS Imagery. *Remote Sens.* **2015**, *7*, 15782–15803. [CrossRef]
62. Kleinn, C.; Corrales, L.; Morales, D. Forest area in Costa Rica: A comparative study of tropical forest cover estimates over time. *Environ. Monit. Assess.* **2002**, *73*, 17–40. [CrossRef]
63. Noi Thanh, P.; Kappas, M. Comparison of Random Forest, k-Nearest Neighbor, and Support Vector Machine Classifiers for Land Cover Classification Using Sentinel-2 Imagery. *Sensors* **2018**, *18*, 18. [CrossRef]
64. Tomppo, E.; Haakana, M.; Katila, M.; Perasaari, J. *Multi-Source National Forest Inventory. Methods and Applications*; Springer: Berlin, Germany, 2008; Volume 18. [CrossRef]
65. Fix, D.; Hodges, J.L. *Discriminatory Analysis, Nonparametric Discrimination: Consistency Properties*; Technical Report 4; USAF School of Aviation Medicine: Randolph Field, TX, USA, 1951.
66. Moeur, M.; Stage, A.R. Most similar neighbor: An improved sampling inference procedure for natural resource planning. *For. Sci.* **1995**, *41*, 337–359.
67. Franco-Lopez, H.; Ek, A.R.; Bauer, M.E. Estimation and mapping of forest stand density, volumen, and cover type using the k-nearest neighbors method. *Remote Sens. Environ.* **2001**, *77*, 251–274. [CrossRef]
68. McRoberts, R.E.; Tomppo, E.O. Remote sensing support for national forest inventories. *Remote Sens. Environ.* **2007**, *110*, 412–419. [CrossRef]

69. Ohmann, J.L.; Gregory, M.J. Predictive mapping of forest composition and structure with direct gradient analysis and nearest-neighbor imputation in coastal Oregon, U.S.A. *Can. J. For. Res.* **2002**, *32*, 725–741. [CrossRef]
70. Wulder, M.A.; Franklin, S.E.; White, J.C.; Linke, J.; Magnussen, S. An accuracy assessment framework for large-area land cover classification products derived from medium-resolution satellite data. *Int. J. Remote Sens.* **2006**, *27*, 663–683. [CrossRef]
71. Halperin, J.; LeMay, V.; Coops, N.; Verchot, L.; Marshall, P.; Lochhead, K. Canopy cover estimation in miombo woodlands of Zambia: Comparison of Landsat 8 OLI versus RapidEye imagery using parametric, nonparametric, and semiparametric methods. *Remote Sens. Environ.* **2016**, *179*, 170–182. [CrossRef]
72. Ramo, R.; García, M.; Rodríguez, D.; Chuvieco, E. A data mining approach for global burned area mapping. *Int. J. Appl. Earth Obs. Geoinf.* **2018**, *73*, 39–51. [CrossRef]
73. Kanevski, M.A.; Pozdnoukhov, V. *Machine Learning for Spatial Environmental Data: Theory Applications and Software*; CRC Press: Boca Raton, FL, USA, 2009; 368p.
74. Roy, D.P.; Huang, H.; Boschetti, L.; Giglio, L.; Yan, L.; Zhang, H.H.; Li, Z. Landsat-8 and Sentinel-2 burned area mapping—A combined sensor multi-temporal change detection approach. *Remote Sens. Environ.* **2019**, *231*, 111254. [CrossRef]
75. Liu, J.; Heiskanen, J.; Maeda, E.E.; Pellikka, P.K.E. Burned area detection based on Landsat time series in savannas of southern Burkina Faso. *Int. J. Appl. Earth Obs. Geoinf.* **2018**, *64*, 210–220. [CrossRef]
76. Giglio, L.; Boschetti, L.; Roy, D.P.; Humber, M.L.; Justice, C.O. The Collection 6 MODIS burned area mapping algorithm and product. *Remote Sens. Environ.* **2018**, *217*, 72–85. [CrossRef]
77. Stroppiana, D.; Bordogna, G.; Carrara, P.; Boschetti, M.; Boschetti, L.; Brivio, P. A method for extracting burned areas from Landsat TM/ETM+ images by soft aggregation of multiple Spectral Indices and a region growing algorithm. *ISPRS J. Photogramm. Remote Sens.* **2012**, *69*, 88–102. [CrossRef]
78. Fornacca, D.; Ren, G.; Xiao, W. Performance of Three MODIS Fire Products (MCD45A1, MCD64A1, MCD14ML), and ESA Fire_CCI in a Mountainous Area of Northwest Yunnan, China, Characterized by Frequent Small Fires. *Remote Sens.* **2017**, *9*, 1131. [CrossRef]
79. Smiraglia, D.; Filipponi, F.; Mandrone, S.; Tornato, A.; Taramelli, A. Agreement Index for Burned Area Mapping: Integration of Multiple Spectral Indices Using Sentinel-2 Satellite Images. *Remote Sens.* **2020**, *12*, 1862. [CrossRef]
80. Singh, M.; Evans, D.; Chevance, J.; Tan, B.S.; Wiggins, N.; Kong, L.; Sakhoeun, S. Evaluating remote sensing datasets and machine learning algorithms for mapping plantations and successional forests in Phnom Kulen National Park of Cambodia. *PeerJ* **2019**, *7*, e7841. [CrossRef]
81. Rishickesh, R.; Shahina, A.; Khan, A.N. Predicting forest fires using supervised and ensemble machine learning algorithms. *Int. J. Recent Technol. Eng.* **2019**, *8*, 3697–3705. [CrossRef]
82. Yao, J.; Raffuse, S.M.; Brauer, M.; Williamson, J.; Bowman, D.M.J.S.; Johnston, H.; Henderson, S.B. Predicting the minimum height of forest fire smoke within the atmosphere using machine learning and data from the CALIPSO satellite. *Remote Sens. Environ.* **2018**, *206*, 98–106. [CrossRef]
83. Ramo, R.; Chuvieco, E. Developing a Random Forest Algorithm for MODIS Global Burned Area Classification. *Remote Sens.* **2017**, *9*, 1193. [CrossRef]
84. Collins, L.; Griffioen, P.; Newell, G.; Mellor, A. The utility of Random Forests for wildfire severity mapping. *Remote Sens. Environ.* **2018**, *216*, 374–384. [CrossRef]
85. Leshem, G. Improvement of Adaboost Algorithm by Using Random Forests as Weak Learner and Using This Algorithm as Statistics Machine Learning for Traffic Flow Prediction. Ph.D. Thesis, Hebrew University, Jerusalem, Israel, 2005.
86. Dos Santos, E.E.; Cruz Sena, N.; Balestrin, D.; Fernandes Filho, E.I.; Marciano da Costa, L.; Bozzi Zeferino, L. Prediction of burned areas using the random forest classifier in the Minas Gerais state. *Floreta e Ambiente* **2020**, *27*, e20180115. [CrossRef]
87. Gibson, R.; Danaher, T.; Hehir, W.; Collins, S. A remote sensing approach to mapping fire severity in south-eastern Australia using sentinel 2 and random forest. *Remote Sens. Environ.* **2020**, *240*, 111702. [CrossRef]
88. Collins, L.; McCarthy, G.; Mellor, A.; Newell, G.; Smith, L. Training data requirements for fire severity mapping using Landsat imagery and random forest. *Remote Sens. Environ.* **2020**, 111839. [CrossRef]
89. Wood, D.A. Prediction and data mining of burned areas of forest fires: Optimized data matching and mining algorithm provides valuable insight. *Artif. Intell. Agric.* **2021**, *5*, 24–42. [CrossRef]
90. Castelli, M.; Vanneschi, L.; Popovic, A. Predicting burned areas of forest fires: An artificial intelligence approach. *Fire Ecol.* **2015**, *11*, 106–118. [CrossRef]
91. Naganathan, H.; Seshaseyee, S.P.; Kim, J.; Chong, W.K.; Chou, J.S. Wildfire predictions: Determining reliable models using fused dataset. *Glob. J. Comput. Sci. Technol.* **2016**, *16*, 29–40.
92. Cortez, P.; Morais, A.A. A data mining approach to predict forest fires using meteorological data. In *New Trends in Artificial Intelligence, Proceedings of the 13th Portuguese Conference on Artificial Intelligence*; Springer: Berlin, Germany, 2007; pp. 512–523.
93. Miranda, P.; Coelho, F.; Tomé, A.R.; Valente, M.A.; Carvalho, A.; Pires, C.; Pires, H.O.; Pires, V.C.; Ramalho, C. 20th century Portuguese climate and climate scenarios. In *Climate Change in Portugal: Scenarios, Impacts and Adaptation Measures (SIAM Project)*; Santos, F.D.K., Ed.; Gradiva: Lisbon, Portugal, 2002; pp. 23–83.
94. Lourenço, L. Tipos de tempo correspondentes aos grandes incêndios florestais ocorridos em 1986 no Centro de Portugal. *Finisterra* **1988**, *23*, 251–270. [CrossRef]

95. Vilar, L.; Camia, A.; San-Miguel-Ayanz, J. A comparison of remote sensing products and forest fire statistics for improving fire information in Mediterranean Europe. *Eur. J. Remote Sens.* **2015**, *48*, 345–364. [CrossRef]
96. Nunes, M.C.S.; Vasconcelos, M.J.; Pereira, J.M.C.; Dasgupta, N.; Alldredge, R.J.; Rego, F.C. Land cover type and fire in Portugal: Do fires burn land cover selectively? *Landsc. Ecol.* **2005**, *20*, 661–673. [CrossRef]
97. Calvo, L.; Santalla, S.; Valbuena, L.; Marcos, E.; Tarrega, R.; Luis-Calabuig, E. Post-fire natural regeneration of a Pinus pinaster forest in NW Spain. *Plant Ecol.* **2008**, *197*, 81–90. [CrossRef]
98. USGS EROS Center. 2019. Available online: https://earthexplorer.usgs.gov/ (accessed on 24 January 2021).
99. Randerson, J.T.; Chen, Y.; van der Werf, G.R.; Rogers, B.M.; Morton, D.C. Global burned area and biomass burning emissions from small fires. *J. Geophys. Res.* **2012**, *117*, G04012. [CrossRef]
100. Pieschke, R.L. *US Geological Survey Distribution of European Space Agency's Sentinel-2 Data*; No. 2017-3026; US Geological Survey: Reston, VA, USA, 2017.
101. Vermote, E. MOD09A1 MODIS/Terra Surface Reflectance 8-Day L3 Global 500 m SIN Grid V006 [Data Set]. NASA EOSDIS Land Processes DAAC 2015. Available online: https://lpdaac.usgs.gov/products/mod09a1v006/ (accessed on 22 November 2020).
102. Thomas, I.L.; Ching, N.P.; Benning, V.M.; D'Aguanno, J.A. A review of multi-channel indices of class separability. *Int. J. Remote Sens.* **1987**, *8*, 331–350. [CrossRef]
103. Richards, J.A.; Jia, X. Feature Reduction. In *Remote Sensing Digital Image Analysis: An Introduction*, 4th ed.; Springer: Berlin/Heidelberg, Germany, 2006; pp. 267–294.
104. Van Niel, T.G.; McVicar, T.R.; Datt, B. On the relationship between training sample size and data dimensionality: Monte Carlo analysis of broadband multi-temporal classification. *Remote Sens. Environ.* **2005**, *98*, 468–480. [CrossRef]
105. Aha, D.W. Artificial Intelligence Review. *Lazy Learn.* **1997**, *11*, 1–6. [CrossRef]
106. McRoberts, R.E. A two-step nearest neighbors algorithm using satellite imagery for predicting forest structure within species composition classes. *Remote Sens. Environ.* **2009**, *113*, 532–545. [CrossRef]
107. Stojanova, D.; Kobler, A.; Ogrinc, P.; Zenko, B.; Dzeroski, S. Estimating the risk of fire outbreaks in the natural environment. *Data Min. Knowl. Discov.* **2012**, *24*, 411–442. [CrossRef]
108. Cariou, C.; Le Moan, S.; Chehdi, K. Improving K-Nearest Neighbor Approaches for Density-Based Pixel Clustering in Hyperspectral Remote Sensing Images. *Remote Sens.* **2020**, *12*, 3745. [CrossRef]
109. Rodríguez-Galiano, V.F.; Ghimire, B.; Rogan, J.; Chica-Olmo, M.; Rigol-Sánchez, J.P. An assessment of the effectiveness of a random forest classifier for land-cover classification. *ISPRS J. Photogramm. Remote Sens.* **2012**, *67*, 93–104. [CrossRef]
110. Hastie, R.; Tibshirani, R.; Friedman, J. *The Elements of Statistical Learning. Data Mining, Inference, and Prediction*, 2nd ed.; Springer: Berlin, Germany, 2009.
111. Quinlan, J.R. *C4.5: Programs for Machine Learning*; Morgan Kaufmann: San Mateo, CA, USA, 1993.
112. Breiman, L.; Friedman, J.H.; Olshen, R.A.; Stone, C.J. *Classification and Regression Trees. Statistics/Probability Series*; Wadsworth & Brooks/Cole Advanced Books & Software: Monterey, CA, USA, 1984.
113. Cutler, D.; Edwards, T.; Beard, K.; Cutler, A.; Hess, K.; Gibson, J.; Lawler, J. Random Forests for Classification in Ecology. *Ecology* **2007**, *88*, 2783–2792. [CrossRef]
114. ICNF 2019. *Defesa da Floresta Contra Incêndios*; Instituto da Conservação da Natureza e das Florestas: Lisboa, Portugal, 2019.
115. Story, M.; Congalton, R.G. Accuracy assessment: A user's perspective. *Photogramm. Eng. Remote Sens.* **1986**, *52*, 397–399.
116. Fawcett, T. An introduction to ROC analysis. *Pattern Recognit. Lett.* **2006**, *27*, 861–874. [CrossRef]
117. Cohen, J.A. Coefficient of agreement for nominal scales. *Educ. Psychol. Meas.* **1960**, *20*, 37–46. [CrossRef]
118. Pereira, J.M.C.; Sá, A.C.L.; Sousa, A.M.O.; Silva, J.M.N.; Santos, T.N.; Carreiras, J.M.B. Spectral characterisation and discrimination of burned áreas. In *Remote Sensing of Large Wildfires in the European Mediterranean Basin*; Chuvieco, E., Ed.; Springer: Berlin, Germany, 1999; pp. 123–138.
119. Nogueira, J.M.P.; Ruffault, J.; Chuvieco, E.; Mouillot, F. Can We Go Beyond Burned Area in the Assessment of Global Remote Sensing Products with Fire Patch Metrics? *Remote Sens.* **2017**, *9*, 7. [CrossRef]
120. Chuvieco, E.; Lizundia-Loiola, J.; Pettinari, M.L.; Ramo, P.M.; Tansey, K.; Mouillot, F.; Laurent, P.; Storm, T.; Heil, A.; Plummer, S. Generation and analysis of a new global burned area product based on MODIS 250m reflectance bands and thermal anomalies. *Earth Syst. Sci. Data* **2018**, *10*, 2015–2031. [CrossRef]
121. Padilla, M.; Stehman, S.V.; Litago, J.; Chuvieco, E. Assessing the temporal stability of the accuracy of a time series of burned area products. *Remote Sens.* **2014**, *6*, 2050–2068. [CrossRef]
122. Roy, D.P.; Boschetti, L. Southern Africa validation of the MODIS, L3JRC and GlobCarbon burned-area products. *IEEE Trans. Geosci. Remote Sens.* **2009**, *47*, 1032–1044. [CrossRef]
123. Ngadze., F.; Mpakairi, K.S.; Kavhu, B.; Ndaimani, H.; Maremba, M.S. Exploring the utility of Sentinel-2 MSI and Landsat 8 OLI in burned area mapping for a heterogenous savannah landscape. *PLoS ONE* **2020**, *15*, e0232962. [CrossRef]
124. Soler, M.; Úbeda, X. Evaluation of fire severity via analysis of photosynthetic pigments: Oak, eucalyptus and cork oak leaves in a mediterranean forest. *J. Environ. Manag.* **2018**, *206*, 65–68. [CrossRef]
125. Sever, L.; Leach, J.; Bren, L. Remote sensing of post-fire vegetation recovery; a study using Landsat 5 tm imagery and NDVI in north-east Victoria. *J. Spat. Sci.* **2012**, *57*, 175–191. [CrossRef]
126. Falkowski, M.J.; Gessler, P.E.; Morgan, P.; Hudak, A.T.; Smith, A.M.S. Characterizing and mapping forest fire fuels using aster imagery and gradient modeling. *For. Ecol. Manag.* **2005**, *217*, 129–146. [CrossRef]

127. Fernández-Manso, A.; Fernández-Manso, O.; Quintano, C. SENTINEL-2A red-edge spectral indices suitability for discriminating burn severity. *Int. J. Appl. Earth Obs. Geoinf.* **2016**, *50*, 170–175. [CrossRef]
128. Silva Junior, J.A.; Pacheco, A.P. Análise do Modelo Linear de Mistura Espectral na Avaliação de Incêndios Florestais no Parque Nacional do Araguaia, Tocantins, Brasil: Imagens EO-1/Hyperion e Landsat-7/ETM+. *Anuário do Instituto de Geociências* **2020**, *43*, 4. [CrossRef]
129. ICNF—Sistema de Gestão de Informação de Incêndios Florestais. 7.º Relatório Provisório de Incêndios Rurais Departamento de Gestão de Fogos Rurais/Divisão de Apoio à Gestão de Fogos Rurais. 2019. Available online: http://www2.icnf.pt/portal/florestas/dfci/Resource/doc/rel/2019/2019-10-10-RPIR-07-01jan-30set.pdf (accessed on 24 January 2021).
130. Oshiro, T.; Pérez, P.; Baranauskas, J. How Many Trees in a Random Forest? *Lect. Notes Comput. Sci.* **2012**, 7376. [CrossRef]
131. Çömert, R.; Matci, D.; Avdan, U. Object based burned area mapping with random forest algorithm. *Int. J. Eng. Geosci.* **2019**, *4*, 78–87. [CrossRef]
132. Meng, Q.; Cieszewski, C.; Madden, M.; Borders, B.K. Nearest Neighbor Method for Forest Inventory Using Remote Sensing Data. *GISci. Remote Sens.* **2007**, *44*, 149–165. [CrossRef]
133. Zhao, M.; Chen, J. Improvement and comparison of weighted k Nearest Neighbors classifiers for model selection. *J. Softw. Eng.* **2016**, *10*, 109–118. [CrossRef]
134. Blanzieri, E.; Melgani, F. Nearest Neighbor Classification of Remote Sensing Images with the Maximal Margin Principle. *IEEE Trans. Geosci. Remote Sens.* **2008**, *46*, 1804–1811. [CrossRef]
135. Veraverbeke, S.; Lhermitte, S.; Verstraeten, W.; Goossens, R. A time-integrated MODIS burn severity assessment using the multi-temporal differenced normalized burn ratio (dNBRMT). *Int. J. Appl. Earth Obs. Geoinf.* **2011**, *13*, 52–58. [CrossRef]
136. Pleniou, M.; Koutsias, N. Sensitivity of spectral reflectance values to different burn and vegetation ratios: A multi-scale approach applied in a fire affected area. *ISPRS J. Photogramm. Remote Sens.* **2013**, *79*, 199–210. [CrossRef]
137. Santana, N.A.; Morales, C.A.S.; da Silva, D.A.A.; Antoniolli, Z.I.; Jacques, R.J.S. Soil Biological, Chemical, and Physical Properties After a Wildfire Event in a Eucalyptus Forest in the Pampa Biome. *Rev. Braileira Ciênc. Solo* **2018**, *42*. [CrossRef]
138. Lambin, E.; Goyvaerts, K.; Petit, C. Remotely-sensed indicators of burning efficiency of savannah and forest fires. *Int. J. Remote Sens.* **2003**, *24*, 3105–3118. [CrossRef]
139. Bastarrika, A.; Chuvieco, E.; Martín, M.P. Mapping burned areas from Landsat TM/ETM+ data with a two-phase algorithm: Balancing omission and commission errors. *Remote Sens. Environ.* **2011**, *115*, 1003–1012. [CrossRef]
140. Sertel, E.; Ugur, A. Comparison of pixel and object-based classification for burned area mapping using SPOT-6 images. *Geomat. Nat. Hazards Risk* **2016**, *7*, 1198–1206. [CrossRef]
141. Lanorte, A.; Danese, M.; Lasaponara, R.; Murgante, B. Multiscale mapping of burn area and severity using multisensor satellite data and spatial autocorrelation analysis. *Int. J. Appl. Earth Obs. Geoinf.* **2013**, *20*, 42–51. [CrossRef]
142. Palomino-Ángel, S.; Anaya-Acevedo, J.A. Evaluation of the causes of error in the MCD45 burned-area product for the savannas of northern South America. *Dyna* **2012**, *79*, 35–44.
143. Shimabukuro, Y.E.; Dutra, A.C.; Arai, E.; Duarte, V.; Cassol, H.L.G.; Pereira, G.; Cardozo, F.D.S. Mapping burned areas of Mato Grosso state brazilian amazon using multisensor datasets. *Remote Sens.* **2020**, *12*, 3827. [CrossRef]
144. Tanase, M.A.; Belenguer-Plomer, M.A.; Roteta, E.; Bastarrika, A.; Wheeler, J.; Fernández-Carrillo, Á.; Tansey, K.; Wiedemann, W.; Navratil, P.; Lohberger, S.; et al. Burned Area Detection and Mapping: Intercomparison of Sentinel-1 and Sentinel-2 Based Algorithms over Tropical Africa. *Remote Sens.* **2020**, *12*, 334. [CrossRef]
145. Schroeder, W.; Ruminski, M.; Csiszar, I.; Giglio, L.; Prins, E.; Schmidt, C.; Morisette, J. Validation analyses of an operational fire monitoring product: The hazard mapping system. *Int. J. Remote Sens.* **2008**, 6059–6066. [CrossRef]
146. Calle, A.; Casanova, J.L.; Romo, A. Fire detection and monitoring using MSG Spinning Enhanced Visible and Infrared Imager (SEVIRI) data. *J. Geophys. Res.* **2006**, *111*, G04S06. [CrossRef]
147. Koltunov, A.; Ustin, S.L. Early fire detection using non-linear multitemporal prediction of thermal imagery. *Remote Sens. Environ.* **2007**, *110*, 18–28. [CrossRef]
148. Amos, C.; Petropoulos, G.P.; Ferentinos, K.P. Determining the use of Sentinel-2A MSI for wildfire burning & severity detection. *Int. J. Remote Sens.* **2019**, *40*, 905–930. [CrossRef]
149. Axel, A.C. Burned Area Mapping of an Escaped Fire into Tropical Dry Forest in Western Madagascar Using Multi-Season Landsat OLI Data. *Remote Sens.* **2018**, *10*, 371. [CrossRef]
150. Zhang, P.; Nascetti, A.; Ban, Y.; Gong, M. An implicit radar convolutional burn index for burned area mapping with Sentinel-1 C-band SAR data. *ISPRS J. Photogramm. Remote Sens.* **2019**, *158*, 50–62. [CrossRef]
151. Chou, Y.H.; Minnich, R.A.; Chase, R.A. Mapping probability of fire occurrence in San Jacinto Mountains, California, USA. *Environ. Manag.* **1993**, *17*, 129–140. [CrossRef]
152. Siljander, M. Predictive fire occurrence modelling to improve burned area estimation at a regional scale: A case study in East Caprivi, Namibia. *Int. J. Appl. Earth Obs. Geoinf.* **2009**, *11*, 380–393. [CrossRef]
153. Dlamini, W.M. Application of Bayesian networks for fire risk mapping using GIS and remote sensing data. *GeoJournal* **2011**, *76*, 283–296. [CrossRef]
154. Lizundia-Loiola, J.; Otón, G.; Ramo, R.; Chuvieco, E. A spatio-temporal active-fire clustering approach for global burned area mapping at 250 m from MODIS data. *Remote Sens. Environ.* **2020**, *236*, 111493. [CrossRef]

remote sensing

MDPI

Article

Land Subsidence Susceptibility Mapping Using Persistent Scatterer SAR Interferometry Technique and Optimized Hybrid Machine Learning Algorithms

Babak Ranjgar [1], Seyed Vahid Razavi-Termeh [1], Fatemeh Foroughnia [2,*], Abolghasem Sadeghi-Niaraki [1,3] and Daniele Perissin [4]

1 Geoinformation Tech. Center of Excellence, Faculty of Geodesy and Geomatics Engineering, K.N. Toosi University of Technology, Tehran 19697, Iran; babakranjgar@email.kntu.ac.ir (B.R.); vrazavi@mail.kntu.ac.ir (S.V.R.-T.); a.sadeghi@sejong.ac.kr (A.S.-N.)
2 Department of Geoscience and Remote Sensing, Civil Engineering and Geosciences Faculty, Delft University of Technology, 2628 CN Delft, Stevinweg, The Netherlands
3 Department of Computer Science and Engineering, Sejong University, Seoul 143-747, Korea
4 RASER Limited, Radar and Software Engineering Research Company, Hong Kong; daniele.perissin@sarproz.com
* Correspondence: f.foroughnia@tudelft.nl; Tel.: +31-1527-8354-6

Abstract: In this paper, land subsidence susceptibility was assessed for Shahryar County in Iran using the adaptive neuro-fuzzy inference system (ANFIS) machine learning algorithm. Another aim of the present paper was to assess if ensembles of ANFIS with two meta-heuristic algorithms (imperialist competitive algorithm (ICA) and gray wolf optimization (GWO)) would yield a better prediction performance. A remote sensing synthetic aperture radar (SAR) dataset from 2019 to 2020 and the persistent-scatterer SAR interferometry (PS-InSAR) technique were used to obtain a land subsidence inventory of the study area and use it for training and testing models. Resulting PS points were divided into two parts of 70% and 30% for training and testing the models, respectively. For susceptibility analysis, eleven conditioning factors were taken into account: the altitude, slope, aspect, plan curvature, profile curvature, topographic wetness index (TWI), distance to stream, distance to road, stream density, groundwater drawdown, and land use/land cover (LULC). A frequency ratio (FR) was applied to assess the correlation of factors to subsidence occurrence. The prediction power of the models and their generated land subsidence susceptibility maps (LSSMs) were validated using the root mean square error (RMSE) value and area under curve of receiver operating characteristic (AUC-ROC) analysis. The ROC results showed that ANFIS-ICA had the best accuracy (0.932) among the models (ANFIS-GWO (0.926), ANFIS (0.908)). The results of this work showed that optimizing ANFIS with meta-heuristics considerably improves LSSM accuracy although ANFIS alone had an acceptable result.

Keywords: land subsidence; Geographic Information System (GIS); InSAR; machine learning algorithm; meta-heuristics; Iran

Citation: Ranjgar, B.; Razavi-Termeh, S.V.; Foroughnia, F.; Sadeghi-Niaraki, A.; Perissin, D. Land Subsidence Susceptibility Mapping Using Persistent Scatterer SAR Interferometry Technique and Optimized Hybrid Machine Learning Algorithms. *Remote Sens.* **2021**, *13*, 1326. https://doi.org/10.3390/rs13071326

Academic Editors: Cristiano Tolomei, Paolo Mazzanti and Saverio Romeo

Received: 26 February 2021
Accepted: 29 March 2021
Published: 31 March 2021

1. Introduction

Land subsidence (LS) is one of the most challenging catastrophic geohazards due to its potential consequences, including damage to infrastructures, power lines, and buildings, causing sinkholes, floods in coastal areas, and soil degradation [1–3]. Land subsidence is a gradual and slow deformation or sudden collapse of the Earth's surface, which is caused by numerous natural and human-induced factors [4–7]. The ground subsiding movement can be the result of natural causes such as floods, ground lithology, dissolution of carbonated rocks (e.g., limestone), sediment compaction, and tectonic motions of faults [8–11]. Further, anthropogenic activities that alleviate these geological factors, including underground excavations (e.g., mining and tunneling), underground resource withdrawal (gas or oil), overloading of the land surface through road construction and extending the built environment [12–14], and most importantly, over-exploitation of underground aquifers [15–17].

In the last few decades, the land subsidence phenomenon has widely increased in Iran [18–20], and therefore, growing research interest has focused on studying this geological problem [2,13,17,21–23]. One of the most important causes of the LS in Iran with an arid and semi-arid environment is the excessive groundwater extraction for agricultural usage [13,24]. Therefore, modeling factors affecting land subsidence and land subsidence susceptibility mapping (LSSM) is vital for the environment, safety, economy, and human well-being.

Remote sensing (RS) and geographic information system (GIS) data and tools have been helpful in land subsidence susceptibility studies in terms of acquiring fine resolution data and analyzing various factors affecting this phenomenon [10,11,25]. Many statistical and probabilistic approaches have been applied in the literature to provide susceptibility maps and monitor subsidence. These methods include the frequency ratio (FR) [26], weight of evidence (WOE) [27], logistic regression (LR) [28], evidential belief function (EBF) models [11], artificial neural networks (ANNs) [29,30], analytical hierarchy processes (AHP) [31,32], multi-criteria decision making (MCDM) models [22], as well as fuzzy logic (FL) [27,30] and adaptive neuro-fuzzy inference systems (ANFIS) [33].

However, these methods are mainly based on human assumptions and need expert knowledge. Recently, GIS-based machine learning algorithms (MLAs) have become a favorite in modeling and analyzing environmental hazards, especially LS. They can cope with data peculiarities, reveal complex relationships between data, and produce high accuracy and close-to-real world results [10,23]. Lee and Park [12] conducted a comparative investigation between the decision tree (DT) algorithm and FR model in estimation of LS and its causing factors. Abdollahi et al. [34] applied a support vector machine (SVM) to predict LSS using water table drawdown and other influential factors. Taravatrooy et al. [35] used a hybrid clustering method based on k-means, genetic optimization, and several soft computing algorithms to examine subsided zones. Tien Bui et al. [10] compared four MLAs (Bayesian logistic regression (BLR), SVM, logistic model tree (LMT), and alternate decision tree (ADT)) in assessing LSS near abandoned mining areas in South Korea. In a study in Kerman, Iran, the random forest (RF) algorithm showed superior capability in LSS mapping [36]. Ebrahimy et al. [23] performed a comparative study using three tree-based MLAs, a boosted regression tree (BRT), RF, and classification and regression tree (CART), for studying land susceptibility in Tasuj plane, Iran. Evaluation results revealed that BRT had the best performance. In another study by Rahmati et al. [37], four tree-based MLAs, a rule-based decision tree (RDT), RF, CART, and BRT, were compared for generating LS hazard maps in Hamedan Province, Iran. The results indicated that RF had the best accuracy amongst the employed methods.

Despite the better performance and accuracy of MLAs, all the above-mentioned approaches are dependent on the availability and precision of the subsidence inventory data, which is a serious challenge in developing countries [2,11,37]. On the other hand, interferometric synthetic aperture radar (InSAR) has been utilized in land displacement measurement and demonstrated promising results with millimetric precision [38,39]. SAR is satellite data so there is no need for time-consuming field survey data acquisition; therefore, it is superior to other approaches such as leveling data and is denser than ground positioning system (GPS) station data. Furthermore, radar data are functional in all-time all-weather conditions, making this a cost-efficient method to obtain land subsidence measurements. Recently, InSAR methods were employed in LSS studies as reliable input data along with other data to achieve finer accuracies [6,40]. In this paper, we used the PS-InSAR method to obtain land subsidence inventory data and utilize them among other subsidence triggering factors for land subsidence susceptibility mapping in the study area. As a novel methodology in LSSM, we used ANFIS optimized with two meta-heuristic algorithms: (1) imperialist competitive algorithm (ICA) and (2) gray wolf optimization (GWO). ANFIS uses hybrid learning of ANN in adjusting its membership functions (MF) with output data [41,42]. Further, by taking advantage of meta-heuristic algorithms, the weight parameters of MFs were optimized. The results of the method were compared using the statistical approach of the root mean square error (RMSE) value. Furthermore, the accuracy of LSSMs was evaluated by the area under the receiver operating characteristic (ROC) curve.

2. Materials and Methods

The methodology applied in this research (summarized in Figure 1) is to generate an updated land subsidence inventory through the PSInSAR technique using a Sentinel-1 SAR dataset spanning the period from 2019 to 2020. Moreover, the aim of this paper is to develop and use an ensemble of ANFIS and meta-heuristic algorithms in modeling land subsidence susceptibility. The main steps of the study are as follows. First, a spatial database was created using generated land subsidence inventory and the layers of the conditioning factors. In the second step, PS-InSAR-derived subsidence inventory data were divided into training (70%) and testing (30%) data. Next, MF parameters were optimized in ANFIS using ICA and GWO meta-heuristic algorithms, and then LSSMs were produced using ANFIS, ANFIS-ICA, and ANFIS-GWO individually. Finally, the produced land susceptibility maps were compared and evaluated using the area under the ROC curves.

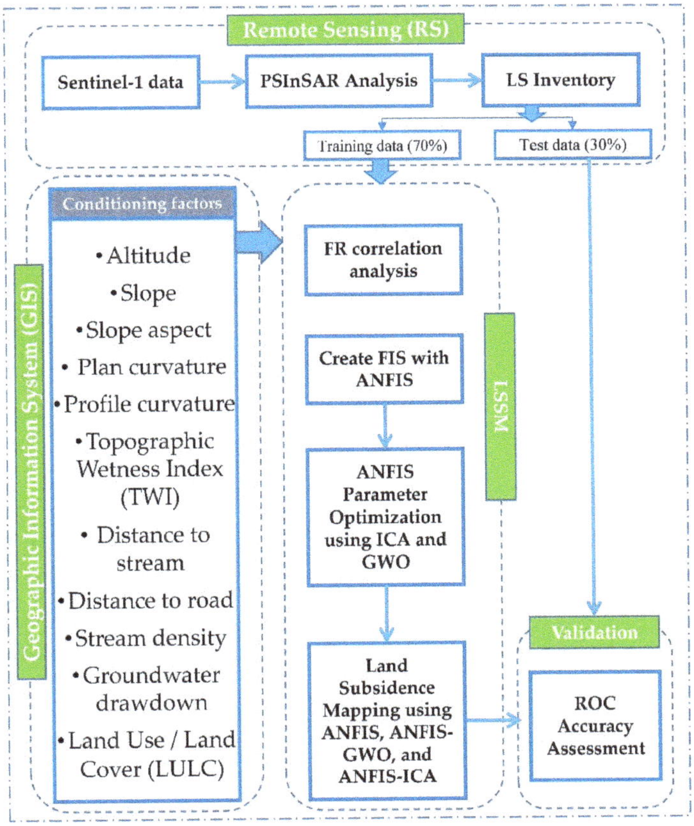

Figure 1. Flowchart of the overall methodology.

2.1. Study Area

The region of interest in this paper is Shahryar, the central city of Shahryar County within 35°35′ to 35°42′ latitudes and 50°59′ to 51°6′ longitudes (Figure 2), with the elevation ranging from 1081 to 1222 m. This county is located west of Tehran, the capital of Iran. In recent years, there has been an increase in population migration to cities near Tehran, including Shahryar, for better jobs and income. According to Iran census data in 2016, the county is the 12th largest in the country with a population of more than 700,000 people. This has become a serious problem in urban environment management and food production. Shahryar County is known for its green and beautiful landscape and the major income of the people in the area originates from gardening and agriculture. Owing to population increase, the demand for food has grown dramatically. Therefore, more illegal wells are dug. As a result, the county has suffered from severe land subsidence (250 to 310 mm/year) due to exhaustion of underground water aquifers and water table dropdown.

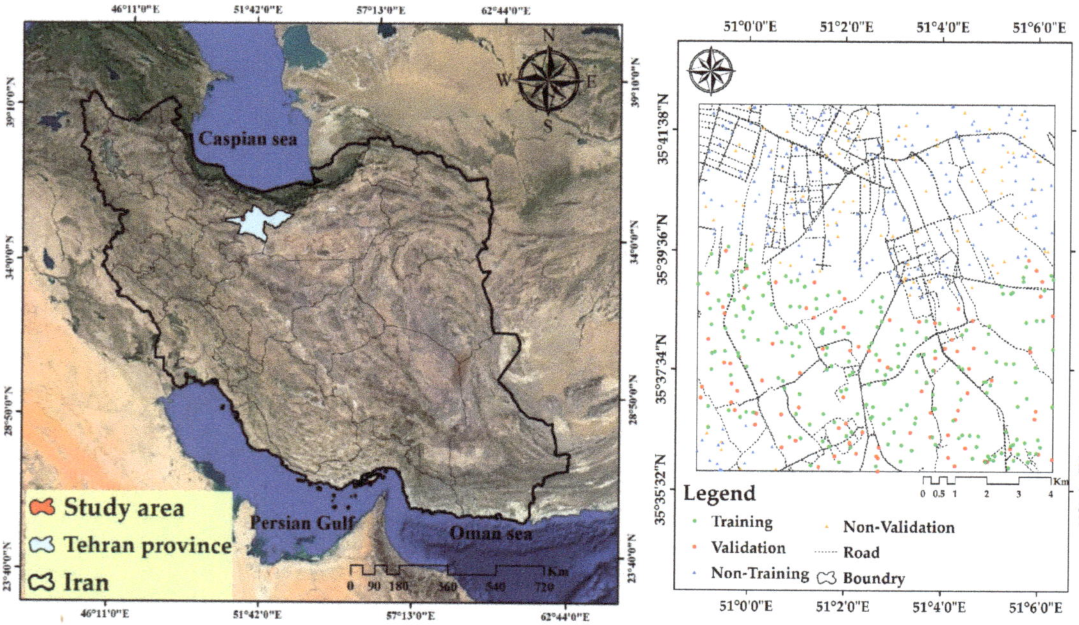

Figure 2. The location of the study area along with the extracted land subsidence inventory.

2.2. Date Used

As outlined earlier, the methodology has two main parts. One is the procedure of generating a land subsidence inventory using the PS-InSAR technique, which needs a satellite SAR dataset. The other is the conditioning factors that are taken into consideration for hazard modeling. In the following section, the process of acquiring and preparing these datasets is discussed.

2.2.1. SAR Data

To generate a land subsidence inventory in order to acquire the training and testing data necessary for LSSM, sentinel-1A single look complex (SLC) SAR data provided by European Space Agency (ESA) were used. In total, 31 SAR images were obtained from January 2019 to January 2020 in ascending track with dual polarization (vertical–vertical and vertical–horizontal) (Table 1).

Table 1. The details of the SAR data used.

Satellite	Acquisition Period	Incidence Angle	Total No.	Polarization
Sentinel-1A	2019/01/02–2020/01/21	~39°	31	VV-VH

2.2.2. Factors Affecting Land Subsidence

To date, there is no single guideline unanimously applied for selecting subsidence-affecting factors. However, based on the literature of LSSM studies carried out in Iran [13,23,33,37], data that were directly and indirectly related to the subsidence-inducing factors are gathered and used for mapping LSS in the study area. These input data include the altitude, slope, aspect, plan curvature, profile curvature, topographic wetness index (TWI), distance to stream, distance to road, stream density, groundwater drawdown, and land use/land cover (LULC) (Table 2). Altitude (Figure 3a) and its derivative factors such as slope (Figure 3b), aspect (Figure 3c), TWI (Figure 3d), plan (Figure 3e), and profile (Figure 3f) are among crucial topo-hydrological criteria of ground subsidence [2]. An advanced space borne thermal emission and reflection radiometer (ASTER) digital elevation model (DEM) was obtained (1 arc second or approximately 30 m resolution) and processed in the GIS environment to produce topography-related factor layers. Land subsidence can cause deformations in the Earth's surface slope and topography [43]. Therefore, altitude and its derivatives were considered because they can directly affect LS. The slope can have a potential impact on runoff infiltration since steep slopes bring about less recharge due to limited infiltration of rainfall [2,37]. As mentioned above, the main cause of land subsidence in Iran is excessive underground water extraction. Undue groundwater extraction results in pore water pressure (PWP) decrease and aquifer compaction increase [44]. Therefore, the well inventory of the study area was acquired from the Iranian department of water resource management (IDWRM) to generate groundwater drawdown (Figure 3k). Further, distance to stream (Figure 3i) and stream density (Figure 3j) were used, as these factors can impact the groundwater level by recharging the groundwater tables [45]. The network of streams was extracted from the ASTER DEM. Another geomorphological influential factor is the ground lithology of the area. However, we could not take the lithological layer into account since this factor did not have substantial variation in the region of interest. Road data were obtained from the open street map (OSM) at the scale of 1:100,000, and distance to a road was calculated in the GIS environment (Figure 3h). Google Earth Engine (GEE) cloud computing, gathering massive volumes of various satellite imagery alongside popular machine learning algorithms, is a suitable platform for analyzing geo big data and monitoring the environment [46,47]. The available 10-m Iran-wide LULC map was used. The map was generated in the GEE platform using Sentinel-1 and Sentinel-2 images and object-based random forest classifier with 95% overall accuracy [48]. All the data were generated or resampled to a 30 m pixel size.

Table 2. The details of the input conditioning factors.

Factors	Source	Scale (Resolution)	Classification Method
Altitude	ASTER DEM	30 × 30	Natural breaks
Slope angle			Natural breaks
Slope aspect			Manual
Plan curvature			Manual
Profile curvature			Natural breaks
TWI			Natural breaks
Distance to river			Manual
Stream density			Natural breaks
Land cover	Sentinel-1 and Sentinel-2	30 × 30	Land cover units
Distance to road	Open street map (OSM)	1:100,000	Manual
Groundwater drawdown	Well inventory of the study area	30 × 30	Natural breaks

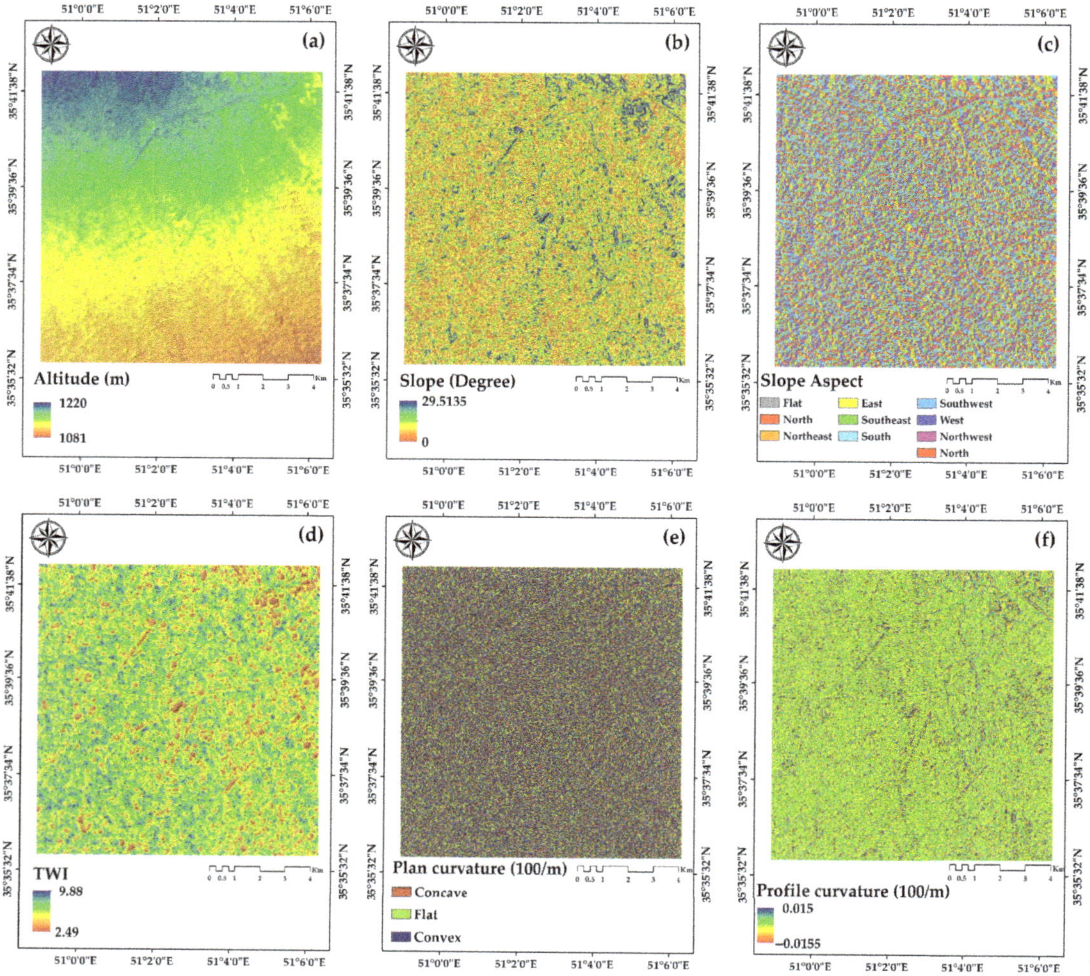

Figure 3. *Cont.*

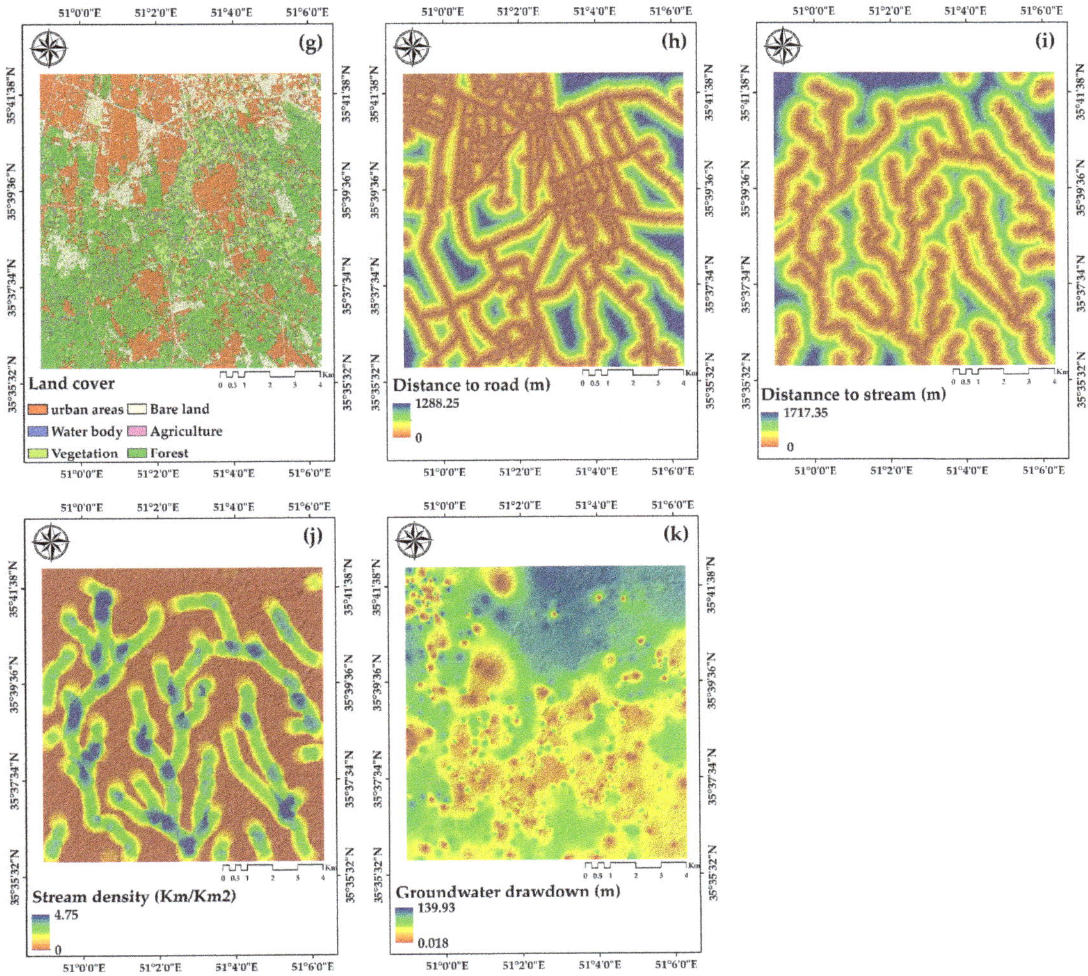

Figure 3. Land subsidence conditioning factors: (**a**) altitude; (**b**) slope; (**c**) slope aspect; (**d**) topographic wetness index (TWI); (**e**) plan curvature; (**f**) profile curvature; (**g**) land cover; (**h**) distance to road; (**i**) distance to stream; (**j**) stream density and (**k**) groundwater drawdown.

3. Methodology

In this section, the methods and models used in different parts of the approach are presented in detail. First, the PS-InSAR technique used for generating land subsidence map of the study area and the LS inventory are discussed. Then, the ANFIS model and evolutionary algorithms, GWO and ICA, are stated. Moreover, FR analysis and the approach to optimize ANFIS using meta-heuristics are presented. Finally, the validation methodology of this paper is outlined.

3.1. *PS-InSAR Technique*

SAR Interferometry (InSAR) is a well-established technique for measuring and monitoring ground deformation with millimetric accuracy. This is mainly based on the phase difference between SAR images acquired at different times and slightly different sensor positions. Time-series InSAR analysis aims to identify coherent image pixels (persistent scatterers (PSs)), which have high phase stability and reflect strong backscatter to the satellite over a long time period. A baseline configuration can determine the set of interferometric image pairs, which is used in the time series analysis. The baseline shows the distance between two images, in terms of antenna position (spatial baseline), acquisition time (temporal baseline), or Doppler centroid (Doppler baseline). The single-master configuration, where each image is co-registered to a unique master image, is the most common one for PSI analysis [49,50]. The master image is chosen in the middle of the 2D spatio-temporal space so that the high coherence of all formed interferometric pairs (interferograms) is guaranteed. The interferogram contains the ground deformation phase component as well as some other distinct contributions, such as atmospheric disturbance, topographic, and flat-Earth terms. These components are removed in the next step from the interferometric phase using an external DEM.

A spatial network is formed using a primary set of points as PS candidates (PSCs) to estimate the Atmospheric Phase Screen (APS) and further densification of PS points. Since the original interferometric phase is wrapped (i.e., phase observations in the $[-\pi, +\pi)$), and it is composed of a large number of phase contributions, the PSCs cannot be selected based on the phase. Thus, the amplitude dispersion index (ADI) based on Equation (1) is used as an approximation of phase stability. A point can be selected as a PSC if it always has a higher amplitude value than a suitable threshold. It was proved that assuming sufficient data images, the phase behavior with the standard deviation (σ_v) lower than a threshold of 0.25 is similar to the trend of ADI. Hence, this index, which represents the phase stability of points, can be used for PSCs selection.

$$D_A = \frac{\sigma_A}{m_A} \simeq \sigma_v \leq 0.25 \tag{1}$$

where σ_A is the standard deviation and m_A is the average amplitude value of each pixel over time. Next, the spatial network is used to estimate the unknown parameters, DEM error (residual topographic phase component), and the deformation rate, along with each connection between two adjacent PSCs through the maximization of a periodogram (Equation (2)) [51]. All PSI methods are based on assumptions regarding the spatial and temporal smoothness of the deformation signal, expressed by a model. Here, the model is considered as a linear deformation trend in time.

$$\xi\left[\Delta v(p_{ij}).\Delta h(p_{ij})\right] = \frac{1}{N}\sum_{s=1}^{N} e^{j\left[\Delta\varphi_{s.k}(\rho_{ij}) - \frac{4\pi}{\lambda}\Delta v(p_{ij})B_{t.s} - \frac{4\pi}{\lambda R \sin\theta}\Delta h(p_{ij})B_{n.s}\right]} \tag{2}$$

where p_{ij} demonstrates the connection between adjacent PSCs p_i and p_j. N is the number of interferograms. The term $\Delta\varphi_{s.k}$ is the double difference interferometric phase in image pairs s and k, while $B_{t.s}$ and $B_{n.s}$ are the temporal and interferometric normal baselines, respectively. θ refers to the incidence angle of the SAR signal.

For each PSC, the average residual phase after correction for the modeled parameters is taken to obtain an estimate for the atmospheric signal in the master image. Then, the atmospheric signal of the slave images as well as phase noise is separated from un-modeled deformation based on high-pass filtering. After APS removal, to increase point density, the second set of PSs is selected using a higher threshold for the ADI criterion. The unknown parameters are re-estimated for all pixels based on another maximization of the periodogram [49]. Eventually, temporal coherence, which is a function of residual phase noise, is used to determine the final PSs which build the land deformation map.

3.2. Adaptive Neuro-Fuzzy Inference System (ANFIS)

Fuzzy inference systems (FIS) are capable of depicting multifaceted processes using the concepts and if-then rules. However, they are incapable of learning [52]. Furthermore, if the number of input variables is too large, then selecting the membership functions and setting the fuzzy rules will become challenging [53]. On the other hand, learning algorithms automatically choose the suitable set of parameters for fuzzy membership functions despite their inability to explain the system under study. Thus, the ANFIS model [54] is a mixture of ANN and FL benefiting from both ANN's computation capability and FL's decision making. The structure of the ANFIS model contains five layers, called adaptive and fixed [55]. The ANFIS model employs the Takagi–Sugeno–Kang fuzzy algorithm in two rules of 'if-then' with two inputs, x and y, and one output f for both as follows [56]:

$$\begin{array}{l} Rule\ 1: if\ x\ is\ A_1\ and\ y\ is\ B_1, \\ then\ \ f_1 = p_1x + q_1y + r_1 \end{array} \tag{3}$$

$$\begin{array}{l} Rule\ 2: if\ x\ is\ A_2\ and\ y\ is\ B_2, \\ then\ \ f_2 = p_2x + q_2y + r_2 \end{array} \tag{4}$$

Each node contains adaptive nodes, and input variables are fuzzified in first layer (Equations (5) and (6)):

$$O_{1,i} = \mu A_{i(x)} \tag{5}$$

$$O_{1,i} = \mu B_{i(y)} \tag{6}$$

where, x and y are the input nodes, A and B are the linguistic variables, and $\mu A_{i(x)}$ and $\mu B_{i(y)}$ are membership functions for that node. The second layer contains fixed nodes denoted as л to compute the strength of the rules. The output of each node is the product of all input signals to that node (Equation (7)):

$$O_{2,i} = W_i = \mu A_{i(x)} \mu B_{i(y)},\ i = 1,\ 2 \tag{7}$$

where W_i is the output for each node.

The third layer encompasses fixed nodes denoted as N. The nodes in this layer are the normalized outputs of the second layer, which are referred to as the normal firepower (Equation (8)):

$$O_{3,i} = w_i = \frac{w_i}{\sum_{j=1}^{2} w_j},\ i = 1,\ 2 \tag{8}$$

All nodes in the fourth layer are adaptive and associated with a node function described by the following equation:

$$O_{4,i} = w_i f_i = w_i(p_i x + q_i x + r_i) \tag{9}$$

where w_i is the normalized firepower of third layer and p_i, q_i, and r_i are node parameters. The parameters of this layer are can be interpreted as the result parameters.

The final layer has only one node denoted as Σ, which represents the summation of all the input signals:

$$O_{5,i} = \sum w_i f_i = \frac{\sum w_i f_i}{\sum w_i} \tag{10}$$

3.3. *Imperialist Competitive Algorithm*

The imperialist competitive algorithm (ICA) is a novel evolutionary algorithm based on human social evolution, developed by Atashpaz-Gargari and Lucas [57]. The ICA belongs to the group of swarm intelligence, which provides a powerful algorithm for solving NP-hard problems through its capability of dealing with continuous optimization. In this algorithm, the primary population is composed of several countries, and they interact with each other to form empires. Assuming the value of the objective function, colonialist and colonial groups are formed based on the existing countries. After ascertaining a colonialist, other countries are randomly allocated to one of the colonizers [57,58]. Every colonialist and its associated colony is called an empire. The algorithm then simulates the competition among imperialists in order to acquire more colonies. The best colonialist typically has more chance to occupy more colonies.

Another way of allocating the colonies to each colonialist is based on their normalized cost, which is calculated via Equation (11):

$$N.T.C_n = max_i\{T.c_i\} - T.c_n, \quad (11)$$

where $T.c_n$ is the empire's total cost n, and $N.T.C_n$ is the total normalized cost value of that empire. Possession eventuality of the colonization competition by each empire is calculated by Equation (12):

$$P_{p_n} = \left| \frac{N.T.C_n}{\sum_{i=1}^{N_{imp}} N.T.C_i} \right| \quad (12)$$

The next phase is to attempt to approach a colonial country to analyze the colonies' cultural and social structures in different political and social layers. The colonies then move to the colonialist country. The colonialist and colony will change their positions; the new colonialist position will continue with the algorithm. The new colonialist country will start applying adjustment to its colonies this time. To calculate the cost function, the total empire cost is given by Equation (13) as follows:

$$T.c_n = f(imp) + \mathfrak{d}mean(f(colony)) \quad (13)$$

where $f(imp)$ is the value of the cost function for the colonialist, $f(colony)$ represents the mean values of the cost function for the colonies, and the constant $\mathfrak{d}$ is considered a value between 0 and 1.

Finally, the cost of each empire is calculated, and the colonies of weak empires are abolished and join to stronger empires. This process of recruitment or competition between colonialists is continued. In the next stage, empires that have lost all their colonies will be eliminated and will join other colonies. The process is repeated until a single universal empire in the globe is built that is very close to the empire with colonial nations [59].

3.4. Grey Wolf Optimization

The novel grey wolf optimization (GWO) algorithm, presented by Mirjalili et al. [60], is an inspiration from the hunting behavior of grey wolves and the social hierarchy in nature. Wolves are social animals that live in packs, and they have a hierarchy consisting of four groups. The leader of each group, the alpha wolf (α), makes decisions about hunting, sleeping, and walking time, and all the other group members must follow its directives. In terms of hierarchy, the other wolves fall into three levels, called beta (β), delta (δ), and omega (ω). The beta wolves at the second level assist the alpha in making decisions and devise them. They are the best candidates for alpha replacement. Another notable characteristic is their group hunting, which can be summarized in four stages: (1) encircling prey, (2) hunting, (3) attacking prey (exploitation), and (4) searching for prey (exploring) [61]. The hunting process (optimization) is led by α, β, and δ wolves, and ω wolves have to abide by these three groups.

1. Encircling prey

In the first stage, the grey wolves encircle and surround the prey during hunting. To define this phase mathematically, the following Equations (14) and (15) are proposed. The parameter D measures the distance between the grey wolf and the prey, and $\vec{X}$ represents the location of the prey:

$$D = \left| \vec{C}.\vec{X}_p(t) - \vec{X}(t) \right| \quad (14)$$

$$\vec{X}(t+1) = \vec{X}_p(t) - \vec{A}.\vec{D} \quad (15)$$

where t denotes the current iteration, and $\vec{X}_p$ and $\vec{X}$ denote the position vectors of the prey and the grey wolves, respectively. The $\vec{A}$ and $\vec{C}$ coefficient vectors are defined as follows:

$$\vec{A} = 2a.r_2 - a \tag{16}$$

$$\vec{C} = 2.\vec{r}_1 \tag{17}$$

where components of a are linearly decreased from 2 to 0 over the course of iterations, and r_1 and r_2 are random vectors between [0,1].

2. Hunting

After the encircling of the prey, the hunting phase is guided by α, β, and δ since they are supposed to have compressive knowledge about the prey's position. This can be computed using following formulas:

$$\vec{D}_a = \left|\vec{C}_1.\vec{X}_a - \vec{X}\right|,\ \vec{D}_\beta = \left|\vec{C}_2.\vec{X}_\beta - \vec{X}\right|,\ \vec{D}_\delta = \left|\vec{C}_3.\vec{X}_\delta - \vec{X}\right| \tag{18}$$

$$\vec{X}_1 = \vec{X}_a - \vec{A}_1.\vec{D},\ \vec{X}_2 = \vec{X}_\beta - \vec{A}_2.\vec{D},\ \vec{X}_3 = \vec{X}_\delta - \vec{A}_3.\vec{D} \tag{19}$$

where $\vec{X}_1$, $\vec{X}_2$, $\vec{X}_3$ denote the position of α, β and δ, wolves respectively. $\vec{A}_1$, $\vec{A}_2$, $\vec{A}_3$ and $\vec{C}_1$, $\vec{C}_2$, $\vec{C}_3$ are the respective coefficient vectors. The position of a grey wolf in the search space can be updated as follows:

$$\vec{X}(t+1) = \frac{\vec{X}_1 + \vec{X}_2 + \vec{X}_3}{3} \tag{20}$$

The other wolves update their positions randomly according to the position of the prey.

3. Attacking prey

The process of hunting ends when the prey stops moving and gray wolves attack the prey. It is important to note that the fluctuation range of $\vec{A}$ is $[-2a, 2a]$, where a is linearly decreased from 2 to 0. The exploration trend happens when $\left|\vec{A}\right| < 1$ and $\left|\vec{C}\right| < 1$. At this moment, the wolves attack the prey.

4. Search for prey

The grey wolves track and chase the prey. The pursuing of the prey is known as the exploration phase in the GWO algorithm [62]. The parameters α, β, and δ have guidance responsibility in this process. If $\left|\vec{A}\right| > 1$, it means the grey wolves are split and distributed in diverse ways for searching of the prey. After finding it, they congregate to attack. The coefficient $\vec{C}$ provides a random weight for the prey while $\left|\vec{C}\right| > 1$ and promotes the exploration phase. In addition, $\vec{C}$ models the natural hindrances in hunting for the grey wolves.

3.5. *Frequency Ratio*

One of the statistical bivariate models is FR, which is widely used in modeling environmental hazards as a geospatial assessment tool for quantifying the potential relationship between dependent and independent variables [63]. The FR value for a certain class from a given factor can be calculated using:

$$FR = \frac{\frac{N_{pix}(X_i)}{\sum_{i=1}^{m} X_i}}{\frac{N_{pix}(X_j)}{\sum_{j=1}^{n} N_{pix}(X_j)}} \tag{21}$$

where $N_{pix}(X_i)$ is the number of pixels in each class of each factor with land subsidence locations. $X.N_{pix}(X_j)$ is the number of pixels of X_j factor, m is the number of classes in the X_i factor, and n is the number of factors in the study area [64].

3.6. ANFIS with Meta-Heuristic Algorithms

In ANFIS, parameter adjustment and the creation of a basic fuzzy system are done by combining traditional methods and then back error propagation. In this research, ICA and GWO were used as meta-heuristic algorithms to enhance the results of the ANFIS system and also to tweak the parameters of membership functions [52,65]. First, using input and target data, the FIS is created by the ANFIS model. Next, the membership functions are optimized and adjusted by meta-heuristic algorithms, and the output for the ANFIS (y) model is computed by [66]:

$$e = t - y \tag{22}$$

$$RSME = \sqrt{mean(e)^2} \tag{23}$$

where t is the target data, y is a function of input data and optimized FIS, and e is the error function that should be minimized. When the final conditions are met with the best output, the optimization process stops; otherwise, the membership function optimization is repeated.

3.7. Validation

In this research, the ROC curves were used for the accuracy assessment of the LSS models employed [6,10,39]. The ROC curve analysis is a common method to evaluate the goodness-of-fit and prediction power of models regarding the area under the curve (AUC) [2,67]. Ranging from 0 to 1, higher AUC values represent more reliable and accurate model performance. According to Yesilnacar [53], the qualitative relationship between AUC and the prediction accuracy of a model can be classified into the following categories: 0.5–0.6 (poor), 0.6–0.7 (average), 0.7–0.8 (good), 0.8–0.9 (very good), and 0.9–1 (excellent).

4. Results

By utilizing the spatial data and subsidence inventory generated and the methods discussed above, the mapping and assessment of land subsidence susceptibility for Shahryar County were conducted. In the following sub-sections, the results of the various parts of the methodology are thoroughly discussed.

4.1. Result of PS-InSAR

The ground deformation rate along the line of sight (LOS) direction during the acquisition time interval was obtained based on the PS-InSAR technique (see Section 3.1). SARPROZ (SAR PROcessing tool by periZ) [51] software was used to implement the PS-InSAR technique in the current research paper. The star configuration of the S-1A SLC time-series data stack is shown in Figure 4. The master image for the dataset was selected based on maximizing the stack coherence [68]. All slave images were co-registered with respect to the master image (2019/09/11). The shuttle radar topography mission (SRTM) DEM was used to remove topographic-related phase components from the interferometric phases. After the selection of 7881 PSCs, a spatial network was created by Delaunay Triangulation, connecting each point to the other. The unknown LOS velocity and DEM error were calculated along with the connections by maximizing a periodogram. All the obtained parameters were then integrated into the absolute values with respect to a reference point, which had no subsidence rate. The atmospheric phase for PSCs can be resampled on the uniform image grid as the APS. With having APS compensated for all slave images, the unknowns were estimated again for more PS points, selected by a lower threshold on ADI to obtain a more dense subsidence map. Figure 5a shows the LOS deformation map. According to Figure 5a, the maximum velocity was about −175 mm/year, which occurred in the southern agricultural part of the region of interest, where more ground-water extraction was observed.

Further, the LOS deformation rates should be decomposed into horizontal and vertical deformation components as it is the inherently vertical movement of the Earth's surface with a slight horizontal displacement. It has been proved that the horizontal deformation is a very small portion of motion compared to the vertical deformation [67,69]. Hence, the LOS deformation could be assumed as negligible and converted simply into the vertical deformation rates using the cosine of the incidence angle of the radar signal. The interpolated land subsidence inventory map was designed based on the vertical velocity deformation map for Shahryar County, depicted in Figure 5b. ADI was more successful in identifying PS points in man-made areas with stable targets than agricultural areas [70]. Since vegetated regions are the main land cover in our case study, an interpolation was applied to the obtained vertical map to extend the deformation information for the whole study area. The inverse distance weighted (IDW) interpolation was used with a weighting power of 2 and neighboring radius of 12 for calculating the vertical velocity interpolation.

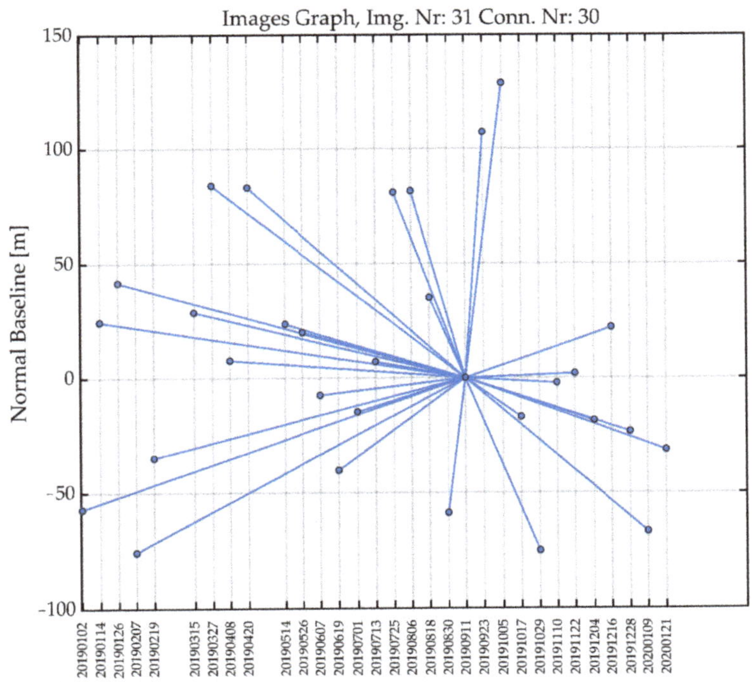

Figure 4. The perpendicular baseline graph for the time-series data stack. The dots represents the images, and the edges denote the interferograms.

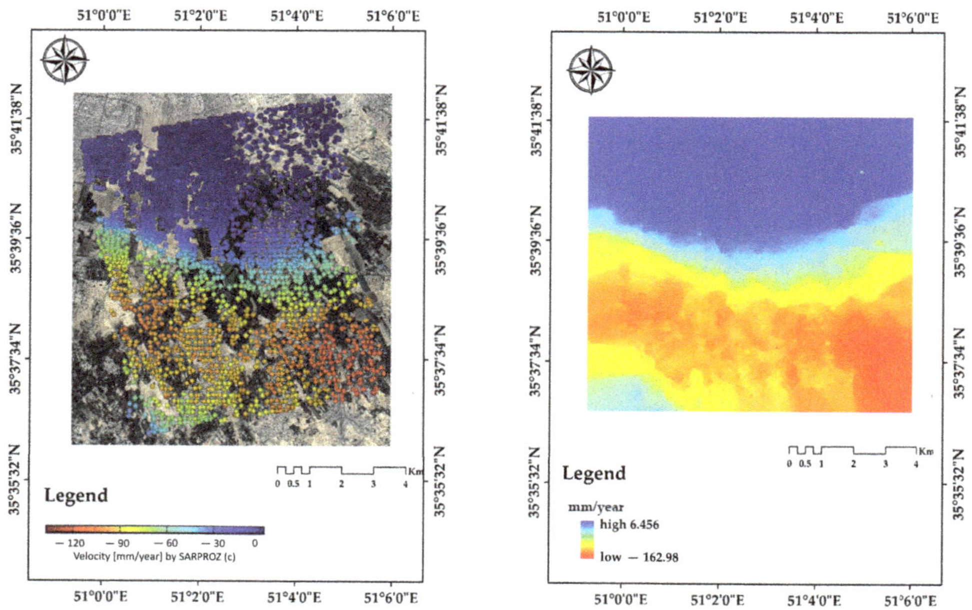

(a) (b)

Figure 5. (**a**) The annual velocity map based on the PSInSAR technique. The map is superimposed on the Google Earth (GE) imagery; (**b**) IDW interpolation raster of the velocity map.

4.2. Result of FR

The results of the FR analysis in identifying the relationship of land subsidence occurrence with the conditioning factors are summarized in Table 3. Two out of five altitude classes had the highest probability (FR > 1.0), with 1119 to 1137 m being the most correlated class with land subsidence, followed by the altitude class lower than 1119 m. The results of the slope angle analysis showed that slopes ranging between 4.5 and 6.8 degrees had the highest FR (1.11). Further, a TWI class lower than 4.84, profile curvature higher than 0.0029, convex plan curvature class, and flat (F) slope aspect had the most influence on LSS for each corresponding factor. Land cover analysis results indicated that forest and urban classes had the highest probability of land subsidence occurrence, with FR values of 1.17 and 1.04, respectively. Distance to a stream of between 50 to 100 m had the highest FR, and the class of lower than 50 m had a considerable correlation; a stream density higher than 2.68 had the highest correlation with land subsidence and the 1.23 to 1.92 class, which also had a considerable FR value. For distance to road, the 0 to 100 m class had the highest FR followed by the 100 to 200 m classes. Finally, groundwater drawdown ranging from 55 to 83 m and from 28 to 55 m had higher impacts on land subsidence occurrence.

Table 3. Relationship between land subsidence occurrence and conditioning factors using the FR model.

FR	No. of Land Subsidence Areas	No. Pixels in the Domain	Class	FR	No. of Land Subsidence Areas	No. Pixels in the Domain	Class
			Distance to stream (m)				**Altitude (m)**
1.18	46	26,180	0–50	1.63	69	32,246	<1119
1.31	43	22,119	50–100	1.78	95	40,783	1119–1137
1.09	37	22,772	100–150	0.907	42	35,442	1137–1157
0.98	27	18,411	150–200	0.092	4	33,287	1157–1179
0.73	57	52,240	>200	0	0	19,088	>1179
			Distance to road (m)				**Slope angle**
1.052	80	51,395	0–100	0.88	57	49,345	0–2.5
1.051	40	25,705	100–200	1.01	71	53,745	2.5–4.5
0.939	28	20,151	200–300	1.11	53	36,493	4.5–6.8
0.905	17	12,692	300–400	1.05	24	17,416	6.8–10.4
0.95	45	32,000	>400	0.99	5	3846	>10.4
			Stream density				**TWI**
0.81	83	68,535	0–0.428	1.2	32	20,419	<4.84
0.99	32	21,673	0.428–1.23	0.97	68	53,623	4.84–5.48
1.3	56	28,947	1.23–1.92	0.94	75	60,506	5.48–6.08
1.03	25	16,307	1.92–2.68	1.029	33	24,560	6.08–7.62
1.5	14	6260	>2.68	0.882	2	1736	>7.62
			Groundwater drawdown (m)				**Profile curvature**
0.76	10	8837	<28	0.73	12	12,501	<−0.015
1.28	80	42,231	28–55	0.97	51	40,102	−0.01
1.31	100	51,222	55–83	0.78	52	50,521	−0
0.52	20	25,857	83–111	1.19	70	44,847	−0
0	0	13,796	>111	1.48	25	12,874	>0.0029
			Slope aspect				**Land cover**
3.7	2	413	F	1.04	62	362,054	Urban areas
1.15	26	17,236	N	0.63	1	9577	Water body
0.66	16	18,326	NE	0.81	22	164,856	Vegetation
1.02	29	21,603	E	0.63	19	183,677	Bare land
1.01	30	22,602	SE	0.55	2	21,954	Agriculture
0.98	31	24,177	S	1.17	104	539,017	Forest
1.05	29	21,085	SW				**Plan curvature**
0.93	23	18,742	W	0.98	69	53,666	Concave
1.1	24	16,662	NW	0.9	62	52,532	Flat
				1.1	79	54,643	Convex

4.3. Result of Hybrid Models

In the course of implementation of the hybrid models, 70% of the land subsidence points (210 locations) were used for training with values 1, and the same number of randomly selected non-subsidence points were taken into account with 0 values in the training phase. For the test dataset, 30% of the subsidence inventory (90 locations) with a value of 1 was used, with 90 randomly assigned points with a value of 0. The training datasets were used to calibrate the weights of the membership functions. The testing dataset was used to evaluate the performance of the trained ANFIS ensemble models. Hybrid models were implemented in MATLAB 2017b software. The parameters used in meta-heuristic algorithms are presented in Table 4. The prediction power of ANFIS and the two hybrid models with the training dataset (target) along with the comparison of the output and target testing dataset is shown in Figure 6.

Table 4. Parameters used in hybrid algorithms.

ICA	GWO
Iteration = 1000 Population = 40 Number of empires = 10 Selection pressure = 1 Assimilation coefficient = 2 Revolution probability = 0.1 Revolution rate = 0.5	Iteration = 1000 Number of wolf groups = 30

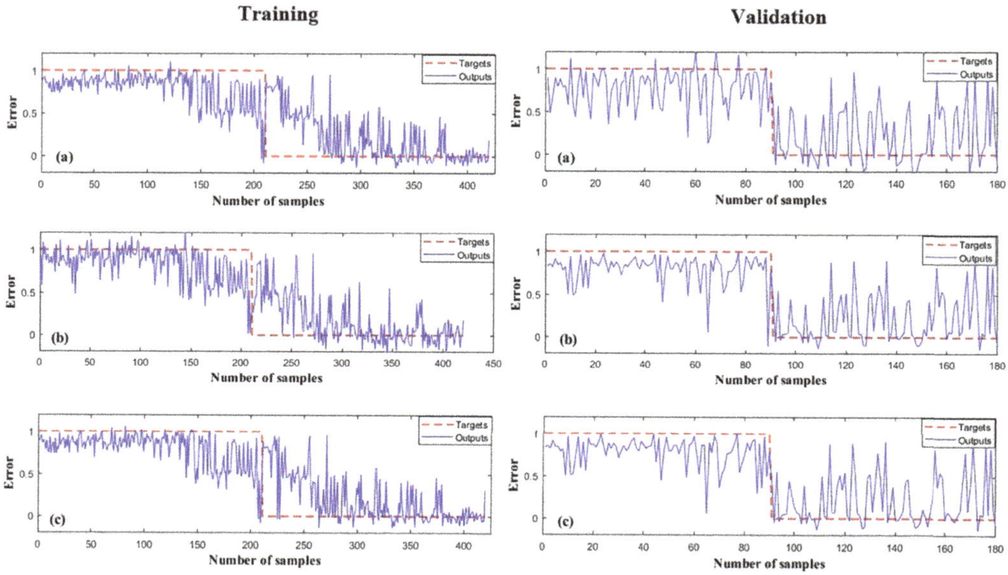

Figure 6. The target and output values for training and validation datasets of (**a**) ANFIS, (**b**) ANFIS-GWO, and (**c**) ANFIS-ICA.

The RMSEs of the training and testing phases were calculated and are shown in Table 5. The two ensemble models enhanced the ANFIS model, and the ANFIS-ICA outperformed the ANFIS-GWO with an RMSE of 0.276 in the training phase and 0.3199 in the validation and testing phase. The ANFIS-GWO yielded an RMSE of 0.313 and 0.3217 in training and validation phases, respectively. Finally, the ANFIS model resulted in 0.323 in training and 0.34 in the validation phase.

Table 5. The comparison of model performance.

Model	RMSE	
	Train	Validation
ANFIS	0.323	0.340
ANFIS-GWO	0.313	0.3217
ANFIS-ICA	0.276	0.3199

The convergence results of the two ANFIS-ICA and ANFIS-GWO ensemble models up to 1000 iterations are shown in Figure 7. ANFIS-ICA had a better convergence value (0.276) than ANFIS-GWO (0.313). The lowest amount of the cost function (RMSE) indicates the best cost and thus the best performance in predicting the results.

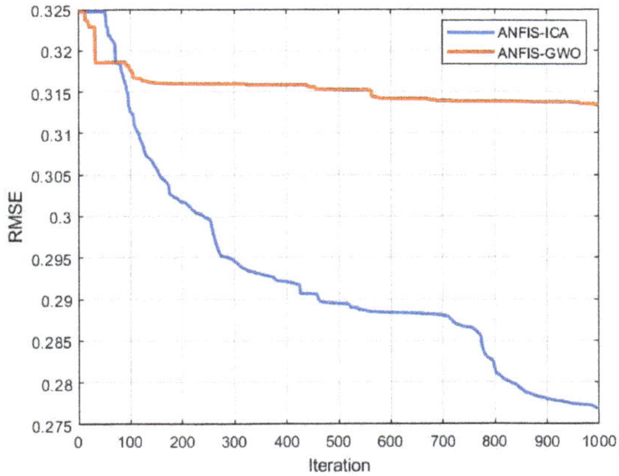

Figure 7. The convergence graph of the objective functions.

4.4. LSSM Using ANFIS and Its Optimized Models

The original ANFIS and its optimized ensembles in this research were trained and used to estimate land subsidence susceptibility in the study area. Susceptibility modelling and estimation were all carried out in MATLAB 2017b and were then exported to ArcGIS 10.3 software to classify and generate the susceptibility maps. Land susceptibility index was classified into five classes, very high, high, moderate, low, and very low, based on a natural break classification scheme [22,41]. Figure 8 presents the generated classified subsidence susceptibility maps obtained from ANFIS, ANFIS-GWO, and ANFIS-ICA. As can be seen, all the output subsidence susceptibility maps are similar and consistent with each other, particularly the ones for ANFIS-ICA and ANFIS-GWO. Moreover, the map based on ANFIS-ICA is much smoother than the others.

4.5. Validation

The ROC curves were calculated for all LSS maps using the test data. Figure 9 demonstrates the comparison of AUC for all the models used. The results showed that the ANFIS-ICA had the highest prediction accuracy (0.932), followed by the ANFIS-GWO (0.926) and ANFIS (0.908). This proves that the combination of the ANFIS model with meta-heuristic algorithms such as GWO and ICA can significantly improve the output land subsidence susceptibility maps in comparison to ANFIS alone.

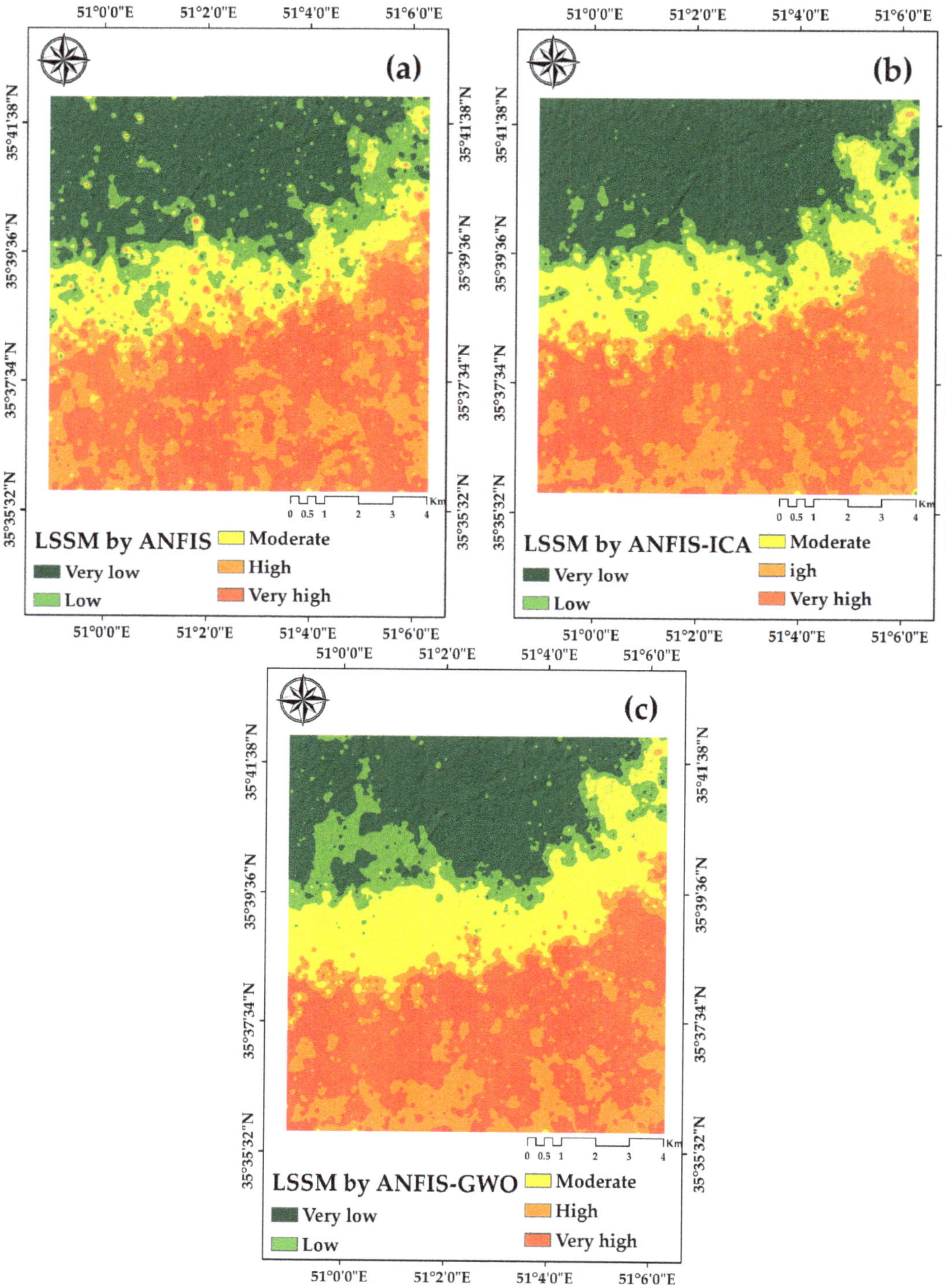

Figure 8. The LSS maps of the study area using (**a**) ANFIS, (**b**) ANFIS-ICA, and (**c**) ANFIS-GWO.

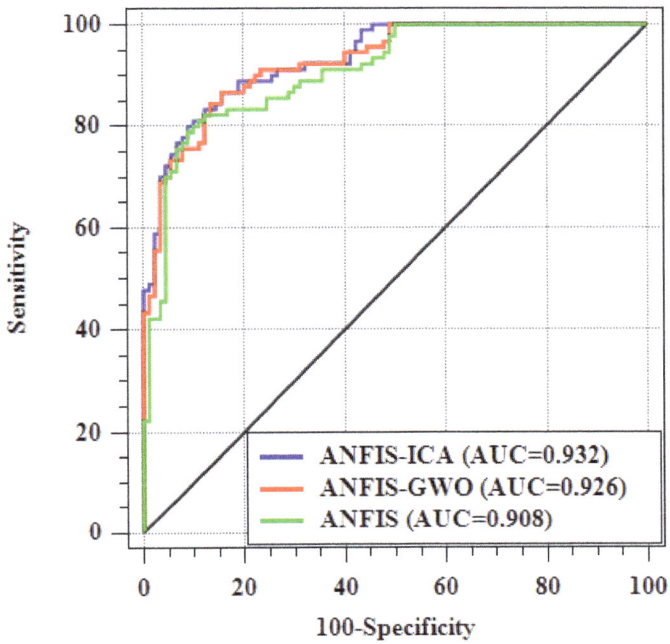

Figure 9. The ROC curves for the LSSMs of the models and their AUC.

5. Discussion

Land subsidence is the slow vertical lowering of the Earth's surface, posing a serious threat to both the environment and human life. Recently, there has been an increasing interest in land subsidence analysis and monitoring in Iran as it is one of the highest subsidence-prone countries [17,18,20,24]. Natural hazard vulnerability analysis using machine learning algorithms (i.e., ANFIS) has shown promising results. Therefore, in this research, the focus was to employ the ANFIS model in combination with meta-heuristics in land subsidence susceptibility mapping.

Land subsidence inventories are necessary for accurate subsidence susceptibility analysis. The use of remote sensing SAR data is suitable for providing subsidence inventories due to their wide availability, independence from fieldwork, time and cost efficiency, frequent repeatability over time, and, especially, high precision [6]. In this work, the PS-InSAR technique with its millimetric precision was employed to determine the subsided areas in the region of interest to form the inventory data used for training and testing the LSS models.

Important conditioning factors for determining land subsidence prone areas were identified and collected based on either the literature or availability of data. The FR model was used to evaluate the correlation and influence of the factors. The results showed that all the factors employed in this paper had a considerable effect on LSS in Shahryar County. Among all the factors, the flat (slope aspect) area had the highest FR value (3.7), indicating high subsidence susceptibility in flat areas. The slope angle is related to the hydro physiographic characteristics that can influence the water infiltration rate and the volume and velocity of the Earth's surface flow [13]. Altitude and groundwater drawdown were the best predictors of land subsidence in this study, followed by stream density and distance to stream. Rahmati et al. and Arabameri et al. [2,37] also found that groundwater drawdown had a greater impact on land subsidence. Other topo-hydrographic factors, such as stream density and distance to stream are indirectly related to LS as they impact groundwater recharge and infiltration [2,40] and, as can be seen in the results, the land areas closer to streams and with a certain stream density were more susceptible to LS. In terms of altitude, the lower lands were more prone to subsidence as the class 1137 to 1119 m and those lower than 1119 m had the highest FR. TWI, plan curvature, and profile curvature are among secondary topographic derivatives indirectly influencing LS [2,40,71]. These factors were not among best predictors of subsidence in the study area, which may be due to smooth and low altitude changes in the study area. The FR analysis showed that a TWI lower than 4.84 was strongly correlated with LS. In a similar study [40], lower TWI values (i.e., 2.54 to 8) have been reported to be more prone to subsidence. Positive and convex plan and profile curvatures had the highest FR value, as reported in [40]. Cropland and urban land cover types exist in lower altitude and flat areas. The main water source of the area is groundwater; therefore, more extraction of water in recent years as a result of population increase has caused the subsidence rate to increase. Previous studies have stressed the impact of groundwater extraction on subsidence occurrence [72,73]. Regarding the distance to road factor, the closer to a road, the greater the land subsidence risk, which can be due to closeness to urban land cover and thus indirectly related to the subsidence phenomenon.

Two novel meta-heuristic algorithms, GWO and ICA, were used to optimize the rules and parameters of the ANFIS model. Both of these evolutionary algorithms belong to swarm intelligence. The results showed that ICA had a slower convergence rate than GWO; however, it had better performance. In order to evaluate the prediction power and accuracy of the models, RMSE and ROC criteria were used. The RMSE is simply based on error assessment, whereas ROC is based on true positive (TP), false positive (FP), true negative (TN), and false negative (FN), which is more appropriate for comparison [42]. ANFIS-ICA had the lowest RMSE in both the training (0.276) and testing (0.3199) phases, followed by ANFIS-GWO and ANFIS alone. According to the AUC-ROC results, the ANFIS-ICA model was more accurate (0.932), followed by the ANFIS-GWO model (0.926) and the ANFIS model (0.908). It can be seen that the use of machine learning algorithms resulted in higher prediction accuracy since ANFIS alone yielded a suitable performance compared to other statistical methods in other studies. It can also be concluded that optimization of the ANFIS algorithm by meta-heuristics improves its results considerably. This was also reported in cases of other applications [64,74]. The results showed that the ICA algorithm was more accurate than the GWO algorithm in optimization of the ANFIS model. The advantages of the ICA algorithm are high convergence speed and the ability to optimize functions with a large number of variables [75]. The GWO algorithm has a small number of disadvantages, including a low solving accuracy, poor local searching ability, and slow convergence rate [60].

The output land subsidence susceptibility maps of the three models were similar and in line with each other. However, the map produced by the ANFIS-ICA was smoother than that of the other two. As could be observed, the high-risk areas were predicted where the groundwater extraction was higher, elevation was lower, and agricultural land use was higher. This is because the main source of income in the study area is agriculture. Further, the population has increased; therefore, food production has stressed the groundwater, the main water source of the area, and thus the land subsidence risk has become higher in those areas. Further, it is evident that the subsidence trend is gradually reaching towards the urban part of the Shahryar County, posing a serious threat to settlements and human life. The generated LSSMs in this paper can benefit authorities and decision-makers to identify subsidence-prone areas regarding environmental and urban management.

6. Conclusions

Land subsidence is an important issue in Iran due to the semi-arid and arid climate and excessive groundwater extraction. Therefore, modeling, simulation, and risk mapping offer valuable knowledge of environmental geohazards. GIS-based predictions have proved to be essential for authorities in terms of planning and decision-making. In this work, we used remote sensing SAR data and the PSInSAR technique to create a land subsidence inventory of the study area as a high-precision tool with a low cost and frequent reproducibility. Since machine learning tools have shown appropriate performance in modeling and mapping hazard susceptibility, the ANFIS model was used in this research to map the land subsidence risk in Shahryar County, Tehran province, Iran. Another objective of this paper was to investigate the effect of optimization of the ANFIS model through meta-heuristics. Two novel evolutionary algorithms, namely, GWO and ICA, were used to create ensemble models. The results of the three models in both training and testing phases were assessed by RMSE. In both phases, ANFIS-ICA had the lowest RMSE, followed by ANFIS-GWO and ANFIS alone. AUC-ROC analysis was also used for model evaluation, and its results indicated that ANFIS-ICA had the best prediction performance (0.932), followed by ANFIS-GWO (0.926) and ANFIS (0.908). To conclude, the results overall showed the applicability of the ANFIS machine learning algorithm in land subsidence susceptibility mapping and the effectiveness of its ensembles with meta-heuristic algorithms. The methodology used is reproducible and can be applied to other regions with different environmental parameters to test the modelling performance. Further studies should be applied using other machine learning and deep learning algorithms to compare their prediction accuracy. In addition, future research can focus on developing risk monitoring and early-warning frameworks.

Author Contributions: Conceptualization, B.R. and S.V.R.-T.; Data curation, B.R. and S.V.R.-T.; methodology, S.V.R.-T., F.F. and B.R.; software, D.P., F.F. and S.V.R.-T.; validation, S.V.R.-T., F.F. and B.R.; formal analysis, B.R., S.V.R.-T., F.F., A.S.-N. and D.P.; investigation, B.R.; resources, F.F.; writing—original draft preparation, B.R.; writing—review and editing, B.R., S.V.R.-T. and F.F.; visualization, S.V.R.-T. and B.R.; supervision, A.S.-N. and D.P.; project administration, F.F., A.S.-N. and D.P. All authors have read and agreed to the published version of the manuscript.

Funding: This research received no external funding.

Acknowledgments: The authors would like to thank the European Space Agency (ESA) for freely providing Sentinel-1 satellite imagery.

Conflicts of Interest: The authors declare no conflict of interest.

References

1. Holzer, T.L.; Galloway, D.L. Impacts of land subsidence caused by withdrawal of underground fluids in the United States. *Hum. Geol. Agents* **2005**, *16*, 87–99.
2. Arabameri, A.; Saha, S.; Roy, J.; Tiefenbacher, J.P.; Cerda, A.; Biggs, T.; Pradhan, B.; Thi Ngo, P.T.; Collins, A.L. A novel ensemble computational intelligence approach for the spatial prediction of land subsidence susceptibility. *Sci. Total Environ.* **2020**, *726*, 138595. [CrossRef] [PubMed]
3. Raspini, F.; Bianchini, S.; Moretti, S.; Loupasakis, C.; Rozos, D.; Duro, J.; Garcia, M. Advanced interpretation of interferometric SAR data to detect, monitor and model ground subsidence: Outcomes from the ESA-GMES Terrafirma project. *Nat. Hazards* **2016**, *83*, 155–181. [CrossRef]
4. Galloway, D.L.; Burbey, T.J. Review: Regional land subsidence accompanying groundwater extraction. *Hydrogeol. J.* **2011**, *19*, 1459–1486. [CrossRef]
5. Xue, Y.-Q.; Zhang, Y.; Ye, S.-J.; Wu, J.-C.; Li, Q.-F. Land subsidence in China. *Environ. Geol.* **2005**, *48*, 713–720. [CrossRef]
6. Bianchini, S.; Solari, L.; Del Soldato, M.; Raspini, F.; Montalti, R.; Ciampalini, A.; Casagli, N. Ground Subsidence Susceptibility (GSS) Mapping in Grosseto Plain (Tuscany, Italy) Based on Satellite InSAR Data Using Frequency Ratio and Fuzzy Logic. *Remote Sens.* **2019**, *11*, 2015. [CrossRef]
7. Shi, X.; Jiang, S.; Xu, H.; Jiang, F.; He, Z.; Wu, J. The effects of artificial recharge of groundwater on controlling land subsidence and its influence on groundwater quality and aquifer energy storage in Shanghai, China. *Environ. Earth Sci.* **2016**, *75*, 195. [CrossRef]
8. Modoni, G.; Darini, G.; Spacagna, R.L.; Saroli, M.; Russo, G.; Croce, P. Spatial analysis of land subsidence induced by groundwater withdrawal. *Eng. Geol.* **2013**, *167*, 59–71. [CrossRef]
9. Stanley, J.-D.; Corwin, K.A. Measuring Strata Thicknesses in Cores to Assess Recent Sediment Compaction and Subsidence of Egypt's Nile Delta Coastal Margin. *J. Coast. Res.* **2012**, *29*, 657–670. [CrossRef]
10. Tien Bui, D.; Shahabi, H.; Shirzadi, A.; Chapi, K.; Pradhan, B.; Chen, W.; Khosravi, K.; Panahi, M.; Bin Ahmad, B.; Saro, L. Land Subsidence Susceptibility Mapping in South Korea Using Machine Learning Algorithms. *Sensors* **2018**, *18*, 2464. [CrossRef]
11. Pradhan, B.; Abokharima, M.H.; Jebur, M.N.; Tehrany, M.S. Land subsidence susceptibility mapping at Kinta Valley (Malaysia) using the evidential belief function model in GIS. *Nat. Hazards* **2014**, *73*, 1019–1042. [CrossRef]
12. Lee, S.; Park, I. Application of decision tree model for the ground subsidence hazard mapping near abandoned underground coal mines. *J. Environ. Manag.* **2013**, *127*, 166–176. [CrossRef] [PubMed]

13. Mohammady, M.; Pourghasemi, H.R.; Amiri, M. Land subsidence susceptibility assessment using random forest machine learning algorithm. *Environ. Earth Sci.* **2019**, *78*, 503. [CrossRef]
14. Fiaschi, S.; Tessitore, S.; Bonì, R.; Di Martire, D.; Achilli, V.; Borgstrom, S.; Ibrahim, A.; Floris, M.; Meisina, C.; Ramondini, M.; et al. From ERS-1/2 to Sentinel-1: Two decades of subsidence monitored through A-DInSAR techniques in the Ravenna area (Italy). *GIScience Remote Sens.* **2017**, *54*, 305–328. [CrossRef]
15. Galloway, D.L.; Jones, D.R.; Ingebritsen, S.E. *Land Subsidence in the United States*; US Geological Survey: Reston, VA, USA, 1999; Volume 1182.
16. Conway, B.D. Land subsidence and earth fissures in south-central and southern Arizona, USA. *Hydrogeol. J.* **2016**, *24*, 649–655. [CrossRef]
17. Mahmoudpour, M.; Khamehchiyan, M.; Nikudel, M.R.; Ghassemi, M.R. Numerical simulation and prediction of regional land subsidence caused by groundwater exploitation in the southwest plain of Tehran, Iran. *Eng. Geol.* **2016**, *201*, 6–28. [CrossRef]
18. Motagh, M.; Walter, T.R.; Sharifi, M.A.; Fielding, E.; Schenk, A.; Anderssohn, J.; Zschau, J. Land subsidence in Iran caused by widespread water reservoir overexploitation. *Geophys. Res. Lett.* **2008**, *35*, L16403. [CrossRef]
19. Dehghani, M.; Valadan Zoej, M.J.; Hooper, A.; Hanssen, R.F.; Entezam, I.; Saatchi, S. Hybrid conventional and Persistent Scatterer SAR interferometry for land subsidence monitoring in the Tehran Basin, Iran. *ISPRS J. Photogramm. Remote Sens.* **2013**, *79*, 157–170. [CrossRef]
20. Tarighat, F.; Foroughnia, F.; Perissin, D. Monitoring of Power Towers' Movement Using Persistent Scatterer SAR Interferometry in South West of Tehran. *Remote Sens.* **2021**, *13*, 407. [CrossRef]
21. Rahmati, O.; Golkarian, A.; Biggs, T.; Keesstra, S.; Mohammadi, F.; Daliakopoulos, I.N. Land subsidence hazard modeling: Machine learning to identify predictors and the role of human activities. *J. Environ. Manag.* **2019**, *236*, 466–480. [CrossRef] [PubMed]
22. Ghorbanzadeh, O.; Feizizadeh, B.; Blaschke, T. An interval matrix method used to optimize the decision matrix in AHP technique for land subsidence susceptibility mapping. *Environ. Earth Sci.* **2018**, *77*, 584. [CrossRef]
23. Ebrahimy, H.; Feizizadeh, B.; Salmani, S.; Azadi, H. A comparative study of land subsidence susceptibility mapping of Tasuj plane, Iran, using boosted regression tree, random forest and classification and regression tree methods. *Environ. Earth Sci.* **2020**, *79*, 223. [CrossRef]
24. Foroughnia, F.; Nemati, S.; Maghsoudi, Y.; Perissin, D. An iterative PS-InSAR method for the analysis of large spatio-temporal baseline data stacks for land subsidence estimation. *Int. J. Appl. Earth Obs. Geoinf.* **2019**, *74*, 248–258. [CrossRef]
25. Chaussard, E.; Wdowinski, S.; Cabral-Cano, E.; Amelung, F. Land subsidence in central Mexico detected by ALOS InSAR time-series. *Remote Sens. Environ.* **2014**, *140*, 94–106. [CrossRef]
26. Wang, Q.; Li, W. A GIS-based comparative evaluation of analytical hierarchy process and frequency ratio models for landslide susceptibility mapping. *Phys. Geogr.* **2017**, *38*, 318–337. [CrossRef]
27. Choi, J.-K.; Kim, K.-D.; Lee, S.; Won, J.-S. Application of a fuzzy operator to susceptibility estimations of coal mine subsidence in Taebaek City, Korea. *Environ. Earth Sci.* **2010**, *59*, 1009–1022. [CrossRef]
28. Zhou, G.; Yan, H.; Chen, K.; Zhang, R. Spatial analysis for susceptibility of second-time karst sinkholes: A case study of Jili Village in Guangxi, China. *Comput. Geosci.* **2016**, *89*, 144–160. [CrossRef]
29. Lee, S.; Park, I.; Choi, J.-K. Spatial prediction of ground subsidence susceptibility using an artificial neural network. *Environ. Manag.* **2012**, *49*, 347–358. [CrossRef]
30. Rafie, M.; Samimi Namin, F. Prediction of subsidence risk by FMEA using artificial neural network and fuzzy inference system. *Int. J. Min. Sci. Technol.* **2015**, *25*, 655–663. [CrossRef]
31. Hu, B.; Zhou, J.; Wang, J.; Chen, Z.; Wang, D.; Xu, S. Risk assessment of land subsidence at Tianjin coastal area in China. *Environ. Earth Sci.* **2009**, *59*, 269. [CrossRef]
32. Rezaei, M.; Yazdani Noori, Z.; Dashti Barmaki, M. Land subsidence susceptibility mapping using Analytical Hierarchy Process (AHP) and Certain Factor (CF) models at Neyshabur plain, Iran. *Geocarto Int.* **2020**, 1–17. [CrossRef]
33. Ghorbanzadeh, O.; Blaschke, T.; Aryal, J.; Gholaminia, K. A new GIS-based technique using an adaptive neuro-fuzzy inference system for land subsidence susceptibility mapping. *J. Spat. Sci.* **2020**, *65*, 401–418. [CrossRef]
34. Abdollahi, S.; Pourghasemi, H.R.; Ghanbarian, G.A.; Safaeian, R. Prioritization of effective factors in the occurrence of land subsidence and its susceptibility mapping using an SVM model and their different kernel functions. *Bull. Eng. Geol. Environ.* **2019**, *78*, 4017–4034. [CrossRef]
35. Taravatrooy, N.; Nikoo, M.R.; Sadegh, M.; Parvinnia, M. A hybrid clustering-fusion methodology for land subsidence estimation. *Nat. Hazards* **2018**, *94*, 905–926. [CrossRef]
36. Pourghasemi, H.R.; Mohseni Saravi, M. Land-Subsidence Spatial Modeling Using the Random Forest Data-Mining Technique. In *Spatial Modeling in GIS and R for Earth and Environmental Sciences*; Elsevier: Amsterdam, The Netherlands, 2019; pp. 147–159.
37. Rahmati, O.; Falah, F.; Naghibi, S.A.; Biggs, T.; Soltani, M.; Deo, R.C.; Cerdà, A.; Mohammadi, F.; Tien Bui, D. Land subsidence modelling using tree-based machine learning algorithms. *Sci. Total Environ.* **2019**, *672*, 239–252. [CrossRef]
38. Tomás, R.; Romero, R.; Mulas, J.; Marturià, J.J.; Mallorquí, J.J.; Lopez-Sanchez, J.M.; Herrera, G.; Gutiérrez, F.; González, P.J.; Fernández, J.; et al. Radar interferometry techniques for the study of ground subsidence phenomena: A review of practical issues through cases in Spain. *Environ. Earth Sci.* **2014**, *71*, 163–181. [CrossRef]

39. Del Soldato, M.; Farolfi, G.; Rosi, A.; Raspini, F.; Casagli, N. Subsidence Evolution of the Firenze–Prato–Pistoia Plain (Central Italy) Combining PSI and GNSS Data. *Remote Sens.* **2018**, *10*, 1146. [CrossRef]
40. Hakim, W.; Achmad, A.; Lee, C.-W. Land Subsidence Susceptibility Mapping in Jakarta Using Functional and Meta-Ensemble Machine Learning Algorithm Based on Time-Series InSAR Data. *Remote Sens.* **2020**, *12*, 3627. [CrossRef]
41. Polykretis, C.; Chalkias, C.; Ferentinou, M. Adaptive neuro-fuzzy inference system (ANFIS) modeling for landslide susceptibility assessment in a Mediterranean hilly area. *Bull. Eng. Geol. Environ.* **2019**, *78*, 1173–1187. [CrossRef]
42. Termeh, S.V.R.; Khosravi, K.; Sartaj, M.; Keesstra, S.D.; Tsai, F.T.-C.; Dijksma, R.; Pham, B.T. Optimization of an adaptive neuro-fuzzy inference system for groundwater potential mapping. *Hydrogeol. J.* **2019**, *27*, 2511–2534. [CrossRef]
43. Oh, H.-J.; Ahn, S.-C.; Choi, J.-K.; Lee, S. Sensitivity analysis for the GIS-based mapping of the ground subsidence hazard near abandoned underground coal mines. *Environ. Earth Sci.* **2011**, *64*, 347–358. [CrossRef]
44. Pacheco-Martínez, J.; Cabral-Cano, E.; Wdowinski, S.; Hernández-Marín, M.; Ortiz-Lozano, J.; Zermeño-de-León, M. Application of InSAR and Gravimetry for Land Subsidence Hazard Zoning in Aguascalientes, Mexico. *Remote Sens.* **2015**, *7*, 17035–17050. [CrossRef]
45. Arabameri, A.; Lee, S.; Tiefenbacher, J.P.; Ngo, P.T.T. Novel Ensemble of MCDM-Artificial Intelligence Techniques for Groundwater-Potential Mapping in Arid and Semi-Arid Regions (Iran). *Remote Sens.* **2020**, *12*, 490. [CrossRef]
46. Amani, M.; Ghorbanian, A.; Ahmadi, S.A.; Kakooei, M.; Moghimi, A.; Mirmazloumi, S.M.; Alizadeh Moghaddam, S.H.; Mahdavi, S.; Ghahremanloo, M.; Parsian, S.; et al. Google Earth Engine Cloud Computing Platform for Remote Sensing Big Data Applications: A Comprehensive Review. *IEEE J. Sel. Top. Appl. Earth Obs. Remote Sens.* **2020**, *13*, 5326–5350. [CrossRef]
47. Amani, M.; Kakooei, M.; Moghimi, A.; Ghorbanian, A.; Ranjgar, B.; Mahdavi, S.; Davidson, A.; Fisette, T.; Rollin, P.; Brisco, B.; et al. Application of Google Earth Engine Cloud Computing Platform, Sentinel Imagery, and Neural Networks for Crop Mapping in Canada. *Remote Sens.* **2020**, *12*, 3561. [CrossRef]
48. Ghorbanian, A.; Kakooei, M.; Amani, M.; Mahdavi, S.; Mohammadzadeh, A.; Hasanlou, M. Improved land cover map of Iran using Sentinel imagery within Google Earth Engine and a novel automatic workflow for land cover classification using migrated training samples. *ISPRS J. Photogramm. Remote Sens.* **2020**, *167*, 276–288. [CrossRef]
49. Ferretti, A.; Prati, C.; Rocca, F. Permanent scatterers in SAR interferometry. *IEEE Trans. Geosci. Remote Sens.* **2001**, *39*, 8–20. [CrossRef]
50. Kampes, B.M. *Radar Interferometry*; Springer: Dordrecht, The Netherlands, 2006.
51. Perissin, D.; Wang, Z.; Lin, H. Shanghai subway tunnels and highways monitoring through Cosmo-SkyMed Persistent Scatterers. *ISPRS J. Photogramm. Remote Sens.* **2012**, *73*, 58–67. [CrossRef]
52. Razavi Termeh, S.V.; Kornejady, A.; Pourghasemi, H.R.; Keesstra, S. Flood susceptibility mapping using novel ensembles of adaptive neuro fuzzy inference system and metaheuristic algorithms. *Sci. Total Environ.* **2018**, *615*, 438–451. [CrossRef]
53. Yesilnacar, E.; Topal, T. Landslide susceptibility mapping: A comparison of logistic regression and neural networks methods in a medium scale study, Hendek region (Turkey). *Eng. Geol.* **2005**, *79*, 251–266. [CrossRef]
54. Jang, J.-S.R. ANFIS: Adaptive-network-based fuzzy inference system. *IEEE Trans. Syst. Man. Cybern.* **1993**, *23*, 665–685. [CrossRef]
55. Tien Bui, D.; Pradhan, B.; Lofman, O.; Revhaug, I.; Dick, O.B. Landslide susceptibility mapping at Hoa Binh province (Vietnam) using an adaptive neuro-fuzzy inference system and GIS. *Comput. Geosci.* **2012**, *45*, 199–211. [CrossRef]
56. Takagi, T.; Sugeno, M. Fuzzy identification of systems and its applications to modeling and control. *IEEE Trans. Syst. Man. Cybern.* **1985**, *SMC-15*, 116–132. [CrossRef]
57. Atashpaz-Gargari, E.; Lucas, C. Imperialist competitive algorithm: An algorithm for optimization inspired by imperialistic competition. In Proceedings of the 2007 IEEE Congress on Evolutionary Computation, Singapore, 25–28 September 2007; pp. 4661–4667.
58. Razavi-Termeh, S.V.; Sadeghi-Niaraki, A.; Choi, S.-M. Gully erosion susceptibility mapping using artificial intelligence and statistical models. *Geomat. Nat. Hazards Risk* **2020**, *11*, 821–844. [CrossRef]
59. Esmaeil, A.G.; Farzad, H.; Ramin, R.; Caro, L. Colonial competitive algorithm: A novel approach for PID controller design in MIMO distillation column process. *Int. J. Intell. Comput. Cybern.* **2008**, *1*, 337–355. [CrossRef]
60. Mirjalili, S.; Mirjalili, S.M.; Lewis, A. Grey Wolf Optimizer. *Adv. Eng. Softw.* **2014**, *69*, 46–61. [CrossRef]
61. Rahmati, O.; Darabi, H.; Panahi, M.; Kalantari, Z.; Naghibi, S.A.; Ferreira, C.S.S.; Kornejady, A.; Karimidastenaei, Z.; Mohammadi, F.; Stefanidis, S.; et al. Development of novel hybridized models for urban flood susceptibility mapping. *Sci. Rep.* **2020**, *10*, 12937. [CrossRef]
62. Chen, W.; Hong, H.; Panahi, M.; Shahabi, H.; Wang, Y.; Shirzadi, A.; Pirasteh, S.; Alesheikh, A.A.; Khosravi, K.; Panahi, S.; et al. Spatial Prediction of Landslide Susceptibility Using GIS-Based Data Mining Techniques of ANFIS with Whale Optimization Algorithm (WOA) and Grey Wolf Optimizer (GWO). *Appl. Sci.* **2019**, *9*, 3755. [CrossRef]
63. Rahmati, O.; Pourghasemi, H.R.; Zeinivand, H. Flood susceptibility mapping using frequency ratio and weights-of-evidence models in the Golastan Province, Iran. *Geocarto Int.* **2016**, *31*, 42–70. [CrossRef]
64. Regmi, A.D.; Devkota, K.C.; Yoshida, K.; Pradhan, B.; Pourghasemi, H.R.; Kumamoto, T.; Akgun, A. Application of frequency ratio, statistical index, and weights-of-evidence models and their comparison in landslide susceptibility mapping in Central Nepal Himalaya. *Arab. J. Geosci.* **2014**, *7*, 725–742. [CrossRef]
65. Razavi-Termeh, S.V.; Khosravi, K.; Sadeghi-Niaraki, A.; Choi, S.-M.; Singh, V.P. Improving groundwater potential mapping using metaheuristic approaches. *Hydrol. Sci. J.* **2020**, *65*, 2729–2749. [CrossRef]

66. Razavi-Termeh, S.V.; Sadeghi-Niaraki, A.; Choi, S.-M. Ubiquitous GIS-Based Forest Fire Susceptibility Mapping Using Artificial Intelligence Methods. *Remote Sens.* **2020**, *12*, 1689. [CrossRef]
67. Pepe, A.; Bonano, M.; Zhao, Q.; Yang, T.; Wang, H. The Use of C-/X-Band Time-Gapped SAR Data and Geotechnical Models for the Study of Shanghai's Ocean-Reclaimed Lands through the SBAS-DInSAR Technique. *Remote Sens.* **2016**, *8*, 911. [CrossRef]
68. Kampes, B.M. Displacement Parameter Estimation Using Permanent Scatterer Interferometry. Ph.D. Thesis, Delft University of Technology, Delft, The Netherlands, September 2005.
69. Ren, H.; Feng, X. Calculating vertical deformation using a single InSAR pair based on singular value decomposition in mining areas. *Int. J. Appl. Earth Obs. Geoinf.* **2020**, *92*, 102115. [CrossRef]
70. Hooper, A.; Segall, P.; Zebker, H. Persistent scatterer interferometric synthetic aperture radar for crustal deformation analysis, with application to Volcán Alcedo, Galápagos. *J. Geophys. Res. Solid Earth* **2007**, *112*, B07407. [CrossRef]
71. Pourghasemi, H.R.; Beheshtirad, M. Assessment of a data-driven evidential belief function model and GIS for groundwater potential mapping in the Koohrang Watershed, Iran. *Geocarto Int.* **2015**, *30*, 662–685. [CrossRef]
72. Ye, S.; Xue, Y.; Wu, J.; Yan, X.; Yu, J. Progression and mitigation of land subsidence in China. *Hydrogeol. J.* **2016**, *24*, 685–693. [CrossRef]
73. Suganthi, S.; Elango, L.; Subramanian, S.K. Microwave D-InSAR technique for assessment of land subsidence in Kolkata city, India. *Arab. J. Geosci.* **2017**, *10*, 458. [CrossRef]
74. Moayedi, H.; Mehrabi, M.; Bui, D.T.; Pradhan, B.; Foong, L.K. Fuzzy-metaheuristic ensembles for spatial assessment of forest fire susceptibility. *J. Environ. Manag.* **2020**, *260*, 109867. [CrossRef]
75. Pourghasemi, H.R.; Razavi-Termeh, S.V.; Kariminejad, N.; Hong, H.; Chen, W. An assessment of metaheuristic approaches for flood assessment. *J. Hydrol.* **2020**, *582*, 124536. [CrossRef]

remote sensing

MDPI

Communication

A New Approach for Identification of Potential Rockfall Source Areas Controlled by Rock Mass Strength at a Regional Scale

Xueliang Wang [1,2,3,*], Haiyang Liu [1,2,3] and Juanjuan Sun [1,2,3]

1 Key Laboratory of Shale Gas and Geoengineering, Institute of Geology and Geophysics, Chinese Academy of Sciences, Beijing 100029, China; liuhaiyang@mail.iggcas.ac.cn (H.L.); sunjuanjuan17@mails.ucas.edu.cn (J.S.)
2 Innovation Academy for Earth Sciences, Chinese Academy of Sciences, Beijing 100029, China
3 University of Chinese Academy of Sciences, Beijing 100049, China
* Correspondence: wangxueliang@mail.iggcas.ac.cn

Abstract: The identification of rockfall source areas is a fundamental work for rockfall disaster prevention and mitigation. Based on the Culmann model, a pair of important indicators to estimate the state of slope stability is the relief and slope angles. Considering the limit of field survey and the increasing requirements for identification over a large area, a new approach using the relief–slope angle relationship to identify rockfall source areas controlled by rock mass strength at a regional scale is proposed in this paper. Using data from helicopter-based remote sensing imagery, a digital elevation model of 10 m resolution, and field work, historical rockfalls in the Wolong study area of Tibet where frequent rockfalls occur are identified. A clear inverse relationship between the relief and slope angles of historical rockfalls enables us to calculate the rock mass strength of the landscape scale by the Culmann model and the relief–slope angle relationship curve. Other parameters used in our proposed approach are calculated by ArcGIS and statistic tools. By applying our approach, the potential rockfall source areas in the study are identified and further zoned into three susceptibility classes that could be used as a reference for a regional rockfall susceptibility study. Using the space partition of historical rockfall inventory, our prediction result is validated. Most of the rockfall source areas (i.e., 71.92%) identified in the validation area are occupied by historical rockfalls, which proves the good prediction of our approach. The dominant uncertainty in this paper is derived from the process of calculating rock mass strength, defining the specific area for searching potential rockfall source areas, and the resolution of the digital elevation model.

Keywords: rockfall source areas; identification; relief; slope angle; rock mass strength; rockfall susceptibility

Citation: Wang, X.; Liu, H.; Sun, J. A New Approach for Identification of Potential Rockfall Source Areas Controlled by Rock Mass Strength at a Regional Scale. *Remote Sens.* **2021**, *13*, 938. https://doi.org/10.3390/rs13050938

Academic Editor: Paolo Mazzanti

Received: 7 February 2021
Accepted: 26 February 2021
Published: 3 March 2021

Publisher's Note: MDPI stays neutral with regard to jurisdictional claims in published maps and institutional affiliations.

1. Introduction

The identification of potential rockfall source areas is the first step in assessing rockfall susceptibility, hazard, risk, and determining rockfall disaster prevention and mitigation [1–7]. However, it is very difficult work to carry out in mountain areas, especially the steep and high-relief slopes, which are not accessible on site. Field investigation provides the most effective method to survey the distribution of potential rockfall source areas at a specific site [8]. Through field investigation, the engineering geology conditions controlling rockfall distribution, including the rock mass strength, orientation of structures, joint density, slope angle, relief, and the activity of tectonic faults, could be carefully studied on site [7,9,10]. Recent technologies, including unmanned aerial vehicles, terrestrial laser scanning, monitoring systems, photogrammetry, and point cloud analysis software tools (e.g., AgiSoft, Photoscan, and Coltop) [11–14], help researchers to acquire detailed information of the above conditions.

For the identification of potential rockfall source areas at a regional scale, the traditional field investigation is not as effective as that at a specific site because of its limited investigation scope and because it consumes much time and human resources. Hence, some researchers have developed regional rockfall susceptibility models based on ArcGIS to estimate the spatial distribution of rockfall using the causal factors of rockfall including lithology, terrain, elevation, faults, rainfall, and earthquakes [15–17]. Based on the results of rockfall susceptibility assessments, the whole area could be zoned into different areas with different susceptible degrees, which provides useful guidance for regional land use and rockfall disaster prevention plans. Alternatively, some researchers have identified rockfall sources at a regional scale by remote sensing interpretation technologies using multi-temporal aerial photos, helicopter-based remote sensing imagery, and high-resolution digital elevation model (DEM) [5,6,18]. Subsequently, regional locations of rockfall source areas could be identified for further rockfall kinematic modeling and predicting regional rockfall hazards [19].

Rock mass strength is thought to be the basic controlling factor of slope stability [20,21]. A rock slope with a low rock mass strength fractured by different types of fractures is prone to rockfall [7,22]. Based on Culmann's two-dimensional slope stability model [23], a hillslope is susceptible to rockfall if its relief, slope angle, or both are larger than the threshold values [24–27]. This means that the relief and slope angles are a pair of important indicators that could be used to identify rockfall source areas on the slopes whose stability is dominantly controlled by the rock mass strength. Until now, previous studies rarely focused on the approaches combining the relief and slope angles to identify rockfall source areas controlled by the rock mass strength at a regional scale [5,28]. Focusing on this issue, comprehensive technologies, including helicopter-based remote sensing imagery, a DEM with 10 m resolution, images from Google Earth, and field work, were adopted in this study. Lastly, a new approach, the procedures, and the application criteria for identifying rockfall source areas at a regional scale are proposed and were applied in the study area.

2. Materials and Methods

The Wolong (WL) area of Tibet where ideal geological conditions exist to investigate the characteristics of rock mass outcrops and the distribution of rockfall source areas was selected for this study (Figure 1). To identify potential rockfall source areas at a regional scale, a new approach combining the relief and slope angles based on the rock mass strength is proposed.

2.1. Study Area

The lithology of the area is mainly diorite and granite with a small component of gneiss [29,30]. Because of the steep terrain and the widely distributed tectonic structures, the main type of slope failure in the WL area is rockfall. Based on the rockfall scars left on the steep slopes and the rockfall deposits distributed widely, rockfalls in the study area occur frequently, and the dominant kinematic failure modes are toppling and planer sliding (Figure 2).

In this study, based on a power grid project, the helicopter-based remote sensing imagery obtained in 2017 and the 10 m resolution DEM over the study area were acquired with the help of the research group State Grid Corporation of China. Using the data of the complete study area, images of Google Earth, and field work, a rockfall inventory including 235 rockfall scars on bedrocks and 109 rockfall deposits was prepared (Figure 1). The rockfall scars were identified based on the fresh bedrock color left on the scars (Figure 2). Because many of the rockfall deposits were removed by the Yarlung Tsangpo River, and some of the rockfalls are adjacent, it was not possible to track each rockfall's deposits. Rockfall deposits at the foot of the slopes were identified based on the shape of the deposit (e.g., pyramid) and the identifiable rockfall blocks (e.g., meters) left on the deposits.

2.2. Framework of Identifying Potential Rockfall Source Areas

By using ArcGIS and statistic tools, our new approach proposed in this paper includes the framework as follows:

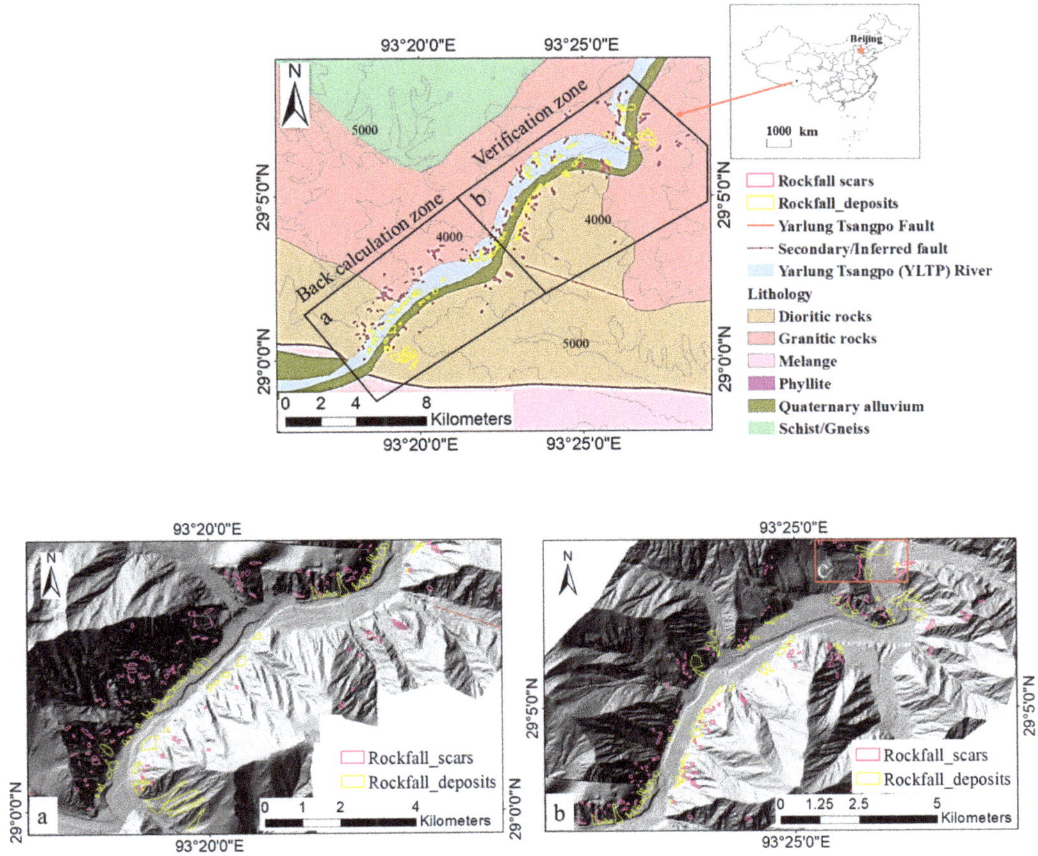

Figure 1. Simplified geology map of the study area. Two zones are separated by the space partition for validation of the prediction result, with (**a**) the back calculation zone and (**b**) the verification zone. (**c**) Sample area presented in Figure 2.

(1) Calculate the rock mass strength parameters (c_1, φ_1) of the bedrock slope at the landscape scale and build the relief–slope (R–S) angle relationship curve in the study area (Figure 3a).

(2) Measure the present relief (H) and slope angle (β) of each specific area (A) (Figure 3b) with the potential rockfall over the slope areas.

(3) Calculate the limit relief (H_c) of the specific area (A) by the Culmann model [23] (Figure 3a). The Culmann model indicates that the relief of the slope is controlled by the rock mass strength, and the slope angle (β) has the following relationship with the limit relief (maximum height) (H_c):

$$H_c = \frac{4C}{\gamma}\frac{sin\beta cos\varphi}{[1-\cos(\beta-\varphi)]} \tag{1}$$

where c is cohesion, γ is the bulk density, and φ is the internal friction angle.

(4) Compare the present relief (H) and the limit relief (H_c) of the specific area to estimate its state of stability. When the present relief of the bedrock is larger than the limit relief (Figure 3b), the bedrock is prone to generate rockfalls. Hence, the area whose relief exceeds its limit relief is identified as the potential rockfall source area.

(5) Using the procedure from steps (1) to (4), all areas of the slopes over the study area are searched, and their states of stability are estimated. Eventually, all the potential rockfall source are identified.

Figure 2. Samples of the rockfall scars and deposits on the helicopter-based remote sensing imagery with the location c in Figure 1b and the historical rockfalls with kinematic failure modes of toppling and planer sliding and their deposits at the foot of the slopes.

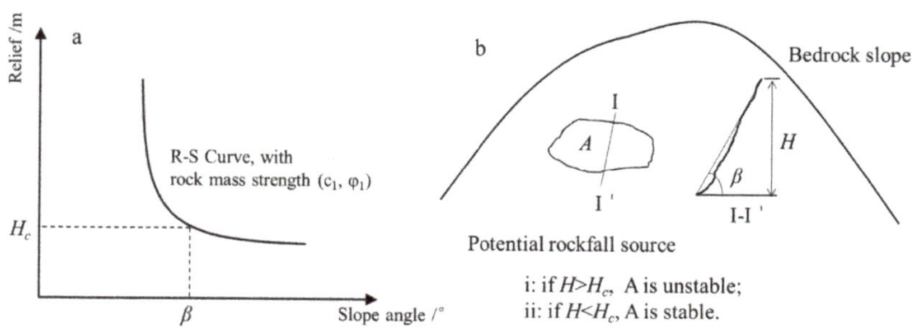

Figure 3. Sketch map describing the approach. (**a**) Slope angle versus relief for hillslope. (**b**) Sketch map of estimating the state of stability for the specific area (A).

Different from the availability of rock mass strength tested in the laboratory, the rock mass strength of bedrocks at the landscape scale is very difficult to test on site because of the lack of suitable approaches [20,21]. Schmidt and Montgomery [24] proposed an approach to estimate the rock mass strength parameters (c, φ) using the relief and slope angles of historical rockfalls with the Culmann model. Previous studies have applied the approach to calculate the parameters of rock mass strength using the relief and slope angles of historical landsides or rockfall scars in some cases [25–27,31]. In reference to previous studies, data of the relief and slope angles of 235 historical rockfall scars were first extracted by ArcGIS. Then, the parameters of the rock mass strength (c, φ) at the landscape scale in our study area were calculated under the precondition that the bedrock relief is controlled by the rock mass strength.

3. Results

The geometrical characteristics of historical rockfalls were analyzed to define the specific area (A) used in the new approach for searching rockfall source areas. Applying our proposed approach, the rockfall source areas were identified in the study area and were zoned into three susceptibility classes.

3.1. Geometrical Characteristics of Historical Rockfalls

Before calculating the relief and slope angles of each specific area, a suitable value of the specific area (A) should be first determined. To define the value of A in this study, the geometrical characteristics of 235 historical rockfalls were first analyzed by ArcGIS. The values of the relief and slope angles and the areas of historical rockfalls were measured separately. According to the statistical results, the relief of historical rockfalls is mainly distributed between 40 and 130 m, the slope angle is generally larger than 45°, and the area of each historical rockfall scar is generally less than 9000 m^2. All three groups of data show Gaussian distribution characteristics (Figure 4). The mean area of historical rockfalls is 5217 m^2, which was adopted as the specific area (A) for searching rockfall source areas of the slopes over the study area.

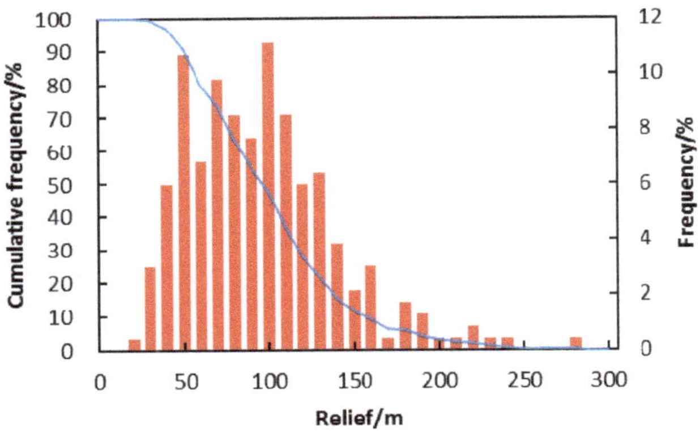

(**a**). Relief frequency statistics

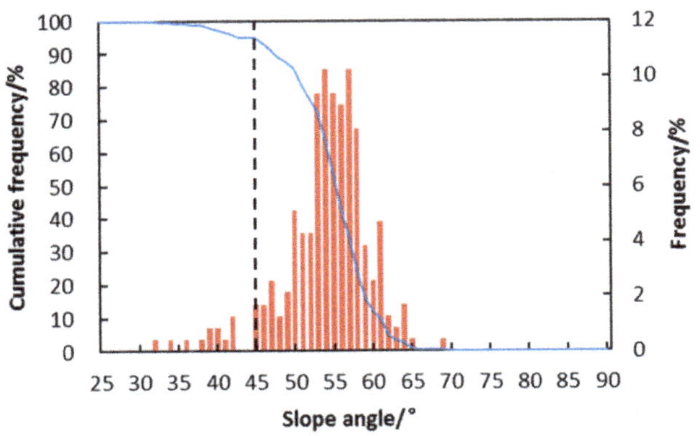

(**b**). Slope angle frequency statistics

Figure 4. *Cont.*

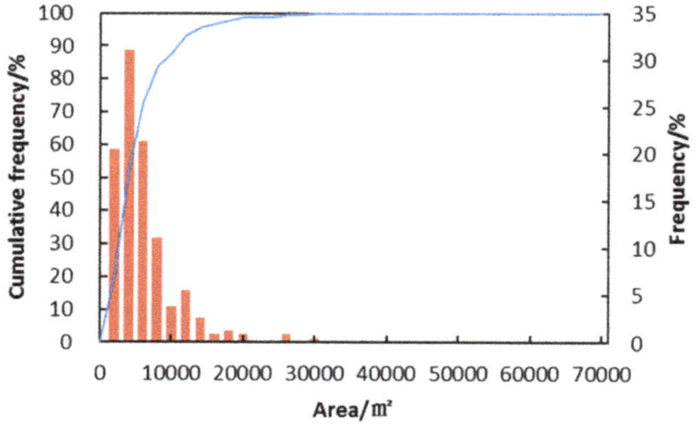

(**c**). Area frequency statistics

Figure 4. The geometrical statistics of historical rockfalls.

3.2. Locations of Identified Rockfall Sources

A clear inverse relationship between the relief and slope angles of 235 historical rockfalls (Figure 5) enabled us to calculate the rock mass strength at the landscape scale in this study. According to previous studies [25–27,31], the minimum and maximum parameters of rock mass strength can be estimated using the lower envelope and the upper envelope of the R–S curves obtained by data of the relief and slope angles of failed slopes. The upper envelope of the R–S curve (Figure 5) represents the maximum strength of the rock mass, and the lower envelope represents the minimum.

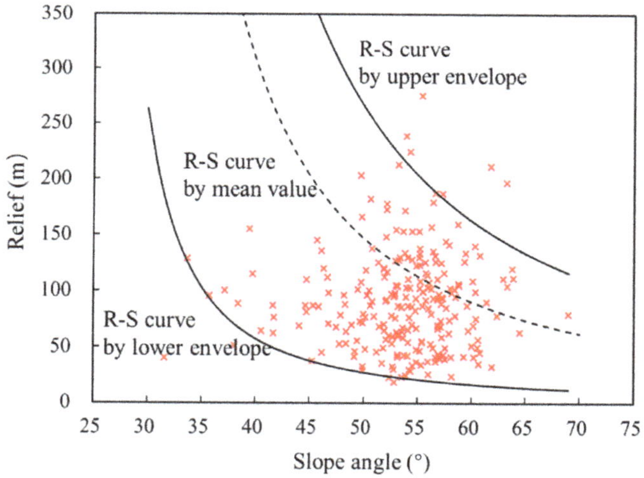

Figure 5. Rock mass strength fitting result and the relief–slope (R–S) angle relationship in the study area.

Applying the results of 235 historical rockfalls to Equation (1) by the optimization algorithm (Figure 5), the minimum and maximum rock mass cohesions (c) in the study area are 28 Pa and 270 kPa, respectively, and both internal friction angles (φ) are 23°. Using the upper envelope and the mean value of the upper envelope and the lower envelope, with the lower envelope corresponding to each slope angle, three R–S relationship curves were built as the threshold to determine if each specific area is stable or unstable (Figure 5).

The R–S curves could be regarded as three different estimates: the aggressive estimate, the moderate estimate, and the conservative estimate, corresponding to the upper envelope, the mean value, and the lower envelope curves, which could be selected by different aims or rockfall disaster prevention and mitigation strategies.

Based on our approach, to identify the rockfall source area is to identify the area whose present relief exceeds the limit relief corresponding to its slope angle (Figure 3). Hence, we calculated the limit relief corresponding to each slope angle (Table 1) using the three R–S curves presented in Figure 5. Considering the actual slope distribution of the historical rockfalls (Figure 4b), we mainly focused on the slopes with angles larger than 45°. Comparing the present relief and the limit relief in each specific area of the slopes in the study area, we obtained the rockfall source areas in ArcGIS (Figure 6). More rockfall source areas are distributed at the lower parts of the slopes whose slope angles are relatively bigger, which is probably affected by the intense incision of the Yarlung Tsangpo River [32].

Table 1. Limit relief of the specific areas corresponding to the slope angle.

Slope Angle (°)	Limit Relief (m)			Slope Angle (°)	Limit Relief (m)			Slope Angle (°)	Limit Relief (m)		
	C-U	C-M	C-L		C-U	C-M	C-L		C-U	C-M	C-L
46.0	35.5	189.1	342.7	61.0	16.2	86.2	156.2	76.0	9.6	50.9	92.3
47.0	33.2	176.7	320.3	62.0	15.6	82.8	150.0	77.0	9.3	49.4	89.5
48.0	31.2	165.8	300.4	63.0	15.0	79.6	144.2	78.0	9.0	47.9	86.9
49.0	29.3	155.8	282.4	64.0	14.4	76.6	138.8	79.0	8.7	46.5	84.3
50.0	27.6	146.9	266.1	65.0	13.9	73.7	133.6	80.0	8.5	45.2	81.9
51.0	26.1	138.7	251.4	66.0	13.4	71.1	128.8	81.0	8.3	43.9	79.6
52.0	24.7	131.3	238.0	67.0	12.9	68.5	124.2	82.0	8.0	42.7	77.3
53.0	23.4	124.6	225.7	68.0	12.4	66.2	119.9	83.0	7.8	41.5	75.2
54.0	22.2	118.4	214.5	69.0	12.0	63.9	115.8	84.0	7.6	40.3	73.1
55.0	21.2	112.7	204.1	70.0	11.6	61.8	111.9	85.0	7.4	39.2	71.1
56.0	20.2	107.4	194.6	71.0	11.2	59.7	108.2	86.0	7.2	38.2	69.2
57.0	19.3	102.5	185.8	72.0	10.9	57.8	104.7	87.0	7.0	37.2	67.3
58.0	18.4	98.0	177.6	73.0	10.5	55.9	101.4	88.0	6.8	36.2	65.5
59.0	17.6	93.8	170.0	74.0	10.2	54.2	98.2	89.0	6.6	35.2	63.8
60.0	16.9	89.9	162.9	75.0	9.9	52.5	95.2	90.0	6.5	34.3	62.2

Note: for the slope angle, the front part is open interval, and latter part is closed interval. For example, 46 represents the range of (45, 46]. C-U: R–S curve by upper envelope; C-M: R–S curve by mean value; C-L: R–S curve by lower envelope.

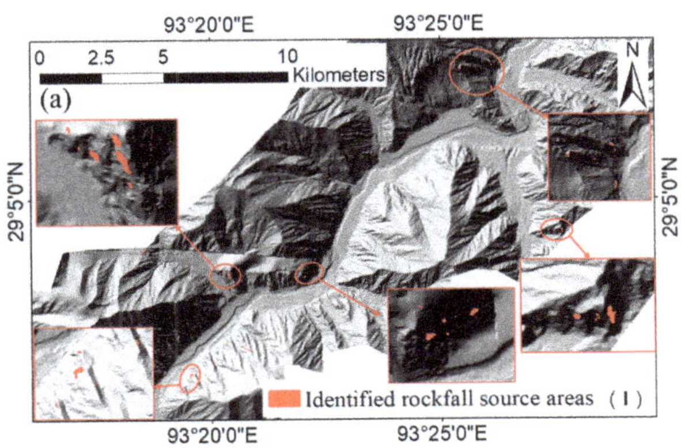

Figure 6. *Cont.*

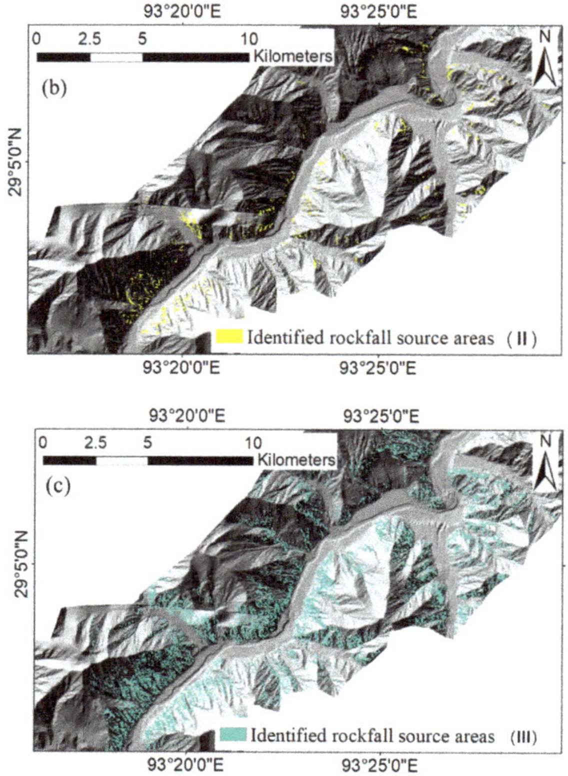

Figure 6. The identification results of the rockfall source areas in the study area. The results of (**a–c**) correspond to high (I), medium (II) and low (III) rockfall susceptibility classes, respectively.

3.3. Zoning Map of Rockfall Susceptibility

In rockfall risk analysis, the rockfall susceptibility assessment is the first step to carry out [16,28,33]. The rockfall susceptibility map helps to highlight the spatial distribution of potentially unstable slopes [34], which is usually zoned into different susceptibility classes, e.g., high, medium, and low susceptibility, to represent different rockfall susceptible degrees of slopes. Based on the sketch in Figure 3, if $H > H_c$, the area A on the slope is unstable, which means it is prone to rockfall in the future. The larger H is than Hc, the higher the possibility of rockfall in area A, and hence the higher rockfall susceptibility of area A. To provide a reference for the susceptibility study following the identification of the rockfall source areas by our approach, the rockfall source areas were zoned into three susceptibility classes, high (I), medium (II), and low susceptibility (III) areas, using the upper envelope, the mean envelope, and the lower envelope in Figure 5. In this way, the regional rockfall susceptibility maps were produced in the study area (Figure 6).

The slope angle and the elevation of all the susceptibility classes were analyzed, and their frequency distributions were obtained. According to the results (Figure 7), the rockfall source areas with different susceptibility classes have different ranges of slope angle. The rockfall source areas within the high susceptibility class are mainly distributed on the slopes with the angles of 60–66°, those of medium susceptibility are distributed on the slopes with the angles of 54–61°, and those of low susceptibility are distributed on the slopes with the angles of 46–55°. However, based on the distribution statistics of elevation in Figure 7, no obvious relationship between elevation and susceptibility classes was observed in this study.

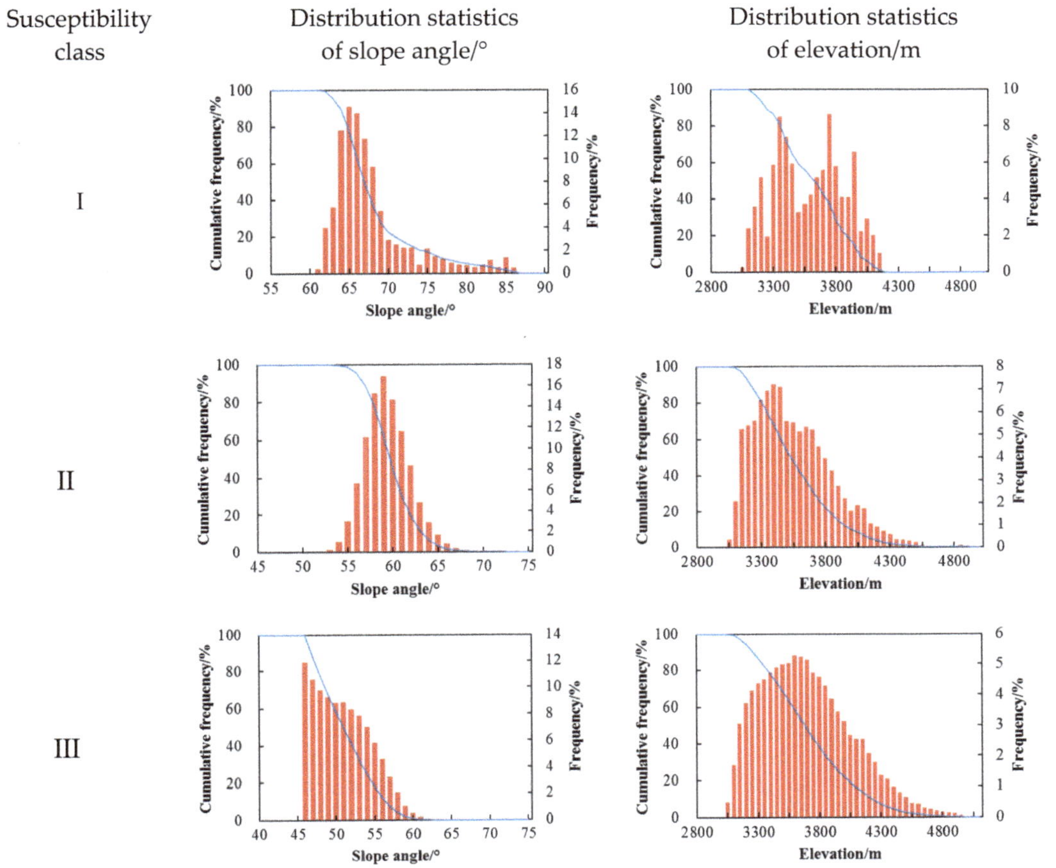

Figure 7. The distribution of slope and elevation of the identified rockfall source areas within different susceptibility classes.

4. Discussion

It is important to validate the prediction results in prediction modeling [35]. Hence, the validation of our proposed new approach is discussed here. The uncertainty in our new approach is mainly derived from the process of calculating the rock mass strength parameters, defining the specific area (A) used for searching the rockfall source areas, and the resolution of the DEM.

4.1. Validation

Proper validation should be based on the comparison between the prediction results and the actual characteristics affected by future rockfalls [16,35]. The selection of approaches depends on the characteristics of dataset. In this study, the space partition to separate our rockfall inventory into two groups was chosen since information was lacking related to the time of the 235 historical rockfalls (Figure 8). To validate the prediction result, the rockfall inventory was partitioned into two groups. One group was used for prediction (Figure 1a) and the other was used for validation (Figure 1b).

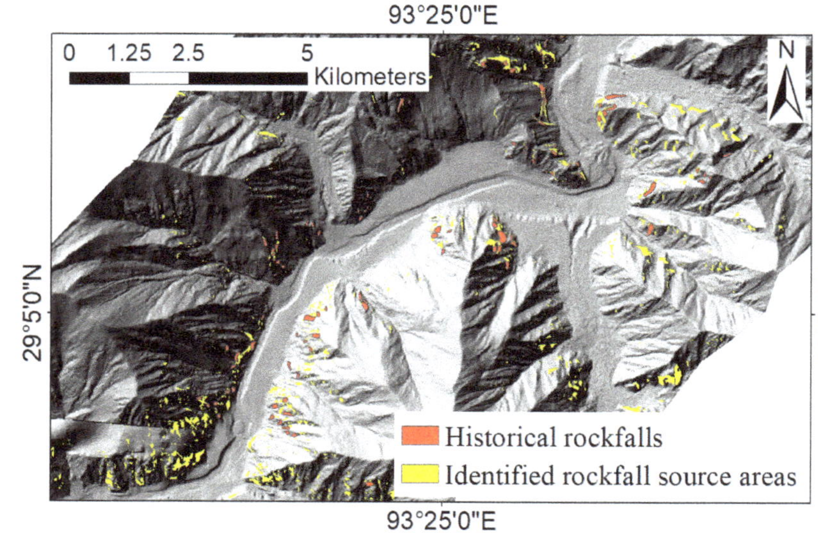

Figure 8. The validation of the prediction result obtained by using one group of historical rockfalls separated by the space partition (Figure 1a).

The first group with 118 historical rockfalls (Figure 1a) was adopted to calculate the rock mass strength and build the R–S relationship curve. According to the fitting result of the selected 118 rockfall sources, the minimum and maximum rock mass cohesions (c) are 28 and 230 kPa, respectively, and both internal friction angles (φ) are 22°. To present a common case of application, we used the mean value of the upper envelope and the lower envelope to build the R–S relationship curve in this paper. Based on the procedure of our approach (Figure 3), the rockfall source areas were obtained. The second group with 117 historical rockfalls (Figure 1b) was used to validate the prediction result (Figure 8). The validation result demonstrated that 117 historical rockfalls occupy 71.92% of the rockfall source areas in the validation area predicted by our new approach, which proves to be a good prediction.

Besides, the prediction result was validated with field work. Most of the rockfall source areas identified by our approach are distributed on slopes with high relief and steep terrain. This is consistent with the observation of rockfall distribution on the maps of DEM and hillshade (Figure 8). Most of the historical rockfall deposits are distributed at the foot of the slopes with identified rockfall source areas, proving that the identified rockfall source areas are distributed rockfalls that occurred in the past, and that unstable slopes are prone to rockfall in the future [19].

4.2. Uncertainty Analysis

An accurate calculation of rock mass strength parameters is the fundamental work in our new proposed approach. For example, for a slope with a specific relief (H) and slope angle (β), the bigger the rock mass strength estimated, the higher the value of the limit relief (Hc) calculated by Equation (1), and hence the bigger the difference between Hc and H. For the stability of the slope, more stable (i.e., the lower rockfall susceptibility) slopes were predicted. In other words, if the rock mass strength parameters are overestimated, the slope would be prone to rockfall with an incorrect prediction of low susceptibility. In our new approach, the more accurate estimated rock mass strength parameters will greatly improve the accuracy of the R–S relationship curve, and hence the prediction results of rockfall source areas. However, it is very difficult to quantitatively estimate the rock mass strength at the landscape. An in situ test is recognized as one of the most reliable methods to obtain rock mass strength [10,20], which proves to be very difficult to carry out in high mountain areas. In this paper, the back analysis for rock mass strength using historical rockfalls based on the Culmann model is adopted, whose accuracy depends on the reliability of the identification, the boundary of historical rockfalls, and the uncertainty in fitting the data.

According to the procedure of our approach and the sketch in Figure 3, the specific area (A) is also important in affecting the accuracy of the prediction result. For example, if a smaller area than the real one was defined to search the potential rockfall source areas, a lower relief (H) than the real value of a potential source area would be obtained. For the same slope angle and corresponding estimated Hc, the area would mistakenly be regarded as more stable (Figure 3) than its real stability state. In this paper, the mean area of historical rockfalls is defined as the specific area (A) for searching the potential rockfall source areas. The definition of the parameter could be further studied in the future.

The resolution of DEM plays an important role in controlling the accuracy of the result [5] because it is the basic data in almost every data process of our new approach. The higher the resolution of the DEM, the higher the accuracy of the values of the relief and slope angles of historical rockfalls and the potential rockfall source areas. However, it is not easy to acquire a high-quality DEM in a large study area currently.

5. Conclusions

The main type of rock mass failure in the Wolong area of Tibet is rockfall. Using data from helicopter-based remote sensing imagery, a DEM with 10 m resolution of the study area, images from Google Earth, and field work, a rockfall inventory including 235 rockfalls scars on bedrocks and 109 rockfall deposits was prepared. According to the statistical results, the relief of historical rockfalls is mainly distributed between 40 and 130 m, the slope angle is generally larger than 45°, and the area of each historical rockfall scar is generally less than 9000 m^2. A clear inverse relationship between the relief and slope angles of historical rockfalls enabled us to calculate the rock mass strength at the landscape scale base on the Culmann model, obtaining the minimum and maximum rock mass cohesions (c) in the study area from 28 to 270 kPa, respectively, and the internal friction angle of 23°.

Required by the actual needs of identification on high and steep slopes, this paper proposes a new approach using the relief–slope angle relationship to identify the rockfall source areas controlled by the rock mass strength on a regional scale. Based on historical rockfalls and a high-resolution DEM, we obtained the parameters used in our proposed approach. By applying our approach, the potential rockfall source areas in the study area were identified and further zoned into three susceptibility classes that could be used as a reference for the study of regional rockfall susceptibility assessment. According to the results, rockfall source areas within the high susceptibility class are mainly distributed on the slopes with the angles of 60–66°, those of medium susceptibility are distributed on the slopes with the angles of 54–61°, and those of low susceptibility are distributed on the slopes with the angles of 46–55°.

By the space partition and the field work, our prediction result was validated. Most of the rockfall source areas (i.e., 71.92%) identified in the validation area are occupied by actual historical rockfalls, which proves the accurate prediction ability of our approach. The locations of the rockfall source areas obtained in this paper could provide reference for actual rockfall disaster prevention and mitigation in the study area. Our proposed approach could be used to identify the rockfall source areas in the regional areas that are not accessible. In the paper, the dominant uncertainty is derived from the process of calculating the rock mass strength parameters, the process of defining the specific area (A) that is used for searching the rockfall source areas, and the resolution of the DEM. Many more studies estimating the rock mass strength at the landscape scale and defining the specific area (A) are necessary in the future.

Author Contributions: Conceptualization, X.W.; methodology X.W., H.L.; formal analysis, H.L., J.S.; writing—original draft preparation, X.W., H.L.; writing—review and editing, X.W.; visualization, H.L.; supervision, X.W.; project administration, J.S.; funding acquisition, X.W. All authors have read and agreed to the published version of the manuscript.

Funding: This work was supported by the Second Tibetan Plateau Scientific Expedition and Research Program (STEP) (Grant No. 2019QZKK0904), the Strategic Priority Research Program of the Chinese Academy of Sciences (Grant No. XDA23090402), the Application of Synthetic Aperture Radar-Based Geological Hazard Analysis Technology on the Strategic Electricity Transmission Passage of Sichuan-Tibet Plateau (Grant No. 52199918000C).

Acknowledgments: We thank the program research group of the State Grid Corporation of China for providing the 10 m resolution topographic DEM data.

Conflicts of Interest: The authors declare no conflict of interest. The funders had no role in the design of the study; in the collection, analyses, or interpretation of data; in the writing of the manuscript; or in the decision to publish the results.

References

1. Mazzoccola, D.F.; Hudson, J.A. A comprehensive method of rock mass characterization for indicating natural slope instability. *Q. J. Eng. Geol. Hydrogeol.* **1996**, *29*, 37–56. [CrossRef]
2. Bunce, C.M.; Cruden, D.M.; Morgenstern, N.R. Assessment of the hazard from rock fall on a highway. *Can. Geotech. J.* **1997**, *34*, 344–356. [CrossRef]
3. Guzzetti, F.; Carrara, A.; Cardinali, M.; Reichenbach, P. Landslide hazard evaluation: A review of current techniques and their application in a multi-scale study, Central Italy. *Geomorphology* **1999**, *31*, 181–216. [CrossRef]
4. Gokceoglu, C.; Sonmez, H.; Ercanoglu, M. Discontinuity controlled probabilistic slope failure risk maps of the Altindag (settlement) region in Turkey. *Eng. Geol.* **2000**, *55*, 277–296. [CrossRef]
5. Loye, A.; Jaboyedoff, M.; Pedrazzini, A. Identification of potential rockfall source areas at a regional scale using a DEM-based geomorphometric analysis. *Nat. Hazards Earth Syst. Sci.* **2009**, *9*, 1643–1653. [CrossRef]
6. Liu, H.; Wang, X.; Liao, X.; Sun, J.; Zhang, S. Rockfall Investigation and Hazard Assessment from Nang County to Jiacha County in Tibet. *Appl. Sci.* **2019**, *10*, 247. [CrossRef]
7. Wang, X.; Frattini, P.; Stead, D.; Sun, J.; Liu, H.; Valagussa, A.; Li, L. Dynamic rockfall risk analysis. *Eng. Geol.* **2020**, *272*, 105622. [CrossRef]
8. Wang, X.; Zhang, L.; Wang, S.; Agliardi, F.; Frattini, P.; Crosta, G.B.; Yang, Z. Field investigation and rockfall hazard zonation at the Shijing Mountains Sutra caves cultural heritage (China). *Environ. Earth Sci.* **2011**, *66*, 1897–1908. [CrossRef]
9. Stead, D.; Wolter, A. A critical review of rock slope failure mechanisms: The importance of structural geology. *J. Struct. Geol.* **2015**, *74*, 1–23. [CrossRef]
10. Saroglou, C.; Qi, S.; Guo, S.; Wu, F. ARMR, a new classification system for the rating of anisotropic rock masses. *Bull. Int. Assoc. Eng. Geol.* **2019**, *78*, 3611–3626. [CrossRef]
11. Jaboyedoff, M.; Metzger, R.; Oppikofer, T.; Couture, R.; Derron, M.H.; Locat, J.; Turmel, D. New insight techniques to analyze rock-slope relief using DEM and 3D-imaging cloud points: COLTOP 3D software. In Proceedings of the 1st Canada-US Rock mechanics Symposium, Vancouver, BC, Canada, 27–31 May 2007; pp. 27–31. [CrossRef]
12. DiMatteo, L.; Romeo, S.; Kieffer, D.S. Rockfall analysis in Alpine area by using a reliable integrated monitoring system: Results from the Ingelsberg slope (Salzburg Land, Austria). *Bull. Eng. Geol. Environ.* **2017**, *76*, 413–420. [CrossRef]
13. Mazzanti, P.; Schilirò, L.; Martino, S.; Antonielli, B.; Brizi, E.; Brunetti, A.; Margottini, C.; Mugnozza, G.S. The Contribution of Terrestrial Laser Scanning to the Analysis of Cliff Slope Stability in Sugano (Central Italy). *Remote Sens.* **2018**, *10*, 1475. [CrossRef]
14. Kovanič, L'.; Blistan, P.; Urban, R.; Štroner, M.; Blišťanová, M.; Bartoš, K.; Pukanská, K. Analysis of the Suitability of High-Resolution DEM Obtained Using ALS and UAS (SfM) for the Identification of Changes and Monitoring the Development of Selected Geohazards in the Alpine Environment—A Case Study in High Tatras, Slovakia. *Remote Sens.* **2020**, *12*, 3901. [CrossRef]
15. Cascini, L. Applicability of landslide susceptibility and hazard zoning at different scales. *Eng. Geol.* **2008**, *102*, 164–177. [CrossRef]
16. Wang, X.L.; Zhang, L.Q.; Wang, S.J.; Lari, S. Regional landslide susceptibility zoning with considering the aggregation of landslide points and the weights of factors. *Landslides* **2014**, *11*, 399–409. [CrossRef]
17. Budetta, P.; De Luca, C.; Nappi, M. Quantitative rockfall risk assessment for an important road by means of the rockfall risk management (RO.MA.) method. *Bull. Int. Assoc. Eng. Geol.* **2016**, *75*, 1377–1397. [CrossRef]
18. Duarte, R.M.; Marquínez, J. The influence of environmental and lithologic factors on rockfall at a regional scale: An evaluation using GIS. *Geomorphology* **2002**, *43*, 117–136. [CrossRef]
19. Frattini, P.; Crosta, G.; Carrara, A.; Agliardi, F. Assessment of rockfall susceptibility by integrating statistical and physically-based approaches. *Geomorphology* **2008**, *94*, 419–437. [CrossRef]
20. Hoek, E. Strength of jointed rock masses. *Géotechnique* **1983**, *33*, 187–223. [CrossRef]
21. Gudmundsson, A. *Rock Fractures in Geological Processes*; Cambridge University Press (CUP): Cambridge, UK, 2011.

22. Francioni, M.; Antonaci, F.; Sciarra, N.; Robiati, C.; Coggan, J.; Stead, D.; Calamita, F. Application of Unmanned Aerial Vehicle Data and Discrete Fracture Network Models for Improved Rockfall Simulations. *Remote Sens.* **2020**, *12*, 2053. [CrossRef]
23. Culmann, C. *Die Graphische Statik*; Meyer and Zeller: Zurich, Switzerland, 1875; 644p.
24. Schmidt, K.M.; Montgomery, D.R. Limits to Relief. *Science* **1995**, *270*, 617–620. [CrossRef]
25. Montgomery, D.R.; Brandon, M.T. Topographic controls on erosion rates in tectonically active mountain ranges. *Earth Planet. Sci. Lett.* **2002**, *201*, 481–489. [CrossRef]
26. Crosta, G.B.; Utili, S.; De Blasio, F.V.; Castellanza, R. Reassessing rock mass properties and slope instability triggering conditions in Valles Marineris, Mars. *Earth Planet. Sci. Lett.* **2014**, *388*, 329–342. [CrossRef]
27. DiBiase, R.A.; Rossi, M.W.; Neely, A.B. Fracture density and grain size controls on the relief structure of bedrock landscapes. *Geology* **2018**, *46*, 399–402. [CrossRef]
28. Crosta, G.B.; Agliardi, F. A methodology for physically based rockfall hazard assessment. *Nat. Hazards Earth Syst. Sci.* **2003**, *3*, 407–422. [CrossRef]
29. Quidelleur, X.; Grove, M.; Lovera, O.M.; Harrison, T.M.; Yin, A.; Ryerson, F.J. Thermal evolution and slip history of the Renbu Zedong Thrust, southeastern Tibet. *J. Geophys. Res. Space Phys.* **1997**, *102*, 2659–2679. [CrossRef]
30. Xu, Z.-Q.; Dilek, Y.; Yang, J.-S.; Liang, F.-H.; Liu, F.; Ba, D.-Z.; Cai, Z.-H.; Li, G.-W.; Dong, H.-W.; Ji, S.-C. Crustal structure of the Indus–Tsangpo suture zone and its ophiolites in southern Tibet. *Gondwana Res.* **2015**, *27*, 507–524. [CrossRef]
31. Burbank, D.W.; Leland, J.; Fielding, E.; Anderson, R.S.; Brozovic, N.; Reid, M.R.; Duncan, C. Bedrock incision, rock uplift and threshold hillslopes in the northwestern Himalayas. *Nat. Cell Biol.* **1996**, *379*, 505–510. [CrossRef]
32. Finnegan, N.J.; Hallet, B.; Montgomery, D.R.; Zeitler, P.K.; Stone, J.O.; Anders, A.M.; Liu, Y.P. Coupling of rock uplift and river incision in the Namche Barwa–Gyala Peri massif, Tibet. *Geol. Soc. Am. Bull.* **2008**, *120*, 142–155. [CrossRef]
33. Fell, R.; Cororninas, J.; Bonnard, C.; Cascini, L.; Leroi, E.; Savage, W.Z. Guidelines for landslide susceptibility, hazard and risk-zoning for land use planning. *Eng. Geol.* **2008**, *102*, 85–98. [CrossRef]
34. Ayalew, L.; Yamagishi, H.; Maruib, H.; Kannoc, T. Landslides in Sado Island of Japan: Part II. GIS-based susceptibility mapping with comparisons of results from two methods and verifications. *Eng. Geol.* **2005**, *81*, 432–445. [CrossRef]
35. Chung, C.F.; Fabbri, A.G. Validation of spatial prediction models for landslide hazard mapping. *Nat. Hazard.* **2003**, *30*, 451–472. [CrossRef]

remote sensing MDPI

Article

Rapid Assessment of Hillslope Erosion Risk after the 2019–2020 Wildfires and Storm Events in Sydney Drinking Water Catchment

Xihua Yang [1,2,*], Mingxi Zhang [2], Lorena Oliveira [3], Quinn R. Ollivier [3], Shane Faulkner [3] and Adam Roff [1,4]

1 New South Wales Department of Planning, Industry and Environment, Parramatta, NSW 2150, Australia; adam.roff@environment.nsw.gov.au

2 School of Life Sciences, Faculty of Science, University of Technology Sydney, Broadway, NSW 2007, Australia; mingxi.zhang@student.uts.edu.au

3 Water New South Wales, Parramatta, NSW 2150, Australia; lorena.oliveira@waternsw.com.au (L.O.); quinn.ollivier@waternsw.com.au (Q.R.O.); shane.faulkner@waternsw.com.au (S.F.)

4 School of Environmental and Life Sciences, University of Newcastle, Callaghan, NSW 2308, Australia

* Correspondence: xihua.yang@environment.nsw.gov.au; Tel.: +61-2-9895-6517

Received: 2 October 2020; Accepted: 17 November 2020; Published: 20 November 2020

Abstract: The Australian Black Summer wildfires between September 2019 and January 2020 burnt many parts of eastern Australia including major forests within the Sydney drinking water catchment (SDWC) area, almost 16.000 km^2. There was great concern on post-fire erosion and water quality hazards to Sydney's drinking water supply, especially after the heavy rainfall events in February 2020. We developed a rapid and innovative approach to estimate post-fire hillslope erosion using weather radar, remote sensing, Google Earth Engine (GEE), Geographical Information Systems (GIS), and the Revised Universal Soil Loss Equation (RUSLE). The event-based rainfall erosivity was estimated from radar-derived rainfall accumulations for all storm events after the wildfires. Satellite data including Sentinel-2, Landsat-8, and Moderate Resolution Imaging Spectroradiometer (MODIS) were used to estimate the fractional vegetation covers and the RUSLE cover-management factor. The study reveals that the average post-fire erosion rate over SDWC in February 2020 was 4.9 Mg ha^{-1} month^{-1}, about 30 times higher than the pre-fire erosion and 10 times higher than the average erosion rate at the same period because of the intense storm events and rainfall erosivity with a return period over 40 years. The high post-fire erosion risk areas (up to 23.8 Mg ha^{-1} month^{-1}) were at sub-catchments near Warragamba Dam which forms Lake Burragorang and supplies drinking water to more than four million people in Sydney. These findings assist in the timely assessment of post-fire erosion and water quality risks and help develop cost-effective fire incident management and mitigation actions for such an area with both significant ecological and drinking water assets. The methodology developed from this study is potentially applicable elsewhere for similar studies as the input datasets (satellite and radar data) and computing platforms (GEE, GIS) are available and accessible worldwide.

Keywords: wildfires; hillslope erosion; satellite imagery; rainfall erosivity; RUSLE

1. Introduction

Because of its mostly hot, dry, and erratic climate, wildfires in Australia and many parts of the world are frequently occurring events during the hotter months. Between 2017 and 2019, severe drought developed across much of eastern and inland Australia including Queensland, New South Wales (NSW), and Victoria, also extending into parts of South Australia and Western Australia. As at late 2019, many regions of Australia were still in significant drought, contributing to water restrictions and extreme fire conditions [1]. The 2019–20 Australian bushfire season, colloquially known as the "Black Summer," began with several serious uncontrolled fires in June 2019, an early start to the wildfire season as it normally starts in early October in NSW [2]. Throughout late 2019 and early 2020 these wildfires both multiplied and combined, to create mega fires that burnt predominantly throughout the southeast of the country, peaking during December–January, and having been since contained and/or extinguished [2–4].

In NSW, the three-year (2017–2019) severe drought had left forests tinder dry, facilitating rapid expansion of fires across the state. Sydney—the nation's most populated city (about 5 million residents)—had been under water restrictions since late 2019, when its dams fell below 45% capacity. The wildfires in NSW between September 2019 to January 2020 (the 2019-20 fires) were unprecedented in their extent and intensity [4]. The NSW Rural Fire Service (RFS) reported that the 2019–2020 fires burnt 5.4 million hectares (including ~11,264 bush or grass fires across 6.7% of the State) and destroyed 2439 homes [3].

Wildfires not only create risk to lives, infrastructure, and properties, but also cause land and ecosystem degradation through increased soil erosion. The subsequent sediment deposition in rivers, lakes, and reservoirs is of great concern for drinking water quality, aquatic habitat, and environmental degradation. When large rainfall events occur in a short period of time, runoff will wash a lot of ash and sediment into waterways and dams. Rain also transports other contaminants, such as building debris, dead animals, and pollutants from fire retardant. Soil sediment, ash, and other contaminants pose a hazard to human health when mobilized in drinking water catchments. It is therefore necessary to quantitatively estimate soil erosion after severe wildfires in order to assess the extent and magnitude of post-fire soil erosion risk and the effectiveness of any rehabilitation or mitigation actions [5].

The Bureau of Meteorology (BoM) recorded 51 mm rainfall in January 2020 and 337 mm in February 2020 over the Sydney drinking water catchment (SDWC) areas. The rainfall events (especially in February) were widespread and helped extinguish the wildfires. However, the rainfall also caused severe runoff and hillslope erosion. As rainfall intensity is the leading agent contributing to greater erosion rates, and the subsequent effects on water quality, it is necessary to obtain timely rainfall information for fire recovery and erosion control practices. The rapid assessment of these risks and timely mitigation actions can significantly reduce risks to public safety, infrastructure, and the environment [5].

Hillslope erosion (including sheet and rill erosion) is the major form of water erosion and the dominant source of sediment in waterways in Australia and many parts of the world [5–7]. Monitoring of hillslope erosion after wildfires may be by direct measurement, such as flumes and sediment traps, or from tipping bucket sampling techniques. However, these techniques are commonly expensive, episodic, and impractical to apply across a large catchment. Observers with different levels of expertise make it difficult to consistently measure or predict long-term soil loss or erosion risk. A common model and consistent datasets are required for reliable soil loss prediction to provide consistent and continuous erosion information for short- and long-term soil and water quality management [5–7].

Many erosion models have been developed to predict soil loss in different regions in the world. Among them, the Universal Soil Loss Equation [8] or the revised USLE (RUSLE) [9] and Water Erosion Prediction Project (WEPP) [10] are widely used to estimate long-term average soil loss rates using rainfall, soil, topography, and land cover and management as inputs. These estimates have been used in long-term planning and soil condition assessment, and have been applied in Australia continent [11,12]. However, the impact of highly variable and extreme rainfall events cannot be estimated using the long-term averages erosion. Severe erosion and sediment transport are often caused by short but strong storm events [5,6].

It is increasingly important to construct an event-based erosion model to predict the extreme erosion risks given the predicted increase in climate variability and fire intensity [13]. Ideally, areas at risk of severe damage should be identified and prioritized for assessment and remediation in the event of fire [14,15], especially for drinking water catchments. Yet, there are few studies and applications of erosion models on estimation of storm event-based rainfall erosivity and erosion. One of the major limiting factors is the lack of rainfall data at high spatial and temporal resolutions, as well as computational capacity of large quantity of spatial data [16,17]. It remains a research challenge to provide quantitative and timely assessments of hillslope erosion after wildland fires during individual storm events to support water and catchment management [5].

Weather radar data have very high temporal resolution (15 min or less) and spatial resolution (1 km or less) with the potential for estimating event-based rainfall erosivity or the 30-min rainfall erosivity index (EI30). Such data sets have recently been applied in event-based erosion modeling to compute a spatial EI30 index [18] and to monitor erosion after wildfires at Warrumbungle National Park in NSW, Australia [5,6]. Its application over a large area or catchment at near real-time of rainfall events is still a research and implementation challenge. The emerging new technologies, such as machine learning, Google Earth Engine (GEE) processing, and high-resolution (spatial, spectral and temporal) satellite remote sensing, can be employed for the efficient implementation of the event-based erosion model, especially when large catchments or regions are concerned [19]. These together can provide timely erosion risk estimation that explicitly link soil loss and sedimentation with vegetation cover and land management, especially in events of wildfires and storms. The improved capability to predict impacts on water quality as a result of wildfires and erosion helps prioritize the mitigation actions in the drinking water catchments.

The aim of this study was to develop a rapid approach to assess the post-fire erosion in near real-time during storm events. We developed an integrated approach using the RUSLE, remote sensing, and Geographical Information System (GIS) to map the potential erosion risk over space and time, with the SDWC as the pilot study area. Water managers were able to use the results to prioritize monitoring points and areas for erosion assessment and interventions following storm events.

2. Materials and Methods

2.1. The Study Area

The SDWC area includes five main catchments and 204 sub-catchments (or drainage units). It extends from north of Lithgow in the upper Blue Mountains, to the source of the Shoalhaven River near Cooma in the south, and from Woronora in the east to the source of the Wollondilly River west of Crookwell. These catchments cover an area of almost 16,000 km^2, about seven times larger than the Australia Capital Territory (ACT) to its southwest (Figure 1). The dam across the Warragamba River forms Lake Burragorang which provides drinking water to Sydney and surrounding regions for more than five million people or 60 per cent of the NSW population.

The annual average rainfall in SDWC is about 841 mm based on the BoM gridded rainfall data [20] over the 30-year period between 1981 and 2010 (or "climate normals") as used in climate maps and statistics in Australia and many parts of the world [21]. The rainfall in 2019 (before and during the 2019–2020 wildfire) is only about 473 mm which was far below the average. The SDWC area experiences significant seasonal variation in monthly rainfall. The months with the highest rainfalls are February (100 mm), November (85 mm), and January (79 mm). The months with least rainfalls are July (55 mm), September (57 mm), and August (59 mm) over the SDWC area.

The dominant land uses are livestock grazing (35%), nature conservation lands or national parks (30%), crown lands and reserves (16%), and others including intensive agriculture, horticulture, mining, and reservoirs (19%) based on the 2017 land use map over the SDWC area [22]. Soils types in SDWC are variable though strongly influenced by lithology and landform. Texture contrast soils with acidic subsoils (Kurosols) dominate the catchments (46%). Sandy soils with minimal soil development (Rudosols and Tenosol) are also common (30%) and often very shallow in steep-sloped terrain. The elevations range from 21 m to 1463 m (a.s.l) with an average slope of 16.7% over the SDWC area. The elevations range from 41 m to 1356 m (a.s.l) with an average slope of 28.7% at the sub-catchments near the Warragamba Dam and 70% of areas being mountainous terrain.

The 2019–2020 fires have severely and extensively burnt major forests within the drinking water catchments for Sydney, about 30% of SDWC areas have been burnt, and 16% of them are in high or extreme severity based on the fire extent and severity mapping (FESM) [23]. The fires were more severe near the Warragamba Dam, where about 81% of designated Special Areas burned (areas with little to no public access), with 30% at high to extreme fire severities (Figure 1). These drinking water catchments are managed by Water New South Wales (WaterNSW) and the New South Wales Department of Planning, Industry and Environment (DPIE).

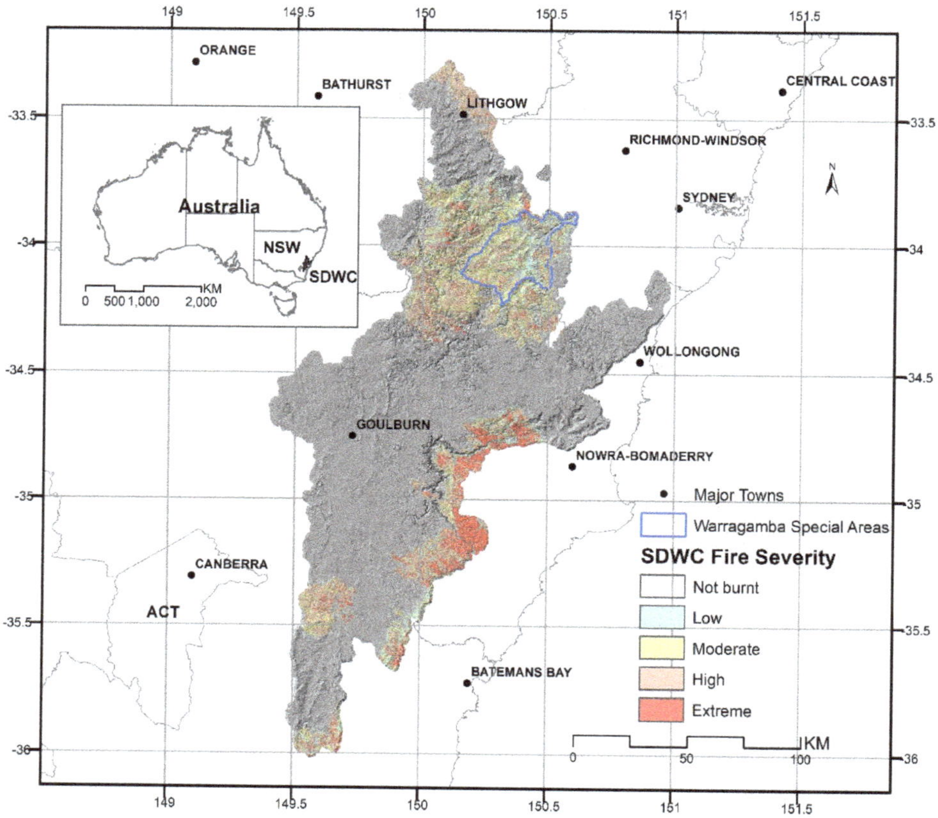

Figure 1. Location of the Sydney drinking water catchment area and the Warragamba Dam (Lake Burragorang) Special Areas with fire severities, New South Wales (NSW), Australia. The background map is the hill-shaded elevation from the 5 m digital elevation model.

2.2. The Datasets

The primary datasets used in this study were radar rainfall data, satellite images (Landsat-8 and Sentinel-2 with cloud cover < 5%), MODIS-derived fractional vegetation cover (FVC), fire severity, soil and water quality data. The Lansat-8 (OLI Level-2 surface reflectance) and Sentinel-2 (Level-2A BOA reflectance images) datasets for the 2019–2020 wildfire period were obtained via the Earth Engine Data Catalog [24]. The FVC datasets, derived from MODIS Nadir BRDF-Adjusted Reflectance product (MCD43A4) Collection 6, include monthly Photosynthetic Vegetation (PV), Non-Photosynthetic Vegetation (NPV), and Bare Soil (BS) at a spatial resolution of 500 m [25,26] were obtained online [27].

Obtaining clear satellite images at high resolution data, for SDWC area is difficult because of clouds and smokes during the 2019–2020 wildfire period. This problem was partially overcome by using Sentinel-2 with a more frequent (5-day) revisiting frequency. The Sentinel-2 multi-spectral instrument (MSI) sensor provides 13 spectral bands with a spatial resolution of 10–20 m in vegetation mapping. In addition, Sentinel-2 offers three new red edge spectral bands, which has the advantage in improving the accuracy for estimating vegetation indices. The blending of the 5-day Sentinel-2 and the 16-day Landsat-8 data made it possible to create time-series vegetation mapping and RUSLE C-factor estimation in high spatial and temporal resolutions.

Radar rainfall can play a significant role in representing the rainfall intensity, especially in areas without a high density of gauge networks [28,29]. Even where rain gauges or pluviograph rainfall stations exist, they are unlikely to replace radar-derived rainfall estimates, because of the high spatial and temporal resolution from radar data. In Australia, BoM produces real-time quality controlled, rainfall estimates (namely Rainfields) and forecasts using radar, rain gauges, and numerical weather prediction models [30]. It converts real-time radar observations of atmospheric reflectivity into quantitative precipitation estimates (QPEs) via several processing steps including quality controlling, cleaning, analyzing, and integrating data from radars and rain gauges in real time, offering improved spatial and temporal resolution in comparison with rain gauges in the areas covered by weather radars. Rainfields (version 3) includes more radars with QPE products and improved resolution, regional mosaic grids at spatial resolution of 1 km^2 and the national radar mosaic at resolution of 2 km^2. In this study, we obtained and used the merged accumulation (Level 2) rainfall for NSW mosaic radar (IDR311MQ, available via FTP to registered users) with the highest quality of QPE at a spatial resolution of 1 km^2 and a temporal resolution of 15 min. These were blended radar and rain gauge rainfall estimates where the calibrated radar estimates were used to add spatial details between the rain gauge locations. Merged rainfall data located at grid points coincident with individual rain gauges are likely to be very similar to gauge observations available at the time of the generation of the product. Rainfields products were stored in NetCDF format and arbitrary coordinate system. We developed automated scripts (R and GIS) to process the Rainfields data including format conversion, re-projection, resampling, and calculation of precipitation. The accumulated precipitation values are calculated by multiplying a scale factor (0.05) and adding an offset (zero in this study) according to the user's guide [30]:

The FESM is a semi-automated fire severity mapping approach in NSW [23] which used a machine learning framework based on the sentinel-2 satellite imagery. The severity map has standardized classes to allow comparison of different fires across the landscape. The FESM severity classes include i) unburnt, ii) low severity (burnt understory, unburnt canopy), iii) moderate severity (partial canopy scorch), iv) high severity (complete canopy scorch, partial canopy consumption), and v) extreme (full canopy consumption). FESM is used in this study for statistical analysis and assessment on impact of fire severity on hillslope erosion.

In addition, recent land use map, soil data, and LiDAR-derived digital elevation models (DEM) [31] were also used to estimate the RUSLE factors (i.e., LS-factor) and statistics. Table 1 summarizes the primary datasets used in this study and their sources.

Table 1. Summary of the primary datasets used in this study.

Dataset	Description	Source	Spatial Resolution	Temporal Resolution
DEM	Digital elevation models (LiDAR [1])	DPIE [2]	5 m	n/a
FESM	Fire Extent and Severity Mapping	DPIE	10 m	monthly
FVC	Fractional vegetation cover	CSIRO [3]	500 m	monthly
Landsat-8	Landsat-8 Operational Land Imager	NASA [4]	30 m	16 days
Landuse	NSW Land use map (2017)	DPIE	polygon	n/a
MODIS	Moderate Resolution Imaging Spectroradiometer	NASA	250–500 m	daily

Table 1. *Cont.*

Dataset	Description	Source	Spatial Resolution	Temporal Resolution
Rainfields	Bureau of Meteorology radar rainfall	BoM [5]	1000 m	15 min
Rainfall	Bureau of Meteorology gridded rainfall	BoM	5000 m	daily
Sentinel-2	Sentinel-2 multi-spectral instrument	ESA [6]	10–60 m	10 days
Soil	Digital soil maps	DPIE	90 m	n/a

[1] LiDAR = Light Detection and Ranging; [2] DPIE = New South Wales Department of Planning, Industry and Environment; [3] CSIRO = Commonwealth Scientific and Industrial Research Organisation Australia; [4] NASA = The National Aeronautics and Space Administration of the USA; [5] BoM = The Bureau of Meteorology, Australia; [6] ESA = European Space Agency.

2.3. Event-Based Erosion Modeling

We integrate near real-time weather radar data, remote sensing, Google Earth Engine (GEE), and GIS to obtain and process timely information for hillslope erosion estimation including event-based rainfall erosivity and FVC or the cover-management factor (or C-factor). Along with other erosion factors in relation to soil and topography, we estimate the hillslope erosion at a specific time (*Ai*) using the modified RUSLE based on [5,9]:

$$A_i = EI_{30i} \times C_i \times K_i \times LS \times P_i \quad (1)$$

where A_i is the computed soil loss per unit area at time *i*, usually in tons per hectare per time unit (Mg ha^{-1} time^{-1}); EI_{30i} is the rainfall–runoff erosivity (R-factor) at time *i* (usually in MJ mm ha^{-1} h^{-1} time^{-1}); K_i is the soil erodibility factor (K-factor, Mg h MJ^{-1} mm^{-1}), a measure of the susceptibility of soil to erosion; *LS* is the slope length and steepness factor (LS-factor, dimensionless); C_i is the cover-management factor (C-factor, 0-1 dimensionless); P_i is the erosion control factor (P-factor, set to 1 for this study); *i* denotes a specific time (i.e., month).

Derivation of the factors required by the RUSLE and its applications in NSW are well described in our previous publications [5–7,32–35]. This study focuses on the recent advancements in weather radar rainfall estimation, satellite estimation of FVC and GEE technology which enable accurate and rapid estimation of some RUSLE factors, specifically for the rainfall erosivity (EI_{30i}) and cover factors (C_i). In addition to these two dynamic factors, the soil erodibility (K_i) factor was estimated based on [35] using updated soil data (soil texture, structure, permeability, and organic matter) for the SDWC area. Using the DEM derived from Light Detection and Ranging (LiDAR) [31], we calculated the LS-factor for the SDWC area with a 5-m spatial resolution based on [33]. The general procedures are illustrated in Figure 2. All RUSLE factors were resampled to a spatial resolution of 30 m before computing the erosion rates (A_i).

2.4. Fractional Vegetation Cover and RUSLE C-Factor Estimation

The RUSLE cover-management factor (C) was estimated from satellite data based on [32] on monthly basis. The satellite data used in this study included MODIS FVC products [25–27], Landsat-8 and Sentinel-2 visible, near-infrared (NIR) and shortwave infrared (SWIR) bands, and Google Earth Engine Burnt Area Mapping (GEEBAM) products including normalized burn ratio (NBR) [36]. The multiple data sources are complementary in spatial (10–500 m) and temporal (daily to 16 days) resolutions providing a means for cross validation.

In this study, Sentinel-2 Level-2A surface reflectance (SR) datasets were queried from the GEE data-pool [24]. An automated GEE script (Sen2cor) was developed for data processing and computation including radiometric calibration, geometric calibration, and atmospheric calibration. We produced the Sentinel-2 SR composites for two periods (period 1: July–August 2019, and period 2: January–February 2020), representing the pre-fire season and post-fire season over SDWC.

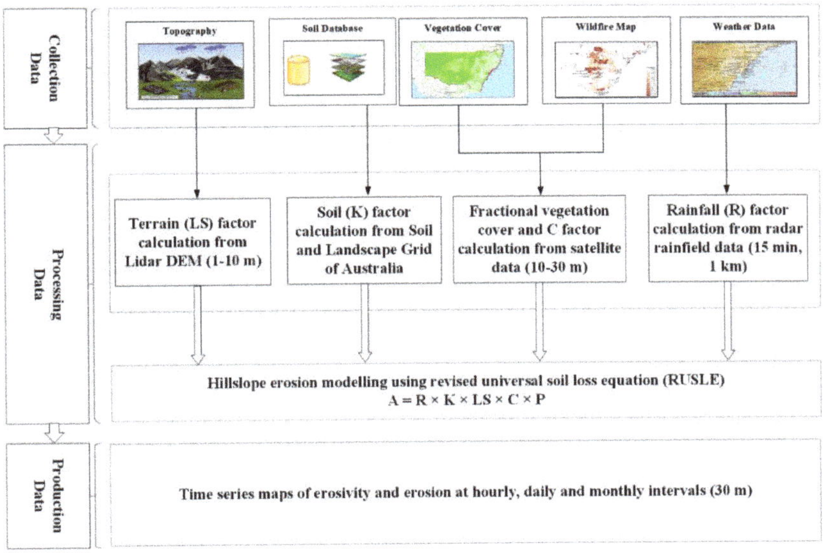

Figure 2. The general procedures of hillslope erosion modeling in this study.

In addition to the MODIS-derived monthly FVC products at a resolution of 500 m [25,26], the PV fraction (fPV) was estimated based on the modified transformed vegetation index (MTVI) [37] using the NIR and the visible bands as presented in Equation (3). The non-photosynthetic vegetation fraction (fNPV) was estimated from the normalized difference senescent vegetation index (NDSVI) using the red and SWIR (shortwave infrared) reflectance as NPV scattering mostly occurs in the SWIR range [38]. The total cover (TC) was calculated as the summary of fPV and fNPV fraction;

$$fPV = 1.7208\left[1.2\left(R_{NIR} - R_{green}\right) - 2.5\left(R_{red} - R_{green}\right)\right] + 0.1004 \quad (2)$$

$$fNPV = 0.82(R_{SWIR} - R_{red})/(R_{SWIR} + R_{red}) + 0.0753 \quad (3)$$

$$TC = (fPV + fNPV)/100 \quad (4)$$

The RUSLE cover and management (C) factor were estimated from the TC as:

$$C_j = \exp\left(-0.799 - 7.74 \times TC_j + 0.0449 \times TC_j{}^2\right) \times EI_j/EI_t \quad (5)$$

where C_j is RUSLE cover-management factor in time j or a given period (e.g., month), TC_j is ground cover (0 to 1) in time j, EI_j is the rainfall erosivity over time j, and EI_t is the total annual rainfall erosivity.

2.5. *Rapid Radar Rainfall Data Processing and Storm Event-Based Rainfall Erosivity Estimation*

Automated scripts in R (version 3.6), a free software environment for statistical computing and graphics, and GIS (ArcGIS version 10.4) were developed to rapidly process radar Rainfields data and rainfall erosivity estimation based on storm events. The processing included the following steps: i) R scripts were developed for batch processing the radar rainfall data and converting NetCDF to Tiff format; ii) the Tiff datasets were then re-projected to Geographic coordinates so that they match with the other datasets; iii) readjust UTC to AEST (Australian East Standard Time = UTC + 10:00 or 11:00 in daylight saving time); and iv) input to ArcGIS for extraction of rainfall accumulation and further calculations for rainfall erosivity and erosion based on RUSLE and our previous studies [5,6].

The EI30 index is commonly used in RUSLE to predict the impact of rainfall events on soil loss [5,18]. For a single storm event, the EI30 is the value of kinetic energy, E in MJ ha^{-1}, multiplied by the peak 30-min rainfall intensity I_{30} (mm hr^{-1}). In this study, E is computed from the composite radar Rainfields data in 15-min intervals from BoM following:

$$E = \sum_{r=1}^{N} e_r \Delta V_r \quad (6)$$

$$e_r = 0.29[1 - 0.72 \exp\left(-a\frac{\Delta V_r}{\Delta t_r}\right)] \quad (7)$$

where $\Delta V_r/\Delta t_r$ is the rainfall intensity (mm hr^{-1}), while ΔV_r refers to rainfall amount during that particular period Δt_r, N is number of 15 min (e.g., N = 2 for 30-min), e_r (MJ ha^{-1} mm^{-1}) means kinetic energy to single storm event, a is an empirical coefficient. The rainfall intensity for interval 30-min (mm hr^{-1}), I_{30} is calculated as

$$I_{30} = P_{30} \times 2 \quad (8)$$

P30 is the maximum 30-min rainfall depth (mm); it is multiplied by 2 to convert to an hourly scale. Peak rainfall amount in 30-min intervals was extracted from radar images at every three 15-min intervals. The accumulated EI30 values in a day were compared with the erosivity estimated from a modified daily rainfall erosivity model [5,6].

2.6. Hillslope Erosion Estimation and Risk Scenario Analysis

With all these RUSLE factors, we estimated the hillslope erosion risk for the period of the 2019–2020 fires on a monthly step and on a storm event basis. The pre- and post-fire erosion rates were estimated using FVC estimated from Sentinel-2, Landsat-8, and MODIS, thus providing cross comparison and validation.

The rainfall erosivity percentiles for each month have been calculated from the period 2000 to 2019 to match with the same period of MODIS-derived FVC time series. The GIS (ArcGIS) "rank" function was used to calculate the percentiles. For example, Percentile 95 is 19th rank, percentile 75 is 15th rank, percentile 55 is 11th rank from the 20-year rainfall data. Using these percentiles, we can work out the likely occurrence of an event. For example, if we have a rainfall value in the 90th percentile, this represent the highest 10% erosion risk for this site, or 90% values will be equal to or below this value. The rainfall erosivity percentiles were further spatially interpolated from 5 km to 30 m (using Spline method) to produce higher resolution surfaces in GIS for RUSLE modeling consistent with other factors [29].

Hillslope erosion rates were further estimated and categorized based on these rainfall erosivity percentiles and the C-factor values related to FESM fire severity classes. We first calculated the mean annual rainfall erosivity for the climate normal period (1981–2010) and used as the basis to estimate the various rainfall erosivity percentiles (e.g., 55%, 75%, 95%,) representing rainfall scenarios. The C-factor values were estimated based on the groundcover levels in association with the fire severity classification. For example, Class 1 or low severity with 75% cover, Class 2 or high severity with 50% cover, Class 3 or extreme severity with 25% cover. These combinations of rainfall erosivity and groundcover represent various possible scenarios of rainfall (amount and intensity) and fire severity classes, thus indicating the likely consequences of erosion risk for any given rainfall and fire regime.

In addition, we also estimated the return period of rainfall and rainfall erosivity for the post-fire storm events by using stationary generalized extreme value (GEV) method [39]. We obtained the annual maxima monthly rainfall, then we fit the GEV to the annual maxima monthly rainfall over the period 1910–2019 using the maximum likelihood method. Then, the parameters of GEV fit were used to plot the return level with the corresponding ±1.96 × standard error for a 95% confidence interval.

2.7. The Normalized Burn Ratio (NBR)

The normalized burn ratio (NBR) and its difference were estimated and used in the GEEBAM program to find out where wildfires in NSW have affected vegetation [36]. GEEBAM used a series of Sentinel-2 images to derive NBR and its difference (dNBR) between the pre-fire and post-fire. NBR is an index designed to highlight burnt areas in large fire zones. The formula is like normalized difference vegetation index (NDVI), except that the formula combines the use of near infrared (NIR) and shortwave infrared (SWIR) wavelengths [40]:

$$NBR = \frac{(NIR - SWIR)}{(NIR + SWIR)} \tag{9}$$

A threshold of dNBR was chosen through visual interpretation to create GEEBAM classes. A higher value of dNBR indicates increased likelihood that the area has burned, while areas with negative dNBR values may indicate regrowth following a fire. GEEBAM was used during the 2019–2020 summer to rapidly predict how severely the tree canopy has burned. It was updated monthly by measuring the change in the color of vegetation after a fire based on NBR. The NBR and fire severity classes were further used in this study to assess the vegetation cover and erosion change before and after the fires.

2.8. Validation of Hillslope Erosion Estimation

The direct assessment of final erosion results was difficult as it was not possible to carry out field measurements on erosion and sediment immediately after the wildfires and storm events, especially during this Covid-19 pandemic period and the accessibility to the mountainous area. As the K-factor and LS-factor are relatively stable and they were validated in our previous studies, the cross validation in this study was focused on the cover-management and the rainfall erosivity factors (refer Equation (1)). These include: i) The cover-management factor estimated from MODIS, Sentinel-2, and Landsat-8; ii) the rainfall erosivity factor estimated from BoM gridded rainfall and radar rainfall. The Nash–Sutcliffe model efficiency coefficient (NSE) and relative error were used to assess the relative accuracy and differences [33]. In addition, normalized burn ratio (NBR) as used in GEEBAM was also used to compare FVC (used in RUSLE).

3. Results

3.1. The Temporal Changes of Rainfall, Erosivity and Erosion at SDWC

Based on the historical rainfall records [20], there were only two months over the past 110 years (1910–2020) that had a rainfall amount greater than the post-fire rainfall in February 2020 at SDWC. They were February 1956 and March 1978; both had an average monthly rainfall amount of 341 mm over the SDWC area. Though the rainfall amount (341 mm) was slightly higher than the February 2020 rainfall (337 mm), the rainfall erosivity in February 2020 was estimated the highest (2187.3 MJ mm ha^{-1} h^{-1} $month^{-1}$). This was because the rainfall in February 2020 was more intense as the storm events were concentrated in just several days (6–13 February), thus more erosive. As shown in Figure 3, the return period for the post-fire rainfall in February 2020 was close to a 1-in-40-year event and the rainfall erosivity a 1-in-257-year event based on the GEV-fitted distribution curve for the 1910–2020 climate.

Table 2 lists the estimated monthly mean rainfall erosivity and erosion at SDWC for the 2019–2020 wildfire period. The monthly mean values on rainfall, erosivity, and erosion are also listed in the table (mean) for comparison and they were estimated from the recent 20-year rainfall to match at the same period of the available MODIS FVC data (2000–2019). The estimated post-fire erosion rate (4.88 Mg ha^{-1} $month^{-1}$) in February 2020 increased more than 30 times compared with the previous months (0.01–0.16 4.88 Mg ha^{-1} $month^{-1}$) and about 10 times compared to the mean erosion rate (0.49 Mg ha^{-1} $month^{-1}$) in the same month for the period 2000–2019. The average rainfall erosivity value over

SDWC in February 2020 was estimated 2187.3 MJ mm ha^{-1} h^{-1}, exceeding the monthly average value of February by about seven times. It was more than five times higher than what is considered critical in soil erosion (i.e., rainfall erosivity values > 500 MJ mm ha^{-1} h^{-1} month^{-1}) [41].

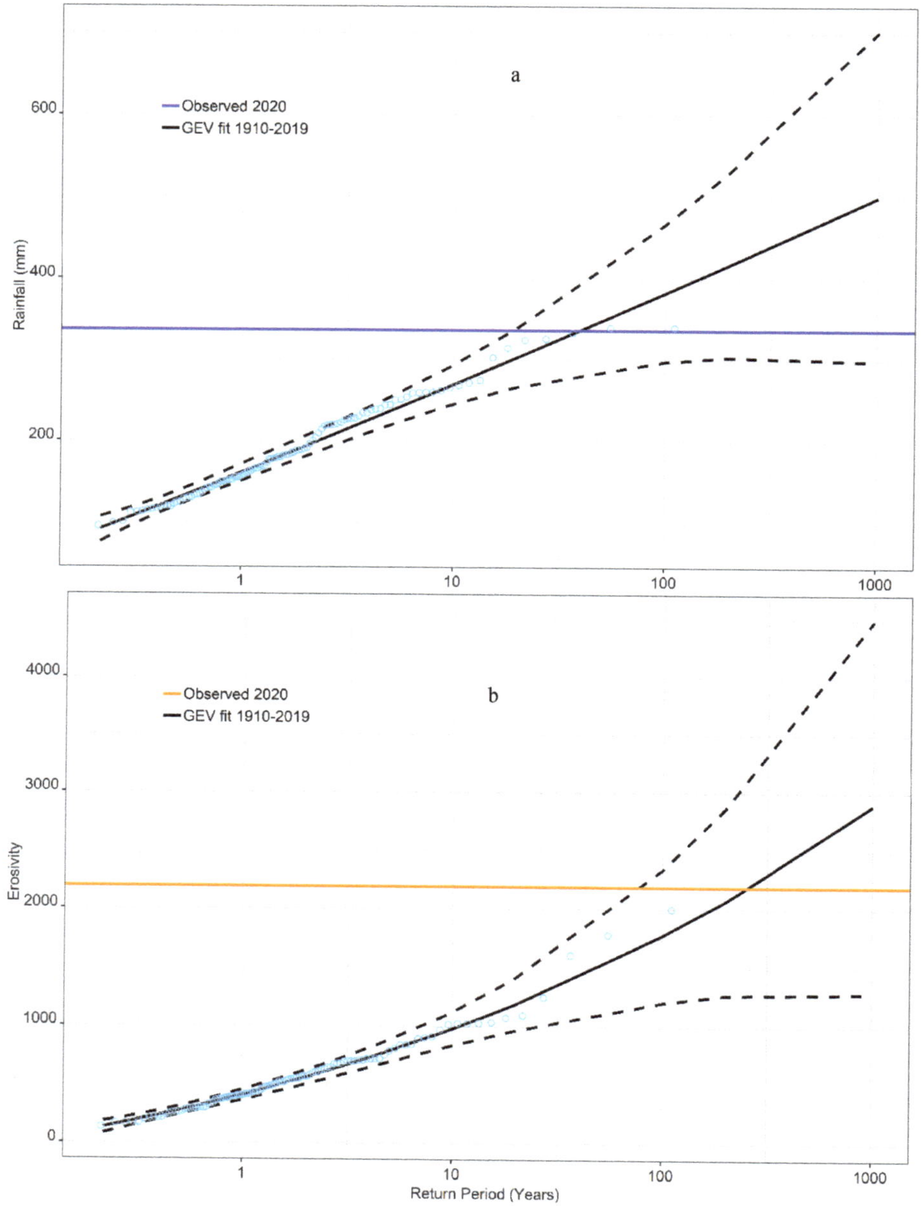

Figure 3. The return periods for the 2019–2020 post-fire in the Sydney drinking water catchment area: (**a**) Rainfall; (**b**) rainfall erosivity. The circles indicate return periods for the annual maximum rainfall based on the 1910–2019 climate; the dash lines are the corresponding ± 1.96 × standard error at the 95% confidence level.

Table 2. Estimated mean monthly erosivity and erosion at Sydney drinking water catchment for the 2019–2020 wildfire period.

Month	FVC (%)	C-factor	Rainfall	Rainfall (Mean)	Erosivity	Erosivity (Mean)	Erosion	Erosion (Mean)
Jul–19	93.5	0.0045	15.9	43	12.0	45.0	0.01	0.04
Aug–19	93	0.0046	22.6	39	16.0	86.5	0.02	0.04
Sep–19	92.5	0.0047	53.1	42	68.7	67.0	0.09	0.05
Oct–19	92.2	0.0048	26.0	53	14.0	92.0	0.02	0.09
Nov–19	90.5	0.0053	18.2	75	12.9	114.7	0.02	0.18
Dec–19	88.8	0.0061	2.6	74	2.0	132.1	0.01	0.17
Jan–20	83.7	0.0081	51.4	78	65.2	167.1	0.16	0.26
Feb–20	86	0.0081	337.2	121	2187.3	332.7	4.88	0.49
Mar–20	90.5	0.007	94.8	88	145.8	105.4	0.24	0.26
Apr–20	93	0.006	53.7	52	58.1	79.2	0.08	0.10
May–20	93	0.006	15.9	37	12.0	68.9	0.07	0.05
Jun–20	93.6	0.006	26	82	13.8	109.5	0.02	0.14

Note: C-factor is unitless; Rainfall in mm month^{-1}; Erosivity in MJ mm ha^{-1} h^{-1} month^{-1}; Erosion in Mg ha^{-1} month^{-1}; the Mean values were calculated for the period 2000–2019.

Figure 4 shows the impacts of total vegetation cover and rainfall on erosion at SDWC for the 2019-20 wildfire period. The sharp increase in the erosion rate in February (Mg ha^{-1} month^{-1}) reflects the combined impacts of wildfire and rainfall on erosion and their relationships.

We further examined the rainfall patterns in February 2020 and found that most erosive rainfall over SDWC area occurred between 6–13 February, accounting for approximately 80% of the total rainfall in that month. With the radar rainfall data at a temporal resolution of 15 min and a spatial resolution of 1000 m, we were able to estimate the EI30 values on hourly and daily basis or storm event basis. Figure 5a shows the estimated daily erosion rate from the radar Rainfields data in January and February 2020. Figure 5b shows an example of the estimated hourly erosion rate from the radar Rainfields data on 9 February, 2020. The radar-based estimation of rainfall erosivity greatly enhanced the temporal and spatial resolutions and potentially more useful for post-fire erosion mitigation.

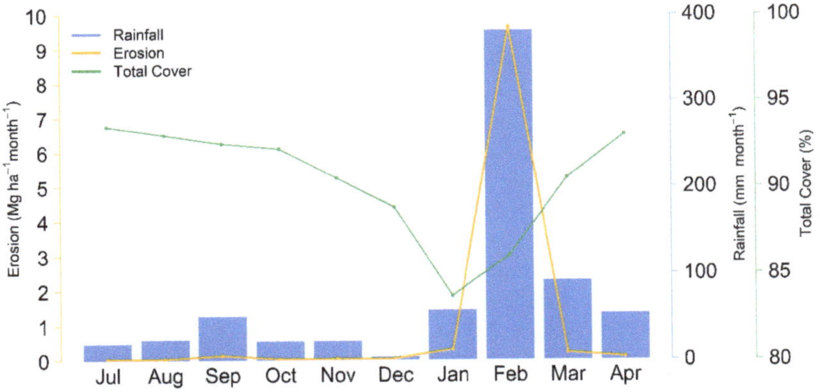

Figure 4. Relationships between erosion and fractional vegetation cover (total cover); and erosion and rainfall at Sydney drinking water catchment during the 2019–2020 wildfire period (July 2019 to April 2020).

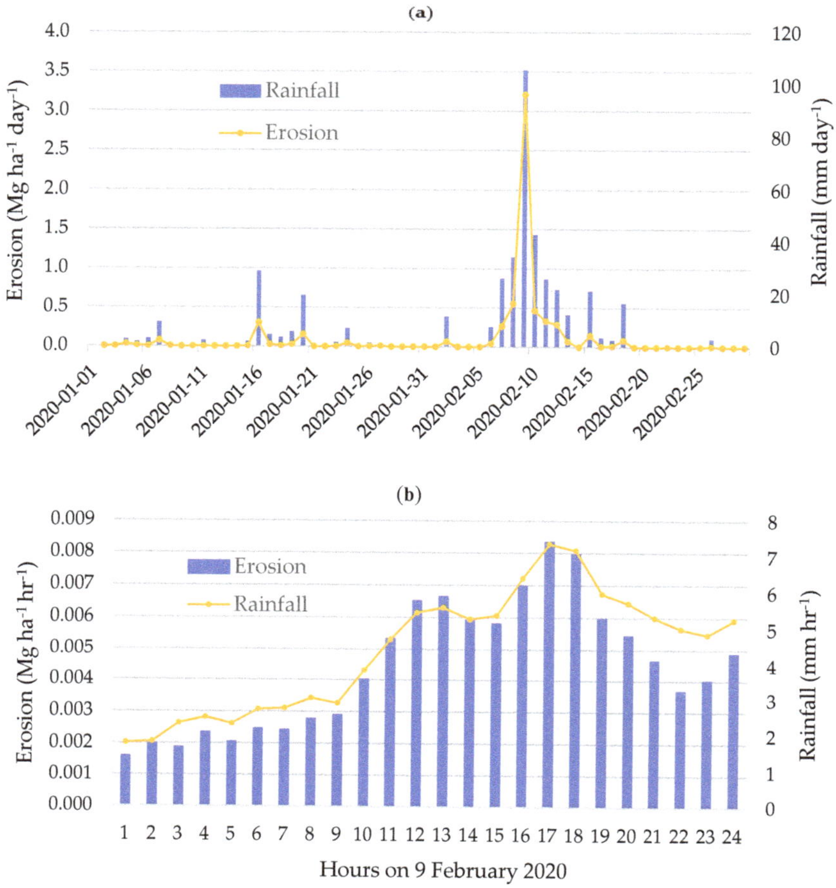

Figure 5. Estimated erosion rate from radar Rainfields data in January and February 2020 at: (**a**) daily step; (**b**) hourly step on 9 February 2020.

3.2. The Spatial Variation of Post-Fire Erosion Risk at SDWC

There was great spatial variation in hillslope erosion rates over the SDWC area. Among the 204 drainage units within SDWC, the post-fire erosion rate (February 2020) ranged from 0.1 to 23.8 Mg ha^{-1} month^{-1} with a mean value of 4.9 and standard deviation of 4.2 (Mg ha^{-1} month^{-1}). Figure 6 shows the average monthly erosion risk in January and February 2020 that represents the effect of post-fire storm events on hillslope erosion across the SDWC area. Most of the highest post-fire erosion risk areas appear to be around the Warragamba Dam (Lake Burragorang) including Wild Dog (ID = 193), Lower Kowmung (ID = 99), and Cedar Ck (ID = 33) as shown in the Figure.

Figure 7 further presents the 10 drainage units with the highest erosion rates over the SDWC area after the 2019–2020 wildfires and storm events in February 2020. The erosion rates range from 13.2 Mg ha^{-1} month^{-1} in Brogers Ck (ID = 24) to 23.8 Mg ha^{-1} month^{-1} in Wild Dog (ID = 193), much higher than the average annual rate in NSW. Linking erosion risk information with these drainage units (the smallest water management units) helps the water management agency (i.e., WaterNSW) to prioritize the monitoring and mitigation actions.

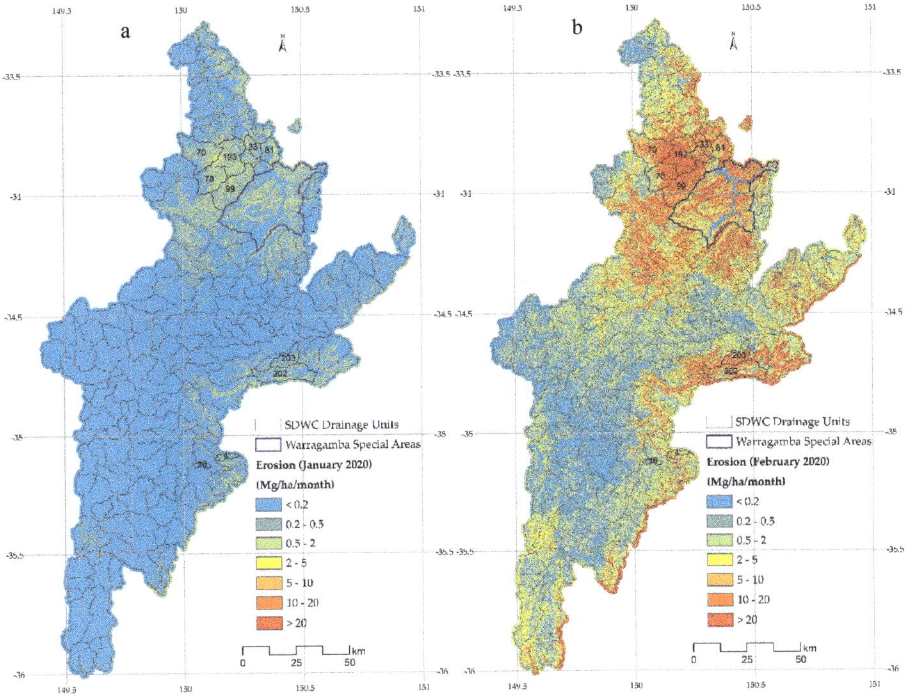

Figure 6. The hillslope erosion in January and February 2020 showing the impacts of fire and storm events on erosion, and the ten high erosion risk sub-catchment (labelled with IDs) across the Sydney drinking water catchment: (**a**) erosion before storm events; (**b**) erosion after storm events.

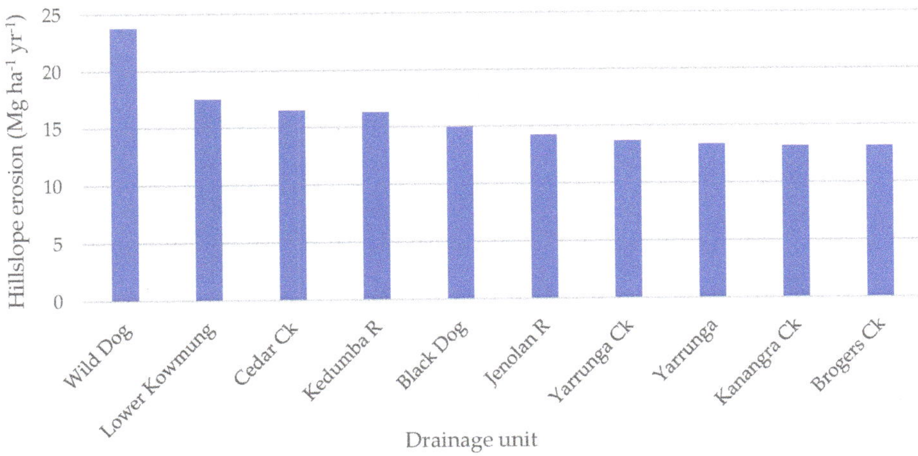

Figure 7. The highest erosion risk drainage units within the Sydney drinking water catchment area after the 2019–2020 wildfires in February 2020.

3.3. The Erosion Risk at Different Fire Severity and Erosivity Scenarios

Erosion rates estimated at the high (75 percentile) and extreme (95 percentile) rainfall erosivity scenarios were compared with the average rainfall erosivity in 2000–2019 (the hillslope erosion modeling period). The vegetation cover levels before (July–August 2019) and after (January–February 2020) the

2019–2020 wildfires were estimated from satellite (Sentinel-2 and Landsat-8) derived FVC. It clearly shows the impact of fire severity on hillslope erosion, and the high fire severity class (canopy fully affected) has an erosion risk about two to four times higher compared to other classes. The erosion could increase up to 450% under extreme rainfall erosivity condition for severely burnt areas. Table 3 summarized the estimated erosion rates under different fire severity classes and rainfall scenarios. Unlike the relative change (%), the actual erosion rates are not always positively correlated to the fire severity classes because of the impacts of other factors such as rainfall, terrain, and soil (refer Equation (1)).

Table 3. Estimated mean erosion rates (Mg ha^{-1} yr^{-1}) at different fire severity classes and rainfall erosivity scenarios at Sydney drinking water catchment area, and the pre- and post-fire changes.

Fire Severity Class	Estimated Mean Erosion Rates (Mg ha^{-1} yr^{-1}) and Change%						
	Average Rainfall Erosivity			High Rainfall Erosivity		Extreme Rainfall Erosivity	
	Pre-Fire	Post-Fire	Change% (%)	Post-Fire	Change%	Post-Fire	Change%
Low	6.0	9.2	52	10.2	68	17.5	190
Moderate	6.1	10.8	76	12.6	105	20.4	232
High	4.5	8.6	91	10.2	126	16.0	254
Extreme	3.5	10.6	202	12.5	255	19.3	448

3.4. Cross Validation and Comparison of Results

Using the relationship between NBR and FVC we estimated the erosion risk directly from NBR using the established methods as presented above, the coefficient of determination (R^2) from the erosion estimates between FVC and NBR reached 0.8346 (Figure 8). It implies that the commonly used NBR index in wildfire studies can be used as a substitute for FVC to estimate the C-factor values and the erosion rates along with other RUSLE factors.

Figure 9 compares the rainfall erosivity and the final erosion estimation from the BoM daily rainfall gridded data (5 km) and the radar Rainfields data (15 min, 1 km). Though the source datasets are very different in resolution and measurement, there is a high correlation between them ($R^2 = 0.7562$), implying that the radar Rainfileds data can be directly used in estimating erosivity.

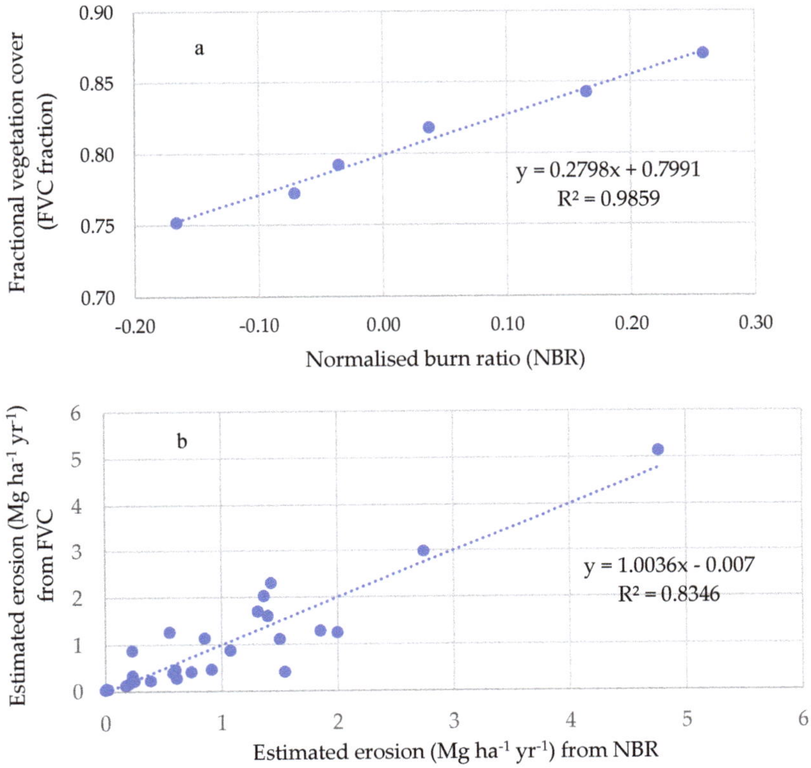

Figure 8. Relationships of fractional vegetation cover (FVC) estimated from MODIS and normalized burn ration (NBR) from Sentinel-2: (**a**) FVC and NBR relationship; (**b**) estimated erosion rates using FVC and NBR.

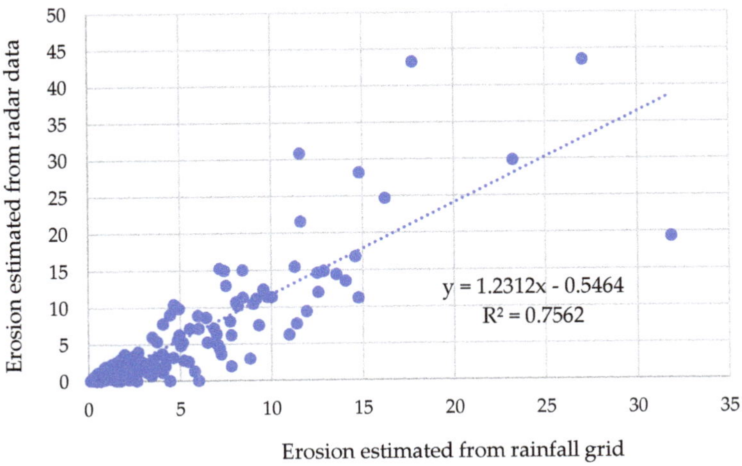

Figure 9. Estimated erosion rates (Mg ha^{-1} month^{-1}) from BoM daily rainfall gridded data and radar Rainfields data for January–February 2020 over Sydney drinking water catchment.

4. Discussion

Vegetation cover (or C-factor) and rainfall (or R-factor) are the two dominant factors affecting the post-fire hillslope erosion. Vegetation cover, which is important in reducing the impact of rainfall, was not significantly lower (about 10%) after the wildfire based on the monthly FVC products [25–27]. Over 80% of the soil surface was still covered by either PV or NPV, both can protect soil from erosion. This agrees well with the study in another national park in Australia [40,41].

Compared to the mean values in February for all years (2000–2020), C-factor value only increased 1.5 times (due to wildfire) but the rainfall erosivity increased about 7 times in February 2020 across the SDWC area. This suggests that rainfall has a far larger impact than groundcover factor, on hillslope erosion after wildfires. This is in line with a similar study in the Blue Mountains catchment using the eWater toolkit (Source) model which reported six times higher sediment load at the downstream outlet after extreme wildfire in 2001 [42]. The post-fire soil erosion is mainly limited by rainfall erosivity which agrees with similar studies [43–45].

There was a prolonged dry period before the 2019–2020 wildfires with less than 300 mm rainfall over the SDWC area for 10 months from April 2019 to January 2020, leading to large quantities of fuel loads. In February 2020 the rainfall amount reached 337 mm month^{-1}, with several intense storm events between 6–13 February. The severe hillslope erosion at SDWC was mainly caused by these extreme rainfall events in February 2020. For example, the erosion rate on a single day (9 February 2020) reached 3.2 (Mg ha^{-1} day^{-1}) which contributed to 65% of the total monthly erosion (4.9 Mg ha^{-1} Month^{-1}) in February 2020.

Despite the increased erosion risk, WaterNSW successfully maintained the supply of safe water to the treatment plants, by proactively managing the water supply system configuration (through sources selection and offtake depth changes) preventing the inflows from the fire-impacted catchment entering the supply during these storm events.

Soil erodibility (K) and slope-steepness (LS) factors are relatively stable compared to the dynamic C-factor and R-factor. Wildfires may alter the soil properties including soil structure, texture, permeability, and soil organic carbon which are all related to the K-factor as discussed in one of our previous studies [46]. The extent of fire effects on these soil properties depends on fire intensity, fire severity, and fire frequency [47] which is complicated and beyond the scope of this study.

The terrain factor also contributed significantly to the post-fire erosion. SWDC area has a rugged terrain compared with many other parts of NSW. The mean slope is about 17% over the SDWC area, and 29% in the Warragamba Dam area, much higher than the state average (6%). Because of the steep terrain, the mean RUSLE LS-factor value at SDWC is 4.7 which is about 2.6 time higher than the State average (1.8). The LS-factor value at the Warragamba Dam area is even higher (7.2) which is about 4 times of the State average. This implies that the erosion risk at our drinking water catchments are likely to be 2.6–4 times higher than the rest of the State regardless of other erosion factors. This also implies the importance of maintaining good vegetation cover and erosion control practices in the drinking water catchment area.

RUSLE, when appropriately used, can produce meaningful information on relative hillslope erosion risk at a given time. The event-based rainfall erosivity estimation relies on the availability of high-resolution rainfall data. The weather radar Rainfield data (15-min, 1 km) are adequate for estimating EI30 index at daily or sub-daily scale. Such data sets will be increasingly important because of the higher likelihood of intense storm events and fire frequency under the changing climate with warmer and drier conditions.

In a GEE environment, various remote sensing data can be searched and used to estimate the FVC and RUSLE C-factor. The widely used NBR in fire severity mapping and FVC has close correlation and can be used as a surrogate to FVC for C-factor estimation in erosion modeling.

As the erosion model has been applied at such high spatial (5–30 m) and temporal (daily or hourly) resolutions, the impacts of fires on soil erosion can be explored at finer landscape scales including individual hydrological catchments, drainage units, or even paddocks. This helps to locate the high erosion risk areas or sediment sources to assist prioritized management practices. Accurate and effective decision-making is best enabled by linking with detailed and timely information on erosion to identify the impacts of rain events after the fires.

However, process-based studies to understand the factors controlling surface runoff and erosion, particularly in relation to aspects of the fire regime are still required to precisely predict the sediment transport and deposition, especially in the waterways.

5. Conclusions

This study developed an innovative yet practical approach for rapid assessment of post-fire hillslope erosion risk using weather radar, remote sensing, GEE, and GIS. The automated scripts running in ArcGIS and GEE allow rapid processing of time-series and event-based spatial data including satellite and weather radar images to estimate the erosion risk for the 2019–2020 wildfires at SDWC.

This study was the first attempt to precisely estimate where and when high erosion risk is likely to occur at daily and/or hourly steps after severe wildland fires across large drinking water catchments, allowing for more accurate estimation of event-based erosion. With these timely estimates of rainfall erosivity and C-factor, along with the existing datasets on soil erodibility and slope-steepness, we were able to deliver rapid assessment of the post-fire hillslope erosion risk and link it with fire severity. These continuous and consistent estimates of erosion rates were used to analyze the erosion risk before and after the 2019–2020 wildfires and the subsequent impacts of rainfall on erosion rates. With these time series datasets, we identified the locations and times of the highest erosion risk. The sub-catchments near Warragamba Dam have the highest erosion risk because of the bushfires and rainfall events.

The rainfall erosivity is the dominant factor affecting hillslope erosion after severe wildfires over the SDWC area. Severe erosion events are often caused by short but intense storm events such as the case in February 2020. High temporal rainfall data are essential in rainfall erosivity modeling. Weather radar Rainfields data are adequate for erosivity or EI30 estimation at catchment and regional scales.

Field observations and measurements of soil erosion are necessary for model validation and improvement. Other relevant water quality models (such as Source developed by eWater Australia) may be used jointly to validate the modeling accuracy and predict likely pollution risk and delivery to the river or lake system after severe wildfires [48]. Over the next two-years we plan to collect more field data on erosion level, ground cover, land management activity, slope and slope length will be gathered along a set of transects at selected trial catchments for further calibration and validation of methods. A hillslope sediment erosion trap network, consisting of moderate and low burn severity, prescribed burn, rainfall event-based, and control sites are planned to be installed at SDWC sub-catchments along with the water quality stations, and maintained by Water NSW. The goal of the traps is to collect post-fire and unburnt erosion rates to validate model estimates. This will help further calibrate and improve the erosion model and link the estimated erosion rates with sediment transport and water quality downstream. The methodology, once fully validated, will be extended to other areas and wildfire events to provide timely and accurate information on erosion water quality immediately after wildfires.

Author Contributions: Conceptualization, X.Y.; methodology, X.Y, M.Z; software, M.Z.; validation, Q.R.O., L.O. and S.F.; formal analysis, X.Y. and M.Z.; writing—original draft preparation, X.Y.; writing—review and editing, X.Y., M.Z., Q.R.O., S.F., and A.R.; visualization, M.Z.; supervision, X.Y. and L.O.; project administration, X.Y. All authors have read and agreed to the published version of the manuscript.

Funding: This research received no external funding.

Acknowledgments: This project was managed through the New South Wales Department of Planning, Industry and Environment (DPIE) and collaborated with Water New South Wales (WaterNSW). Many DPIE and WaterNSW staff, particularly soil surveyors, contributed to this project and their effort is greatly appreciated. We especially thank the contribution and discussion from Lisa Hamilton, John Bickmore, Charity Mundava, Marlene Van Der Sterren from WaterNSW, and Linda Henderson, Jonathan Gray and Mark Young from DPIE. We also thank

Bureau of Meteorology for providing the radar Rainfields data, and CSIRO (Juan Guerschman) for providing the MODIS-derived fractional vegetation cover data.

Conflicts of Interest: The authors declare no conflict of interest.

References

1. Bureau of Meteorology (BoM) Drought Knowledge Centre. Available online: http://www.bom.gov.au/climate/drought/knowledge-centre (accessed on 23 September 2020).
2. New South Wales Rural Fire Service. "Unprecedented Season Breaks All Records" (PDF). Bush Fire Bulletin. Sydney: NSW Rural Fire Service 42 (1). Available online: https://www.rfs.nsw.gov.au/__data/assets/pdf_file/0007/174823/Bush-Fire-Bulletin-Vol-42-No1.pdf (accessed on 12 August 2020).
3. New South Wales Rural Fire Service. Bush Fire Danger Period and Fire Permits. Available online: https://www.rfs.nsw.gov.au/fire-information/BFDP (accessed on 23 September 2020).
4. Boer, M.M.; de Dios, V.R.; Bradstock, R.A. Unprecedented burn area of Australian mega forest fires. The article. *Nat. Clim. Change* **2020**, *10*, 1–2. [CrossRef]
5. Yang, X.; Zhu, Q.; Tulau, M.; McInnes-Clarke, S.; Sun, L.; Zhang, X.P. Near real-time monitoring of post-fire erosion after storm events: A case study in Warrumbungle National Park, Australia. *Int. J. Wildland Fire* **2018**, *27*, 413–422. [CrossRef]
6. Zhu, Q.; Yang, X.; Yu, B.; Tulau, M.; McInnes-Clarke, S.; Nolan, R.H.; Yu, Q. Estimation of Estimation of event-based rainfall erosivity from radar after wildfire. *Land Degrad. Dev.* **2018**, *30*, 33–48. [CrossRef]
7. Yang, X. State and trends of hillslope erosion across New South Wales, Australia. *Catena* **2020**, *186*, 104361. [CrossRef]
8. Wischmeier, W.H.; Smith, D.D. Predicting rainfall erosion losses, a guide to conservation planning. In *Agricultural Handbook*; US Department of Agriculture: Washington, DC, USA, 1978; Volume 537, 58 p.
9. Renard, K.G.; Foster, G.R.; Weesies, G.A.; McCool, D.K.; Yoder, D.C. Predicting soil erosion by water: A guide to conservation planning with the Revised Universal Soil Loss Equation (RUSLE). In *Agricultural Handbook*; US Department of Agriculture: Washington, DC, USA, 1997; Volume 703, pp. 1–251.
10. Flanagan, D.C.; Nearing, M.A. *USDA Water Erosion Prediction Project: Hillslope Profile and Watershed Model Documentation*; NSERL Report number 10; USDA-ARS National Soil Erosion Research Laboratory: West Lafayette, IN, USA, 1995.
11. Lu, H.; Prosser, I.P.; Moran, C.J.; Gallant, J.C.; Priestley, G.; Stevenson, J.G. Predicting sheetwash and rill erosion over the Australian continent. *Aust. J. Soil Res.* **2003**, *41*, 1037–1062. [CrossRef]
12. Teng, H.F.; Viscarra, R.A.; Shi, Z.; Behrens, T.; Chappell, A.; Bui, E. Assimilating satellite imagery and visible-near infrared spectroscopy to model and map soil loss by water erosion in Australia. *Environ. Model. Softw.* **2016**, *77*, 156–167. [CrossRef]
13. Bradstock, R.; Penman, T.; Boer, M.; Price, O.; Clarke, H. Divergent responses of fire to recent warming and drying across south-eastern Australia. *Glob. Change Biol.* **2014**, *20*, 1412–1428. [CrossRef]
14. Moody, J.A.; Ebel, B.A. Infiltration and runoff generation processes in fire-affected soils. *Hydrol. Proc.* **2013**, *28*, 3432–3453. [CrossRef]
15. Nyman, P.; Smith, H.G.; Sherwin, C.B.; Langhans, C.; Lane, P.N.J.; Sheridan, G.J. Predicting sediment delivery from debris flows after wildfire. *Geomorphology* **2015**, *250*, 173–186. [CrossRef]
16. Yin, S.; Xie, Y.; Liu, B.; Nearing, M.A. Rainfall erosivity estimation based on rainfall data collected over a range of temporal resolutions. *Hydrol. Earth Syst. Sci.* **2015**, *19*, 4113–4126. [CrossRef]
17. Veihe, A.; Rey, J.; Quinton, J.N.; Strauss, P.; Sancho, F.M.; Somarriba, M. Modeling of event-based hillslope erosion in Costa Rica, Nicaragua and Mexico: Evaluation of the EUROSEM model. *Catena* **2001**, *44*, 187–203. [CrossRef]
18. Risal, A.; Lim, K.J.; Bhattarai, R.; Yang, J.E.; Noh, H.; Pathak, R.; Kim, J. Development of web-based WERM-S module for estimating spatially distributed rainfall erosivity index (EI30) using RADAR rainfall data. *Catena* **2018**, *161*, 37–49. [CrossRef]
19. Dymond, J.R.; Vale, S.S. An event-based model of soil erosion and sediment transport at the catchment scale. *Geomorphology* **2018**, *318*, 240–249. [CrossRef]
20. Bureau of Meteorology (BoM) Maps and Gridded Spatial Data. Available online: http://www.bom.gov.au/climate/data-services/maps.shtml (accessed on 14 November 2020).

21. Bureau of Meteorology (BoM) Average Annual, Seasonal and Monthly Rainfall. Available online: http://www.bom.gov.au/jsp/ncc/climate_averages/rainfall (accessed on 14 November 2020).
22. Department of Planning, Industry and Environment (DPIE) Dataset NSW Landuse 2017. Available online: https://datasets.seed.nsw.gov.au/dataset/nsw-landuse-2017 (accessed on 14 November 2020).
23. New South Wales Fire Extent and Severity Mapping (FESM). Available online: https://datasets.seed.nsw.gov.au/dataset/fire-extent-and-severity-mapping-fesm (accessed on 26 August 2020).
24. Google Developers Earth Engine Data Catalog. Available online: https://developers.google.com/earth-engine/datasets/catalog/ (accessed on 14 November 2020).
25. Guerschman, J.P.; Scarth, P.F.; McVicar, T.R.; Renzullo, L.J.; Malthus, T.J.; Stewart, J.B.; Rickards, J.E.; Trevithick, R. Assessing the effects of site heterogeneity and soil properties when unmixing photosynthetic vegetation, non-photosynthetic vegetation and bare soil fractions from Landsat and MODIS data. *Remote Sens. Environ.* **2009**, *161*, 12–26. [CrossRef]
26. Guerschman, J.P.; Hill, M.J. Calibration and validation of the Australian fractional cover product for MODIS collection 6. *Remote Sens. Lett.* **2018**, *9*, 696–705. [CrossRef]
27. CSIRO Fractional Cover Datasets. Available online: https://eo-data.csiro.au/remotesensing/v310/australia/monthly/cover/ (accessed on 14 November 2020).
28. Hossain, F.; Anagnostou, E.N.; Dinku, T.; Borga, M. Hydrological model sensitivity to parameter and radar rainfall estimation uncertainty. *Hydrol. Proc.* **2004**, *18*, 3277–3291. [CrossRef]
29. Sidman, G.; Guertin, D.P.; Goodrich, D.C.; Unkrich, C.L.; Burns, I.S. Risk assessment of post-wildfire hydrological response in semiarid basins: The effects of varying rainfall representations in the KINEROS2/AGWA model. *Int. J. Wildland Fire* **2016**, *25*, 268–278. [CrossRef]
30. Bureau of Meteorology (BoM) Rainfields Rainfall Estimates and Forecasts. Available online: http://www.bom.gov.au/australia/radar/about/using_rainfall_accumulations.shtml (accessed on 3 April 2020).
31. Geoscience Australia Digital Elevation Model (DEM) of Australia Derived from LiDAR 5 Metre Grid. Available online: https://ecat.ga.gov.au/geonetwork/srv/eng/catalog.search#/metadata/89644 (accessed on 14 November 2020).
32. Yang, X. Deriving RUSLE cover factor from time-series fractional vegetation cover for soil erosion risk monitoring in New South Wales. *Soil Res.* **2014**, *52*, 253–261. [CrossRef]
33. Yang, X. Digital mapping of RUSLE slope length and steepness factor across New South Wales. *Soil Res.* **2015**, *53*, 216–225. [CrossRef]
34. Yang, X.; Yu, B.F. Modeling and mapping rainfall erosivity in New South Wales, Australia. *Soil Res.* **2015**, *53*, 178–189. [CrossRef]
35. Yang, X.; Gray, J.; Chapman, G.; Zhu, Q.; Tulau, M.; McInnes-Clarke, S. Digital mapping of soil erodibility for water erosion in New South Wales, Australia. *Soil Res.* **2017**, *56*, 158–170. [CrossRef]
36. GEEBAM Google Earth Engine Burnt Area Map (GEEBAM). Available online: https://datasets.seed.nsw.gov.au/dataset/google-earth-engine-burnt-area-map-geebam (accessed on 26 August 2020).
37. Qi, J.; Marsett, R.; Heilman, P.; Biedenbender, S.; Moran, M.S.; Goodrich, D.C.; Weltz, M. RANGES improves satellite based information and land cover assessments in Southwest United States. *Eos Trans. Am. Geophys. Union* **2002**, *83*, 601–606. [CrossRef]
38. Yang, X.; Zhang, X.P.; Lv, D.; Yin, S.Q.; Zhang, M.X.; Zhu, Q.G.Z.; Yu, Q.; Liu, B.Y. Remote sensing estimation of the soil erosion cover-management factor over China's Loess Plateau. *Land Degrad. Dev.* **2020**, 1–14. [CrossRef]
39. Coles, S. *An Introduction to Statistical Modeling of Extreme Values*; Springer: London, UK, 2001; ISBN 1852334592. 208p.
40. Hammill, K.; Bradstock, R. Remote sensing of fire severity in the Blue Mountains: Influence of vegetation type and inferring fire intensity. *Int. J. Wildland Fire* **2006**, *15*, 213–226. [CrossRef]
41. Rufino, R.L. Evaluation of the erosive potential of rain for the State of Paraná: Second approximation. *R. Bras. Ci. Solo* **1986**, *10*, 279–281.
42. Wallbrink, P.; English, P.; Chafer, C.; Humphreys, G.; Shakesby, R.; Blake, B.; Doerr, S. Impacts on water quality by sediments and nutrients released during extreme wildfires. In *CSIRO Land and Water Client Report*; CSIRO Land and Water: Canberra, Australia, 2004.

43. Townsend, S.A.; Douglas, M.M. The effect of a wildfire on stream water quality and catchment water yield in a tropical savanna excluded from fire for 10 years (Kakadu National Park, North Australia). *Water Res.* **2004**, *38*, 3051–3058. [CrossRef]
44. Cawson, J.G.; Sheridan, G.J.; Smith, H.G.; Lane, P.N.J. Surface runoff and erosion after prescribed burning and the effect of different fire regimes in forests and shrublands: A review. *Int. J. Wildland Fire* **2012**, *21*, 857–872. [CrossRef]
45. Shakesby, R.A.; Wallbrink, P.J.; Doerr, S.H.; English, P.M.; Chafer, C.J.; Humphreys, G.S.; Blake, W.H.; Tomkins, K.M. Distinctiveness of wildfire effects on soil erosion in south-east Australian eucalypt forests assessed in a global context. *For. Ecol. Manag.* **2007**, *238*, 347–364. [CrossRef]
46. Tulau, M.J.; Yang, X.; McAlpine, R.; Veeragathipillai, M.; Karunaratne, S.; Zhang, M.; Karunaratne, S.; McInnes-Clarke, S.; Young, M. Impacts of a wildfire on soil organic carbon in Warrumbungle National Park, Australia. *Proc. Linn. Soc. N. S. W.* **2020**, *141*, S209–S227.
47. Li, X.; Li, Q.J.; Chen, H.Y. Effects of a wildfire on selected physical, chemical and biochemical soil properties in a Pinus massoniana forest in South China. *Forests* **2014**, *5*, 2947–2966.
48. eWater Australia. eWater Source—Australia's National Hydrological Modeling Platform. Available online: https://ewater.org.au/products/ewater-source (accessed on 23 September 2020).

Publisher's Note: MDPI stays neutral with regard to jurisdictional claims in published maps and institutional affiliations.

remote sensing

Technical Note

On the Geomagnetic Field Line Resonance Eigenfrequency Variations during Seismic Event

Mirko Piersanti [1,*], William Jerome Burger [2,3], Vincenzo Carbone [4], Roberto Battiston [2,5], Roberto Iuppa [2,5] and Pietro Ubertini [1]

1 INAF-IAPS, Via del Fosso del Cavaliere, 00133 Rome, Italy; pietro.ubertini@inaf.it
2 Trento Institute for Fundamental Physics and Applications (TIFPA), Via Sommarive, Povo, 38122 Trento, Italy; william.burger@tifpa.infn.it (W.J.B.); roberto.battiston@unitn.it (R.B.); roberto.iuppa@unitn.it (R.I.)
3 Centro Ricerche Enrico Fermi (CREF), Via Panisperna, 00100 Rome, Italy
4 Physics Department, Universitá della Calabria, Ponte Pietro Bucci, Rende, 87036 Cosenza, Italy; vincenzo.carbone@fis.unical.it
5 Department of Physics, University of Trento, Via Sommarive, Povo, 38122 Trento, Italy
* Correspondence: mirko.piersanti@inaf.it

Abstract: In this paper, we report high statistical evidence for a seismo–ionosphere effects occurring in conjunction with an earthquake. This finding supports a lithosphere-magnetosphere coupling mechanism producing a plasma density variation along the magnetic field lines, mechanically produced by atmospheric acoustic gravity waves (AGWs) impinging the ionosphere. We have analysed a large sample of earthquakes (EQ) using ground magnetometers data: in 28 of 42 analysed case events, we detect a temporary stepwise decrease (Δf) of the magnetospheric field line resonance (FLR) eigenfrequency (f^*). Δf decreases of ∼5–25 mHz during ∼20–35 min following the time of the EQ. We present an analytical model for f^*, able to reproduce the behaviour observed during the EQ. Our work is in agreement with recent results confirming co-seismic direct coupling between lithosphere, ionosphere and magnetosphere opening the way to new remote sensing methods, from space/ground, of the earth seismic activity.

Keywords: earthquake; coseismic effects; field line resonance; acoustic gravity waves; lithosphere-magnetosphere coupling

Citation: Piersanti, M.; Burger, W.J.; Carbone, V.; Battiston, R.; Iuppa, R.; Ubertini, P. On the Geomagnetic Field Line Resonance Eigenfrequency Variations during Seismic Event. *Remote Sens.* **2021**, *13*, 2839. https://doi.org/10.3390/rs13142839

Academic Editors: Paolo Mazzanti and Saverio Romeo

Received: 8 June 2021
Accepted: 15 July 2021
Published: 19 July 2021

1. Introduction

The study of the physical process connected to the preparation and onset of an earthquake is a topic of increasing interest among the scientific community, also in view of the societal impact of these phenomena. One of the challenges of these studies is to identify physical phenomena which can be directly connected, without ambiguity, with the earthquake geographical location and time window. Most of the evidence in the literature is, indeed, of a statistical nature, while event based, causal observations of the connection among ground, ionosphere and magnetosphere are much more difficult to be convincingly demonstrated. Regarding the statistical evidence, one of the the most interesting and promising result is related to electromagnetic and ionospheric disturbances occurring before and during seismic activities. Examples of these results are the experimental investigation of the lithosphere-ionosphere-magnetosphere coupling [1–3] with the observation of "anomalous" pulses of electromagnetic (EM) emissions in the frequency interval between a few Hz and up to few tens of kHz, as well as the more recent observations of changes of the density of the charged trapped particles registered by satellites [4,5]. More recently, investigations of earthquake preparation phenomena using data registered by the DEMETER satellite provided statistical evidence for spectral damping of VLF (very low frequency) radio signals at F-region altitudes and within a radius of 1000–5000 km from the earthquake epicenter, about 0–3 weeks before the event [6]. In addition, using DEMETER electric and magnetic field data, Bertello et al. [7] found an EM wave at ∼300 Hz propagating 2 days before the L'Aquila 2009 earthquake event. In order to better understand the physical processes present in the lithosphere-atmosphere-magnetosphere interactions, many studies focused on the disturbances induced on the atmospheric electric field [8], on the anomalous geomagnetic pulsations [9,10], as well as on other anomalous disturbances in the ionosphere [11] and magnetosphere [12,13]. Searches for possible seismo-ionospheric effects were also performed before the earthquake by the satellite Interkosmos-19 operating at F-region altitudes, providing for the first time evidence for an increase of both the intensity of VLF noise, in the frequency range between 140 Hz to 15 kHz, and for disturbances of the electron density at a distance from the epicentres up to a few 1000 km [3,14]. More recently, Carbone et al. [15], Piersanti et al. [16] started the development of an analytical model of the coupling between lithosphere-atmosphere-ionosphere-magnetosphere to be submitted to detailed experimental verification (M.I.L.C.). In the M.I.L.C. model the coupling during active seismic conditions is described by the onset of atmospheric and ionospheric EM and particle anomalies: a first successful test of the model was the analysis of the 2018 Bayan EQ, when a series of correlated phenomena were detected both by ground sensors and by low earth orbiting satellites (∼500 km) around the time of EQ occurrence. The authors explained and modelled the experimental observations as due to the generation of an acoustic gravity wave (AGW) induced by the EQ which mechanically perturbed the ionospheric medium causing both an EM emissions and plasma waves. Interestingly, the model predicts a clear decrease of the magnetospheric FLR f^* in concomitance of the EQ occurrence, which has also been observed. This phenomenon was never reported before in the literature and it is particularly interesting, since it represents a direct, unambiguous evidence of the connection between the lithosphere and the magnetosphere, which can be used both for the analysis of coseismic as well as of precursor phenomena. Following the result on the 2018 Bayan EQ, we started a systematic study of this phenomena using 42 EQ in the time span from 2001-07-17 and 2020-08-31. This paper presents the result of this study, in which we analyze the f^* variations using ground magnetometers observations, and we explain the results of these experimental observations with an analytical model describing the f^* behaviour during active seismic conditions.

2. Data and Methods

The ground magnetometers used for the present analysis come from both INTERMAGNET and SUPERMAG magnetometer array networks, which are consortium of observatories guaranteeing a common standard data release to the scientific community, leading to possible comparison among measurements at different observation points. In our analysis we have used 1 s time resolution data.

To evaluate the f^*, we studied the cross-space spectrum [17] between the North-South magnetic field components observed at two geomagnetic observatories close enough to the EQ epicenter location (see Table 1). It is well known that, at the eigenfrequency of a field line centered between two neighboring stations having almost the same magnetic longitude, the phase difference maximizes [17,18]. Waters et al. [19] showed that the patterns of the maximum phase differences in the cross-phase spectrograms were observed consistently from day to day in the dayside region over baselines of about 100 km in the magnetic meridian. Green et al. [18] also reported that, among the several methods that determine the resonant frequency, the phase shift is least affected by geologic inhomogeneity and consistently defines the resonant frequencies.

Table 1. Characteristics of the EQ events analyzed from 2001 to 2020. The X indicates a FLR eigenfrequency variation. The - indicates the absence of a FLR eigenfrequency variation. The NA indicates the impossibility to evaluate the FLR eigenfrequency. M is the earthquake magnitude. The parameters of the earthquake are provided by USGS data catalog (https://earthquake.usgs.gov/, accessed on: 16 July 2021).

FLR	Date	UTC Time	K_p	M	Latitude	Longitude	Region
X	17/07/2001	14.50.57	0	6.3	3.061° S	148.180° E	Bismarck Sea
X	27/11/2002	00.17.20	1	5.4	12.279° N	120.753° E	Philippines
X	12/12/2003	08.07.30	1	5.2	0.110° S	123.991° E	Indonesia
X	28/01/2004	07.41.04	1	5.7	4.931° S	153.584° E	New Guinea
-	09/02/2006	05.44.30	2	6.2	4.810° S	133.063° E	Indonesia
X	17/05/2006	01.21.26	1	6.0	3.743° S	144.305° E	New Guinea
X	24/06/2006	00.03.07	1	6.3	3.071° S	127.183° E	Indonesia
X	16/09/2007	01.20.38	2	6.4	2.763° S	101.106° E	Indonesia
-	26/10/2007	16.34.47	0	6.0	3.271° S	143.763° E	New Guinea
X	14/11/2007	17.44.04	2	5.7	23.215° S	70.526° W	Chile
-	25/07/2008	20.11.07	1	6.5	5.808° S	146.658° E	New Guinea
X	11/09/2008	00.00.02	1	6.6	1.885° N	127.363° E	Indonesia
-	19/12/2008	00.34.58	2	6.8	20.372° N	146.339° E	Mariana Islands
X	06/01/2009	19.56.25	2	6.0	0.566° S	132.784° E	Indonesia
X	16/02/2009	00.33.36	2	6.1	3.664° S	149.608° E	Bismarck Sea
X	02/03/2009	00.03.39	1	6.5	1.105° S	119.868° E	Indonesia
X	25/07/2009	18.41.58	2	5.8	1.869° N	97.020° E	Indonesia
X	15/10/2009	03.34.28	1	6.0	1.111° N	85.322° W	Ecuador
-	24/02/2008	04.36.29	2	6.5	3.741° S	101.986° E	Indonesia
NA	07/06/2008	19.10.48	2	5.0	3.552° S	140.851° E	Indonesia
-	02/07/2008	00.08.31	2	5.2	12.451° N	44.202° W	Mid-Atlantic
X	07/02/2008	23.16.41	1	5.3	17.558° N	144.922° E	Mariana Islands
-	19/12/2006	12.48.16	2	6.0	2.458° N	98.000° E	Idonesia
X	16/11/2009	18.34.24	0	5.2	19.556° S	70.365° W	Chile
NA	11/01/2009	14.03.49	1	5.6	6.388° S	147.423° E	New Guinea
NA	11/01/2009	14.15.54	1	5.0	0.769° S	133.506° E	Indonesia
X	16/09/2008	21.47.14	2	5.7	17.438° N	73.915° E	India
X	24/05/2003	01.46.06	1	5.9	14.428° N	53.813° E	Owen region
-	14/11/2007	18.55.49	2	5.1	22.670° S	70.292° W	Chile
NA	26/10/2007	16.34.47	1	5.6	3.271° S	143.7630 E	New Guinea
X	22/11/2003	09.30.03	1	5.1	13.281° N	57.466° E	Arabic Sea
X	12/03/2008	01.32.34	2	6.0	1.934° N	132.519° E	Indonesia
X	02/02/2013	14.17.33	1	6.9	42.8° N	143.27° E	Japan
-	25/10/2013	17.10.16	2	7.1	37.194° N	144.66° E	Japan
X	06/10/2017	07.59.32	1	6.2	37.325° N	144.02° E	Japan
X	08/01/2019	12.39.31	2	6.3	30.526° N	131.113° E	Japan
X	18/06/2019	13.22.22	0	6.4	38.563° N	139.504° E	Japan
X	27/07/2019	18.31.07	1	6.3	33.015° N	137.413° E	Japan
X	19/04/2020	20.39.08	2	6.3	38.858° N	141.99° E	Japan
-	21/11/2016	20.58.47	1	6.9	38.296° N	141.642° E	Japan
X	05/08/2018	11.58.00	0	6.5	8.28° S	116.4° E	Indonesia
X	25/04/2015	06.45.21	2	6.6	28.18° N	84.72° E	Nepal

3. FLR Frequency Behaviour during Seismic Events

We have evaluated the FLR frequency behaviour for 42 low latitudes EQs (below 39° of geographical latitude) in the time span from 2001-07-17 and 2020-08-31. A part of these EQ (first 32 events) belongs to the sample selected by Battiston et al. [4] using POES satellite data. All the EQs have been chosen as result of a cross-check with the planetary geomagnetic K_p index [20] in order to exclude any possible f^* variation of solar origin. Table 1 summarizes the results. First of all, the K_p index ranges between 0 and 2, indicating that any possible f^* variation of solar origin can be reasonably neglected. The EQ are characterized by a magnitude (M) greater than 5. Then, we found 28 cases out of 42 in which there is a clear variation of the estimated f^* (indicated with the X). In 4 cases it was not possible to correctly evaluate f^* because of the post-sunset occurrence of the EQ (indicated with the NA). In fact, as explained in Menk et al. [21], the determination of the FLR frequency usually fails in the nightside regions. Finally, no f^* variations has been detected for 10 case events (indicated with −).

Figures 1 and 2 show four examples of FLR eigenfrequency time dependence in a time window around the EQ occurrence (red dashed line) using the cross-phase spectrogram. Colours are representative of the phase difference between the two stations selected for the f^* evaluation.

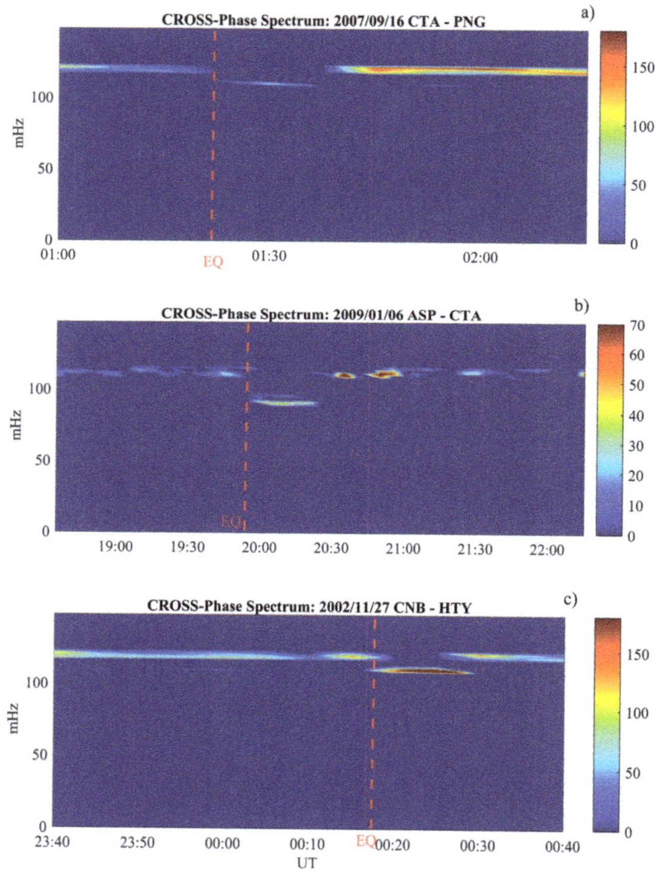

Figure 1. The cross-phase dynamical spectrograms between two low-latitude ground stations near the earthquake epicenter: panel (**a**) Sumatra 16 September 2009 EQ; panel (**b**) Indonesia 6 January 2009 EQ; panel (**c**) Philippines 11 November 2002 EQ. Each spectrum has been evaluated over a 1 h interval. Spectra have been smoothed both in time and frequency domains (7 frequency bands and 15 temporal bands). The red vertical line represents the earthquake occurrence time. In each panel the top caption reports the INTERMAGNET ground station codices used for the evaluation of the dynamical cross-phase spectrogram. The color-bar represents the phase difference in degrees between the equatorward and poleward ground magnetometer.

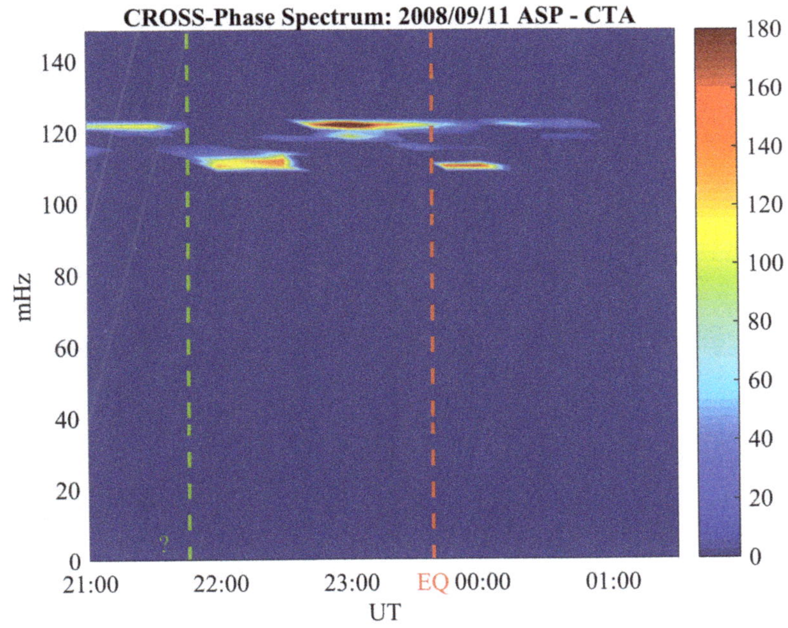

Figure 2. The cross-phase dynamical spectrograms for the Indonesia 11 September 2008 earthquake. Each spectrum has been evaluated over one hour interval. The spectra have been smoothed both in time and frequency domains (7 frequency bands and 15 temporal bands). The red vertical line represents the earthquake occurrence time. The green dashed line shows the occurrence of f^* decrease ~2 h before the EQ main shock. The top caption reports the INTERMAGNET ground station codices used for the evaluation of the dynamical cross-phase spectrogram. The color-bar represents the phase difference in degrees between the equatorward and poleward ground magnetometer.

As expected, at low latitudes the FLR eigenfrequency is around 110 mHz [19,22,23]. For each event, 2 ± 1 min after the earthquake occurrence (red dashed vertical line) there is a clear decrease of f^*. In fact, the upper (a), the middle (b) and the lower panels (c) show variations of ~ -10 mHz, ~ -25 mHz and ~ -12 mHz, respectively. The time duration of such variation is ~15 min for the first (panel a) and the third (panel c) event, and ~30 min for the second event (panel b). It is worth highlighting here that in 97% the FLR frequency variation were characterized by a single decrease coincident with the EQ occurrence, while in the remaining 3% was featured by a double f^* reduction as reported in Figure 2. In fact, in addition to the decrease of the FLR eigenfrequency at the moment of the EQ occurrence (red dashed vertical line), a clear reduction of f^* is also visible less than two hours before (green vertical dashed line). The variation is of ~ -10 mHz both for the coseismic f^* decrease as well as for the precursor f^* decrease, while the time duration is ~40 min for the precursor phenomenon and ~25 min for the coseismic phenomenon.

Figure 3 shows the statistical analysis of the EQ events characterized by a FLR decrease in terms of frequency variation (δf) and relative time duration (δT). It can be easily seen that the typical δT of the eigenfrequency decrease (panel a) is between 25 and 35 min. On the other hand, δf on average shows variations of ~10 mHz. Figure 3c) shows the probability density ($d\mathcal{P}$) of δf as a function of δT. $d\mathcal{P}$ has been estimated constructing bivariate histograms and using a kernel density estimator (e.g., [24]) with the following bin sizes: δf, 3 mHz; δT, 3 min. It results that the co-seismic FLR eigenfrequency variation is characterized by a frequency decrease of 12 ± 3 mHz and a time duration of 36 ± 3 min.

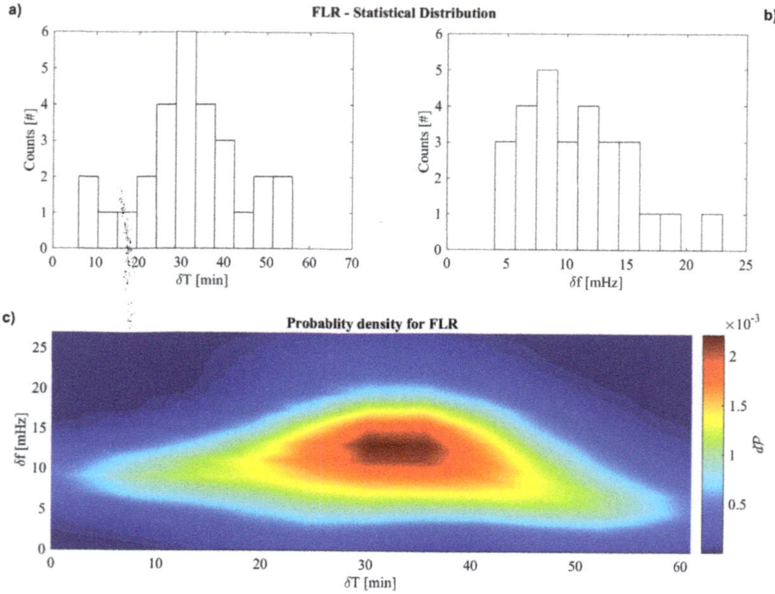

Figure 3. Eigenfrequency variation (panel (**a**)) and relative time duration (panel (**b**)) data distributions of the FLR. Panel (**c**) shows the probability density distribution of the eigenfrequency variation as a function of the relative time duration.

4. Discussion

In order to provide a quantitative explanation to the observed FLR eigenfrequency variations, we modelled the f^* behaviour during the occurrence of an EQ. It is well known that a geomagnetic field line, with both ends fixed in the ionosphere can be sketched as a string whose frequency depends on both the magnetic field geometry and the plasma density along the field line [19,21,25,26]. Following the approach of Singer et al. [27], we have evaluated f^* for an arbitrary magnetic field geometry, starting from the Magnetohydrodynamic (MHD) equations related to a stationary EM wave. By referring to the model reported in Appendix A, we have numerically solved the Equation (A8) using the magnetic equator as reference point $V_A(s_0) = V_A(eq)$ (V_A being the Alfvén speed). We have used the IGRF (International Geomagnetic Reference Field) model [28] for the internal Earth's magnetic field, the T01 model [29,30] for the external part of the Earth's magnetic field and a radial power law dependence for the plasma mass density, $\rho/\rho_{eq} = (r/r_{eq})^{-3}$ [26]. The boundary conditions of fixed footpoints have been established at some level in the ionosphere, i.e., at $h = 120$ km altitude corresponding to the E-layer, where the Alfvén wave is assumed to be perfectly reflected [17]. Finally, the values of the eigenfrequency f^* have been obtained through Equation (A9) of Appendix A.

Figure 4 shows the modelled diurnal eigenfrequency behaviour of a field line footprinted at $\lambda_{mag} = 20°$ (λ_{mag} being the magnetic latitude). Around noon, we modified the plasma density at the footprint of the field line using a gradient pressure (∇p_{den}) which produces a density variation of 15% lasting for about 10 min. Such ∇p_{den} is the result of the application of the M.I.L.C. model to an EQ characterized by a magnitude $M_{EQ} = 6.5$, a PGA = 0.6 g and a Δt = 20 s (PGA and Δt being the Peak Ground Acceleration and time duration of the EQ, respectively). The M.I.L.C. model is based on the assumption that an EQ creates an acoustic gravity wave, which propagates through the atmosphere. The pressure gradient induced by the AGW causes local instability in the ionospheric plasma density distribution, giving rise to both plasma and EM waves propagating up to the magnetosphere. In general, it is well known that the concurring contribution of the EM wave energy and/or of the plasma density variation produces a change in the local FLR eigen-frequency [19,31,32]. Indeed, Figure 4 shows that our model is able to produce a clear f^* collapse in correspondence to the plasma density gradient. This result is the consequence of Equation (A9) according to which any variation in the local magnetic field and/or in the local plasma density produces a corresponding change in f^*. It is important to remind here that the use of the IGRF and the T01 model in solving Equation (A8) produces a 5% maximum error in the evaluation of f^* [23].

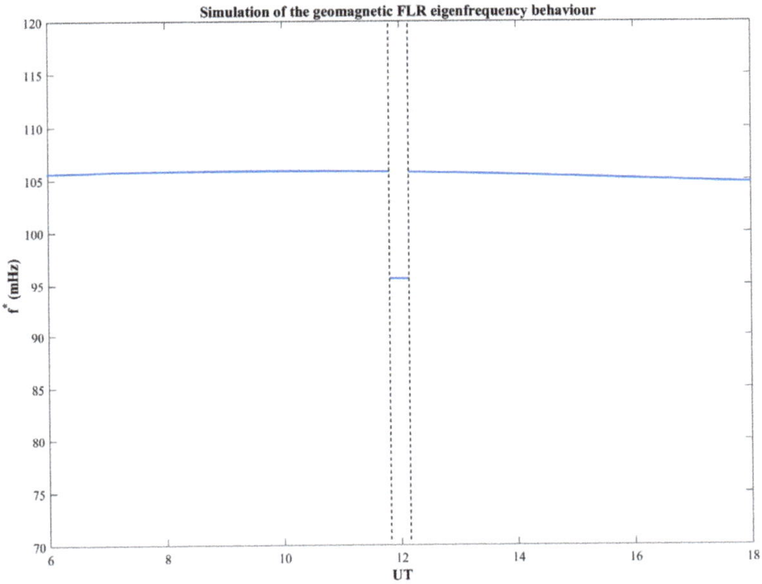

Figure 4. As an example, in the Figure, is shown the simulated behaviour of the FLR eigen-frequency obtained by our modeled for a potential EQ occurred at 20° magnetic latitude.

The result displayed in Figure 4 is consistent with the experimental FLR eigenfrequency behaviour detected in correspondence of an EQ. In fact, both Figure 1 and 2 show a sudden decrease of f^* of ~10 mHz, 2 ± 1 min after the EQ occurrence lasting for 20–30 min. Such result completely agree with the probability density bi-variate distribution in Figure 3c). However, we need to stress here that at low magnetic latitudes ($0° \leq \lambda_{mag} \leq 30°$) the geomagnetic field line is almost completely surrounded by the ionosphere. As a consequence any alteration in the ionospheric plasma density induces a variation in the corresponding eigenfrequency. Consequently, we do interpret the f^* changes observed in our 28 EQ events (see Table 1) as caused by the ionospheric plasma density variation induced by the emission of a co-seismic AGW leading to a pressure gradient [15].

Finally, in the case of the absence of a co-seismic AGW emission, no possible f^* variations can be detected (10 case event, see Table 1). Such a hypothesis is confirmed by Carbone et al. [15], showing that the atmospheric fluctuations excited by a generic seismic event on the top of the first layer of the atmosphere can be evanescent. In fact, depending on the characteristic parameters of the EQ (length of the fault, peak ground acceleration strong time duration and so on), a the propagation of the AGW up to the ionosphere can be prevented. In order to confirm such hypothesis, for these events, we analyzed the vertical atmospheric temperature profiles using the approach described in Piersanti et al. [16] to catch for possible AGW injection. Here, we display the analysis of the 19/12/2006 Sumatra EQ, since the remaining nine case events show similar results.

Figure 5a shows the atmospheric vertical temperature profile (T) as obtained from ERA5, which is the 5th generation atmospheric data set produced by the European Centre for Medium-Range Weather Forecasts [33]. The temperature fluctuations (T'), evaluated as the difference between T and its 2 km moving average, show the expected minimum and maximum at the tropopause ($\sim$18 km) and the stratopause [34], respectively. A similar behaviour can be found in both the $Brunt - Väisälä$ frequency (N^2) and the potential energy density (E_P) value [35]. The lack of any possible wave behaviour in T' confirms the absence of AGW (and reference therein [36]) injected at the moment of the EQ occurrence. As a consequence, we can reasonably affirm that the missing of AGW prevents any possible variation of ionospheric plasma density distribution leading to the FLR eigenfrequency variation.

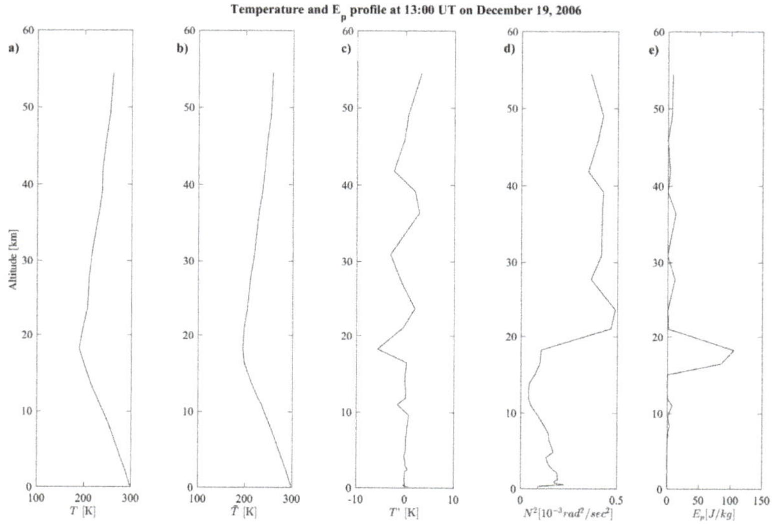

Figure 5. Example of AGW analysis. Vertical profiles of: (**a**) temperature; (**b**) background temperature; (**c**) temperature deviation, (**d**) square term of $Brunt - Väisälä$ frequency, and (**e**) potential energy at 13:00 UT on 19 December 2006.

Finally, it is worth noticing that the variation of f^* does not show any dependence on earthquake magnitude (not shown). Such result agrees with Carbone et al. [15], who demonstrated that the emission of a non-evanescent AGW, generating FLR variation, does not depend on the individual earthquake parameter alone, but on both the combination of the length of the fault, the PGA, the time duration of the EQ, etc (see dispersion relation in Carbone et al. [15]), and the local atmospheric scale height.

5. Conclusions

In the last 20 years, many investigations focused on the possible identification of magnetospheric perturbations directly connected to earthquake occurrence ([37] and reference therein). This paper presents the first evidence, via observation and modelling, of changes in magnetospheric FLR eigenfrequency associated to the EQ occurrence, demonstrating a causal connection between seismic phenomena and space based observables . We have analyzed more than 40 low latitudes EQ from 2000 to 2020, during quiet solar condition in order to search for magnetospheric signal associate to seismic activity. In 28 events, we found a clear sudden decrease of the magnetospheric FLR eigenfrequencies, while in 10 cases we did not find any f^* variation. The proposed explanation is that the plasma density at the footprint of the field line magnetically connected to the EQ location was modified of about $\Delta\rho \simeq 15\%$, by a gradient pressure fluctuations ∇p_{den} induced by a propagation of an AGW emitted during the EQ occurrence [15,16]. At low latitudes the magnetospheric field lines are fully surrounded by the ionosphere and the FLR eigenfrequencies, depending on both the magnetic field and the plasma density along the field line [19,31], is expected to decrease [23,26,32]. On the other hand, the possible explanation of the null f^* variation observed in 10 case events, can be found in the lack of the vertical propagation of the AGW (evanescent) up to the ionosphere, as predicted by the Carbone et al. [15] analytical model.

In is interesting to note that the FLR decrease observed in one case some hours before the EQ occurrence (Figure 2) could be due to various reasons, such as high-level seismic activity (especially for events characterized by a sequence of foreshocks before the main shocks), or to the outflow of radioactive gases (e.g., due to radon decay) by the Earth's surface [37,38]. Indeed, both these phenomena would be able to generate changes in atmospheric temperature and hence AGW formation (e.g., [39]). A similar result was found in Piersanti et al. [16] who found a decrease of f^* 5 h before the EQ occurrence. They explained their observation in terms of the M.I.L.C. model, pointing out that any AGW can produce a variation of the ionospheric plasma density distribution (such as travelling ionospheric disturbances [40]) which in turns changes the Alfvén velocity along the field line giving rise to a change of the FLR eigenfrequency [19].

In conclusion, our results confirm analytically the direct coupling among lithosphere, ionosphere and magnetosphere during active seismic conditions, supporting the models introduced by [16,37] and opening the way to new remote sensing methods combining space and ground sensing of the earth seismic activity.

Author Contributions: M.P. writing—original draft preparation, formal analysis, methodology, investigation and Conceptualization; W.J.B. writing—review and editing, methodology, investigation; V.C. writing—review and editing, validation; R.B. writing—review and editing, methodology and coordination; R.I. data curation and validation; P.U. writing—review and editing and project administration. All authors have read and agreed to the published version of the manuscript.

Funding: This research received no external funding.

Institutional Review Board Statement: Not applicable.

Informed Consent Statement: Not applicable.

Data Availability Statement: Ground geomagnetic observatory data are provided by INTERMAGNET (www.intermagnet.org, accessed on 16 July 2021), and SUPERMAG (https://supermag.jhuapl.edu), accessed on 16 July 2021; Earthquake parameters comes from USGS data catalog (https://earthquake.usgs.gov/), accessed on 16 July 2021.

Acknowledgments: The authors thank the national institutes that support INTERMAGNET for promoting high standards of the magnetic observatory practice (www.intermagnet.org, accessed on 14 July 2021). The authors thank the provision of magnetometer data from SUPERMAG (https://supermag.jhuapl.edu, accessed on 14 July 2021). The parameters of the earthquake are provided by USGS data catalog (https://earthquake.usgs.gov/, accessed on 14 July 2021). M. Piersanti, R. Battiston and R. Iuppa thank the Italian Space Agency for the financial support under the contract ASI "LIMADOU Scienza+" n° 2020-31-HH.0. M. Piersanti thanks the ISSI-BJ project "The electromagnetic data validation and scientific application research based on CSES satellite" and Dragon 5 cooperation 2020-2024 (ID. 59236).

Conflicts of Interest: The authors declare no conflict of interest.

Abbreviations

The following abbreviations are used in this manuscript:

AGW	Acoustic Gravity Wave
EM	Electromagnetic
EQ	Earthquake
FLR	Field Line Resonance
IGRF	International Geomagnetic Reference Field
MHD	Magnetohydrodynamic
M.I.L.C.	Magnetosphere Ionosphere Lithosphere Coupling
VLF	Very Low Frequency

Appendix A. Field Line Resonance Eigenfrequency Equation

The external perturbation pressure δP_{AGW} produced by the AGW, impinging on the ionospheric plasma from below, is able to modify the density field thus inducing a dynamics of plasma which, in turns, produces a perturbed magnetic field $\mathbf{b} = \mathbf{B} - \mathbf{B}_0$ of the background magnetic field $\mathbf{B}_0$. As an order of magnitude estimate, from the Ampere law $|\nabla \mathbf{b}| \sim \beta \nabla \delta \rho$, being $\delta \rho$ the fluctuating density and β is the plasma parameter. As a consequence, by using a relation between density and pressure fluctuation, for example the adiabatic gas law, it can be found $|\nabla \mathbf{b}| \sim (\beta \rho_0^2 / \gamma P_0 B_0) \nabla \delta P_{AGW}$, being ρ_0 and P_0 some reference values for density and pressure, and γ is the adiabatic index.

The corresponding low-frequency dynamics of the plasma can be roughly described by the ideal dissipationless MHD equations

$$\begin{aligned} \rho \left[\frac{\partial \mathbf{v}}{\partial t} + (\mathbf{v} \cdot \nabla) \mathbf{v} \right] &= \mathbf{j} \times \mathbf{B} - \nabla P \\ \frac{\partial \mathbf{B}}{\partial t} &= \nabla \times \mathbf{E} \end{aligned} \tag{A1}$$

where $\mathbf{v}$ is the fluctuation velocity, $\mathbf{j}$ the current density, $\mathbf{E}$ the electric field fluctuations and P the internal pressure. The electric field fluctuations can be obtained by the Ohm's law by neglecting the Hall term and the electron pressure gradient because we are interested at the low-frequency evolution of plasma corresponding to scales much greater than the Larmor radius, so that $\mathbf{E} = -\mathbf{v} \times \mathbf{B}$. Furthermore, we are dealing with a low-β plasma [41], so that dynamical processes occurring within the ionospheric plasma cannot significant alter the background magnetic field, so that the internal pressure gradient ∇P can be neglected in the momentum Equation (A1). It is worthwhile to note that the same approximations are usually used to describe the plasma dynamics of low-β laboratory plasma, for example confined in Reversed Field Pinch devices (e.g., [42]).

To model the eigenfrequencies f^* and amplitudes of low-frequency transverse waves, we use a linear model from Equation (A1), which describes the linear dynamics of fluctuations, namely

$$\begin{aligned} \rho_0 \mu_0 \frac{\partial \mathbf{v}}{\partial t} &\simeq (\nabla \times \mathbf{b}) \times \mathbf{B}_0 \\ \frac{\partial \mathbf{b}}{\partial t} &\simeq \nabla \times (\mathbf{v} \times \mathbf{B}_0) \end{aligned} \tag{A2}$$

where we used the Ampere's law $\nabla \times \mathbf{B} = \mu_0 \mathbf{j}$. Note that, by considering the plasma dynamics generated by AGW, the perturbed magnetic field can be viewed as generated by a small displacement ξ of the plasma [43], not by a compression of the field lines, so that $\mathbf{b}$ lye along the field line and produces the force in the momentum equation. The linear model (A2) neglects the background current density $\mathbf{j}_0$ related to the background magnetic field [29,30]. In fact, as an order of magnitude estimate, the background current density $|\mathbf{j}_0| \simeq 1.8 \times 10^{-10}$ A/m^2 results ten times lower than the current density $|\mathbf{j}| \sim b/l_{\parallel} \simeq 10^{-9}$ A/m^2 ($l_{\parallel}$ is the scale length along the fluctuating magnetic field direction). Let us consider a geometry where the z-axis is directed between the field lines. However, both footpoints of a field line are fixed in the ionosphere, so that the displacement and the separation of adjacent lines can be described by a function $h(x, y, z)$. From the last Equation (A2), using $\mathbf{v} = \partial\xi/\partial t$ along the z-direction, we get the perturbed magnetic field

$$\mathbf{b} = \nabla \times (\xi \mathbf{e}_z \times \mathbf{B}_0) \quad \text{(A3)}$$

(being $\mathbf{e}_z$ the unit vector in the direction z) apart from a constant which can be cast to zero without losing generality. Then, from the momentum equation we obtain a relation for the displacement along the z direction

$$\rho_0\mu_0 \frac{\partial^2 \xi}{\partial t^2} = (\mathbf{B}_0 \cdot \nabla)[(\mathbf{B}_0 \cdot \nabla)\xi] \quad \text{(A4)}$$

By introducing the Alfvén velocity $\mathbf{V}_A = \mathbf{B}_0/\sqrt{\rho_0\mu_0}$ and the normalized displacement $\xi' = \xi/h$ we finally obtain the wave equation for the displacement

$$\frac{\partial^2 \xi'}{\partial t^2} = V_A^2(\mathbf{e}_0 \cdot \nabla)\left[h(\mathbf{e}_0 \cdot \nabla)\xi' + \xi'(\mathbf{e}_0 \cdot \nabla)h\right] \quad \text{(A5)}$$

where $\mathbf{e}_0$ represents the unit vector along the background magnetic field.

If we consider now the ansatz where ξ' behaves as $e^{i\omega t}$, under the hypothesis that the field curvature is smooth enough so the function h is slowly variable, from Equation (A5) we obtain

$$(\mathbf{e}_0 \cdot \nabla)[(\mathbf{e}_0 \cdot \nabla)\xi'] + \left[(\mathbf{e}_0 \cdot \nabla)\ln h^2\right][(\mathbf{e}_0 \cdot \nabla)\xi'] + \frac{\omega^2}{V_A^2}\xi' \simeq 0 \quad \text{(A6)}$$

Introducing the coordinate s along the field line $(\mathbf{e}_0 \cdot \nabla) = \ell^{-1}\partial/\partial s$, where ℓ is the characteristic length of the field line between two ionospheric footpoints, say using the coordinate system where $0 \leq s \leq 1$, we obtain the characteristic value wave equation

$$\frac{\partial^2 \xi'}{\partial s^2} + P(s)\frac{\partial \xi'}{\partial s} + \left(\frac{\omega^2 \ell^2}{V_A^2}\right)\xi' = 0 \quad \text{(A7)}$$

being $P(s) = \partial \ln h^2/\partial s$, a unknown function which depends on the coordinate along the field line. The characteristic frequencies $f^\star$ we are looking for, correspond to the characteristic values ω, which can be obtained once the Sturm-Liouville Equation (A7) is solved supplied by appropriate boundary conditions, for example the condition of fixed footpoints $\xi'(s_\star) = (\partial\xi'/\partial s)_{s=s_\star} = 0$ at both $s_\star = 0, 1$.

To solve the characteristic value equation we can introduce a unknown eigenvalue λ by modifying the equation as

$$\frac{\partial^2 \xi'}{\partial s^2} + P(s)\frac{\partial \xi'}{\partial s} + \lambda\left[\frac{V_A^2(s_0)}{V_A^2(s)}\right]\xi' = 0 \quad \text{(A8)}$$

where $V_A(s_0)$ is the value of the Alfvén speed in a point s_0. The solution of Equation (A8) gives us the eigenvalue λ compatible with both the boundary conditions and a fixed value of $V_A(s_0)$. Finally, by a comparison of (A7) and (A8), the characteristic frequencies results to be

$$f^\star = \frac{V_A^2(s_0)\sqrt{\lambda}}{\ell} \quad \text{(A9)}$$

It can be possible to analytically solve Equation (A8) for some particular geometries, by making explicit the function $P(s)$ and estimate the value of $V_A(s_0)$. For example in a dipole field the azimuthal field line displacement, is proportional to $h \simeq r\sin\theta$ [27], corresponding to a toroidal mode. However, we aimed to a direct comparison with real observations of the eigenfrequencies f^*, and this necessarily requires a numerical integration of Equation (A8), because we need the exact knowledge of the function $P(s)$.

References

1. Gokhberg, M.B.; Morgounov, V.A.; Yoshino, T.; Tomizawa, I. Experimental measurement of electromagnetic emissions possibly related to earthquakes in Japan. *J. Geophys. Res.* **1982**, *87*, 7824–7828. [CrossRef]
2. Gokhberg, M.B.; Pilipenko, V.A.; Pokhotelov, O.A. On the seismic precursors within the ionosphere. *Izv. Acad. Sci. USSR Ser. Physics Earth* **1983**, *10*, 17–21.
3. Larkina, V.I.; Nalivayko, A.V.; Gershenzon, N.I.; Gokhberg, M.B.; Liperovskiy, V.A.; Shalimov, S.L. Observation of VLF emission related with seismic activity on the Intercosmos-19 satellite. *Geomagn. Aeron.* **1993**, *23*, 684–687.
4. Battiston, R.; Vitale, V. First evidence for correlations between electron fluxes measured by NOAA-POES satellites and large seismic events. *Nucl. Phys. B Proc. Suppl.* **2013**, *243–244*, 249–257. [CrossRef]
5. Sgrigna, V.; Carota, L.; Conti, L.; Corsi, M.; Galper, A.M.; Koldashov, S.V.; Murashov, A.M.; Picozza, P.; Scrimaglio, R.; Stagni, L. Correlations between earthquakes and anomalous particle bursts from SAMPEX/PET satellite observations. *J. Atmos. Sol. Terr. Phys.* **2005**, *67*, 1448–1462. [CrossRef]
6. Molchanov, O.; Rozhnoi, A.; Solovieva, M.; Akentieva, O.; Berthelier, J.J.; Parrot, M.; Lefeuvre, F.; Biagi, P.F.; Castellana, L.; Hayakawa, M. Global diagnostics of the ionospheric perturbations related to the seismic activity using the VLF radio signals collected on the DEMETER satellite. *Nat. Hazard Earth Sys.* **2006**, *6*, 745–753. [CrossRef]
7. Bertello, I.; Piersanti, M.; Cidi, M.; Diego, P.; Ubertini, P. Electromagnetic field observations by the DEMETER satellite in connection with the L'Aquila earthquake. *Ann. Geophys.* **2018**, *36*, 1483–1493. [CrossRef]
8. Gokhberg, M.B.; Kustov, A.V.; Liperovsky, V.A.; Liperovskaya, R.K.; Kharin, E.P.; Shalimov, S.L. About disturbances in F-region of ionosphere before strong earth-quakes. *Izvestiya Acad. Sci. USSR Ser. Physics Earth* **1988**, *4*, 12–20.
9. Fraser-Smith, A.C.; Bernardi, A.; McGill, P.R.; Ladd, M.; Helliwell, R.; Villard, O.G., Jr. Low-frequency magnetic field measurements near the epicenter of the Ms 7.1 Loma Prieta earthquake. *Geophys. Res. Lett.* **1990**, *17*, 1465–1468. [CrossRef]
10. Gogatishvili, I.M. Geomagnetic precursors of intense earthquakes in the spectrum of geomagnetic pulsations with frequencies of 1–0.02 Hz. *Geomagn. Aeron.* **1984**, *24*, 697–700.
11. Kolokolov, L.E.; Liperovskaya, E.V.; Liperovsky, V.A.; Pokhotelov, O.A.; Mararovsky, A.V.; Shalimov, S.L. Sudden diffusion of sporadic E-layers in the mid-latitude ionosphere during the earthquake preparation. *Izvestiya RAN Earth Phys.* **1992**, *7*, 105–113.
12. Parrot, M. Statistical study of ELF/VLF emissions recorded by a low-altitude satellite during seismic events. *J. Geophys. Res.* **1994**, *99*, 23339. [CrossRef]
13. Serebryakova, O.N.; Bilichenko, S.V.; Chmyrev, V.M.; Parrot, M.; Ranch, J.L.; Lefeuvre, F.; Pokhotelov, O.A. Electromagnetic ELF radiation from earthquakes regions as observed by low-altitude satellites. *Geophys. Res. Lett.* **1992**, *19*, 91. [CrossRef]
14. Migulin, V.V.; Larkina, V.I.; Molchanov, O.A.; Nalivaiko, A.V.; Gokhberg, M.B.; Pilipenko, V.A.; Liperovsky, V.A.; Pokhotelov, O.A.; Shalimov, S.L. Detection of earthquake influence on the ELF/VLF emissions at the upper ionosphere. *Preprint IZMIRAN* **1982**, *25*, 2390.
15. Carbone, V.; Piersanti, M.; Materassi, M.; Battiston, R.; Lepreti, F.; Ubertini, P. A mathematical model of Lithosphere-Atmospherecoupling for seismic events. *Sci. Rep. Nat.* **2021**. [CrossRef]
16. Piersanti, M.; Materassi, M.; Battiston, R.; Carbone, V.; Cicone, A.; D'Angelo, G.; Diego, P.; Ubertini, P. Magnetospheric–Ionospheric–Lithospheric Coupling Model. 1: Observations during the 5 August 2018 Bayan Earthquake. *Remote Sens.* **2020**, *12*, 3299. [CrossRef]
17. Waters, C.L.; Menk, F.W.; Fraser, B.J. Low latitude geomagnetic field line resonances: Experiment and modeling. *J. Geophys. Res.* **1994**, *99*, 547. [CrossRef]
18. Green, A.W.; Worthington, E.W.; Baransky, L.N.; Fedorov, E.N.; Kurneva, N.A.; Pilipenko, V.A.; Shvetzov, D.N.; Bektemirov, A.A.; Philipov, G.V. Alfven field line resonances at low latitudes (L = 1.5). *J. Geophys. Res.* **1993**, *98*, 15693–15699. [CrossRef]
19. Waters, C.L.; Samson, J.C.; Donovan, E.F. Variation of plasmatrough density derived from magnetospheric field line resonances. *J. Geophys. Res.* **1996**, *101*, 24737–24745. [CrossRef]
20. Matzka, J.; Bronkalla, O.; Tornow, K.; Elger, K.; Stolle, C. Geomagnetic Kp index. V. 1.0. GFZ Data Services. 2021. Available online: https://dataservices.gfz-potsdam.de/panmetaworks/showshort.php?id=escidoc:5216888 (accessed on 16 July 2021).
21. Menk, F.W.; Waters, C.L. *Magnetoseismology: Ground-Based Remote Sensing of Earth's Magnetosphere*; Wiley: Hoboken, NJ, USA, 2013.
22. Menk, F.W.; Waters, C.L.; Fraser, B.J. Field line resonances and waveguide modes at low latitudes: 1. Observations. *J. Geophys. Res.* **2000**, *105*, 7747–7761. [CrossRef]
23. Vellante, M.; Piersanti, M.; Pietropaolo, E. Comparison of equatorial plasma mass densities deduced from field line resonances observed at ground for dipole and IGRF models. *J. Geophys. Res.* **2014**, *119*. [CrossRef]
24. Martinez, W.L.; Martinez, A.R. *Computational Statistics Handbook with MATLAB*; Chapman and Hall/CRC: Boca Raton, FL, USA, 2002.
25. Rankin, R.; Tikhonchuk, V.T. Dispersive shear Alfvén waves on model Tsyganenko magnetic field lines. *Adv. Space Res.* **2001**, *28*, 1595. [CrossRef]

26. Vellante, M.; Piersanti, M.; Heilig, B.; Reda, J.; Corpo, A.D. Magnetospheric plasma density inferred from field line resonances: Effects of using different magnetic field models. In Proceedings of the 2014 XXXIth URSI General Assembly and Scientific Symposium (URSI GASS), Beijing, China, 16–23 August 2014; pp. 1–4. [CrossRef]
27. Singer, H.J.; Southwood, D.J.; Walker, R.J.; Kivelson, M.G. Alfven wave resonances in a realistic magnetospheric magnetic field geometry. *J. Geophys. Res.* **1981**, *86*, 4589. [CrossRef]
28. Thébault, E.; Finlay, C.C.; Beggan, C.D.; Alken, P.; Aubert, J.; Barrois, O.; Bertr, F.; Bondar, T.; Boness, A.; Brocco, L.; et al. International Geomagnetic Reference Field: The 12th generation. *Earth Planet Sp.* **2015**, *67*, 79. [CrossRef]
29. Tsyganenko, N.A. A model of the magnetosphere with a dawn-dusk asymmetry, 1, Mathematical structure. *J. Geophys. Res.* **2002**, *107*. [CrossRef]
30. Tsyganenko, N.A. A model of the near magnetosphere with a dawn-dusk asymmetry, 2, Parameterization and fitting to observations. *J. Geophys. Res.* **2002**, *107*. [CrossRef]
31. Menk, F.W.; Kale, Z.; Sciffer, M.; Robinson, P.; Waters, C.L.; Grew, R.l.; Clilverd, M.; Mann, I. Remote sensing the plasmasphere, plasmapause, plumes and other features using ground-based magnetometers. *J. Space Weather Space Clim.* **2014**, *4*, A34. [CrossRef]
32. Piersanti, M.; Villante, U.; Waters, C.; Coco, I. The 8 June 2000 ULF wave activity: A case study. *J. Geophys. Res.* **2012**, *117*. [CrossRef]
33. Hennermann, K. ERA5 Data Documentation. In Copernicus Knowledge Base. 2017. Available online: https://confluence.ecmwf.int/display/CKB/ERA5+data+documentation (accessed on 19 October 2017).
34. Tsuda, T.; Murayama, Y.; Nakamura, T.; Vincent, R.A.; Manson, A.H.; Meek, C.E.; Wilson, R.L. Variations of the gravity wave characteristics with height, season and latitude revealed by comparative observations. *J. Atmos. Terr. Phys.* **1994**, *56*, 555–568. [CrossRef]
35. Tsuda, T.; Nishida, M.; Rocken, C.; Ware, R.H. A global morphology of gravity wave activity in the stratosphere revealed by the GPS occultation data (GPS/MET). *J. Geophys. Res.* **2000**, *105*, 7257–7273. [CrossRef]
36. Yang, S.-S.; Asano, T.; Hayakawa, M. Abnormal gravity wave activity in the stratosphere prior to the 2016 Kumamoto earthquakes. *J. Geophys. Res. Space Phys.* **2019**, *124*. [CrossRef]
37. Pulinets, S.A.; Ouzounov, D.P. Lithosphere–atmosphere–ionosphere coupling (LAIC) model—An unified concept for earthquake precursors validation. *J. Asian Earth Sci.* **2011**, *41*, 371–382. [CrossRef]
38. Hayakawa, M.; Kasahara, Y.; Nakamura, T.; Hobara, Y.; Rozhnoi, A.; Solovieva, M.; Molchanov, O.; Korepanov, V. Atmospheric gravity waves as a possible candidate for seismo-ionospheric perturbation. *J. Atmo. Electr.* **2011**, *31*, 129–140. [CrossRef]
39. Hayakawa, M.; Kasahara, Y.; Nakamura, T.; Muto, F.; Horie, T.; Maekawa, S.; Hobara, Y.; Rozhnoi, A.A.; Solovieva, M.; Molchanov, O.A. A statistical study on the correlation between lower ionospheric perturbations as seen by subionospheric VLF/LF propagation and earthquakes. *J. Geophys. Res.* **2010**, *115*, A09305. [CrossRef]
40. Hocke, K.; Schlegel, K. A review of atmospheric gravity waves and travelling ionospheric disturbances. *Ann. Geophys.* **1996**, *14*, 1996. [CrossRef]
41. Stubbe, P.; Hagfors, T. The Earth's ionosphere: A wall-less plasma laboratory. *Surv. Geophys.* **1997**, *18*, 57–127. [CrossRef]
42. Cappello, S.; Escande, D.F. Bifurcation in viscoelastic MHD: The Hartmann Number and the Reversed Field Pinch. *Phys. Rev. Lett.* **2000**, *85*, 3838–3841. [CrossRef]
43. Cummings, W.D.; O'Sullivan, R.J.; Coleman, P.J. Standing Alfvén waves in the magnetosphere. *J. Geophys. Res.* **1969**, *74*, 778–793. [CrossRef]

MDPI
St. Alban-Anlage 66
4052 Basel
Switzerland
Tel. +41 61 683 77 34
Fax +41 61 302 89 18
www.mdpi.com

Remote Sensing Editorial Office
E-mail: remotesensing@mdpi.com
www.mdpi.com/journal/remotesensing

www.ingramcontent.com/pod-product-compliance
Lightning Source LLC
Chambersburg PA
CBHW061546220926
51850CB00005B/268
9783036568324